SPECTRAL ANALYSIS IN GEOPHYSICS

FURTHER TITLES IN THIS SERIES

1. *F.A. VENING MEINESZ*
THE EARTH'S CRUST AND MANTLE

2. *T. RIKITAKE*
ELECTROMAGNETISM AND THE EARTH'S INTERIOR

3. *D.W. COLLINSON, K.M. CREER and S.K. RUNCORN*
METHODS IN PALAEOMAGNETISM

4. *M. BÅTH*
MATHEMATICAL ASPECTS OF SEISMOLOGY

5. *F.D. STACEY and S.K. BANERJEE*
THE PHYSICAL PRINCIPLES OF ROCK MAGNETISM

6. *L. CIVETTA, P. GASPARINI, A. RAPOLLA and G. LUONGO (Editors)*
PHYSICAL VOLCANOLOGY

Developments in Solid Earth Geophysics
7

SPECTRAL ANALYSIS IN GEOPHYSICS

by

MARKUS BÅTH

Seismological Institute
University of Uppsala, Sweden

ELSEVIER SCIENTIFIC PUBLISHING COMPANY
Amsterdam – Oxford – New York 1974

ELSEVIER SCIENTIFIC PUBLISHING COMPANY
335 JAN VAN GALENSTRAAT
P.O. BOX 211, AMSTERDAM, THE NETHERLANDS

AMERICAN ELSEVIER PUBLISHING COMPANY, INC.
52 VANDERBILT AVENUE
NEW YORK, NEW YORK 10017

LIBRARY OF CONGRESS CARD NUMBER: 74-77576

ISBN: 0-444-41222-0

WITH 101 ILLUSTRATIONS AND 54 TABLES

PRINTED IN THE NETHERLANDS

PREFACE

Spectral analysis methods play a dominant role in present-day geophysical research. The purpose of this book is to introduce the reader into the methods for spectral analysis, not the least all its pitfalls, and to give applications to geophysics. Among the latter, seismology will occupy the major part, where spectroscopy has probably scored its greatest success, but applications also to meteorology, oceanography, gravity and geomagnetism will be given a corresponding examination. All through the book, the emphasis will be on methods, and results will be mentioned only briefly and for the rest covered by references. Chapters 1–6 deal with general methodology with special reference to geophysical applications, while Chapters 7–10 are devoted to special geophysical fields.

The presentation assumes a background of the reader corresponding to a B.Sc. or M.Sc. degree in some geophysical branch, with a corresponding knowledge of applied mathematics and some insight into statistics. Although primarily addressed to geophysicists, the book could be used by anybody with a corresponding background and with an interest and a need for spectral methods, as the first six chapters also have a general applicability.

In working up the material, I have aimed at serving a number of related but different purposes: as a university textbook for geophysicists, as an introduction and a guide to research workers in the field, as an assistance to workers in several fields of practical or applied geophysics, especially earthquake engineering, geophysical prospection, nuclear explosion detection, and as a handbook, especially by including as complete references as possible up to around 1972. In selecting references, mainly those dealing with observational spectra have been included, and only those which are easily available to the reader at larger libraries or geophysical institutes, excluding numerous theoretical papers, unpublished manuscripts and institute reports with more limited circulation. The aim of the book is not only to convey to the reader an ability to read professional papers and books in this field without difficulty, but also and primarily to lead to an ability for the reader to perform his own spectral analyses. In other words, both education, research and practical use have been constantly in my mind during the preparation of the book.

The need for a review of geophysical spectroscopy, especially in seismology, became obvious in our institute around 1965. But it was not until later that a major step forward was taken by us in this direction, especially by Dr. Hamid N.

Al-Sadi. As a partial fulfillment of his degree requirements at Uppsala University (1969–1972), he made a review on seismological spectroscopy. Even though completely re-worked and above all very much extended, his draft served as a useful assistance, at least in some parts of the initial stage of the present work. I like to express my sincere thanks to Dr. Al-Sadi. Also, I like to thank my colleague, Asst. Professor Ota Kulhánek, for several helpful discussions, my institute colleagues Klaus Meyer and Rutger Wahlström for assistance on special points, Miss Ulla Hjelmqvist, Uppsala, for directing the typing work, Mrs Ester Dreimanis, Uppsala, for drafting the figures, and my wife Ingrid Båth for continuous encouragement and assistance in an often pressing work. The Swedish Natural Science Research Council is gratefully acknowledged for a supporting grant and the Elsevier Scientific Publishing Company for its appreciated collaboration in the preparation of the book.

Seismological Institute
Uppsala, Sweden

MARKUS BÅTH
August, 1973

CONTENTS

LIST OF SYMBOLS

a_n, b_n	Fourier coefficients
$a(\omega)$, $b(\omega)$	Fourier cosine and sine transforms, respectively
a	1/time-constant, arbitrary constant
$a(t)$	acceleration
A, $A(t)$, $A(\omega)$	amplitude, displacement
b	fault length
$B(\theta) = \|B(\theta)\| \mathrm{e}^{i\Psi(\theta)}$	source spatial function
c	phase velocity, also used for arbitrary constant
cps, cpm, cph, cpd, cpy	cycles per second, minute, hour, day, year, respectively (cps = Hz = Hertz)
$C(\omega)$	crustal transfer function
$C_r(\omega, r)$	transfer function of receiver crust
$C_s(\omega)$	transfer function of source crust
$\mathrm{C}_{11}(\tau)$	autocorrelation
$\mathrm{C}_{12}(\tau)$	cross-correlation
dB	decibel
D	directivity
e	base of natural logarithm system
E	total energy (seismic wave energy, tsunami energy)
E, N, Z	east–west, north–south and vertical component, respectively
$E_{11}(\omega) = E(\omega)$	power
$E_{12}(\omega)$	cross-power
Ex	expectance
$f(t)$	function of time
$f_1(t) \star f_2(t)$	convolution of $f_1(t)$ with $f_2(t)$
$F(\omega)$	Fourier transform of $f(t)$
$F^*(\omega)$	complex conjugate of $F(\omega)$
$\|F(\omega)\|$	absolute value of $F(\omega)$
$f(t) \leftrightarrow F(\omega)$	Fourier pair
g, $g(x, y)$	acceleration of gravity
$g(t)$	filter output for an input $f(t)$
$G(\omega)$	Fourier transform of $g(t)$
$G(r)$, $G(\Delta)$	geometrical spreading, referring to amplitude
h	depth, focal depth, layer thickness
$h(t)$	impulse response of a filter
$H(\omega)$	transfer function of a filter, also used for horizontal component
$\|H(\omega)\|^2$	power transfer function
i	$\sqrt{-1}$, also angle of incidence
$i(t)$	instrumental impulse response
$I(\omega)$	instrumental transfer function

Im	imaginary part
k (k_x, k_y, k_z)	wavenumber $= 2\pi/L$
K	coupling
k, i, j, l, m, n, M, N	integers
ln	natural logarithm
log	logarithm to the base 10
$\log_2$	logarithm to the base 2
L	wavelength
m	body-wave magnitude
M	surface-wave magnitude
$M(\omega, r)$	mantle transfer function
$n(t)$	noise time function
$N(\omega)$	Fourier transform of $n(t)$
$p(t)$	pressure time function
$P(\omega)$	pressure frequency function
$\bar{P}(\omega)$	modified periodogram
$P_{12}(\omega)$	co-spectrum
Pr	probability
q	normalization factor in expression for $I(\omega)$
$q(\Delta, h)$	calibration function in magnitude formula
Q	quality factor
$Q_{12}(\omega)$	quad-spectrum
r	distance (along ray path)
r, θ, z	cylindrical coordinates
r_0	earth's radius
R, $R(\omega)$	reflection coefficient
Re	real part
$s(t)$	source time function, also used for signal time function
$S(\omega)$	Fourier transform of $s(t)$
t	time
$\bar{t}$	quefrency
T	period, fundamental period, record length
$\bar{T}$	temperature, also used for average T
$u(t)$	Heaviside unit step
U	group velocity
v	cyclic frequency $= 1/T$
V, $v(t)$, $V(\omega)$	velocity in general
V_P, V_S	velocity of P and S, respectively
V_F	fault propagation velocity
$w_j(x)$	Walsh function
$w(t)$	time window
$W(\omega)$	spectral window, Fourier transform of $w(t)$
W	explosion yield
x, y, z	rectangular coordinates
α, β	constants
γ	coefficient of spectral slope
$\gamma_{12}(\omega)$	coherence
$\Gamma(k)$	topography
δ	colatitude
$\delta(t)$	Dirac delta function

Δ	epicentral distance
Δt	digitizing interval
Δt_e	equivalent width of $f(t)$
Δx	difference in x
ε	longitude
θ, ψ	direction, azimuth
$\varkappa$	attenuation coefficient, referring to amplitude
Λ	triangular function
μ	number of degrees of freedom
ν	integer
ξ, η	spatial lags
Π	rectangular function
ρ	density
σ	standard deviation
σ^2	variance
τ	time lag
Φ, φ	phase angle
Ψ	phase angle of $B(\theta)$
ω	angular frequency $= 2\pi\nu$
ω_N, ν_N	Nyquist frequency

Remark. The parallel use of two frequency measures has to be especially noted. ω is angular frequency and measures radians per second ($2\pi/T$), whereas ν is cyclic frequency and measures cycles per second ($1/T$). ω is mostly used in our theoretical expressions, whereas ν appears when an observed spectrum is dealt with. They should not be mixed with each other.

Chapter 1

METHODS IN WAVEFORM AND SPECTRA STUDIES

The present chapter is essentially an introduction to this book. The presently used techniques for analysis of geophysical records and converting them into spectra require an extensive mathematical development and the availability of large electronic computers. Early analyses were by necessity less sophisticated and worked mostly with direct measurements from records. These will be reviewed in this chapter, as they constitute a background to the recent developments. The modern analytical and computational methods, which will occupy the following chapters of this book, have been paralleled by some non-analytical methods, providing spectra directly. They will also be reviewed in this chapter. To fix ideas better, our discussion will be mainly focussed on seismograph records, even though our comments are equally applicable to almost any type of geophysical records.

1.1 ANALYSIS OF SEISMOGRAMS AND OTHER GEOPHYSICAL RECORDS

The wave field generated by a seismic energy source (earthquake or explosion) is a function both of space and time. A complete representation of this function is provided by the ensemble of seismograph recordings around the earth.

We can represent the wave field by a function such as $f(x, y, z, t)$, where f represents the wave (amplitude, power, or other property), x,y,z are the spatial coordinates and t is time. It is, however, inconvenient to deal with such a function, and it is easier to consider cases where one or several of the variables are kept constant. We may mention a few examples of this:

(1) $f(t)$ represents the seismograph recording at a given station with coordinates x,y,z.

(2) $f(x,y,z)$ represents a stationary, i.e. time-independent, wave field. Such fields are not generated by earthquakes or explosions, as these are transient phenomena. But by artificial means it is possible to have a continuously acting source with constant power, e.g. a pumping station, which generates a stationary wave field. In nature, microseisms represent quasi-stationary wave fields. Alternatively, $f(x,y,z)$ represents the field at just one instant of time.

(3) $f(x,t)$ represents seismograph recordings along a profile of stations, extending in a given direction from the source.

The space covered by observations is by necessity limited to the earth's surface or anyway very near to it. Each seismograph station provides the time history of the seismic waves arriving at that station. With two-dimensional recording generally used, either on paper or on magnetic tape, one axis is time and the other the amplitude in one spatial direction. In a few exceptional installations, both axes are spatial, for instance, north–south and east–west components etc. Three such seismographs, each one for the three combinations of the three components, serve the purpose of displaying the particle motion more directly. Excluding such apparatuses, the dominating feature of seismographs is to use time as one of the two coordinates. In other words, seismographs show the wave motion in the time domain. This fact has been of great significance in the development of seismic wave analysis. The attention has become most concentrated on time-dependent characteristics, such as arrival times, travel times and velocities, i.e. kinematical properties of the wave motion. In addition, the other recorded coordinate, generally the amplitude with its corresponding period has attracted some attention. Because of lower measuring reliability and greater difficulty of interpretation, amplitudes contributed less to the early development of seismology than time measurements did. Whereas it was customary already in the beginning of instrumental seismology to combine arrival-time measurements from a number of stations for computation of epicenters and origin times, it took many years until a corresponding ensemble treatment was applied to amplitude measurements. It was not until the 1930's in connection with the introduction of magnitude scales, that greater attention was paid to amplitudes.

The time, amplitude and period measurements just referred to are usually made in a simple and quick way visually by means of an ordinary measuring scale. In this book we shall learn how seismology has proceeded beyond these limits, extracting more information from detailed analyses of the waveform (in the time domain) and of the corresponding spectra (in the frequency domain).

Thus, in seismogram analysis we can distinguish the following two approaches: (1) phase readings: arrival times, amplitudes, periods; and (2) waveform and spectra analysis.

The two methods differ not only in the ease of measurement, but also in the extent to which they lend themselves to an easy interpretation. In general, the seismogram and its properties are the result of the source action (earthquake, explosion, or other), of the traversed medium (path properties) and of the receiving seismograph characteristics. A seismogram property is most apt to be used for interpretational work if it is easy to measure reliably and if it depends on only one or a few of the factors mentioned. This is true of the arrival time and travel time. It can be measured directly from a record in an accurate way and it depends only on the medium properties. But even simple measurements of amplitudes from a seismic record are more complicated. Partly is the calculation into ground amplitude less accurate than a time reading, as it involves an accurate knowledge of seismograph response characteristics, partly does the amplitude depend both on source and path properties. Simple amplitude measurements have mostly been used for magnitude calculations, that is, after correction for path properties (distance), they give a measure of the source energy. Simple amplitude measurements have the advantage of being easy and quick to acquire and reliable enough for magnitude determination, but they do not represent all the information which can be extracted from a record. Studies of waveforms and spectra (item 2 above) go far beyond this step, but they involve much more of computational work and interpretational difficulties.

In the development of seismology it is the arrival time readings that have given the most reliable foundations for our present knowledge of the earth's internal structure. The seismic wave velocities are the internal properties known with the highest accuracy and directly deduceable from the travel times. In contrast, most of the simple amplitude readings have been used for deducing source properties. The new techniques of waveform and spectra studies are able to supplement the simple methods in an efficient way. By suitable combination of observations it is possible to isolate one or another of the many properties influencing an observed waveform or spectrum, and thus to study this property in more detail than the simple techniques are able to. This is intimately connected with the fact that item 1 represents point readings in the records, whereas for item 2 the whole wave groups enter into the analysis.

The waveform and spectra analysis of geophysical records lagged behind the simple time-domain readings. The reasons were partly of a practical nature, especially the lengthy calculations needed for spectral estimations, partly of a theoretical nature, concerning explanation of observed waveforms and spectra in terms of source, path and receiver properties. During the last twenty years, these difficulties have been removed, with the consequence that spectral methods are nowadays among the most significant techniques in geophysical research. This is among other witnessed by the recent enormous increase in geophysical literature, which deals with spectra and their interpretation.

1.2 ANALYSIS IN PRE-COMPUTER TIME

Trying to make a brief sketch of the historical development of spectral analysis, we can state first that in the true sense of the word, spectral analysis of geophysical phenomena hardly began before large electronic computers came into more general use, i.e. within about the last twenty years. But this development has some important precursors, by which one tried to present data not only in a simple time-domain style.

1.2.1 Harmonic analysis

The most important by far of these precursors is the harmonic analysis or

Fourier series expansion of a given time series of data. The possibility to resolve a given curve into a series of sine and cosine functions, i.e. harmonic functions, has rendered the name *harmonic analysis* to this computational technique. By the nature of this analysis, it is applied only to phenomena with at least one well-defined fundamental period, or at least a suspected fundamental period. Meteorology offers outstanding examples of extensive harmonic analysis of various meteorological elements, which is natural considering their well-known annual and diurnal periods. On the other hand, meteorological turbulence, being apparently non-periodic, is usually not analyzed in this way, but is an area where modern spectral–analysis methods have scored enormous success. Other phenomena with well-defined periods are the tides of the earth. For example, the oceanic tides have for many years been subjected to harmonic analysis, especially for prediction purposes. Also within seismology, harmonic analysis has been used for a number of problems, such as investigations of periodicities of earthquakes (mostly with discouraging results) or of annual variation of microseisms (being better defined due to their relation to meteorological phenomena). On the other hand, harmonic analysis of seismograph records has hardly been done, but has been reserved as another area where spectral methods have proved extremely useful.

The developments up to the time around World War II were mainly concentrated on determinations of periodicities of various phenomena, as well as of the statistical significance of the various periods found. This development has been extensively discussed by several writers, notably Stumpff (1937). His book covers the field up to that time, and gives also 319 references to earlier literature in the field of periodicity investigations.

Besides harmonic analysis, there are a number of other early attempts to analyze data, beyond just time-domain readings. Among these, we shall briefly describe some of the most important attempts, i.e. analyses of frequencies of occurrence and of amplitude–period relationships. Waveform analysis of seismograph records also belongs to the earlier era, but is a field which has been renewed with modern techniques.

1.2.2 Frequency of occurrence

Frequency of occurrence is a statistical measure that is often used, and has been so also in the past, because of the relative ease by which it generally can be determined. The distribution of frequency can be displayed in relation to any parameter of interest. Such distributions are often referred to as *histograms*, and displayed as a step curve. In general they offer no difficulties to construct, only that precautions are taken to have large enough samples, in addition to have samples representative for the phenomenon under investigation. Even though such representations are not spectra, they may nevertheless be termed spectra, taking this word in a more general sense of distribution.

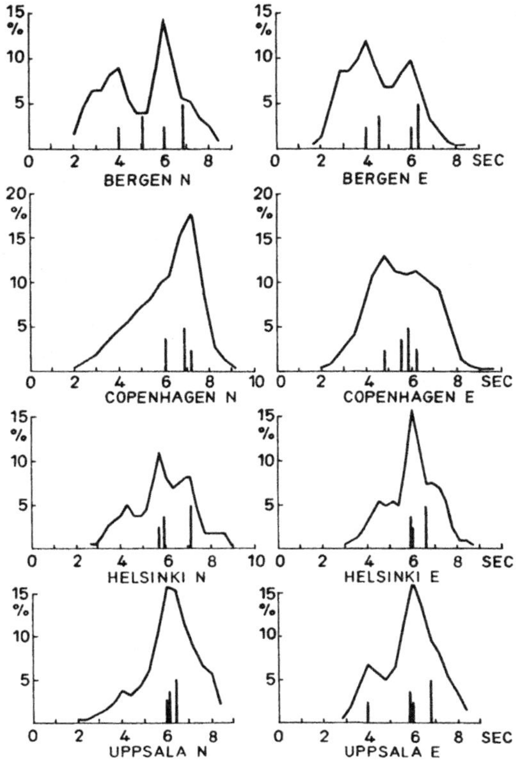

Fig.1. Period spectra of microseisms for some seismograph stations in Fennoscandia for March 23, 1949, around 06^h GMT. The vertical bars indicate calculated periods: the shortest gives period of frequency maximum, i.e. the mode, the next longer bar gives the mean period, and the longest bar gives the period of amplitude maximum. After Båth (1953).

Instead of dwelling on numerous examples of such distributions which can be found in the geophysical literature, let me just give one example. Fig.1 from Båth (1953) shows the frequency distribution of microseismic periods, measured at four Fennoscandian stations for a 30-minute interval on March 23, 1949. The frequencies of occurrence are expressed in percentages, and each single "period spectrum" is based on over 200 measurements made directly on the time-domain records. The different curves are comparable, as the seismograph characteristics are nearly the same and as the microseismic source is the same for all stations, with a cyclone near the north coast of Norway dominating the weather situation and with the microseismic origin located at or near the whole Atlantic coast of Norway. For this and a few other situations, a comparative study of simultaneous period spectra at the four stations enabled us to investigate the influence of several factors on these spectra, especially various effects of the distance to the microseismic source. Quite generally, a single distribution curve is not very informative, but becomes so by comparison with a number of other distributions sampled under the same or comparable conditions. The period spectra shown in Fig.1 would have about the same shape as the amplitude–period spectra (i.e. true spectra) only if the amplitudes were proportional to the respective frequencies of occurrence.

The latter condition is approximately true for resonance phenomena. For these, peaks in the "period spectra" (frequency of occurrence of various periods) are located at the same periods as the amplitude maxima in the amplitude–period spectra. If only readings of such peak periods are relevant, then the period spectra can be used as a substitute for reading them from true (amplitude–period) spectra. The advantage of the period spectra is their relative ease of determination as compared to the true spectra, especially when more sophisticated computer techniques are not available. A typical example is offered by (vertical) resonance vibrations of crustal layers, excited in various ways by seismic waves. The peak periods are then characteristic for each locality and permit a calculation of layer thicknesses (see further section 7.1), as well as a classification of different localities for earthquake engineering purposes. The water-level oscillations (seiches) of lakes or semi–enclosed water basins are another example, where the period spectra may serve as a useful substitute for true amplitude–period spectra.

1.2.3 Amplitude–period graphs

An amplitude–period graph, showing the distribution of amplitude over all the periods or frequencies involved in a studied phenomenon, is in fact nothing but a spectrum. It is therefore closer to the main subject of our study than the plots of frequency of occurrence dealt with in the preceding section. The reason that we still include amplitude–period graphs in this discussion of pre-computer time achievements in spectral analysis is that such graphs were earlier usually determined by direct time-domain readings, i.e. readings of amplitudes and appertaining periods directly from records. The measurements are thus quite simple and straightforward and no theory is required for the construction of such a spectrum. Also the calculations can easily be made on any desk computer.

The earlier geophysical literature offers many examples where this technique has been used. There is no purpose in trying to give a full account of this here, and it may suffice with one example, concerning microseisms. Båth (1949, pp.60–63) studied amplitude–period relations of 3–8 sec microseisms recorded at Uppsala, Sweden. Fig.2 shows an average curve, recalculated with log(amplitude) and log(cyclic frequency) as coordinates. The points represent average amplitudes for given integer period values, and the averaging has been made over numerous cases, although not to the same number for all periods involved. The general amplitude decrease towards increasing frequencies is very obvious from these measurements, a result which has been confirmed many times by later more sophisticated spectral studies (section 9.3.3). Two reference slopes (γ) have been

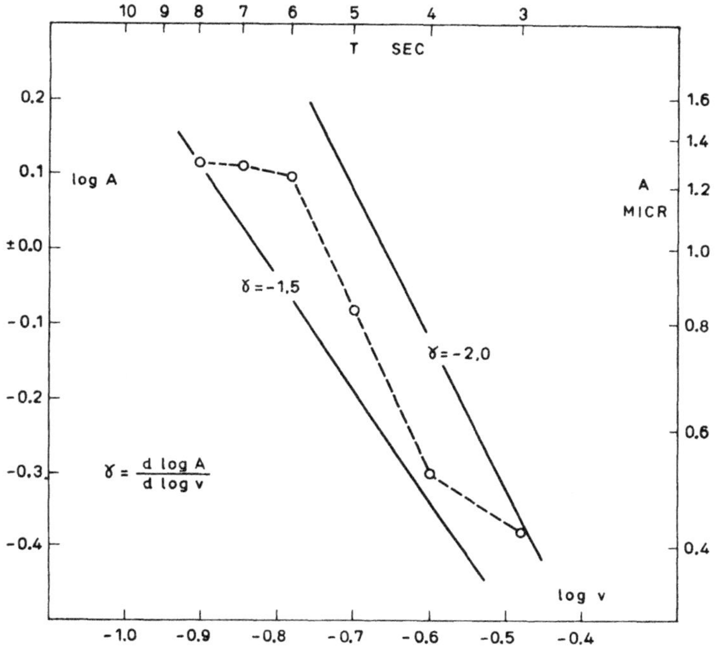

Fig.2. Amplitude–frequency relations of microseisms obtained as an average of numerous measurements on medium-period seismographs at Uppsala.

put into Fig.2, and it is to be observed that these refer to amplitudes. For power spectra, more generally used in this connection, the slope will be twice as large. The spectrum shown in Fig.2 represents average conditions, i.e., it may be considered as an average spectrum. For individual cases, it would be generally much more difficult to arrive at a reliable spectrum by this simple time-domain method. Another example from geophysics amenable to a similar treatment consists of ocean surface waves by simultaneous observations of wave heights and periods.

The amplitude–period representation found much useful application already at an early stage in investigating resonance vibrations both of buildings and other structures as well as of surface layers. In such experiments, the body under investigation is subjected to artificial vibrations of given frequencies and the excited amplitudes are measured. A direct plot of amplitude versus period thus yields the needed information on resonance frequencies.

Instead of using the amplitude directly measured, some function of it can be calculated and plotted versus period. This is done for example in *periodogram methods*, of which there exist several versions. Such methods have mostly been used in a search for dominant periods in records as a preparation for harmonic analysis. In a simple form, the periodogram analysis consists in measuring recorded amplitudes at equal spacing, then averaging the amplitudes at intervals of varying lengths, and plotting the average amplitudes versus the respective interval lengths. Dominant periods will show up as maxima in such a plot.

1.3 WAVEFORM ANALYSIS AND SYNTHESIS

From direct observations of seismograms it soon became clear that the seismic waves exhibit a variety of different types. But it also appeared obvious that certain source regions repeatedly give the same or similar types, distinct from the type obtained from other source regions. This discovery led to the prospects of using the wavetype both for source identification and for structural studies.

Waveform analysis represents a research method which goes beyond the simple time measurements on seismic records, especially by including the whole wavegroup into the study. Analysis of waveforms or waveshapes of seismic waves works entirely in the time domain, with given records. The aim of such studies is in general to associate at any given station certain observed wave characteristics with each epicentral region. Such work is therefore based on the experience that such characteristics exist, depending on similarities partly in the earthquake mechanism, partly in the propagation paths, for any epicentral region and any given station. When such characteristics have been found by analysis of a greater material of records, they can partly serve the purpose of preliminary epicenter locations and source identifications, partly contribute to studies of source mechanism and path properties.

As an alternative to accumulating characteristics for many regions from records at one or a few stations, it is naturally possible to make comparative studies on worldwide seismograph networks of waveforms from a number of events. The latter procedure will probably yield more information. The double or multiple onsets of P with one or several small precursors followed after a few seconds by a larger onset could certainly benefit from an investigation of the last-mentioned type. Knowledge of the multiple P-waves would be of importance not only to epicentral calculations but also to our apprehension of focal mechanism and wave propagation (multiple shocks and/or multiple propagation paths).

Extensive waveform studies were made earlier in seismology (Vesanen, 1942, 1946), and they seem to have met with some renewed interest recently. Waveform analysis can be made on any seismic wave, and for a discussion it is suitable to differ between body waves and surface waves.

1.3.1 Body waves

That certain regions nearly always are accompanied by specific features is a well-known phenomenon. For example, at Swedish stations, earthquakes in the North Atlantic have generally more long-period double onsets within the P-group, usually less sharp than from the Asiatic side. Likewise, Greek earthquakes have mostly double onsets, 4 to 5 sec apart. Mexican earthquakes produce exceptionally long-period phases, and so on. No doubt, body waves carry the imprints both of source mechanism and path properties, and it may often be a difficult task to separate the two.

On the other hand, it has to be made clear that waveforms of body waves do not alone permit a location and identification of events. Even though mechanisms may be fairly similar within any one epicentral region, there are frequently individual deviations, which will affect the waveforms. This is well-known from numerous focal mechanism studies. As one example, I may mention that in a comparative study of over forty years of Uppsala Wiechert records of numerous intermediate-depth earthquakes in Hindu Kush, I found only two shocks, separated by many years (1922 and 1939), which were so similar that one was like a true copy of the other. The others all showed variations, even though their main features were similar. As another example, it may be mentioned that it is no great problem to find P-phases from deep-focus earthquakes which look exactly like P-phases from nuclear explosions. Clearly, a separation exclusively from the P-wave form is not possible, at least not if we restrict ourselves to just one station.

1.3.2 Surface waves

The shape of the surface waves, especially their dispersion, is an effect of the propagation path, both its length and its structure, and the source has no influence on this phenomenon. The source may exert its influence on the surface waves mainly in the relative amplitudes of different components, i.e. their spectral composition, and in the relative excitation of Love and Rayleigh waves, which is a function of source mechanism and direction to station. Also the initial phase at the source influences the observed phases. From personal experience, I have found the surface-wave shape to be an extremely sensitive indicator whether given phenomena are of identical origin or not. The simple method used is to trace one of the wave trains on a piece of transparent paper and to place this on top of the other record. If every detail in two records coincides, then they are certainly of the same origin. The method is very sensitive to variations in distance (the amount of which can be judged from dispersion curves), but also to azimuth, because of the azimuthal variation of crustal structure (especially the uneven distribution of oceanic and continental structures). For preliminary and quick location of events, the surface-wave method appears superior to the body-wave method, especially if for the station investigated, a "waveform library" has been accumulated. Moreover, the surface-wave method can be applied also to relatively weak events, as soon as a dispersed wave train can be seen, whereas in such cases P and S are usually too small for any trustworthy waveform analysis. Because of the dependence of surface-wave shapes on the initial phase, the method is best applicable to cases with the same initial phase, i.e. explosive sources (Von Seggern, 1972). But if the method is used essentially to test identity of events, then it can be used more generally; any differences found could then depend either on slightly different epicenters, different depths or different initial phases.

A counterpart to classification of events by their waveforms would be a classification by means of spectra of waves. Apparently, no systematic study has been made of the relative potentialities of these two approaches for classification of seismic events. A piece of information of interest in this respect can be found in Toksöz and Ben-Menahem (1964), who demonstrate very great similarity between amplitude spectra of Rayleigh waves from atmospheric explosions at a given point and recorded at a given station. Different stations, even relatively close, give quite different spectra for the same explosion, depending upon path influences. For earthquakes, greater variability is to be expected.

1.3.3 Synthetic seismograms

There is also another aspect of seismic waveform analysis to be emphasized, i.e. analysis of records in the time domain. The analysis referred to above is mostly concerned with classification of waveforms, based on descriptive analyses without much effort to really explain observed features. This is especially true for body waves. For surface waves, dispersion studies have naturally contributed to explanations of observed waveforms. The other approach, I am now referring to, is based on a mathematical analysis, consisting in the deduction of a so-called *synthetic seismogram*, based on certain assumptions about source, path and receiver properties. Comparison of synthetic and observed waveforms will then enable successive modifications of the theoretical assumptions so as to reach close agreement. We shall list briefly the most important points in the development of synthetic-seismogram studies.

(1) Probably the first theoretical seismogram was constructed by H. Lamb already in 1904 (see Båth, 1968, chapter 12, for a full description). Due to a number of simplifying assumptions, this seismogram is fairly far from really observed seismograms, but his effort, commonly called *Lamb's problem*, has played an enormous role in seismology. Several later theoretical developments of this problem have been made.

(2) Next we like to mention Ricker (1940, 1943, 1944, 1953), who developed the *wavelet theory of seismogram structure*. Starting from a sharp pulse at the source, corresponding to an explosion, he calculated its changes of shape during propagation, essentially due to modification because of viscosity, which involved dissipation. The seismogram is explained as constituted of a number of overlapping wavelets, whose shape, width and amplitude vary with distance from the source. These efforts were essentially directed towards application in seismic

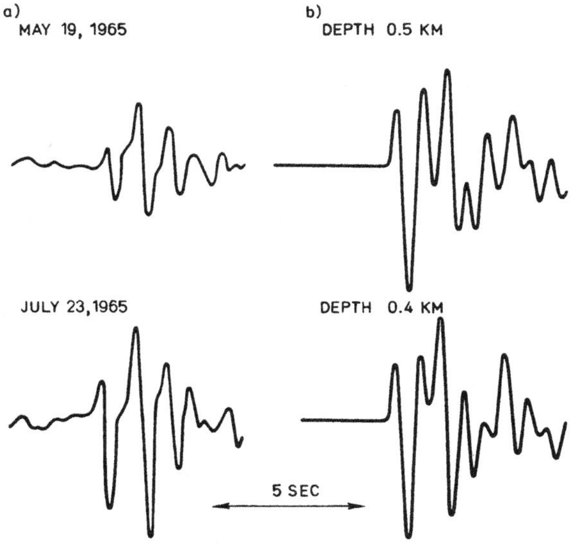

Fig.3. Vertical-component P-wave records from Umeå, Sweden, of a) Nevada Test Site underground nuclear explosions compared with b) corresponding synthetic seismograms. After Kogeus (1968).

prospecting work, where tests showed good agreement with theory. Further application especially to ScS–waves was made by Ōtsuka (1963).

(3) Renewed interest concerning synthetic seismograms, with still more realistic earth models, arose during the 1960's, partly in connection with efforts to discriminate underground nuclear explosions and earthquakes, partly in connection with more detailed investigations of source functions of earthquakes after the development of the moving-source theory. These studies concern both near and distant events and are essentially restricted to the time domain. Among a number of such studies, I like to mention Kogeus (1968), Beaudet (1970) and Hasegawa (1971a), who synthesized records of underground explosive sources (Fig.3), and Berckhemer and Jacob (1968), Mikumo (1969, 1971, 1972), Bollinger (1968, 1970) and Kanamori (1970a, b), who were able to make deductions about earthquake source properties from P-wave forms and from synthetic surface waves, respectively. Jensen and Ellis (1971) developed methods for computation of synthetic P- and S-waves, using linear systems theory, and A. Douglas et al. (1972) constructed seismograms both for P– and Rayleigh waves, both for explosions and earthquakes. Also the internal structure of the earth has been investigated by the synthetic seismogram method, by Helmberger and Wiggins (1971) by means of body waves. Already earlier, methods had been developed for calculation of synthetic surface waves by Satô (1960), with application to experimental flexural waves, and by Aki and Nordquist (1961), with application to earthquake-

generated Rayleigh waves. Likewise, Chander et al. (1968) synthesized PL–waves. The synthetic–seismogram method has also become of great significance to seismic prospecting, especially by providing a means of studying the P-wave forms for reflection and transmission through thinly stratified multilayered media (Berzon and Ratnikova, 1971).

(4) Probably the theoretically most attractive recent developments of synthetic–seismogram construction and waveform analysis are based on superposition of the normal modes of the earth's vibrations, which represent all kinds of seismic waves. Among several notable contributors to this important field, I like to mention Y. Satô in Japan and Z. Alterman in Israel with their collaborators. Recent, extensive reviews of this field of study have been given by Landisman et al. (1970) and Alterman and Loewenthal (1972).

Finally, it should be observed that the more recent developments of waveform analysis, especially those mentioned under 3 and 4 above, would not have been possible without access to large electronic computers, as distinct from developments discussed in earlier sections of this chapter.

Even though waveform analysis may have had its greatest success in analyses of signals from earthquakes and explosions, it is clear that it will also prove useful in numerous other fields of geophysics:

(1) Microseisms generally exhibit typical forms, mostly of quite different periods, depending upon source generation and source distance.

(2) Atmospheric pressure waves exhibit typical waveforms, partly due to dispersion, as for seismic surface waves. Due to the time variation of the atmosphere, use of waveforms in the atmospheric case is rendered more difficult than for seismic surface waves.

(3) Meteorological records often exhibit typical waveforms, as for example a thermogram and a barogram during a cold-front passage.

(4) The same is true for records of many other phenomena, as for example of ocean surface waves and swell or of geomagnetic storms, etc.

An interpreter's detailed knowledge of such waveforms and characteristic features of his records is of immense help in evaluations, especially for quick, preliminary estimates of sources, their nature and distance, for any recorded phenomenon. If he tries to apply waveform analysis to seismic records, a combined study of body and surface waves will then lead to the safest result.

1.4 FREQUENCY AND SPECTRAL ANALYSIS

The analysis methods discussed hitherto are characterized by their restriction to the time domain. The term frequency is seldom met with in the early literature, and when needed it is substituted by the period. Wherever there is a function of any physical phenomenon which fluctuates in time and/or space, the rate of fluctuation, i.e. the frequency or the wavenumber, becomes a significant parameter in defining the function. The seismic signal is a typical example of such a function.

The frequency (and/or wavenumber) is in many respects a more significant and more useful independent variable to use than just the time (and/or space coordinates). Transforming a given record into the frequency domain does not mean the addition of anything new but only a rearrangement of the given data in a different order, i.e. arranged according to frequency instead of according to time sequence. The significance of such a representation is above all clear from the fact that most geophysical phenomena are expressed theoretically in frequency-dependent form. Therefore, this form of presentation is more directly related to the phenomena under study. Moreover, a number of mathematical operations, e.g. those of filtering, are easier to apply in the frequency domain than in the time domain. Another advantage of the spectral analysis is that the whole signal shape is used in the analysis, whereas the common time-domain measurements on seismograms are point measurements, as for instance, of the arrival time of the first onset, its displacement direction, or of the maximum amplitude in a wave group. Analyses of whole wave groups are no doubt bound to yield much more information than just point measurements in the time domain. The transform of a record into the frequency (and/or wavenumber) domain is termed its *spectrum*.

The background of the word *spectrum* is both linguistic and historical. From the language point of view, spectrum is derived from the Latin verb *specio* = "see", something that one sees, i.e. "appearance", "form", "apparition", especially of a visionary and/or imaginary character (Dr. S. Hedberg, Uppsala, personal communication). In the history of science, it is to be noted that Sir Isaac Newton in 1671 used the word spectrum as a Latin loan-word in this sense.

In a general sense, a spectrum expresses nothing other than a function, i.e. the dependence of some quantity on one or several independent variables. In science, the term spectrum has a more restricted implication. But still it represents the dependence of some function on one or several independent variables. The function may be amplitude or power or any other property, and the independent parameter is generally frequency and/or wavenumber.

Spectrum as applied in geophysics (and in other fields of science) is a statistical quantity with a definite mathematical expression, derived from time or space functions by certain transformations. The spectrum is in general a complex function, represented by the following two forms, both used in the geophysical literature:

(1) sum of real and imaginary parts:

$$F(\omega) = a(\omega) - ib(\omega)$$

(2) product of real and complex parts:

$$F(\omega) = |F(\omega)|\, e^{i\Phi(\omega)}$$

where:

$$|F(\omega)| = [a^2(\omega) + b^2(\omega)]^{\frac{1}{2}}$$

$$\Phi(\omega) = \tan^{-1}\left[-\frac{b(\omega)}{a(\omega)}\right] + 2n\pi$$

$$n = 0, \pm 1, \pm 2, \ldots$$

In the expressions above, $F(\omega)$ is the function whose variation with the independent variable ω is described. That is, $F(\omega)$ is the spectrum. If $a(\omega)$ and $b(\omega)$ are amplitudes, then $|F(\omega)|$ is the *amplitude spectrum* and $\Phi(\omega)$ the corresponding *phase spectrum*. The independent variable is usually frequency (ω or v) or wavenumber (k). In the expressions above, there is only one independent variable ω. This is termed one-dimensional spectral analysis. But it is possible to extend this concept to two or more variables, for example, using wavenumbers in two spatial directions, k_x and k_y, as independent variables.

The conversion from a geophysical record (in the time domain) to a spectrum (in frequency domain) requires both an advanced mathematical discussion and large computers for its practical performance. This will constitute the main content of this book. But in addition there are some non-analytical methods developed, by which a spectrum can be directly recorded. Such methods have only found a limited application, but in view of their importance they will be reviewed in section 1.6.

There is now an enormously large literature on spectral analysis from various points of view. Due to its wide range of applicability, the field has attracted scientists from many different branches. There are many basic theoretical treatments which fall within the realm of pure mathematics. Being an essentially statistical treatment of data, it is also quite natural that statistical journals contain much on spectral analysis. From the field of application, we have to mention engineering, especially electrical and electronics engineering, where spectra became a common tool earlier than in geophysics. The presentations by different scientists are rather different, depending upon whether they approach this field from the pure mathematical point of view, or from statistics or from engineering applications viewpoints. In the present book we shall be concerned with spectral applications in geophysics, especially seismology. Therefore, our approach will be more on the practical side, but so much of basic theory will be incorporated that the student will get a full understanding of the various operations to be made in order to obtain and to interpret a spectrum.

Geophysical applications of Fourier transforms comprise both their use in theoretical discussions, including the solution of differential equations (Båth, 1968, pp.220–227), and their use in handling observations, to deduce Fourier spectra. The theoretical applications, in general being quite straight-forward, will not be dealt with in this book. On the other hand, the observational applications need consideration of a large number of influencing factors. These will constitute the main subject of this volume.

Any observational study consists essentially of three stages: (1) acquisition of data; (2) handling of data; and (3) conclusions from data.

In this book, we shall be almost exclusively concerned with item 2, and then only concerning calculation of spectra and related quantities, as correlation functions. Numerous examples from various branches of geophysics will be given which clearly demonstrate in what ways spectral studies have enriched geophysical investigations, by looking at phenomena in another domain.

1.5 SPECTRA VERSUS TIME-DOMAIN ANALYSIS

The advantage of spectra is that they work in the frequency (or wavenumber) domain, where the independent variable (frequency or wavenumber) provides for a reliable and unique check that comparisons of different records are referred to the same value of this parameter. Measuring only in the time domain, we may in many cases not be sure that the comparison has been adequately made, as time alone does not provide such a unique control that we are comparing exactly the same thing in two records. This remark concerns all phenomena where frequency (wavenumber) is of decisive significance. In addition to being more informative, the frequency-domain presentations are often simpler to handle computationally.

However, in spite of this unambiguity of frequency-domain measurements versus time-domain measurements, it should not be overlooked that there are cases where fully unambiguous comparisons of records are possible already in the time domain. From my own experience in seismology, I may mention two examples:

(1) Records of an identical source (same location, same mechanism) at the same station, e.g. records at one and the same station of atmospheric nuclear explosions over Novaya Zemlya (Båth, 1962).

(2) Records of surface noise propagating downwards along a vertical profile of seismometers, at least in some cases (Båth, 1966a).

In both these cases from my experience, it has been perfectly possible to compare all records in each group and identify them, wave by wave, even wiggle by wiggle. In such exceptional cases, spectral analysis is practically superfluous, because the amplitude ratio between corresponding points in two or more records will bear a constant ratio to any spectral ratio which can be formed. Moreover, the time-domain measurements provide such information with the same accuracy at a fraction of the work needed for a spectral analysis.

It should nevertheless be observed that such identical records are much more an exception than a rule, and this means that in other cases, which constitute the majority in geophysical research, spectral analysis has a definite advantage in unambiguity over the simple time-domain studies.

1.6 NON-ANALYTICAL (ANALOGUE) METHODS FOR SPECTRAL DETERMINATION

In present time, most spectra are calculated on large computers. This is the analytical method, against which we have to consider certain non-analytical methods. They were partly developed already early in geophysical history, in pre-computer time, partly recent developments have been made, which conveniently supplement computer methods. Nowadays, such methods may have justification only when large amounts of data should be handled in some special way, where the job to construct a special analyzer would be repaid in time-saving. In the following review we shall differ between three groups of non-analytical spectrographs or spectral analyzers and we shall concentrate mostly on recent developments.

1.6.1 Resonators and filter methods

This approach has two aspects. The earlier and simpler one is to use a series of well-tuned seismographs, each of which will record within a limited spectral band, equivalent to mechanical filtering. Instead of using a multiple instrumental set-up at each place, it is often more advantageous to use only one or a few broad-band instruments with magnetic-tape recording, which then can be passed through variable filters or a battery of filters (low-pass, band-pass, high-pass), usually electrical. This constitutes the second approach, which is the most common one to-day. The output will give the spectral amplitude or (if squared) the energy corresponding to the center frequency of each filter setting. The narrower the filter bands are, the more details will naturally come out in the spectrum.

Frequency analysis, essentially based on this technique, has been applied by Milne (1959) to T-phases, using a single band-pass filter of variable center frequency, Grossling (1959) to records of underground explosions, also using a variable narrow-band filter, Willis and Johnson (1959) and Willis (1963a, b, c) to records of quarry and underwater shots, Kato et al. (1966) to records of geomagnetic pulsations, Tsujiura (1966, 1967, 1969) to records of body phases from near and distant earthquakes, Lambert et al. (1972) to records of seismic surface waves, calling the resulting spectra *group spectra*. Besides permitting the construction of spectra, band-pass filtering in the time domain frequently leads to higher time accuracy in the reading of phase arrivals. As demonstrated for instance by Willis and Johnson (1959) and Willis (1964), in records of near explosions, some phases become readable only in low-frequency outputs, while they are practically unreadable in high-frequency outputs or in the original broad-band records.

The reader will find an extensive treatment of resonators and their spectral properties in the book by Kharkevich (1960). Spectral analyzers nowadays available on the instrumental market, are generally based upon narrow-band electrical filtering.

1.6.2 Sound spectrograph or acoustical spectrograph

The basic principles of this spectrograph are based on instruments used in the frequency analysis of sound waves. Developments of sound spectrographs about a quarter century ago are well described by Koenig et al. (1946). The sound is transformed into electric analogue form via a microphone and recorded in a three-dimensional plot: horizontal axis is time, vertical axis is frequency and the energy or intensity is shown by varying darkness of the plot. Whereas the time and frequency axes could be read accurately, the energy could not be read quantitatively in a reliable way from earlier constructions. Thanks to improvements by various methods, reliable quantitative energy measures were made possible by Koenig and Ruppel (1948) and Kersta (1948). The latter produced time sections, i.e. two-dimensional cuts for given instants, showing energy (or amplitude) versus frequency, just like an ordinary spectrum.

With the three-dimensional presentation, a vertical line clearly gives the intensity–frequency spectrum at any chosen moment, whereas a horizontal line shows the intensity–time variation for any given frequency. As compared to representation by a single intensity–frequency spectrum, the sound spectrograph produces more detailed information, which also allows for the non-stationarity (i.e. variability with time) of the signal; in other words, it provides for *time-varying or instantaneous spectra* (cf. section 3.6.5).

Another potentiality of this method is its flexibility. As soon as a signal is available in electrical analogue form, then electrical filters and other devices provide for a great variability in the presentation, also for a wide range of frequencies and of resolution, both in time and frequency. The method is available without modification to any signal, e.g. seismic signals, which can be obtained in electrical analogue form.

Khudzinskii and Melamud (1957) describe a frequency analyzer, mainly for exploration seismology, which is a modified version of the Rodman analyzer, originally developed for analysis of sound waves. Their paper gives also a review of earlier Russian frequency analyzers, going back to Galitzin's days (1913).

With an electronic spectrograph, developed from R. K. Potter's sound spectrograph (magnetic-tape record with a heterodyne type of analyzer, recording the output on spark-sensitive facsimile paper), Ewing et al. (1959, 1961a, b) and Landisman et al. (1962) analyzed transient seismic signals, claiming discovery of P-, PKP- and possibly S-wave dispersion, while Oliver and Page (1963) studied long-period microseisms. In principle, the given record is in electrical analogue form (on magnetic tape, mounted on a drum), which is fed through an electrical circuit whose band-pass filtering is varied, by which successive spectral components of the given record are filtered out. The method provides for high resolution of composite transient signals, both surface and body waves, and is thus superior

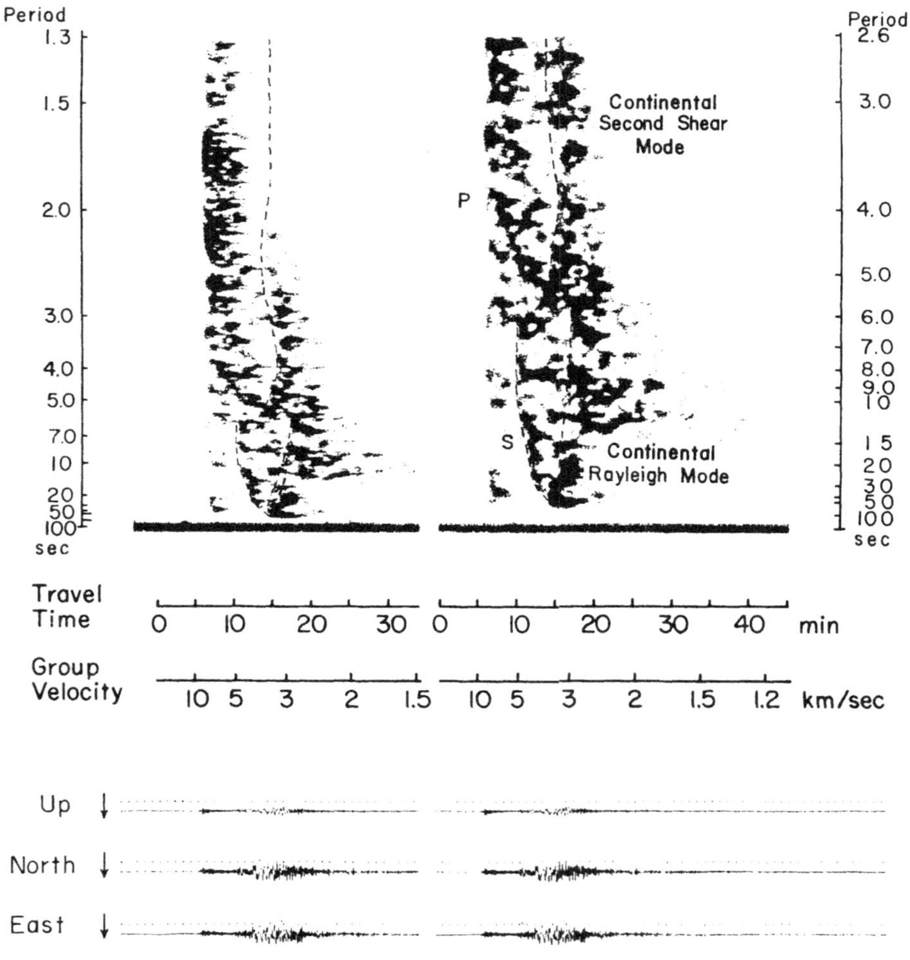

Fig.4. Sonagraph analysis of the Oaxaca, Mexico, earthquake of December 12, 1951, as recorded at Pasadena. After Ewing et al. (1959) by due permission of the authors.

to the simple peak-and-trough method and to Fourier analysis. The latter, like the methods described under 1.6.1, provide time-averages, whereas the spectrograph instead displays the spectrum as a function of time. The great advantage of displaying frequency versus time (i.e. time-varying spectra) is to demonstrate clearly the temporal development of the phenomenon under investigation. An example is shown in Fig.4. A further modification of this spectrograph, providing for continuous recording, was developed and used by Walker et al. (1964) for microseism studies. Application of the sonagraph method to spectral analysis of seismic waves from microearthquakes (section 8.1.6) is reported by Terashima (1968) and Srivastava et al. (1971).

P. F. Smith et al. (1955) describe a method, using a heterodyne-type sonic analyzer, for direct recording of a continuous amplitude-frequency spectrum of sound in the ocean, suitable for investigating oceanic noise, such as of meteorological, biological and volcanic origin. More application of sound spectrographs to underwater sound recording is reported by Schevill et al. (1962), R. H. Johnson and Norris (1972), and others. Among applications of sonagram analysis we mention also investigations of the daily variation of geomagnetic pulsations of short period (Hirasawa and Nagata, 1966; Kawamura, 1970). A new and considerably improved version of the sound spectrograph was described by Presti (1966).

1.6.3 Light spectrograph or optical spectrograph

Just as sound spectrographs require the signal to be in electrical analogue form, the optical spectrographs require the signal to be in optical-intensity analogue form. Such records may be *variable-area records*, which have been blackened to one side of the record trace, or they may be *variable-density records*, in which the photographic density is proportional to recorded amplitudes. Presentations of these kinds display reflected waves across a set of records much clearer than the ordinary wiggly-trace photographic records do.

An optical scanning method, developed by Jackson (1964, 1965), uses the monochromatic light diffraction caused by opaque obstacles. The principle of the method is obvious from Fig.5, where the set-up is compared with the corresponding one used in physics for analyzing light of unknown composition. In the seismic application, the light is known (monochromatic, preferably laser), whereas the seismogram represents the unknown structure to be analyzed. The Fraunhofer diffraction is analogous to the Fourier transform, as the diffraction process is dependent on the sizes of the elements in the diffraction screen. A spherical lense produces a two-dimensional transform and a cylindrical lense a one-dimensional transform. For further details, see Arsac (1966), Porcello et al. (1969) or textbooks in physics, for example Crawford (1965, pp.481–495). Like the sound spectrograph,

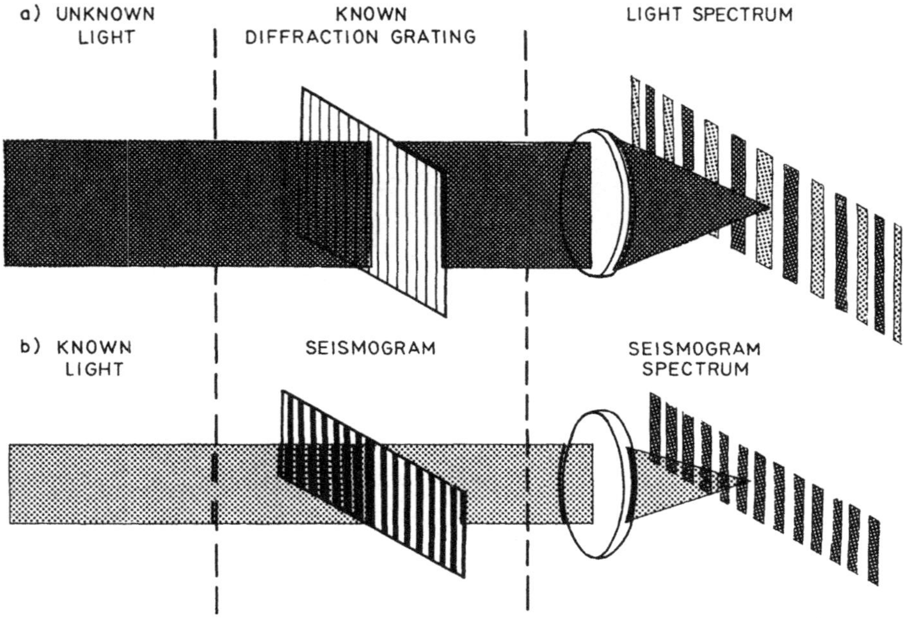

Fig.5. Comparison of a) optical spectroscopy with b) diffraction analysis of seismograms. After Jackson (1965).

the optical spectrograph gives a three-dimensional presentation of the wave structure, with the same coordinates.

This method of analysis has found application to seismic prospection records, but cannot be applied to ordinary teleseismic records. The method compares favourably with digital and analogue computer methods, especially as it permits the operator to inspect and to modify the process at every point much easier. The method permits a number of operations to be made on the spectra, such as various

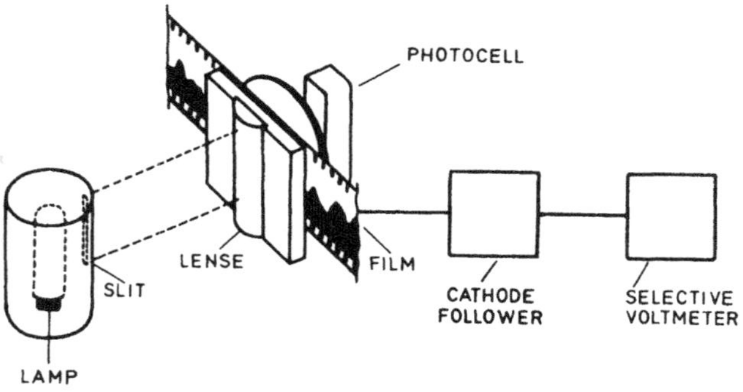

Fig.6. Photomechanical analyzer. After O. Kulhánek (personal communication, 1972).

kinds of filtering, especially elimination of multiple reflections and noise, correlations, etc. Well-written descriptions of the method have been published by Dobrin et al. (1965) and Grau (1966). Other methods, based on geometrical optics, also permit calculation of correlations and convolution (Jackson, 1968).

The same principle has found application for obtaining two-dimensional Fourier spectra of the sea surface from photographs of sun glitter, i.e. sunlight reflected from the ocean surface (Stilwell, 1969). In the transform plane, the radius to any point is proportional to the wavenumber and its direction indicates the direction of wave propagation. The optical density is related to the ocean-wave spectral amplitude.

Several alternative forms of optical spectralizers exist (see for example Manley, 1945, p.219; Howell et al., 1959; Saito, 1960). Essentially, a photomechanical wave analyzer (Fig.6) consists of a photomechanical transducer and a measuring instrument. The transducer transforms a graphical record of seismic vibrations into electrical form. The portion of the record to be studied is photographically transferred onto a 35-mm film. The lower or the upper portion of the film is blackened so that the trace itself forms the border between the opaque and transparent part, thus producing a variable-area record. The film is passing with a constant speed between an illuminated slit and a photocell. By properly selecting the width of the slit, the illuminated area of the photocathode, and thus also the photocell current, are at any moment proportional to the instantaneous amplitude of the analyzed trace. The photocell current is led through auxiliary components to a selective voltmeter where the spectral amplitudes can be read directly. If the speed of the film can be gradually changed within a prescribed range, then the selective property of the measuring instrument is no longer necessary. The frequency range of the transducer depends on the width of the slit. Towards higher frequencies a narrower slit is required. Saito (1960) uses a combination of a photomechanical analyzer and a sonagraph recorder for determination of spectra of geomagnetic pulsations.

1.6.4 Other analogue methods

Analogue methods, as distinct from digital methods, start generally from a graphical representation of the curve $f(t)$ to be analyzed. A large number of different mechanical, electrical and optical analyzers have been developed for this purpose, too many to review here and nearly all of them only of historical interest today. For instance, A. A. Michelson constructed in 1898 a harmonic analyzer, which made it possible to determine the first 80 Fourier components of a given curve. It could also be used as a harmonic synthesizer. Stumpff (1937) presented old methods of this kind in great detail. Barber (1961) has also reviewed a great number of such methods for determination of Fourier coefficients as well as for filtering. Such methods exhibit much ingenuity in their construction, but are of relatively little use in the present computer era. They earlier found application to the analysis of meteorological time series as well as other periodic phenomena, but were more seldom used for analysis of seismic waveforms. It is also to be noted that waveform analysis found wide application in mechanical and electrical engineering much earlier than in geophysics.

Among other special methods for wave analysis, we may mention the *envelope method* (Manley, 1945), applicable only to particularly simple waveforms, sometimes encountered in engineering applications. Examples of the envelope method applied to seismic surface waves of Rayleigh type can be found in a paper by Cleary and Peaslee (1962).

A number of different apparatuses were earlier constructed not only for Fourier analysis, but also for such related calculations as of auto- and cross-correlations, etc. See, for instance, Shima (1962). Also such calculations are nowadays generally taken over by digital computers, and therefore we do not see any need to deal with such various constructions in this book.

1.7 ELECTROMAGNETIC VERSUS MECHANICAL SPECTRA

It is instructive to compare the electromagnetic spectra, well-known from physics, with mechanical spectra, to be dealt with in this book. By mechanical spectra we denote spectra of mechanical quantities, as distinct from spectra of electromagnetic phenomena. Mechanical spectra may refer to observations of seismic waves, meteorological turbulence, ocean surface waves, variations of gravity or geomagnetic elements, just to give a few examples. A comparison can be summarized in the following points.

(1) *Method of generation.* Electromagnetic spectra or optical spectra (being a narrower range) are produced in laboratory physics by refraction of light through a prism, which produces a line and/or a band spectrum. These correspond to the frequency spectra in the mechanical case. Diffraction patterns in optics are a counterpart to the wavenumber spectra in the mechanical case; these will not be considered further in the present review.

(2) *History.* The optical spectra have much longer history, the developments starting already in 1859 with the physicists R. W. Bunsen and R. Kirchhoff. For mechanical spectra, it is not easy to give just one name with equal dominance, but the name of J. W. Tukey and the year 1949 seem a valid counterpart. The relatively late development of mechanical spectral methods is connected with the advent of high-speed electronic computers, even though some early developments of spectral methods are due to A. Schuster during the first decade of the 20th century. We do not consider in this connection harmonic analysis, which has a much longer history than mechanical spectra, and can be traced back to J. Fourier. Later development of harmonic analysis is marked by such names as A. Schuster,

H. and Y. Labrouste, D. Brunt, F. Vercelli, J. Bartels, A. T. Doodson, and others.

(3) *Spectral parameters.* Optical spectra display power (intensity) against frequency. Mechanical spectra use a much greater variety of presentations, including also phase spectra.

(4) *Type of spectra.* Optical spectra are line and/or band spectra, depending upon the state of the radiating matter, whether atoms or molecules. Mechanical spectra may be line spectra or continuous spectra, and this depends partly on the source: a periodic source gives a line spectrum and a random process gives a continuous spectrum, partly and essentially on the method of analysis: the Fourier series method yields line spectra and the Fourier transform method continuous spectra.

(5) *Precision.* Optical spectra, obtained under well-controlled laboratory conditions, represent a higher degree of precision than mechanical spectra. Among the latter, it is practically only spectra of the earth's free oscillations that can stand a comparison with optical spectra in resolution and accuracy.

(6) *Range.* Electromagnetic spectra span over an enormously larger range than mechanical spectra. The spectra meet in general at periods of approximately 0.1–0.01 sec, from where the electromagnetic spectrum extends down to periods of around 10^{-24} sec. Even if we would take a mechanical period from the oldest geological era until now, the electromagnetic spectrum extends over periods of five more orders of magnitude. Among geophysical variations, probably the observed or inferred periods of geomagnetic elements cover the largest range, from about 10^{-4} sec to 10^{11} sec (section 10.1.1). See Fig.7.

(7) *Explanation of spectra.* For optical spectra, obtained under well-defined laboratory conditions, almost all effort goes into explanation and identification

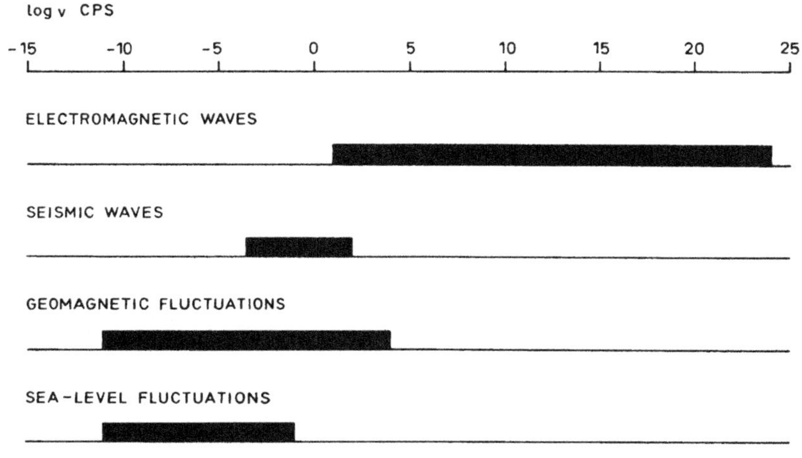

Fig.7. Frequency ranges of some geophysical phenomena compared to the electromagnetic spectrum.

of observed lines and bands, whereas relatively little is said about the influence of experimental conditions. For mechanical spectra, however, there is much concern about methods of calculating spectra, by any procedure, and there is at least as much effort to be spent on spectral calculation as on spectral interpretation. This state of affairs probably explains why in calculation of mechanical spectra, practically nothing can be learnt from the methods to obtain optical spectra.

Chapter 2

FOURIER SERIES AND FOURIER TRANSFORMS

Fourier spectra are dealt with extensively in numerous textbooks. It is not the intention here to recapitulate this in full. Our purpose is rather to direct attention to basic theorems concerning Fourier series and Fourier transforms, which will be necessary for an understanding of the following chapters of this book. The following summary is sufficient for most geophysical applications. For the reader's convenience we include pertinent proofs of formulas, but we have followed the degree of mathematical rigour that is considered sufficient in applied mathematics. We hope that by the end of this chapter the reader will have got a clear understanding of the relation between an analytic function (in the time domain) and its spectrum (in the frequency domain).

In order to get well acquainted with mathematical operations in this as well as in following chapters, the reader is recommended to apply theorems to simple functions, such that the results can be easily understood, at least intuitively. This will give a better insight into the theorems and an ability to trust their application in more complicated cases, such as usually represented by observational data. Also in order to test computer programs, it is a recommended rule to apply the programs to simple functions, for which the results are known or can be easily found analytically.

The following presentation is essentially based on the books by Manley (1945), Jennison (1961), Papoulis (1962), Bracewell (1965), Panter (1965), Lanczos (1966), Arsac (1966), Båth (1968), Jenkins and Watts (1968) and Stuart (1969). Besides providing basic theory, several of these books contain numerous applications of spectral analysis to problems in optics, acoustics, electronics, electrical engineering and information transmission. The books by Davenport and Root (1958) and by Jenkins and Watts (1968) contain extensive treatments from the statistical side. See also Van Isacker (1961), Jenkins (1961a) and Hannan (1966). The reader will also find much valuable material in the books by Bendat (1958) and by Bendat and Piersol (1971), also to a great extent from the statistical side, as well as in Kharkevich (1960) and Hamming (1962). A number of special problems are dealt with in the multi-author book edited by Harris (1967). Numerous useful exercises can be found in the book by Grau (1966). A recent book by Otnes and Enochson (1972) is also recommended, especially for the numerous valuable computational instructions it contains. In addition, there are numerous other books dealing with Fourier theory more from the viewpoint of pure mathematics.

Among these, it may here suffice to refer to the books by Lighthill (1958) and Wiener (1959).

2.1 FOURIER SERIES

Joseph Fourier (1768–1830), a French engineer and mathematician, discovered his theorem in solving problems in heat conduction. He presented his theory to the French Academy of Sciences on December 21, 1807. Even though other mathematicians had proposed closely related theorems, and even though Fourier's own contribution to this field may be subject to discussion, it is nevertheless his name which has been connected with this development in all books over the world.

According to Fourier's theorem, any function $f(t)$ satisfying certain restrictions, can be expressed as a sum of an infinite number of sinusoidal terms. Dirichlet in 1829 formulated restrictions under which the theorem is valid. Briefly stated the *Dirichlet conditions* are as follows:

(1) $f(t)$ should be periodic, i.e. $f(t) = f(t + 2\pi)$, where 2π is the period. If $f(t)$ is not periodic, but defined over a finite range, the sum of the sinusoidal terms will still converge to $f(t)$ over the defined range. Outside the range, the sum will represent repetition of $f(t)$.

(2) $f(t)$ should be at least sectionally continuous, with at most a finite number of discontinuities and finite jumps.

(3) $f(t)$ should possess a finite number of maxima and minima.

(4) The integral

$$\int_{-\pi}^{\pi} f(t)\,\mathrm{d}t$$

should be convergent (which follows from (2)). A frequently occurring, alternative formulation of the conditions is that the integral

$$\int_{-\infty}^{\infty} |f(t)|\,\mathrm{d}t$$

should be finite, where $|f(t)|$ is measured from its mean value.

The Dirichlet conditions appear in slightly different formulations in different treatises, but in essence they correspond to the formulation above. It is to be noted that these conditions for $f(t)$ are *sufficient* but not all of them are necessary. This means that the Fourier theorem is valid for a much wider class of functions. This was demonstrated by Fejér in 1904, who by a different summation procedure showed that the only requirement is that $f(t)$ is integrable. However, this extension is more of a purely mathematical interest than of physical significance. All known functions encountered in the physical world fulfill the more restrictive Dirichlet conditions.

2.1.1 Theoretical development of Fourier series

According to the Fourier theorem, a function $f(t)$ having a fundamental period of 2π and satisfying Dirichlet's conditions, can be represented by the following infinite *Fourier series:*

$$f(t) = a_0 + \sum_{n=1}^{\infty} (a_n \cos nt + b_n \sin nt) \qquad [1]$$

where a_0, a_n, and b_n are constants.

The constants are determined by multiplying [1] in turn by $\cos(0)t$, $\cos nt$ and $\sin nt$ and integrating over the period length 2π with respect to t, using the orthogonality properties of the sine and cosine functions, that is:

$$\int_{-\pi}^{\pi} \sin mt \sin nt \, dt = \int_{-\pi}^{\pi} \cos mt \cos nt \, dt = \begin{cases} \pi \text{ for } m = n \\ 0 \text{ for } m \neq n \end{cases}$$

$$\int_{-\pi}^{\pi} \sin mt \cos nt \, dt = 0 \qquad \text{for all } m \text{ and } n$$

where m and n are integers.

Integration as well as differentiation can be carried out term by term of the Fourier series [1] for all physically realizable functions. By the integration procedures we get:

$$\left.\begin{aligned} a_0 &= \frac{1}{2\pi} \int_{-\pi}^{\pi} f(t) \, dt \\ a_n &= \frac{1}{\pi} \int_{-\pi}^{\pi} f(t) \cos nt \, dt \\ b_n &= \frac{1}{\pi} \int_{-\pi}^{\pi} f(t) \sin nt \, dt \end{aligned}\right| \qquad [2]$$

For the general case when the fundamental period is T, we get the following expressions for the Fourier coefficients and the Fourier series:

$$
\left.
\begin{aligned}
a_0 &= \frac{1}{T}\int_{-T/2}^{T/2} f(t)\,\mathrm{d}t \\
a_n &= \frac{2}{T}\int_{-T/2}^{T/2} f(t)\cos\frac{2n\pi t}{T}\,\mathrm{d}t \\
b_n &= \frac{2}{T}\int_{-T/2}^{T/2} f(t)\sin\frac{2n\pi t}{T}\,\mathrm{d}t \\
f(t) &= a_0 + \sum_{n=1}^{\infty}\left(a_n\cos\frac{2n\pi t}{T} + b_n\sin\frac{2n\pi t}{T}\right)
\end{aligned}
\right| \qquad [3]
$$

It can be proved (see literature referred to in the introduction to this chapter) that the coefficients a_n and b_n converge to zero as n increases to infinity, provided $f(t)$ is integrable. We have here measured $f(t)$ from an arbitrary base-line (level a_0), but we could equally well measure $f(t)$ from a zero base-line ($a_0 = 0$). The latter is the most usual procedure, used in the rest of this book, unless otherwise mentioned. The deductions in this and the following sections are not influenced by the use of an arbitrary base-line.

The coefficients a_n and b_n are called *Fourier coefficients* and their calculation is called *Fourier analysis* or *harmonic analysis*, more specially *waveform analysis*. The expansion [1] is valid for all kinds of functions $f(t)$, fulfilling the Dirichlet conditions, whether real or complex, only that the corresponding Fourier coefficients will be real or complex, respectively. For a physical time-series, like a geophysical record, $f(t)$ is real, and then all coefficients are also real.

It is seen from [2] and [3] that a_0 is the average of $f(t)$ over the integration interval and that a_n and b_n are twice the averages of the modulated functions $f(t)$ cos nt and $f(t)$ sin nt, respectively. With a terminology adopted from electricity theory, the mean value a_0 is called the *d.c.* (direct-current) *component* of $f(t)$ and a_n, b_n (with zero mean value) are called the *a.c.* (alternating-current) *components* of $f(t)$. The integration limits $-\pi$ to $+\pi$ can be exchanged for the limits 0–2π, as is easily shown by virtue of the periodicity of $f(t)$ and of the trigonometric functions. Equations [2] or their equivalents [3] are called the *Cauchy integrals*.

A special theorem, called *Riemann's theorem*, states that the expansion [1] is unique, i.e., there is only one set of values of the coefficients. We can see this in the following way. If we had two alternative sets of coefficients, we could write the right-hand side of [1] in two ways, which should be identically equal for all values of t. Now, this can only happen if respective coefficients in the two sets are all equal. This theorem means that if we were able to determine the coefficients in [1] in some other way than by the method outlined above, the result would still be the same. In fact, the least-square method offers such a possibility. Each observation of $f(t)$ gives an equation of the type [1] with the coefficients as unknowns. If we have more observations than coefficients to determine, then the least-square method is applicable to their determination. Carrying through such a calculation leads to exactly the same formulas [2] and [3] as above, which would also be expected by virtue of the Riemann theorem. For more detail, see Manley (1945, p.172) or Panofsky and Brier (1958, p.219). Extending the range of n also to negative values, it is possible by application of Euler's theorem to express [1] in complex form. This is left as an exercise (see, for instance, Stuart, 1969, p.15).

2.1.2 Application of Fourier series expansion to an analytic function

Concerning expansion into Fourier series, we may distinguish between two situations:

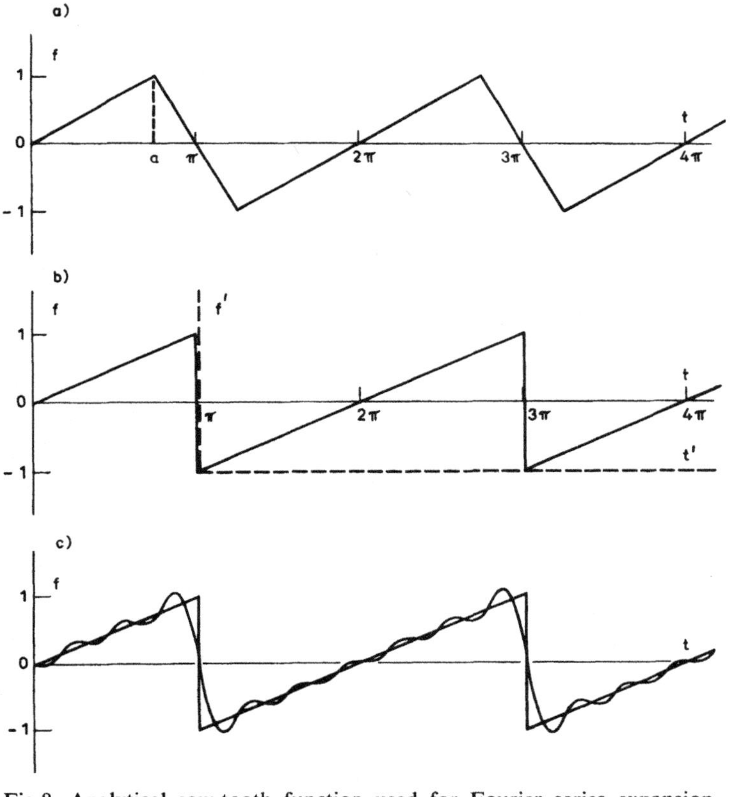

Fig.8. Analytical saw-tooth function used for Fourier series expansion. c) illustrates Gibbs' phenomenon.

(1) The function $f(t)$ to be analyzed is analytic, i.e., is given in a mathematical form.

(2) The function $f(t)$ is given only graphically or numerically, but not analytically. This case happens as a rule within geophysics, where $f(t)$ stands for observational data. Treatment of this will occupy us considerably in following chapters.

In the present section, we shall be concerned with case 1, and we assume the function $f(t)$ to be of the saw-tooth form (Fig.8a):

$$\left.\begin{aligned} f(t) &= t/a && \text{for } 0 \leq t \leq a \\ f(t) &= (\pi - t)/(\pi - a) && \text{for } a \leq t \leq \pi \end{aligned}\right| \qquad [4]$$

A waveform of the saw-tooth type is said to be *skew-symmetrical* or *anti-symmetrical* about the points $t = 0$ and $t = \pi$, which means that $f(t)$ is an odd function around these points. Then, only sine terms will appear in the Fourier expansion. From [2.3] and [4] we find:

$$b_n = \frac{2}{\pi}\int_0^\pi f(t) \sin nt \, dt = \frac{2}{a\pi}\int_0^a t \sin nt \, dt$$

$$+ \frac{2}{(\pi - a)\pi}\int_a^\pi (\pi - t) \sin nt \, dt = \frac{2}{n^2 a(\pi - a)} \sin na \qquad [5]$$

using partial integration. Taking now the special case that $a = \pi$ (Fig.8b) the expression [5] for b_n becomes 0/0. Calculating the limit by l'Hospital's rule, we find:

$$b_n = -2/n\pi \cdot \cos n\pi$$

i.e.:

$$\left.\begin{aligned} b_n &= 2/n\pi && \text{for } n \text{ odd} \\ b_n &= -2/n\pi && \text{for } n \text{ even} \end{aligned}\right| \qquad [6]$$

Obviously, the amplitudes b_n decrease as $1/n$ with increasing n. It can be proved that this is a general property of functions with simple discontinuities.

By [1] the Fourier series expansion of $f(t)$ then becomes:

$$f(t) = \frac{2}{\pi}\left(\frac{\sin t}{1} - \frac{\sin 2t}{2} + \frac{\sin 3t}{3} - \ldots\right) \qquad [7]$$

In a somewhat different coordinate system f' and t' (Fig.8b), related to f and t by:

$$\left.\begin{aligned} f' &= f/2 + 1/2 \\ t' &= t - \pi \end{aligned}\right| \qquad [8]$$

the expansion becomes, after dropping the primes:

$$f(t) = \frac{1}{2} - \frac{1}{\pi}\left(\frac{\sin t}{1} + \frac{\sin 2t}{2} + \frac{\sin 3t}{3} + \ldots\right) \qquad [9]$$

Fig.8c shows a plot of $f(t)$ from [9], including terms up to sin $6t/6$. As we see, the approximation to the saw-tooth curve is worst near the discontinuities of $f(t)$, at $t = 0$ and $t = 2\pi$. This is a general behaviour, referred to as *Gibbs' phenomenon*. At the discontinuities themselves, the series for $f(t)$ assumes values half-way between the two adjacent $f(t)$, i.e. $1/2[f(t+) + f(t-)]$, which is also a general property of Fourier expansions. In this case, the values at the discontinuities become 1/2. It is clear from [9] that, for instance, $f(0) = f(2\pi) = 1/2$, no matter how many terms are included in the expansion. It can be proved that at discontinuities of $f(t)$ always a finite discrepancy will remain irrespective of the number of terms included, whereas for continuous portions of $f(t)$, the approximation will be the better the more terms are included in the series expansion.

Gibbs' phenomenon can be understood as follows. The error from including only a finite number n of terms in $f(t)$ is as follows:

$$\text{Error} = f_n(t) - f(t) = \frac{1}{2} - \frac{1}{\pi}\,\mathrm{Si}\left(n + \frac{1}{2}\right)t \qquad [10]$$

considering the definition of the sine integral Si(x):

$$\mathrm{Si}(x) = \int_0^x \frac{\sin x}{x}\,\mathrm{d}x$$

and the integral:

$$\int_0^\infty \frac{\sin x}{x}\,\mathrm{d}x = \frac{\pi}{2}$$

and replacing the sum in [9] by its corresponding integral. Thus, the error depends only on $(n + 1/2)\,t$. This means that for a given value of this quantity, e.g. corre-

TABLE I

Convergence in approximating the saw-tooth function by a Fourier series expansion

Number of harmonics n	Independent variable t	$(n + 1/2)t$	Error $f_n(t) - f(t)$
Finite, arbitrary	$0, 2\pi, \ldots$	$0, (n + 1/2)2\pi, \ldots$	$\pm 1/2$
Infinite	constant, arbitrary	infinite	0
Arbitrary	arbitrary	constant, corresponding to maximum error	± 0.09

sponding to a maximum error, this error will not diminish with increasing n, i.e. with increasing number of terms. Only t will be smaller for increasing n, keeping $(n + 1/2)\ t$ and the error constant. For the saw-tooth curve, this constant error amounts to about 9% of the step in $f(t)$.

On the other hand, if we keep t constant and let n approach infinity, the error [10] will clearly approach zero. For $t = 0, 2\pi, \ldots$ we find an error of $\pm 1/2$, in agreement with the result obtained above. For better review the different cases are summarized in Table I (see also Fig.8c). These circumstances have important application to the analysis of geophysical records, as we shall see in later chapters.

2.1.3 Application of Fourier series expansion to empirical curves

As mentioned already in section 1.2.1, harmonic analysis or Fourier series expansion of observational data has played a great role in the development of periodicity investigations. In pre-computer time, harmonic analysis completely dominated this field of research. The analysis was effectuated either by special apparatuses (section 1.6.4) or by computational schemes, which facilitated the desk calculations (see for example Manley, 1945). Nowadays, such methods have nearly completely been replaced by electronic computers, mostly with digitized data as input. In working with digitized data, we need to substitute our integral formulas given above for the Fourier coefficients by the corresponding discrete summation formulas. This will be done in section 4.5.2. We shall not dwell on the older methods, and in this section we shall limit ourselves to a few comments.

In the example given in 2.1.2, we have derived the Fourier series expansion from general formulas. These are naturally applicable in all cases amenable to Fourier series expansion. However, in special problems it is advantageous to work with formulas developed particularly for the problem under consideration. One example is the determination of lunar daily geophysical variations (Malin and Chapman, 1970). Another example of applied Fourier series expansions (in space) is offered by geological structures (folds, etc.), which are amenable to

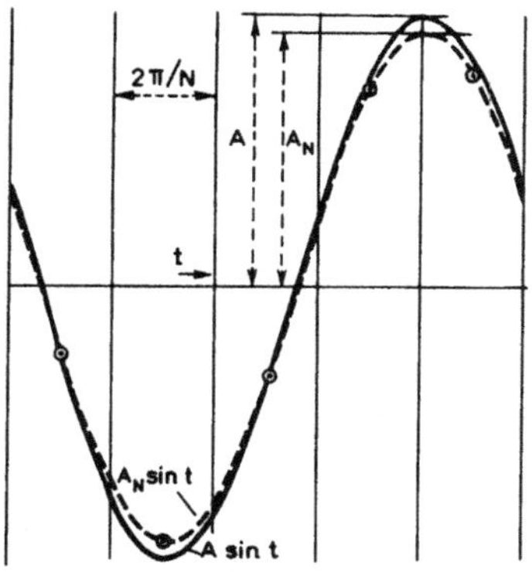

Fig.9. Sketch demonstrating that lower amplitudes are obtained for a sine curve deduced from mean values. After Malin and Chapman (1970).

simplified expressions (Stabler, 1968; Hudleston, 1973, the latter with numerous additional references). See further section 3.5.3.

As emphasized by Malin and Chapman (1970) numerous geophysical series expansions are made with mean values over certain intervals of time. This entails lower amplitudes (Fig.9) than what should be obtained for a larger number of observations closely spaced. The relation between the amplitude A_N of a sine curve with N observations spread equally on the whole cycle and the true amplitude A is obtained by equalizing the areas under the curves:

$$A = A_N \frac{\pi/N}{\sin(\pi/N)} = A_N \frac{1}{\text{sinc}(1/N)} \qquad [11]$$

For $N = 12$ (monthly mean values for annual variation), the correction factor is 1.01152, for $N = 24$ (hourly mean values for daily variation), it is 1.00286, and for $N = \infty$, it is $= 1$. If there is only one observation for the whole range of a period, i.e. $N = 1$, the factor becomes ∞, but then $A_N = 0$, and A is indeterminate. Obviously, $N = 2$ is the lowest number of observations per period (cf. section 4.3.2), and then the factor is $\pi/2 = 1.57080$. The formula [11] holds whenever mean values are used, either in Fourier series expansion or in Fourier transforms.

2.2 FOURIER INTEGRAL AND FOURIER TRANSFORM

2.2.1 Derivation of the Fourier integral

Replacing the argument t of the integrals [3] by the dummy integration variable λ and substituting the expressions [3] for the constants a_0, a_n and b_n in the original Fourier series [1] we get:

$$f(t) = \frac{1}{T}\int_{-T/2}^{T/2} f(\lambda)\,\mathrm{d}\lambda + \sum_{n=1}^{\infty}\left[\left\{\frac{2}{T}\int_{-T/2}^{T/2} f(\lambda)\cos\frac{2n\pi\lambda}{T}\mathrm{d}\lambda\right\}\cos\frac{2n\pi t}{T}\right.$$
$$\left. + \left\{\frac{2}{T}\int_{-T/2}^{T/2} f(\lambda)\sin\frac{2n\pi\lambda}{T}\,\mathrm{d}\lambda\right\}\sin\frac{2n\pi t}{T}\right] \qquad [12]$$

The infinite series on the right-hand side of [12] converges to $f(t)$ if t is not a discontinuity and to $1/2[f(t+) + f(t-)]$, if t is a discontinuity. From [12] we find immediately that $f(T/2) = f(-T/2)$, i.e., that the values of $f(t)$ at the two ends of the interval coincide. This is often referred to as a boundary condition for $f(t)$. However, this condition can be dispensed with on the ground that we can have a finite discontinuity between the two ends, by Dirichlet conditions. Then, the value to be assigned to the end points is:

$$f(\pm T/2) = 1/2[f(T/2) + f(-T/2)]$$

However, fulfillment of the boundary condition entails more rapid convergence of the Fourier series. Then this can be achieved by modifying the original function $f(t)$ by adding another function such that the boundary condition is fulfilled (Lanczos, 1966, p.101). Corresponding theoretical boundary conditions hold for all derivatives of $f(t)$ as seen by differentiation of [12].

The Fourier series in [12] reduces to:

$$f(t) = \frac{1}{T}\int_{-T/2}^{T/2} f(\lambda)\,\mathrm{d}\lambda + \sum_{n=1}^{\infty}\frac{2}{T}\int_{-T/2}^{T/2} f(\lambda)\cos\left[\frac{2n\pi}{T}(t-\lambda)\right]\mathrm{d}\lambda \qquad [13]$$

The factor $\cos\,(2n\pi/T)\,(t - \lambda)$ appearing in the second integral in [13] is termed the *Dirichlet kernel.* Let:

$$\omega_n = 2n\pi/T$$

$$\omega_{n-1} = 2(n-1)\pi/T$$

$$\omega_n - \omega_{n-1} = 2\pi/T = \Delta\omega$$

By substitution in [13], we get:

$$f(t) = \frac{\Delta\omega}{2\pi}\int_{-T/2}^{T/2} f(\lambda)\,d\lambda + \sum_{n=1}^{\infty}\frac{\Delta\omega}{\pi}\int_{-T/2}^{T/2} f(\lambda)\cos[\omega_n(t-\lambda)]d\lambda \qquad [14]$$

Now if we let $T \to \infty$, the following changes take place:
(1) The integral

$$1/T \int_{-T/2}^{T/2} f(\lambda)\,d\lambda \quad \text{vanishes, since} \quad \int_{-T/2}^{T/2} f(\lambda)\,d\lambda$$

is convergent as $f(\lambda)$ satisfies Dirichlet's conditions.

(2) The increment $\Delta\omega$ becomes very small and in the limit $\Delta\omega$ can be represented by $d\omega$. The digitally increasing ω_n becomes the continuous variable ω.

(3) The summation is converted into an integral with the limits 0 and ∞. As a consequence, [14] becomes in the limit:

$$f(t) = \frac{1}{\pi}\int_{0}^{\infty} d\omega \int_{-\infty}^{\infty} f(\lambda)\cos[\omega(t-\lambda)]\,d\lambda \qquad [15]$$

This is the *Fourier integral.*

2.2.2 Derivation of the Fourier transform

Introducing the *cosine and sine transforms* by the following definitions:

$$\left.\begin{aligned} a(\omega) &= \int_{-\infty}^{\infty} f(\lambda)\cos\omega\lambda\,d\lambda \\ b(\omega) &= \int_{-\infty}^{\infty} f(\lambda)\sin\omega\lambda\,d\lambda \end{aligned}\right| \qquad [16]$$

and defining the function $\Phi(\omega)$ by the equation:

$$\sin\Phi(\omega) = -b(\omega)/[a^2(\omega) + b^2(\omega)]^{\frac{1}{2}}$$

which implies:

$$\cos\Phi(\omega) = a(\omega)/[a^2(\omega) + b^2(\omega)]^{\frac{1}{2}}$$

$$\tan \Phi(\omega) = -b(\omega)/a(\omega) \quad \text{and} \quad \Phi(-\omega) = -\Phi(\omega) \qquad [17]$$

we get from [15] by expanding the cosine function under the integral sign and introducing $a(\omega)$ and $b(\omega)$:

$$f(t) = \frac{1}{\pi}\int_{0}^{\infty} [a(\omega)\cos \omega t + b(\omega)\sin \omega t]\mathrm{d}\omega$$

Introducing $\Phi(\omega)$, we get:

$$f(t) = \frac{1}{\pi}\int_{0}^{\infty} [a^2(\omega) + b^2(\omega)]^{\frac{1}{2}} [\cos \Phi(\omega)\cos \omega t - \sin \Phi(\omega)\sin \omega t]\mathrm{d}\omega =$$

$$= \frac{1}{2\pi}\int_{0}^{\infty} [a^2(\omega) + b^2(\omega)]^{\frac{1}{2}}\, \mathrm{e}^{i[\omega t + \Phi(\omega)]}\,\mathrm{d}\omega + \frac{1}{2\pi}\int_{0}^{\infty} [a^2(\omega) + b^2(\omega)]^{\frac{1}{2}}\, \mathrm{e}^{-i[\omega t + \Phi(\omega)]}\,\mathrm{d}\omega =$$

$$= \frac{1}{2\pi}\int_{0}^{\infty} [a^2(\omega) + b^2(\omega)]^{\frac{1}{2}}\, \mathrm{e}^{i[\omega t + \Phi(\omega)]}\,\mathrm{d}\omega + \frac{1}{2\pi}\int_{-\infty}^{0} [a^2(\omega) + b^2(\omega)]^{\frac{1}{2}}\, \mathrm{e}^{i[\omega t + \Phi(\omega)]}\,\mathrm{d}\omega =$$

$$= \frac{1}{2\pi}\int_{-\infty}^{\infty} F(\omega)\,\mathrm{e}^{i\omega t}\,\mathrm{d}\omega \qquad [18]$$

Here we have:

$$F(\omega) = |F(\omega)|\,\mathrm{e}^{i\Phi(\omega)} = [a^2(\omega) + b^2(\omega)]^{\frac{1}{2}}\,\mathrm{e}^{i\Phi(\omega)} =$$

$$= [a^2(\omega) + b^2(\omega)]^{\frac{1}{2}}\,[\cos \Phi(\omega) + i\sin \Phi(\omega)] =$$

$$= a(\omega) - i\,b(\omega) = \int_{-\infty}^{\infty} f(t)\,\mathrm{e}^{-i\omega t}\,\mathrm{d}t \qquad [19]$$

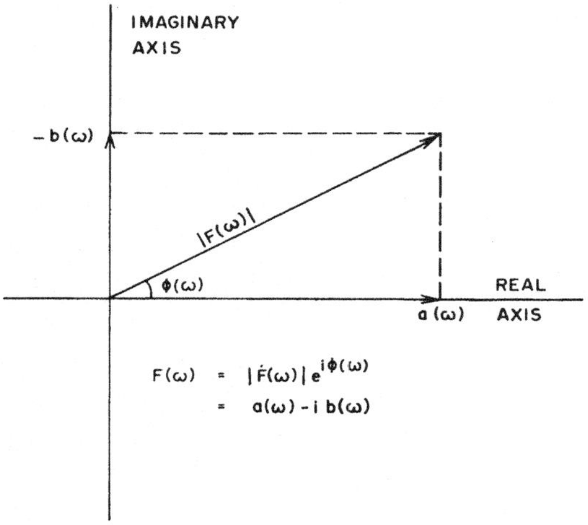

Fig.10. Graphical demonstration of the components of the Fourier spectrum $F(\omega)$.

introducing the expressions [16] for $a(\omega)$ and $b(\omega)$. A corresponding graphical representation is shown in Fig.10.

In summary, we have thus derived the following important set of formulas:

(1) *Fourier transform* (also referred to as *complex Fourier transform* or *Fourier spectrum*):

$$F(\omega) = \int_{-\infty}^{\infty} f(t)\, \mathrm{e}^{-i\omega t}\, \mathrm{d}t \qquad [20]$$

$$F(\omega) = a(\omega) - i\, b(\omega) = |F(\omega)|\, \mathrm{e}^{i\Phi(\omega)} \qquad [21]$$

$$\left.\begin{array}{ll} |F(\omega)| = [a^2(\omega) + b^2(\omega)]^{\frac{1}{2}} & \text{is the } \textit{amplitude spectrum} \\ \Phi(\omega) = \tan^{-1}[-b(\omega)/a(\omega)] + 2n\pi & \text{is the } \textit{phase spectrum} \\ n = 0, \pm 1, \pm 2, \ldots & \end{array}\right| \qquad [22]$$

$-\Phi(\omega)$ is sometimes referred to as the *phase-lag spectrum*. $a(\omega)$ and $b(\omega)$ are given by [16].

(2) *Inverse Fourier transform:*

$$f(t) = \frac{1}{2\pi} \int_{-\infty}^{\infty} F(\omega)\, \mathrm{e}^{i\omega t}\, \mathrm{d}\omega \qquad [23]$$

Equation [20] expresses the *Fourier analysis* of $f(t)$, and [23] expresses the *Fourier synthesis* of $f(t)$, i.e. the synthesis of the various spectral components $F(\omega)$ into the original function $f(t)$. The functions $f(t)$ and $F(\omega)$ are said to form a *Fourier pair*, and their dualism will be denoted as $f(t) \leftrightarrow F(\omega)$. The equations [20] and [23] express a one-to-one correspondence between these two functions. Unlike an ordinary coordinate transformation, the Fourier transformation does not express a point-by-point correspondence in the two domains, but only a curve-by-curve correspondence. The formulas [20] and [23] have general validity for any kind of functions $f(t)$ and $F(\omega)$, i.e. whether real, imaginary or complex, whether even, odd or a combination, only provided Dirichlet's conditions are fulfilled. The Fourier transform is a special kind of *integral transforms* (Båth, 1968, p.211). The equations [20] and [23] are obviously integral equations. For instance, [20] is a *Fredholm integral equation of the first kind*, and $e^{-i\omega t}$ is its *kernel*. The inversion formula [23] is the solution of this integral equation, and vice versa. Cf. Båth (1968, p.319).

By [20], $F(\omega)$ corresponds to an average of $f(t)\ e^{-i\omega t}$ over the interval of integration. In observational series, this interval is by necessity limited in extent. By virtue of the orthogonal properties of the trigonometric functions, the factor $e^{-i\omega t}$ acts like an operator, picking out from $f(t)$ only components with frequency ω. In other words, $F(\omega)$ is an average of those components of $f(t)$, which have the frequency ω. As $F(\omega)$ is referred to unit frequency interval, this quantity is called "density", or more specifically *spectral density* for $F(\omega)$ and *amplitude density* for $|F(\omega)|$. From a dimensional point of view, $F(\omega)$ has the same dimension as $f(t)$. For example, if $f(t)$ is a displacement, i.e., has the dimension of length, also $F(\omega)$ has the dimension of length, and $|F(\omega)|^2$ has the dimension of $(\text{length})^2$.

Alternatively, it is possible to refer the spectral density to unit time instead of unit frequency interval. This is effectuated by multiplying $F(\omega)$ by the number of cycles per second (v) or any proportional quantity, like ω. Thus $\omega F(\omega)$ is spectral density referred to unit time interval.

In [23] the integration extends over all frequencies from $-\infty$ to $+\infty$. Negative frequencies have no physical meaning, but the formulation [23] is preferred for mathematical reasons. This formulation permits application of several efficient mathematical methods. For instance, integration in the complex plane with application of the residue theorem (Båth, 1968) permits evaluation of integrals of the form [20] in closed form. But in physical applications we have to revert to real, positive values of ω.

2.2.3 Corresponding properties of $f(t)$ and $F(\omega)$

By considering symmetry properties of $f(t)$ and $F(\omega)$ we may from [20] and [23] easily deduce some useful relations. Any function $f(t)$ can be split unambiguously into its even and odd parts:

$$f(t) = 1/2[f(t) + f(-t)] + 1/2[f(t) - f(-t)] = f_e(t) + f_o(t) \quad [24]$$

where $f_e(t) = f_e(-t)$ is the even or symmetrical part, and $f_o(t) = -f_o(-t)$ is the odd or anti-symmetrical (skew-symmetrical) part. In general, excepting physical conditions, $f(t)$ is a complex function. Splitting $f(t)$ also into its real Re and imaginary Im parts we have from [24]:

$$f(t) = \mathrm{Re}\, f_e(t) + i\, \mathrm{Im}\, f_e(t) + \mathrm{Re}\, f_o(t) + i\, \mathrm{Im}\, f_o(t) \quad [25]$$

Introducing this expression into the formula [20] for $F(\omega)$, we get:

$$F(\omega) = 2 \int_0^\infty [\mathrm{Re}\, f_e(t) + i\, \mathrm{Im}\, f_e(t)] \cos \omega t \, \mathrm{d}t - 2i \int_0^\infty [\mathrm{Re}\, f_o(t) + i\, \mathrm{Im}\, f_o(t)] \sin \omega t \, \mathrm{d}t \quad [26]$$

The equations [25] and [26] can be immediately extended to the corresponding expressions for $f(-t)$, $f^*(t)$, $f^*(-t)$, and their transforms. The equations [25] and [26] permit us to tell the properties of $F(\omega)$ from the properties of any given function $f(t)$, if it is real or imaginary, even or odd, or exhibits any combination between these cases. The same deductions can easily be performed for $f(-t)$, $f^*(t)$, $f^*(-t)$. For example, $f^*(t) \leftrightarrow F^*(\omega)$. As such derivations do not present any special problems, it is left to the student as an exercise to make a systematic

TABLE II

Corresponding properties of $f(t)$ and $F(\omega)$

$f(t)$	$F(\omega)$
Real and even	real and even
Real and odd	imaginary and odd
Imaginary and even	imaginary and even
Imaginary and odd	real and odd

study. One has only to observe that the oddness or evenness of $F(\omega)$ is determined only by the factors containing ω, i.e. $\cos \omega t$ and $\sin \omega t$. Table II summarizes a few examples which can serve as a start for a more extensive study.

Applying these considerations to a geophysical record or to any physically realizable time-curve $f(t)$, we have to note that $f(t)$ is pure real and, as a consequence of [26], $F(\omega)$ is complex. Thus we have:

Real time function or real time series $f(t)$ ↗ real Fourier series $f(t)$ ↘ complex Fourier transform $F(\omega)$.

Moreover, we find immediately in this case, i.e. for real $f(t)$ and only in this case, that:

$$F(\pm\omega) = F^*(\mp\omega) \qquad [27]$$

Moreover, we see that in this case the real part of $F(\omega)$ is even and the imaginary part is odd. A function fulfilling these conditions is called *Hermitian*. This is therefore a property fulfilled by all geophysical spectra. Conversely, if $F(\omega)$ has this property, then $f(t)$ is real. The Hermitian property of geophysical spectra as well as of all physical spectra is of great significance to bear in mind. It is schematically illustrated in Fig.11.

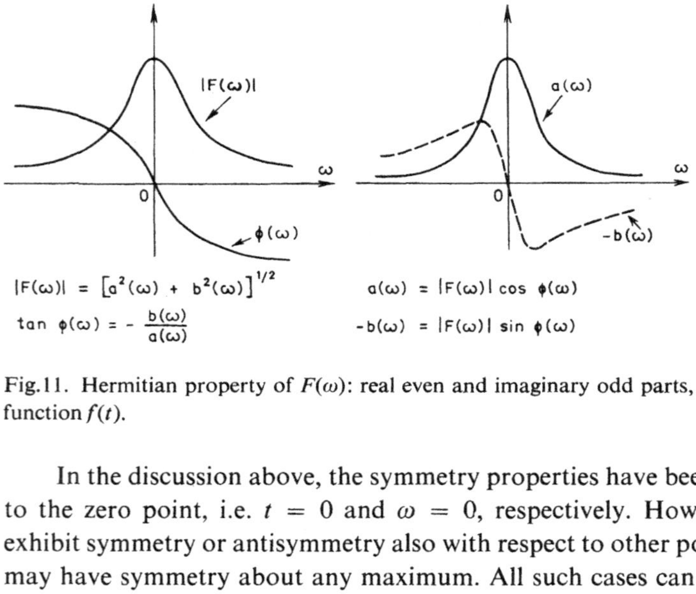

Fig.11. Hermitian property of $F(\omega)$: real even and imaginary odd parts, as obtained from a real function $f(t)$.

In the discussion above, the symmetry properties have been defined in relation to the zero point, i.e. $t = 0$ and $\omega = 0$, respectively. However, functions may exhibit symmetry or antisymmetry also with respect to other points, e.g. a spectrum may have symmetry about any maximum. All such cases can be reduced to those above by substituting $t_0 \pm t$ and $\omega_0 \pm \omega$, respectively, where t_0 and ω_0 correspond to points of symmetry or antisymmetry.

2.2.4 Co-spectrum and quad-spectrum

The real part $a(\omega)$ and the imaginary part $b(\omega)$ of $F(\omega)$, introduced by [16], are also used individually and have special names:

$$a(\omega) = \int_{-\infty}^{\infty} f(t) \cos \omega t \, dt$$

$$b(\omega) = \int_{-\infty}^{\infty} f(t) \sin \omega t \, dt \qquad [28]$$

$a(\omega)$ is the Fourier cosine transform of $f(t)$ and is called *co-spectrum*. $b(\omega)$ is the Fourier sine transform of $f(t)$ and is called *quadrature spectrum* or *quad-spectrum* (cf. Munk et al., 1959). The expression for $ib(\omega)$, entering $F(\omega)$, is obviously:

$$ib(\omega) = \int_{-\infty}^{\infty} f(t) \, e^{i(\pi/2)} \sin \omega t \, dt \qquad [29]$$

i.e., $ib(\omega)$ is the sine transform of the function $f(t)$ after a phase change of $\pi/2$ has been applied to each harmonic of $f(t)$. Each harmonic component of $f(t)$ can be written as $a_n \cos nt + b_n \sin nt$ (see [1]). Increasing nt by $\pi/2$ yields $-a_n \sin nt + b_n \cos nt$. In a vector diagram with a_n and b_n as axes, the latter vector is rotated $\pi/2$ in relation to the first one. This explains the name quadrature.

$a(\omega)$ and $b(\omega)$ correspond to the Fourier coefficients a_n and b_n in equation [3] as follows:

$$a(\omega) \rightarrow 1/2 \, a_n; \qquad b(\omega) \rightarrow 1/2 \, b_n \qquad [30]$$

This is easily seen:

$$a(\omega) = \int_{-\infty}^{\infty} f(\lambda) \cos \omega\lambda \, d\lambda = \int_{-\infty}^{\infty} f(t) \cos \frac{2n\pi}{T} t \, dt \rightarrow$$

$$\rightarrow \lim_{T \to \infty} \frac{1}{T} \int_{-T/2}^{T/2} f(t) \cos \frac{2n\pi}{T} t \, dt = \frac{a_n}{2}$$

2.3 BASIC THEOREMS FOR THE FOURIER TRANSFORM

If $f(t)$ and $F(\omega)$ form a Fourier pair, i.e. if $f(t) \leftrightarrow F(\omega)$, then these two functions are related by the Fourier transform, see [20], and its corresponding inverse transform, see [23]. These equations permit deductions of a number of theorems, of which the more important will be reported in this section.

2.3.1 Linearity theorem, superposition theorem, or addition theorem

If $f_1(t) \leftrightarrow F_1(\omega)$ and $f_2(t) \leftrightarrow F_2(\omega)$, then $[a_1 f_1(t) \pm a_2 f_2(t)] \leftrightarrow [a_1 F_1(\omega) \pm a_2 F_2(\omega)]$, where a_1 and a_2 are constants.

Proof. From [23], we have:

$$a_1 f_1(t) \pm a_2 f_2(t) = a_1/2\pi \int_{-\infty}^{\infty} F_1(\omega)\, e^{i\omega t}\, d\omega \pm a_2/2\pi \int_{-\infty}^{\infty} F_2(\omega)\, e^{i\omega t}\, d\omega =$$

$$= 1/2\pi \int_{-\infty}^{\infty} [a_1 F_1(\omega) \pm a_2 F_2(\omega)]\, e^{i\omega t}\, d\omega$$

that is:

$$[a_1 f_1(t) \pm a_2 f_2(t)] \leftrightarrow [a_1 F_1(\omega) \pm a_2 F_2(\omega)] \qquad [31]$$

It is readily shown that this theorem can be extended to any (finite) sum of signals and their respective spectra:

$$\sum_{n=1}^{m} a_n f_n(t) \leftrightarrow \sum_{n=1}^{m} a_n F_n(\omega) \qquad [32]$$

Thus, the spectrum of the sum of a number of signals is equal to the sum of their individual spectra. This is a practically useful result, as it may permit the calculation of spectra in an easier way, if the given function $f(t)$ can be expressed as a sum over simpler functions. Applications of this rule to cases where $f(t)$ is expressed as a Fourier sum and as a step curve (approximating any given curve) can be found in Kharkevich (1960, pp.210–213).

2.3.2 Duality theorem or symmetry theorem

If $f(t) \leftrightarrow F(\omega)$, then $F(\pm t) \leftrightarrow 2\pi\, f(\mp \omega)$.

Proof. In the Fourier transform formula [20], we exchange t and ω:

$$F(t) = \int_{-\infty}^{\infty} f(\omega)\, e^{-i\omega t}\, d\omega = \int_{-\infty}^{\infty} f(-\omega)\, e^{i\omega t}\, d\omega = 2\pi \frac{1}{2\pi} \int_{-\infty}^{\infty} f(-\omega)\, e^{i\omega t}\, d\omega$$

(substituting $-\omega$ for ω) from which we find:

$$F(t) \leftrightarrow 2\pi f(-\omega)$$

That the transform of $F(-t)$ is $2\pi f(\omega)$ is proved in a fully analogous way. Combining the two formulas into one, we have:

$$F(\pm t) \leftrightarrow 2\pi f(\mp \omega) \qquad [33]$$

Thus, if $F(\omega)$ is the spectrum of a signal $f(t)$, then the spectrum of $F(\pm t)$ is $2\pi f(\mp\omega)$ and the signal of $f(\pm\omega)$ is $(1/2\pi)\ F(\mp t)$.

2.3.3 Time-scaling theorem, reciprocal-spreading theorem, or similarity theorem

If $f(t) \leftrightarrow F(\omega)$ and a is a real constant, then $f(at) \leftrightarrow (1/|a|)\ F(\omega/a)$.
Proof. Starting from the Fourier inversion formula [23] we have:

$$f(at) = \frac{1}{2\pi} \int_{-\infty}^{\infty} F(\omega)\, \mathrm{e}^{i\omega a t}\, \mathrm{d}\omega = \frac{1}{2\pi} \int_{-\infty}^{\infty} \frac{1}{a} F\left(\frac{\omega}{a}\right) \mathrm{e}^{i\omega t}\, \mathrm{d}\omega$$

that is:

$$f(at) \leftrightarrow \frac{1}{a} F\left(\frac{\omega}{a}\right)$$

Similarly, it is shown that:

$$f(-at) \leftrightarrow \frac{1}{a} F\left(-\frac{\omega}{a}\right)$$

whence the two formulas can be summarized into the following:

$$f(at) \leftrightarrow \frac{1}{|a|} F\left(\frac{\omega}{a}\right) \qquad [34]$$

valid for both positive and negative a.

According to this formula, an extension (contraction) of the time scale by a factor a implies a contraction (extension) of the frequency scale by the same factor, while the area under the spectral curve is preserved. This reflects a fundamental relation between the time and frequency domains, to which we shall return in sections 4.3.4 and 5.3.3.

2.3.4 Time-shifting theorem

If $f(t) \leftrightarrow F(\omega)$, then $f(t \pm a) \leftrightarrow e^{\pm ia\omega} F(\omega)$, where a is a real constant. *Proof.* The Fourier inversion formula [23] gives immediately:

$$f(t \pm a) = \frac{1}{2\pi} \int_{-\infty}^{\infty} F(\omega)\, e^{i\omega(t \pm a)}\, d\omega = \frac{1}{2\pi} \int_{-\infty}^{\infty} F(\omega)\, e^{\pm i\omega a}\, e^{i\omega t}\, d\omega$$

that is:

$$f(t \pm a) \leftrightarrow e^{\pm ia\omega} F(\omega) \qquad [35]$$

Obviously, when a time function is shifted by a constant time segment, its amplitude spectrum remains unchanged while its phase spectrum changes by an amount determined by the time shift. This is of course immediately clear if instead we consider the expansion of $f(t)$ into a Fourier series.

2.3.5 Frequency-shifting theorem

If $F(\omega) \leftrightarrow f(t)$, then $F(\omega \pm \omega_0) \leftrightarrow e^{\mp i\omega_0 t} f(t)$, where ω_0 is a real constant. *Proof.* The Fourier transform formula [20] gives immediately:

$$F(\omega \pm \omega_0) = \int_{-\infty}^{\infty} f(t)\, e^{-i(\omega \pm \omega_0)t}\, dt = \int_{-\infty}^{\infty} f(t)\, e^{\mp i\omega_0 t}\, e^{-i\omega t}\, dt$$

that is:

$$F(\omega \pm \omega_0) \leftrightarrow e^{\mp i\omega_0 t} f(t) \qquad [36]$$

Expanding $e^{\mp i\omega_0 t}$ in [36] by Euler's formula and then applying [31], we find immediately the following two relations, named *modulation theorems:*

$$\left.\begin{aligned} f(t) \cos \omega_0 t &\leftrightarrow \frac{1}{2} [F(\omega + \omega_0) + F(\omega - \omega_0)] \\ f(t) \sin \omega_0 t &\leftrightarrow \frac{i}{2} [F(\omega + \omega_0) - F(\omega - \omega_0)] \end{aligned}\right| \qquad [37]$$

As an example from geomagnetism, Coleman and Smith (1966) apply [37] to explain a frequency splitting found by Shapiro and Ward (1966) as due to amplitude modulation. Modulation (both of amplitude, frequency and phase) plays a great role in communication engineering. An extensive discussion of its spectral properties is given by Kharkevich (1960).

2.3.6 Derivation theorem

If $f(t) \leftrightarrow F(\omega)$, then $f'(t) \leftrightarrow i\omega\, F(\omega)$ and $F'(\omega) \leftrightarrow -it\, f(t)$.

Proof. Differentiating the inverse transform formula [23] under the integral sign, we get:

$$f'(t) = \frac{1}{2\pi}\frac{\mathrm{d}}{\mathrm{d}t}\int_{-\infty}^{\infty} F(\omega)\,\mathrm{e}^{i\omega t}\,\mathrm{d}\omega = \frac{1}{2\pi}\int_{-\infty}^{\infty} i\omega\, F(\omega)\,\mathrm{e}^{i\omega t}\,\mathrm{d}\omega$$

that is:

$$f'(t) \leftrightarrow i\omega\, F(\omega) \qquad [38]$$

The derivative of a function emphasizes the higher-frequency components in the spectrum in relation to the lower-frequency components and eliminates any zero-frequency (or d.c.) components. In addition, there is a phase shift of $\pi/2$ for all components.

The second part of the theorem is derived in a fully analogous way by differentiating the transform formula [20] under the integral sign, which results in:

$$F'(\omega) \leftrightarrow -it\, f(t) \qquad [39]$$

Both results can be immediately extended to any higher-order derivative, simply by continuing the differentiation:

$$\left.\begin{aligned} f^{(n)}(t) &\leftrightarrow (i\omega)^n\, F(\omega) \\ F^{(n)}(\omega) &\leftrightarrow (-it)^n\, f(t) \end{aligned}\right| \qquad [40]$$

where n is the order of differentiation.

From [40] we can easily derive the following relation between the differential operators:

$$\left(\frac{d^n}{dx^n} - x^n\right)$$

in the time (t) and the frequency (ω) domains, respectively:

$$\left(\frac{d^n}{dt^n} - t^n\right) f(t) \leftrightarrow (-1)^{(n/2)+1} \left(\frac{d^n}{d\omega^n} - \omega^n\right) F(\omega) \qquad [41]$$

where $n = 1, 2, 3, \ldots$ Note especially that for $n = 2$, the operator is exactly the same in the two domains. The same happens for all $n = 2 + 4m$, where $m = 0, 1, 2, 3, \ldots$

More generally, the derivation theorem may be applied to the solution of ordinary or partial differential equations with constant coefficients. For example, given the following equation with known $g(t)$:

$$a_N \frac{\mathrm{d}^N f(t)}{\mathrm{d}t^N} + a_{N-1} \frac{\mathrm{d}^{N-1} f(t)}{\mathrm{d}t^{N-1}} + \ldots + a_1 \frac{\mathrm{d}f(t)}{\mathrm{d}t} + a_0 f(t) = g(t)$$

or more compact:

$$\sum_{n=0}^{N} a_n \frac{\mathrm{d}f^n(t)}{\mathrm{d}t^n} = g(t) \qquad [42]$$

the derivation theorem yields, when applied to the transform of this equation:

$$F(\omega) \sum_{n=0}^{N} a_n (i\omega)^n = G(\omega) \qquad \text{or:} \qquad F(\omega)\, H(\omega) = G(\omega) \qquad [43]$$

In differential equation theory, this equation is referred to as the *auxiliary equation* (or *subsidiary equation*) to the given differential equation (cf. Båth, 1968, pp.218–246). The sought function $f(t)$ is then obtained from:

$$f(t) = \frac{1}{2\pi} \int_{-\infty}^{\infty} F(\omega)\, \mathrm{e}^{i\omega t}\, \mathrm{d}\omega = \frac{1}{2\pi} \int_{-\infty}^{\infty} \frac{G(\omega)\, \mathrm{e}^{i\omega t}}{\sum_{n=0}^{N} a_n (i\omega)^n}\, \mathrm{d}\omega \qquad [44]$$

2.3.7 Integration theorem

If $f(t) \leftrightarrow F(\omega)$, then $\int_{-\infty}^{\infty} f(t)\, \mathrm{d}t \leftrightarrow \frac{1}{i\omega} F(\omega)$.

Proof. According to [20], the Fourier transform of:

$$\int_{-\infty}^{\infty} f(t)\, \mathrm{d}t \qquad \text{is} \qquad \int_{-\infty}^{\infty} \left[\int_{-\infty}^{\infty} f(t)\, \mathrm{d}t\right] \mathrm{e}^{-i\omega t}\, \mathrm{d}t$$

By partial integration, this expression can be developed as follows:

$$\left[\int_{-\infty}^{\infty} f(t)\,\mathrm{d}t\right]\left[-\frac{1}{i\omega}\,\mathrm{e}^{-i\omega t}\right]_{-\infty}^{\infty} + \frac{1}{i\omega}\int_{-\infty}^{\infty} \mathrm{e}^{-i\omega t} f(t)\,\mathrm{d}t = \frac{1}{i\omega}\,F(\omega)$$

as the first term vanishes because of Dirichlet's conditions. Thus, it results that:

$$\int_{-\infty}^{\infty} f(t)\,\mathrm{d}t \leftrightarrow \frac{1}{i\omega}\,F(\omega) \qquad [45]$$

which should be proved. This constitutes a counterpart to the derivation theorem (2.3.6).

Just as the derivation theorem is useful in solving differential equations (section 2.3.6), the integral theorem may prove useful in transforming a given equation into an algebraic form in the frequency domain.

Another useful application of the derivation and integration theorems refers to the case when one record is the derivative or the integral of another. For example, we have the following common case:

displacement	$f(t) \leftrightarrow F(\omega)$	
velocity	$v(t) = f'(t) \leftrightarrow i\omega\, F(\omega) = \omega\, F(\omega)\, \mathrm{e}^{i(\pi/2)}$	[46]
acceleration	$a(t) = f''(t) \leftrightarrow -\,\omega^2\, F(\omega) = \omega^2\, F(\omega)\, \mathrm{e}^{i\pi}$	

Likewise, we could have started from an acceleration record and applied the integration theorem. Much effort has gone into integration methods of accelerograms, at least earlier. But obviously, passing over the frequency domain may offer an advantageous solution, because the operations, corresponding to integration and derivation in the time domain, are so simple and straight-forward in the frequency domain. Essentially, the integration or derivation problem has been recast into the problem of spectral calculation and its inversion.

2.3.8 General comments on the Fourier transform theorems

It is to be noted that in the theorems given here, the spectrum appears in the form $F(\omega)$, i.e. without separation of amplitude and phase spectra. By [21], the amplitude spectrum is $|F(\omega)|$ and the phase spectrum is $\Phi(\omega)$. Table III summarizes some of the theorems given above, where the amplitude and phase spectra are given separately. These examples may suffice to demonstrate the procedure. In any given case, it is just needed to write an obtained spectral function in the shape $|G(\omega)|\,\mathrm{e}^{i\Psi(\omega)}$ and then to identify $|G(\omega)|$ and $\Psi(\omega)$, respectively, with the given expression.

TABLE III

Examples of relations between time function, amplitude spectrum, and phase spectrum; a and ω_0 are real constants

Section	Time function	Amplitude spectrum	Phase spectrum
	$f(t)$	$\|F(\omega)\|$	$\Phi(\omega)$
2.3.1	$a\,f(t)$	$a\|F(\omega)\|$	$\Phi(\omega)$
2.3.2	$F(\pm t)$	$2\pi\|f(\mp\omega)\|$	$n\pi$ ($n = 0, \pm 1, \pm 2, \ldots, f$ real)
2.3.3	$f(at)$	$\frac{1}{\|a\|}\left\|F\left(\frac{\omega}{a}\right)\right\|$	$\Phi\left(\frac{\omega}{a}\right)$
2.3.4	$f(t \pm a)$	$\|F(\omega)\|$	$\Phi(\omega) \pm a\omega$
2.3.5	$f(t)\,e^{\pm i\omega_0 t}$	$\|F(\omega \mp \omega_0)\|$	$\Phi(\omega \mp \omega_0)$
[1]	$f_1(t) \star f_2(t)$	$\|F_1(\omega)\,F_2(\omega)\|$	$\Phi_1(\omega) + \Phi_2(\omega)$
2.3.6	$f'(t)$	$\omega\|F(\omega)\|$	$\Phi(\omega) + \pi/2$
[1]	$\mathrm{d}/\mathrm{d}t\,[f_1(t) \star f_2(t)]$	$\omega\|F_1(\omega)\,F_2(\omega)\|$	$\Phi_1(\omega) + \Phi_2(\omega) + \pi/2$
2.3.7	$\int_{-\infty}^{\infty} f(t)\,\mathrm{d}t$	$\frac{\|F(\omega)\|}{\omega}$	$\Phi(\omega) - \frac{\pi}{2}$

[1] Theorems relating to convolution (★) will be proved in section 3.2.

Concerning the theorems derived in this section, it is to be noticed that they are valid for the Fourier transform and its inversion as we have defined them, i.e. for infinite extent both in t and ω. Sometimes, when only a limited time section is involved, it may be appropriate to use the *finite Fourier transform*, defined as:

$$F(\omega) = \int_{-\pi}^{\pi} f(t)\,e^{-i\omega t}\,\mathrm{d}t \tag{47}$$

However, the corresponding inversion formula will have its integral extending from $-\infty$ to ∞ over ω. Therefore, the theorems above cannot without special test be applied to the finite Fourier transform. For example, the symmetry theorem (2.3.2) cannot be directly applied to the finite transform. Finite transforms will be of use in discussion of data windows in section 4.4. They are also useful in solving differential equations (cf. Båth, 1968, pp.223–227).

2.4 CALCULATION OF FOURIER TRANSFORMS OF ANALYTIC FUNCTIONS

2.4.1 Transcription between different systems for Fourier transform and its inversion

The calculation of the Fourier transform $F(\omega)$ for any given function $f(t)$

has two different aspects:

(1) $f(t)$ is an analytically given function. In this case the calculation consists in purely mathematical operations, using [20].

(2) $f(t)$ represents a series of observations.

Case 2 is of the greatest significance in geophysical applications, and will be dealt with in following chapters. In this section, we shall be concerned with case 1, which is important in order to get a thorough familiarity with Fourier transforms. This case has also great significance in many theoretical considerations in geophysics and especially in seismology. Moreover, theoretical cases such as those presented in this section, will have a practical value for testing any computer program, before this is applied to actual geophysical records or any other kind of observations.

Numerous tabulations of Fourier transforms can be found both in mathematical tables and in textbooks. In using any of these, it is very important to be clear about the definitions they are using for the inversion and transform formulas. There is quite a variety of such expressions, and the corresponding expressions for the transforms will be different. We can summarize all the various forms in the following formulas, where k, m, n are to be specified in each case:

$$\left.\begin{aligned} f(t) &= \frac{1}{(2\pi)^{k_1}} \int_{-\infty}^{\infty} F(\omega) \exp[(-1)^{m_1} (2\pi)^{n_1} i\omega t] \, d\omega \\ F(\omega) &= \frac{1}{(2\pi)^{k_2}} \int_{-\infty}^{\infty} f(t) \exp[(-1)^{m_2} (2\pi)^{n_2} i\omega t] \, dt \end{aligned}\right| \qquad [48]$$

Writing the transform equations in this compact way, we have to observe that the parameter, especially ω, is only a dummy variable to which no special physical interpretation can be attached. Numerical values of k, m, n are summarized in Table IV for the most commonly occurring systems. System 1 is the one adopted in this book. The table gives also the appearance of the Fourier transform in the different systems for any given function $f(t)$, where $F(\omega)$ is as defined in our system. For instance, a transform found in system 5 can be turned into system 1 by dividing the frequency by 2π. The table permits easy transcription from any system into any other system. Obviously, a given time function $f(t)$ can have its spectrum expressed in a variety of ways. But, all in all, the different expressions for the Fourier transform in Table IV differ only in the scales used, either for the abscissa (frequency) or the ordinate (spectrum F).

In general, one finds more extensive tables of Laplace transforms in the literature than of Fourier transforms. Conversion from Laplace to Fourier transforms may therefore appear to be useful. However, in such conversions we have to consider both the region of convergence of the Laplace transform and if it is

TABLE IV

Rewiew of different inversion formulas and transforms, with reference to [48]

System	k_1	m_1	n_1	k_2	m_2	n_2	Fourier transform	
1	1	2	0	0	1	0	$F_1(\omega) = F(\omega)$	$F(\omega) = F_1(\omega)$
2	1	1	0	0	2	0	$F_2(\omega) = F(-\omega)$	$F(\omega) = F_2(-\omega)$
3	1/2	2	0	1/2	1	0	$F_3(\omega) = \frac{1}{\sqrt{2\pi}} F(\omega)$	$F(\omega) = \sqrt{2\pi}\, F_3(\omega)$
4	1/2	1	0	1/2	2	0	$F_4(\omega) = \frac{1}{\sqrt{2\pi}} F(-\omega)$	$F(\omega) = \sqrt{2\pi}\, F_4(-\omega)$
5	0	2	1	0	1	1	$F_5(\omega) = F(2\pi\omega)$	$F(\omega) = F_5\left(\frac{\omega}{2\pi}\right)$
6	0	2	−1	0	1	−1	$F_6(\omega) = F\left(\frac{\omega}{2\pi}\right)$	$F(\omega) = F_6(2\pi\omega)$
7	—	—	—	[1]	1	0	$F_7(\omega) = \omega F(\omega)$	$F(\omega) = \frac{F_7(\omega)}{\omega}$

[1] $1/(2\pi)^{k2} = \omega$

unilateral (integral extending from 0 to $+\infty$) or bilateral (integral extending from $-\infty$ to $+\infty$). As a consequence, the conversion formulas may be rather complicated (see Papoulis, 1962, p.172) and not very practical.

2.4.2 Practical rules for calculation of Fourier transforms

The following hints may be useful to consider in the calculation of a Fourier transform $F(\omega)$ of a given analytic time function $f(t)$:

(1) Straight-forward integration (real t), using the expression for $F(\omega)$, is the most immediate operation. An extensive table of infinite integrals proves useful to have at hand.

(2) Frequently, integrals are encountered which do not exist according to usual analysis, but interpreted as distributions or generalized functions they may still have a definite expression (Papoulis, 1962, pp.269–282). The most useful among such formulas are the following ("transforms in the limit"):

$$\int_0^\infty \sin \omega t \, dt = \frac{1}{\omega} \qquad \omega \neq 0$$

$$\int_0^\infty \cos \omega t \, dt = \pi\delta(\omega) \qquad [49]$$

(3) Contour integration in the complex plane often proves useful. For example, replacing time t by the imaginary variable $z = it$, the transform formula becomes:

$$F(\omega) = \frac{1}{i} \int_{-i\infty}^{i\infty} f\left(\frac{z}{i}\right) e^{-\omega z} \, dz \qquad [50]$$

which can be solved by contour integration in the complex z-plane. Students not familiar with this method are referred to textbooks in applied mathematics (e.g. Båth, 1968).

(4) Often it is possible to express the given function $f(t)$ in terms of other functions, for which the transforms are known or can be more easily calculated.

(5) The theorems given in section 2.3 are extremely useful for deducing additional formulas. In fact, a small set of transforms, such as given below (see Table V), can be used for many other functions by applying the theorems.

(6) It is always valuable to check that the transform has the right structure corresponding to a given time function, depending upon if the latter is even or odd, real or complex, following the rules laid down in section 2.2.3.

(7) Sometimes, it is most convenient first to expand $f(t)$ in a Fourier series, and then to transform this series term by term. It is often instructive to compare $F(\omega)$ with a Fourier series expansion of $f(t)$. The two correspond to each other, $F(\omega)$ giving the continuous spectrum and the series expansion of $f(t)$ giving the line spectrum.

(8) Useful formulas for calculation of Fourier transforms and inverses can naturally be developed for functions fulfilling certain requirements. For example, Liou (1964) developed such formulas for piecewise linear functions, which can also be used approximately for functions which can be reasonably well represented by linear segments.

(9) Transforms can often be found in textbooks and mathematical tables. Extensive tables of Fourier transforms were given by Campbell and Foster (1948) and Erdélyi (1954). For further references, see Bracewell (1965, p.23). In using tables of transforms it is important to see what definitions have been used for the transform and to transcribe the results accordingly (Table IV).

(10) As tables of Laplace transforms are usually more abundant than those of Fourier transforms, it would appear a suitable method to calculate the latter from tables of Laplace transforms. There are methods for this (see Papoulis, 1962, pp.172–173); however, owing to the precautions to be taken, such recalculations are not very practical, as emphasized in section 2.4.1.

2.4.3 Examples of Fourier transform calculation

In Table V we have collected a number of Fourier transforms, corresponding

TABLE V

Fourier transforms; $a > 0$ is a real constant

No.	$f(t)$	$F(\omega)$
1	rectangular function: $\Pi(t) = \begin{cases} 1 & \lvert t \rvert \le 1/2 \\ 0 & \lvert t \rvert > 1/2 \end{cases}$	$\dfrac{\sin(\omega/2)}{\omega/2} = \operatorname{sinc}\dfrac{\omega}{2\pi}$
2	Fourier kernel: $\dfrac{\sin at}{\pi t} = \dfrac{a}{\pi}\operatorname{sinc}\dfrac{at}{\pi}$	$\Pi\left(\dfrac{\omega}{2a}\right)$
3	filtering (interpolating) function (sine cardinal pulse): $\dfrac{\sin \pi t}{\pi t} = \operatorname{sinc} t$	$\Pi\left(\dfrac{\omega}{2\pi}\right)$
4	$\Pi(t - t_0) + \Pi(t + t_0)$	$2 \operatorname{sinc}\dfrac{\omega}{2\pi}\cos \omega t_0$
5	$\Pi(t)\cos \omega_0 t$	$\dfrac{\sin[(\omega - \omega_0)/2]}{\omega - \omega_0} + \dfrac{\sin[(\omega + \omega_0)/2]}{\omega + \omega_0} = \dfrac{1}{2}\left(\operatorname{sinc}\dfrac{\omega - \omega_0}{2\pi} + \operatorname{sinc}\dfrac{\omega + \omega_0}{2\pi}\right)$
6	triangular function: $\Lambda(t) = \begin{cases} 1 - \lvert t \rvert & \lvert t \rvert \le 1 \\ 0 & \lvert t \rvert > 1 \end{cases}$	$\operatorname{sinc}^2\dfrac{\omega}{2\pi}$
7	Fejér kernel: $\operatorname{sinc}^2 t$	$\Lambda\left(\dfrac{\omega}{2\pi}\right)$
8	sign function: $\operatorname{sgn} t = \begin{cases} 1 & t > 0 \\ -1 & t < 0 \end{cases}$	$-\dfrac{2i}{\omega}$
9	Dirac delta function (unit-area impulse): $\delta(t) = 0 \quad t \neq 0$; $\int_{-\infty}^{\infty} \delta(t)\,\mathrm{d}t = 1$	$1\ [= u(\lvert\omega\rvert)]$
10	$1\ [= u(\lvert t \rvert)]$	$2\pi\,\delta(\omega)$
11	$\delta(t - t_0)$	$e^{-i\omega t_0}$
12	$e^{i\omega_0 t}$	$2\pi\,\delta(\omega - \omega_0)$

TABLE V (continued)

No.	$f(t)$	$F(\omega)$
13	$\cos \omega_0 t$	$\pi[\delta(\omega + \omega_0) + \delta(\omega - \omega_0)]$
14	$\sin \omega_0 t$	$i\pi[\delta(\omega + \omega_0) - \delta(\omega - \omega_0)]$
15	$\sum_{n=-\infty}^{\infty} \delta(t - nt_0)$	$\frac{2\pi}{t_0} \sum_{n=-\infty}^{\infty} \delta\left(\omega - \frac{2n\pi}{t_0}\right)$
16	Heaviside unit-step function: $u(t) = \begin{cases} 1 & t > 0 \\ 0 & t < 0 \end{cases}$	$\pi\,\delta(\omega) - \frac{i}{\omega}$
17	$u(t)\,e^{i\omega_0 t}$	$\pi\,\delta(\omega - \omega_0) - \frac{i}{\omega - \omega_0}$
18	$u(t) \cos \omega_0 t$	$\frac{\pi}{2}[\delta(\omega - \omega_0) + \delta(\omega + \omega_0)] + \frac{i\omega}{\omega_0^2 - \omega^2}$
19	$u(t) \sin \omega_0 t$	$-\frac{i\pi}{2}[\delta(\omega - \omega_0) - \delta(\omega + \omega_0)] + + \frac{\omega_0}{\omega_0^2 - \omega^2}$
20	$u(t)\,\Lambda(2t)$	$\frac{1}{4}\left[\frac{\sin(\omega/4)}{\omega/4}\right]^2 - \frac{i}{\omega}\left[1 - \frac{\sin(\omega/2)}{\omega/2}\right] = \frac{1}{4}\operatorname{sinc}^2\frac{\omega}{4\pi} - \frac{i}{\omega}\left(1 - \operatorname{sinc}\frac{\omega}{2\pi}\right)$
21	$\|t\|$	$-2/\omega^2$
22	$1/t$	$-i\pi \operatorname{sgn} \omega$
23	$u(t)\,e^{-at}$	$(a - i\omega)/(a^2 + \omega^2)$
24	Laplace function: $e^{-a\|t\|}$	$\frac{2a}{a^2 + \omega^2}$
25	Gauss function: e^{-at^2}	$\sqrt{\frac{\pi}{a}}\,e^{-\omega^2/4a}$
26	$t\,e^{-at^2}$	$-\frac{i\omega}{2a}\sqrt{\frac{\pi}{a}}\,e^{-\omega^2/4a}$
27	Berlage function: $u(t)\,t\,e^{-at} \sin \omega_0 t$	$\frac{2\omega_0(a + i\omega)}{[(a + i\omega)^2 + \omega_0^2]^2}$

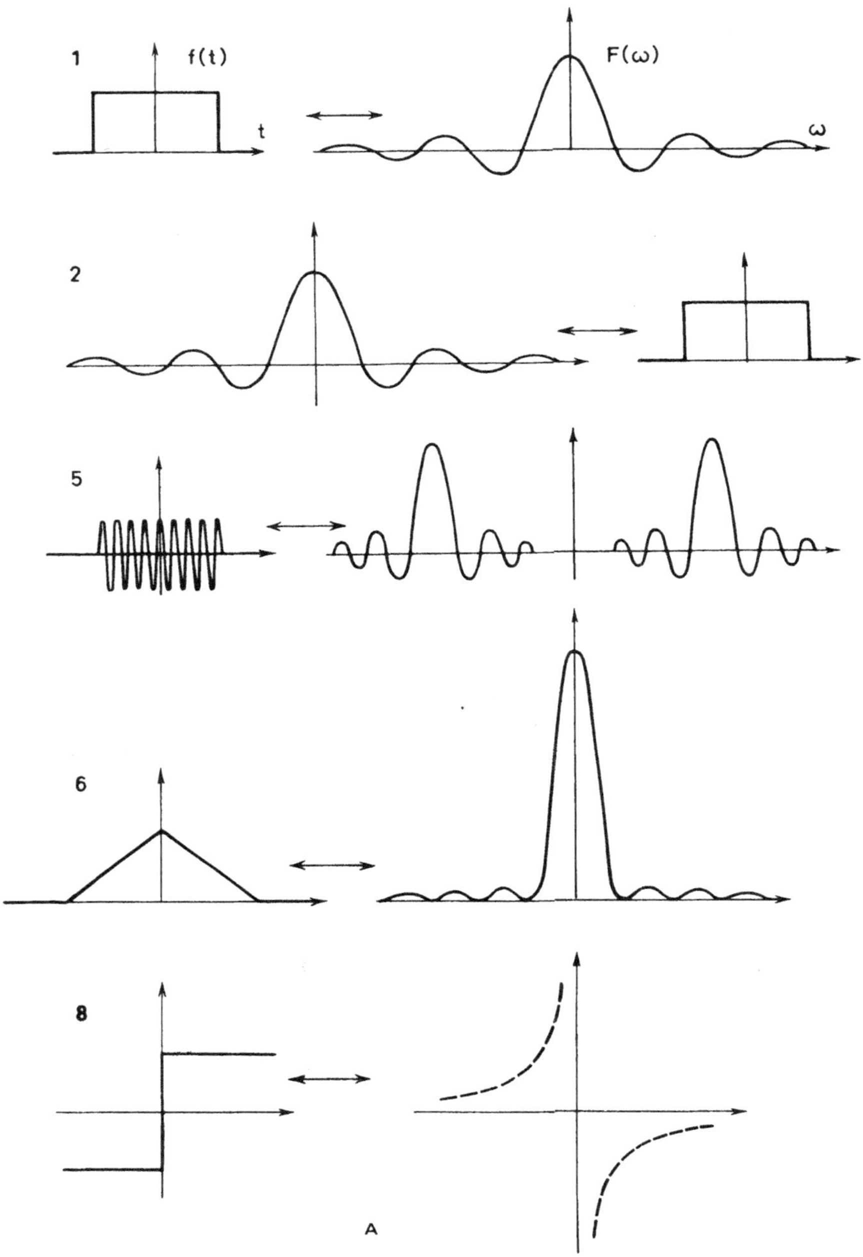

Fig.12A. Fourier transforms of selected analytic time functions. Numbers refer to the listing in Table V. Full lines: real parts, dashed lines: imaginary parts. Scales arbitrary. Modified and redrafted from Papoulis (1962).

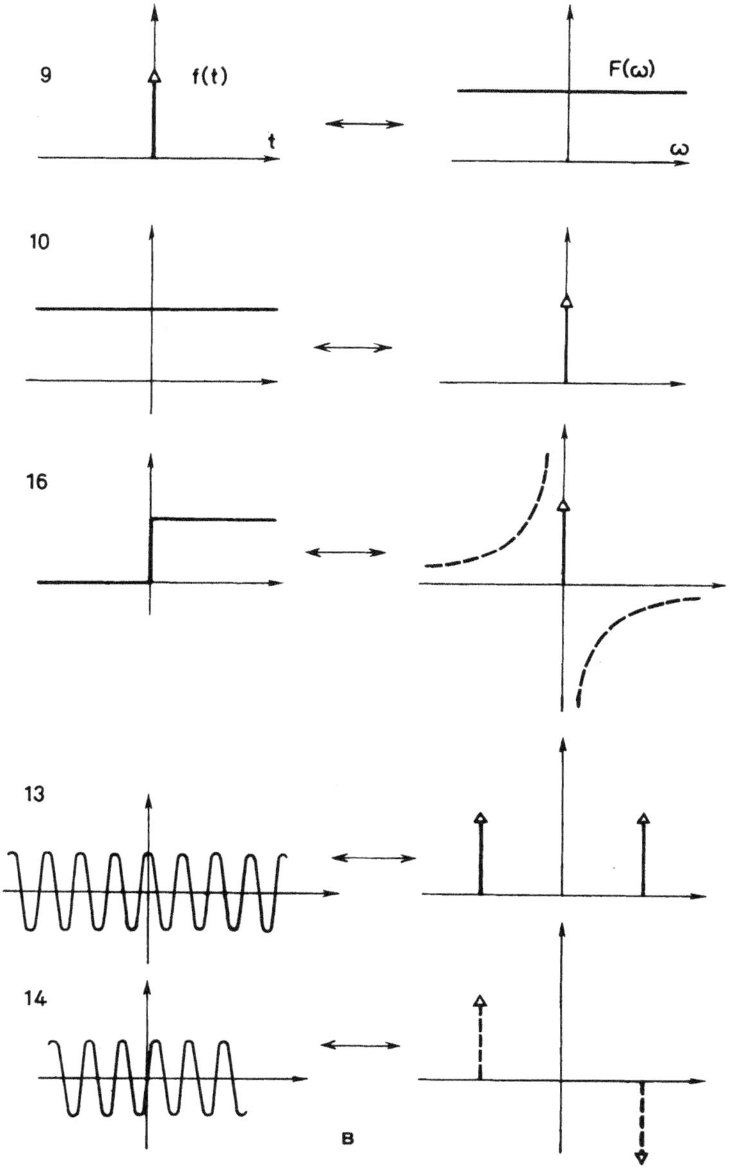

Fig.12B. See legend Fig.12A.

to system 1 in Table IV. It is suggested as an exercise to the student to work through the examples in Table V and to trace all functions graphically. Below we give only some hints for the different examples, which will facilitate the derivations.

(1) We start with the rectangular function $\Pi(t)$. The formula 1 follows

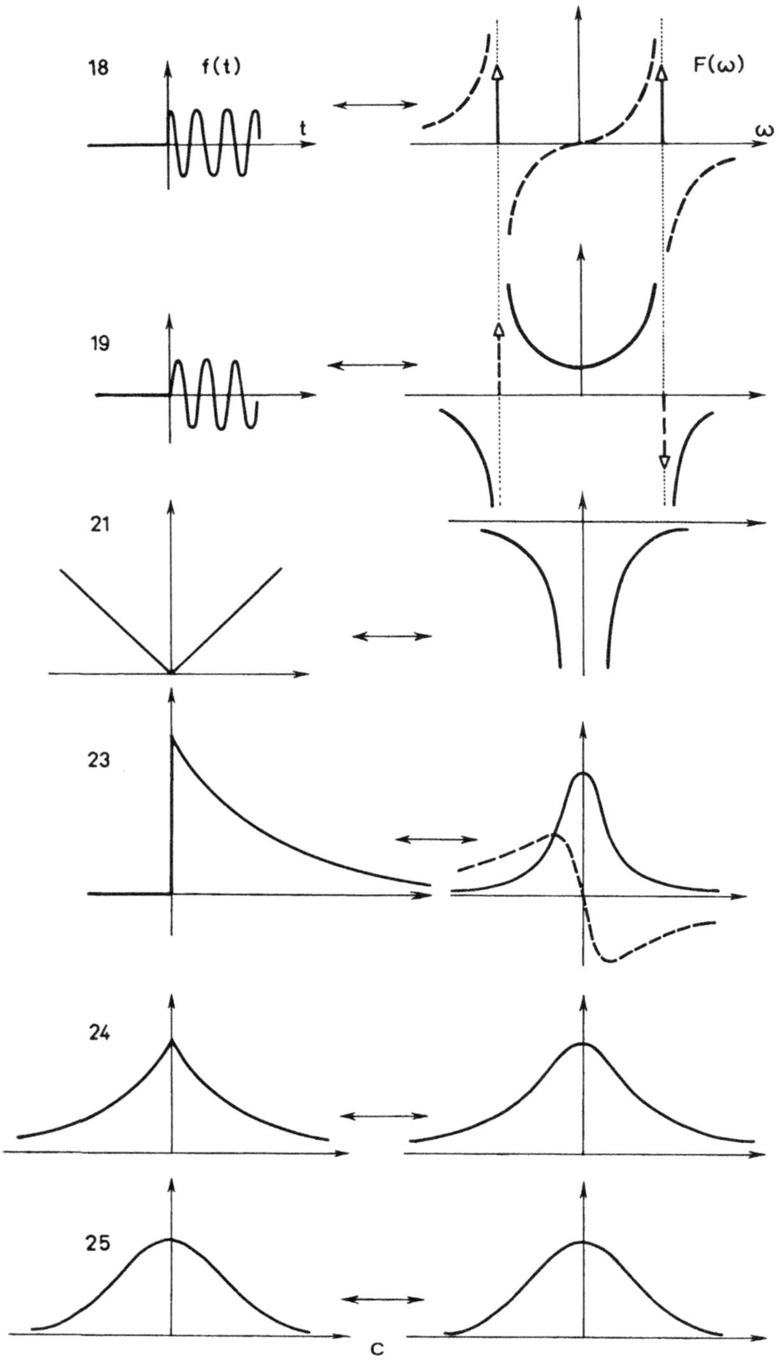

Fig.12C. See legend Fig.12A.

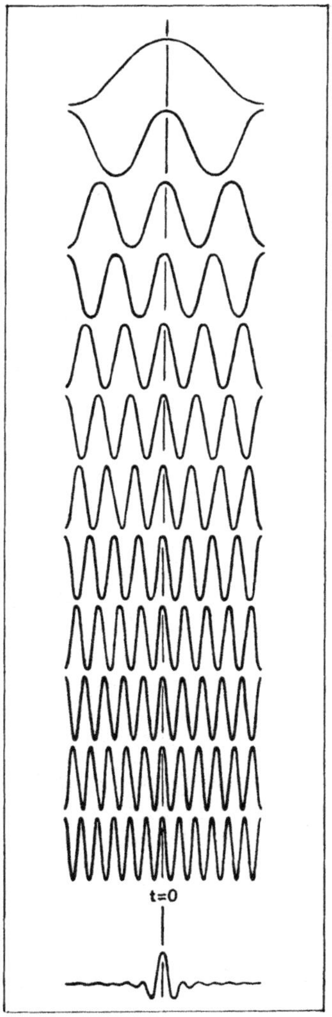

Fig.13. Approximation to the synthesis of the waveform corresponding to a rectangular amplitude spectrum and a zero phase spectrum. After Anstey (1964). More exactly, the sinc-function would correspond to cosinusoids covering continuously the range from zero frequency to the limiting frequency $\omega = a$, all of the same amplitude ($= 1$).

immediately from application of the expression for $F(\omega)$ and observing that the range of integration over t extends from $-1/2$ to $+1/2$, the integrand being zero for all other t. See Fig.12.

(2) From 1 by applying first the symmetry theorem (2.3.2) and then the time-scaling theorem (2.3.3). A clear pictorial presentation of this transform is given in Fig.13 from Anstey (1964).

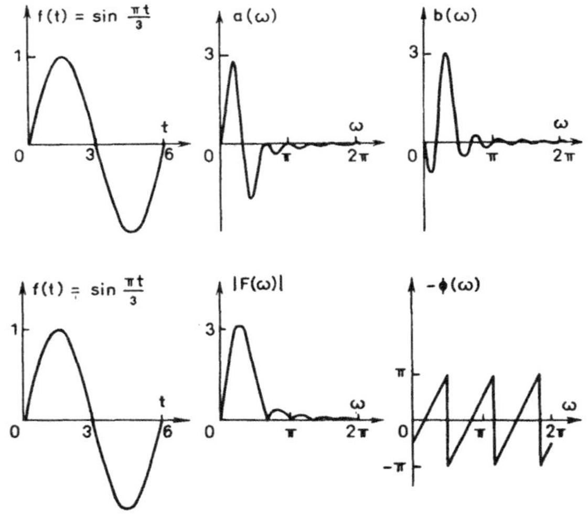

Fig.14. Fourier cosine and sine transforms, amplitude and phase spectra derived from one cycle of a given sine wave. Modified and redrafted after Huang (1966).

(3) A special case of 2, for $a = \pi$. Instead of sinc t, sometimes the notation dif t is used (dif from diffraction), defined in the same way.

(4) From 1 by applying the time-shifting theorem (2.3.4) twice and the linearity theorem (2.3.1).

(5) From 1, by application of the modulation theorem (2.3.5). Note that while the unlimited function cos $\omega_0 t$ has a line spectrum (lines at frequencies ω_0 and $-\omega_0$; see example 13 below), the truncated cosine function has a spectrum covering all frequencies, though mostly concentrated to ω_0 and $-\omega_0$. For purposes of illustration, Fig.14, adopted from Huang (1966) shows a related, though different function with its Fourier coefficients, amplitude and phase spectra.

(6) Derived by straight-forward integration. The triangular function $\Lambda(t)$ has contributions only for $-1 \leq t \leq 1$, and the integration interval thus reduces to this range.

(7) From 6 by the symmetry theorem (2.3.2) and the time-scaling theorem (2.3.3).

(8) Calculating the transform of sgn t by integration in the usual way, we are led to an infinite integral which does not exist:

$$\int_0^\infty \sin \omega t \, dt$$

However, interpreted as a distribution, it has a definite expression, see [49].

(9) Obtained immediately by application of the following formula valid for $\delta(t)$:

$$\int_{-\infty}^{\infty} \delta(t) f(t)\, \mathrm{d}t = f(0) \qquad [51]$$

See Båth (1968, p.248). In the Fourier transform $f(t) = \mathrm{e}^{-i\omega t}$ and $f(0) = 1$. The frequency-independent energy distribution is a characteristic of *white noise*. Applying the derivation theorem, section 2.3.6, to the transform $\delta(t) \leftrightarrow 1$ we get, $\delta'(t) \leftrightarrow i\omega$ or generally:

$$\delta^{(n)}(t) \leftrightarrow (i\omega)^n \qquad [52]$$

(10) From 9 by the symmetry theorem (2.3.2). Note that:

$2\pi\, \delta(\omega) = \delta(\omega/2\pi)$ or in general $\delta(a\omega) = (1/|a|)\, \delta(\omega)$

(11) From 9 by the time-shifting theorem (2.3.4).

(12) From 9 by the frequency-shifting theorem (2.3.5). The transform of $\mathrm{e}^{i\pi t}$ is found immediately as a special case.

(13 and 14) From 10 by the modulation theorems (2.3.5). As a check, combining 13 and 14 we get 12. The transforms of $\cos \pi t$ and $\sin \pi t$ are found as special cases.

(15) This is an infinite sequence of unit pulses, t_0 apart (n integer). A relatively simple way to find its transform is first to expand $f(t)$ in a Fourier series, using [13]:

$$f(t) = \frac{1}{t_0} + \frac{2}{t_0} \sum_{n=1}^{\infty} \cos \frac{2n\pi t}{t_0} \qquad [53]$$

The transform of this expression is found from examples 10 (first term) and 13 (the sum), and applying the linearity theorem (2.3.1). Thus, the so-called *Dirac comb* transforms as a comb, i.e. also as an infinite sequence of pulses, each with an amplitude factor $2\pi/t_0$ and located $2\pi/t_0$ apart along the frequency axis. As an exercise, try to deduce 15 from 11.

(16) Noting that $u(t)$ can be written as:

$$u(t) = 1/2 + 1/2 \operatorname{sgn} t \qquad [54]$$

we can apply 8 and 10 and the linearity theorem (2.3.1). Alternatively, we can carry out a straight-forward integration over $u(t)$. We are then faced with semi-infinite integrals over sine and cosine functions, where we apply [49]. It is worth noting a simple relation between $\Pi(t)$ and $u(t)$, i.e.:

$$\Pi(t) = u(t + 1/2) - u(t - 1/2) \qquad [55]$$

Transforming the expression on the right-hand side, using the time-shifting theorem (2.3.4), we again find the transform of $\Pi(t)$, as given above under 1. It is also of interest to see the connection between our formulas 16 and 9. We know that $\delta(t) = u'(t)$ (Båth, 1968, p.250). Applying the derivation theorem (2.3.6) to our formula 16, we get:

$$\delta(t) = u'(t) \leftrightarrow i\omega[\pi\delta(\omega) - i/\omega] = 1$$

in agreement with case 9.

(17) From 16 by the frequency-shifting theorem (2.3.5).

(18 and 19) From 16 by the modulation theorems (2.3.5). As a check, 18 and 19 combined give 17.

(20) By direct integration, noting that the integrand is different from zero only for $0 < t < 1/2$.

(21) The transform of $|t|$ can be obtained by equalizing the following integral (for proof, see Papoulis, 1962, p.277):

$$|t| = -\frac{1}{\pi}\int_{-\infty}^{\infty} \frac{\cos \omega t}{\omega^2}\, d\omega \qquad [56]$$

and the inversion formula for $f(t)$, considering that $F(\omega)$ is real and even, as $|t|$ is real and even.

(22) Derived by straight-forward integration over $1/t$ and applying the integral formula:

$$\int_0^{\infty} \frac{\sin \omega t}{t}\, dt = \begin{cases} \pi/2 & \text{for } \omega > 0 \\ 0 & \text{for } \omega = 0 \\ -\pi/2 & \text{for } \omega < 0 \end{cases} \qquad [57]$$

Bracewell (1965, p.130) solves this problem by contour integration in the complex t-plane, which also forms a useful exercise. The result also follows from 8 by the symmetry theorem (2.3.2).

(23) By straight-forward integration. Note that integration need only be extended from 0 to $+\infty$, as there is no contribution for $t < 0$.

(24) We write:

$$e^{-a|t|} = e^{-at}\, u(t) + e^{at}\, u(-t)$$

The transform of the first term is given in 23 and of the second term we find it from 23 by exchanging $-t$ for t and using the symmetry theorem (2.3.2) in the form $f(-t) \leftrightarrow F(-\omega)$. Then the result follows immediately by the linearity theorem (2.3.1).

(25) By direct integration over e^{-at^2} (cf. Bracewell, 1965, p.130) and using the integral formula:

$$\int_{-\infty}^{\infty} e^{-at^2}\,dt = \sqrt{\pi/a} \qquad [58]$$

The Gaussian function usually appears under the following form:

$$f(t) = \frac{1}{\sigma\sqrt{2\pi}}\, e^{-t^2/2\sigma^2} \qquad [59]$$

where σ = standard deviation and the average is assumed to be zero. The Fourier transform of [59] is by the same method found to be:

$$e^{-\sigma^2\omega^2/2}$$

In this book, we shall stick to the simple form $f(t) = e^{-at^2}$, where $a = 1/2\sigma^2$. Formulas corresponding to [59] can be immediately written down in every case, as soon as those deduced from e^{-at^2} are given. The transform of $e^{-\pi t^2}$ is obtained as a special case, putting $a = \pi$. We see that the transform of the Gaussian curve is also Gaussian.

(26) By direct integration, guided by the procedure applied in 23.

(27) First derive the transform of te^{-at}, which can be done by direct integration, then apply the modulation theorem (2.3.5).

2.4.4 Some general comments on Fourier transforms

Besides the hints given for the solution of the problems in section 2.4.3, it should be pointed out that many of these are examples where the theorems in section 2.3 are applicable. For instance, the derivation theorem (2.3.6) can be tested on those examples for which one time function is the derivative of the other,

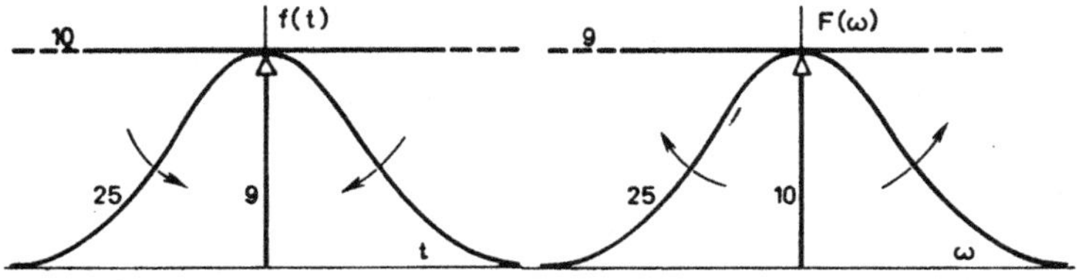

Fig.15. Schematic demonstration of the reciprocal nature of the functions $f(t)$ and $F(\omega)$. The numbers refer to the examples in Table V.

e.g. the pairs 13-14, 18-19 and 25-26. In the case 18-19 one has to apply the formula $u'(t) = \delta(t)$. Sometimes, the transform of a function may not be readily calculable, whereas the transform of some of its derivatives may be much easier to obtain. Then it is advisable to get first a derivative of the given function, then get the transform of the derivative, and finally, by the derivation theorem (2.3.6), get the transform of the originally given function. An example from gravity is given by Sharma and Geldart (1968). This procedure is generally applicable, to all theorems listed in section 2.3. As a further exercise, it is suggested to split the expressions for $F(\omega)$ in Table V into their amplitude spectra and phase spectra (cf. Table III).

The reciprocal nature of the two functions $f(t)$ and $F(\omega)$ is illustrated schematically in Fig.15 for the two most extreme cases, examples 9 and 10, and for an intermediate case, example 25. A function which exhibits the same form in the two domains, like the Gaussian function, is said to be *self-reciprocal*. Fig.15 also illustrates the relations between integration limits in the two domains: 10 infinite t–finite ω, 25 infinite t–infinite ω, 9 finite t–infinite ω. There is no possibility to have finite ranges both in t and ω. A review is given in Table VI.

It is especially instructive to compare $F(\omega)$ for three cases involving $\cos \omega_0 t$:

Case 5 with a limited cosine function: $F(\omega)$ becomes an infinite band.

Case 13 with a doubly infinite cosine function: $F(\omega)$ becomes a limited line spectrum.

Case 18 with a semi-infinite cosine function: $F(\omega)$ consists both of an infinite band and a limited line spectrum.

A pure line spectrum, corresponding to a Fourier series expansion, exists only for a fully periodic function, whereas in the other cases "end effects" of the time function lead to band spectra.

TABLE VI

Examples from Table V regarding relations between corresponding integration intervals of $f(t)$ and $F(\omega)$ and the character of $F(\omega)$

$F(\omega)$	$f(t)$	
	infinite	finite (truncated)
Infinite:		
band spectrum	8, 16–19[1], 21–27	1, 4–6, 9, 11, 20
line spectrum	15	no[2]
Finite:		
band spectrum	2, 3, 7	no[2]
line spectrum	10, 12–14, 16–19[1]	no[2]

[1] Semi-infinite.

[2] No such cases are possible.

It is also worth noting that if $F(\omega)$ has a sharp cut-off, the frequency of $f(t)$ is equal to the cut-off frequency of $F(\omega)$. Examples of this are 2, 3, 7, 10, 12–14. Conversely, if $f(t)$ has a sharp cut-off at some value of t, the corresponding $F(\omega)$ exhibits a "frequency" (factor of ω), called "quefrency", which is equal to the cut-off in the time domain. Examples of this are 1, 4–6, 9, 11, 20.

The practical case, encountered in the treatment of any kind of observational data, corresponds to a finite range in t and an infinite range in ω. Again for practical reasons, the spectrum can only be depicted within a limited range of ω, covering the most significant part of the spectrum.

With reference to Fig.12 there is one further point to be emphasized as an aid in thinking in the two domains. Without knowledge of relations between the two domains, it would be easy to make the mistake that a spike in $f(t)$, example 9, would correspond to a sudden action in the frequency domain; on the contrary, it corresponds to a constant action covering all frequencies. However, this becomes clear by considering Fig.13. In order to produce a spike in the time domain, it is necessary to add cosinusoids of all frequencies, from zero to infinite frequency. Conversely for example 10. This means that a narrow spectrum $F(\omega)$ centered at $\omega = 0$, corresponds to a long duration of the source activity and that a broad spectrum $F(\omega)$ corresponds to a short duration of the source activity. This is important to bear in mind in efforts to explain spectra.

2.5 MULTI-DIMENSIONAL FOURIER SERIES AND TRANSFORMS

2.5.1 Mathematical forms for the multi-dimensional case

The theory developed in this chapter has been concerned with one-dimensional Fourier series and transforms, generally time and frequency domains, as being most significant in geophysical applications. However, the theory can be immediately extended to any number of dimensions both for the Fourier series and the Fourier transform.

For instance, a *two-dimensional Fourier series* expansion with two spatial coordinates x and y can be easily obtained by an extension of [1]:

$$f(x) = a_0 + \sum_{n=1}^{\infty} (a_n \cos nx + b_n \sin nx)$$

In this case, all entering quantities, i.e. $f(x)$, a_0, a_n, b_n, which correspond to one line y = constant, are also functions of y, i.e.:

$$f(x,y) = a_0(y) + \sum_{n=1}^{\infty} [a_n(y) \cos nx + b_n(y) \sin nx]$$

The functions $a_0(y)$, $a_n(y)$ and $b_n(y)$ can be expanded in Fourier series in y, corresponding to [1], and the resulting formula can be written as follows:

$$f(x,y) = \sum_{m=0}^{\infty} \sum_{n=0}^{\infty} (a_{mn} \cos mx \cos ny + b_{mn} \cos mx \sin ny + c_{mn} \sin mx \cos ny + d_{mn} \sin mx \sin ny) \qquad [60]$$

where a_{mn}, b_{mn}, c_{mn}, d_{mn} are constants. Note that by extending the summation to $m = 0$, $n = 0$, the constant term is included in the sum as well as the single terms in cos and sin. Equation [60] represents the two-dimensional Fourier series expansion of the function $f(x,y)$.

Such expansions have found application to two-dimensional fields of anomalies of various kinds, such as gravimetric and geomagnetic (Tsuboi and Fuchida, 1937, 1938; Nagata, 1938; Morelli and Mosetti, 1961; Morelli and Carrozzo, 1963; Hahn, 1965, and in a restricted manner by Parkinson, 1971) and to seismic delay times in crustal and upper-mantle exploration (Raitt, 1969).

Among other harmonic expansions, the *spherical harmonic expansions* deserve special attention, not the least for their indispensable importance in studies of geophysical phenomena of global extent. They provide a simple means of expressing the distribution of a given function over a spherical surface (Chapman and Bartels, 1940, 1951). For a mathematical development of spherical harmonic functions, we refer to textbooks in applied mathematics (for example, Båth, 1968, pp.170–173). The spherical harmonics consist of a combination of sine-cosine and Legendre functions, where the longitudinal dependence is expressed by a Fourier series and the latitudinal one by Legendre functions:

$$f(\varepsilon,\delta) = \sum_{n=0}^{N} \sum_{m=0}^{n} (A_n^{(m)} \cos m\varepsilon + B_n^{(m)} \sin m\varepsilon) P_n^m(\cos \delta) \qquad [61]$$

with ε = longitude, δ = colatitude, P_n^m (cos δ) = the associated Legendre polynomial of the first kind of degree n and order m. In earlier analyses, in pre-computer time, it was customary first to perform a harmonic analysis for each latitude (for example, see Van Mieghem, 1961) and then to fit the result by Legendre functions of latitude. In more recent computations a direct fit is attempted to area means or point values by least-square methods. A summary of geophysical applications is given in section 2.5.4. A four-dimensional dependence could also be envisaged in case of spherical harmonic expansions, cf. [61], by letting the coefficients depend on depth z and time t: $A_n^{(m)}(z,t)$, $B_n^{(m)}(z,t)$, and thus representing the function $f(\varepsilon,\delta,z,t)$.

Three-dimensional Fourier transform formulas read as follows (k_x and k_y are wavenumbers in the two spatial directions x and y):

$$F(k_x,k_y,\omega) = \int_{-\infty}^{\infty}\int_{-\infty}^{\infty}\int_{-\infty}^{\infty} f(x,y,t)\,\mathrm{e}^{-i(k_x x + k_y y + \omega t)}\,\mathrm{d}x\,\mathrm{d}y\,\mathrm{d}t$$

$$f(x,y,t) = 1/(2\pi)^3 \int_{-\infty}^{\infty}\int_{-\infty}^{\infty}\int_{-\infty}^{\infty} F(k_x,k_y,\omega)\,\mathrm{e}^{i(k_x x + k_y y + \omega t)}\,\mathrm{d}k_x\,\mathrm{d}k_y\,\mathrm{d}\omega \qquad [62]$$

Introducing the space–time vector $\mathbf{X}(x,y,t)$ and the wavenumber–frequency vector $\boldsymbol{\Omega}(k_x,k_y,\omega)$, we can write [62] also in the following vectorial form:

$$F(\boldsymbol{\Omega}) = \int_{-\infty}^{\infty} f(\mathbf{X})\,\mathrm{e}^{-i\boldsymbol{\Omega}\cdot\mathbf{X}}\,\mathrm{d}\mathbf{X}$$

$$f(\mathbf{X}) = 1/(2\pi)^3 \int_{-\infty}^{\infty} F(\boldsymbol{\Omega})\,\mathrm{e}^{i\boldsymbol{\Omega}\cdot\mathbf{X}}\,\mathrm{d}\boldsymbol{\Omega} \qquad [63]$$

There is no need to transform all the variables in one operation. For instance, from [62.1] we see immediately the following two cases of partial transformation: Only time–frequency transformation:

$$F(x,y,\omega) = \int_{-\infty}^{\infty} f(x,y,t)\,\mathrm{e}^{-i\omega t}\,\mathrm{d}t \qquad [64]$$

Only space–wavenumber transformation:

$$F(k_x,k_y,t) = \int_{-\infty}^{\infty}\int_{-\infty}^{\infty} f(x,y,t)\,\mathrm{e}^{-i(k_x x + k_y y)}\,\mathrm{d}x\,\mathrm{d}y \qquad [65]$$

Similar partial transformation is naturally equally applicable to the inverse [62.2].

The formulas can be immediately generalized to any number of dimensions, only noting the factor $1/(2\pi)^n$ on the right-hand side of the inversion formula, where n = number of dimensions. The set of equations [62] has applications in geophysics, especially to array stations, where observations are not only made in the time domain but also spatially (x,y). See for example Green et al. (1966). All theorems developed above in this chapter as well as those developed later, can be extended relatively easily to multi-dimensional Fourier transforms.

Note that in the two-dimensional transform, structures in the wavenumber

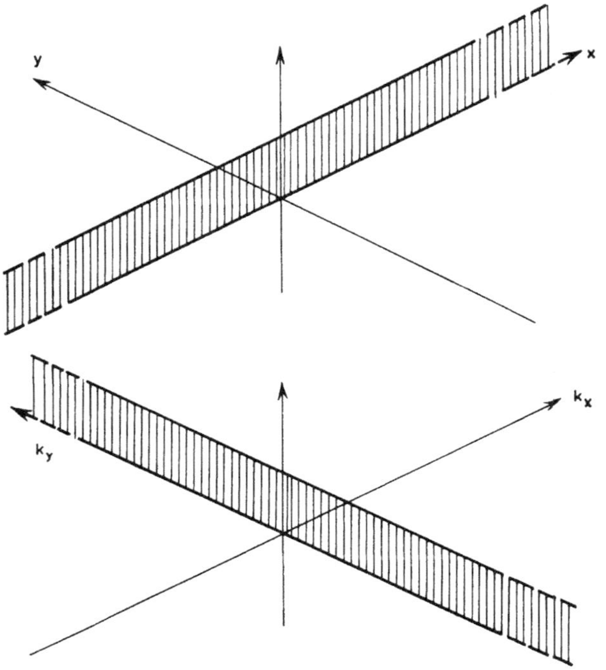

Fig.16. Rotation by 90° of a structure in the two-dimensional wavenumber domain as compared with the space domain. For a body of finite length and thickness, the correspondence holds for any direction; in the degenerate case shown in this figure, only for the coordinate axes.

domain appear rotated 90° as compared to the space domain. The simplest way to illustrate this is by means of a vertical sheet of unit height along the x-axis. Along the x-axis, this transforms into a δ-function (example 10, section 2.4.3). Along the y-axis the given structure means a δ-function, and this transforms into a sheet in the wavenumber domain (example 9, section 2.4.3). As Fig.16 illustrates, the structure appears rotated 90°. This can naturally be extended with proper modifications to more complex configurations (see, e.g. Bracewell, 1965, pp.246–247).

2.5.2 Circular symmetry – Hankel transform

A special interest is attached to the two-dimensional case with circular symmetry. Then, the Fourier transforms are equivalent to the Hankel transform (see Bracewell, 1965, pp.244–248, or Båth, 1968, p.216). This is easily seen as follows. With reference to Fig.17 we have:

$$F(k_x,k_y) = F(k,\psi) = \int_{-\infty}^{\infty}\int_{-\infty}^{\infty} f(x,y)\,\mathrm{e}^{-i(k_x x+k_y y)}\,\mathrm{d}x\,\mathrm{d}y =$$

$$= \int_0^\infty \int_0^{2\pi} f(r)\, e^{-ikr(\cos\psi \cos\theta + \sin\psi \sin\theta)}\, r\, dr\, d\theta =$$

(for circular symmetry)

$$= \int_0^\infty r f(r)\, dr \underbrace{\int_0^{2\pi} e^{-ikr\cos(\theta-\psi)}\, d\theta}_{= 2\pi J_0(kr)} = 2\pi \int_0^\infty r f(r)\, J_0(kr)\, dr$$

(see Båth, 1968, p.123)

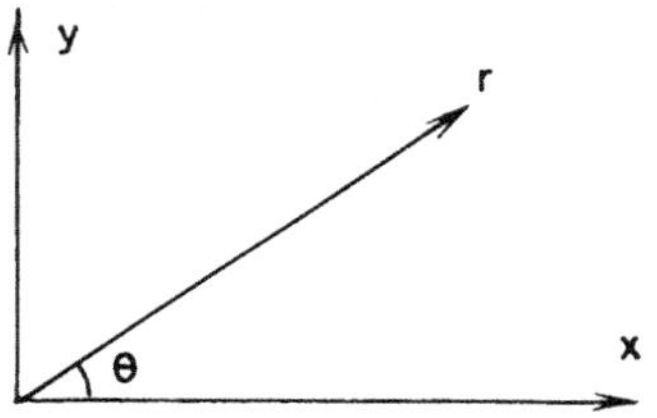

$x = r \cos \theta$

$y = r \sin \theta$

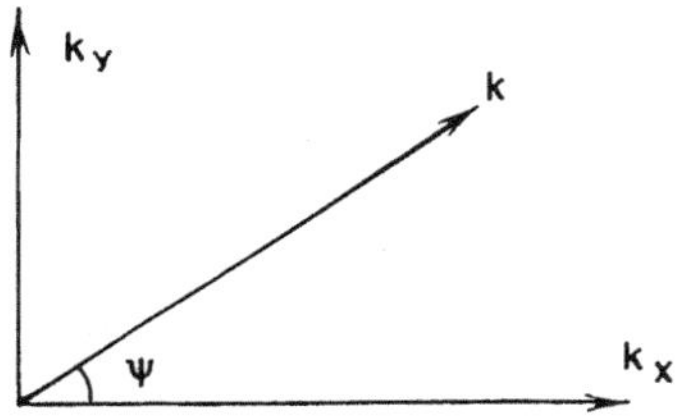

$k_x = k \cos \Psi$

$k_y = k \sin \Psi$

Fig.17. Coordinates used in treatment of two-dimensional circular symmetry.

$J_0(kr)$ is a zero-order Bessel function. The transform is a zero-order Hankel transform (Båth, 1968, p.211, pp.216–218), and is sometimes referred to as the *Hankel spectrum* (Haubrich and McCamy, 1969). In summary, the transform and its inverse are:

$$F(k) = 2\pi \int_0^\infty r f(r)\, J_0(kr)\, dr$$

$$f(r) = 1/2\pi \int_0^\infty k\, F(k)\, J_0(kr)\, dk \qquad [66]$$

2.5.3 Geophysical applications of multi-dimensional transforms

The potentiality of the multi-dimensional Fourier analysis in resolving many problems has been realized more and more in recent time. Let us take a few examples from geophysics:

(1) A 4-dimensional case is encountered, for instance, in meteorology by combining temperature readings along two horizontal profiles (x,y) and a vertical profile (z) and letting time (t) be the fourth variable. The same is true for any time-varying three-dimensional distribution, as of every meteorological or oceanographic parameter. Another example would be a 3-dimensional seismological array. A 4-dimensional dependence in the case of spherical harmonic expansions was envisaged in section 2.5.1.

(2) A 3-dimensional case (x,y,t) is encountered in analyses of the shape of the ocean surface, as due to ocean surface waves, swell, ocean tides, variation in atmospheric pressure, wind, etc. The same variables (x,y,t) enter into a surface seismological array. Another 3-dimensional case (x,y,z) is represented by the distribution of physical parameters in the earth's interior, allowing for lateral inhomogeneity (i.e., including x and y) but assuming stationary conditions (i.e., excluding t as variable). In case the observations cover larger areas or volumes, other coordinates, especially spherical, will be more suitable than the rectangular ones.

(3) A 2-dimensional case (x,t) is represented by any record versus time that is made along a profile in the x-direction. Linear profiles of any kind of measurements belong here, e.g. linear seismological arrays or a linear profile of temperature measurements. A spatial 2-dimensional case (x,y) is represented by the earth's topography, where z and t do not enter. Here again, spherical harmonics are more apt, especially if we want to cover larger portions of the earth. The two-dimensional transform $f(x,y) \leftrightarrow F(k_x,k_y)$ is very much used both in seismology (array-station records), oceanography (ocean surface waves), and in gravity and geomagnetism, as we shall see in later chapters. Applying only partial transform to $f(x,t) \leftrightarrow F(x,\omega)$, we can display $F(x,\omega)$ by isolines in a rectangular coordinate system with distance x and frequency ω as axes. This method has been applied by Willis and DeNoyer (1966) to P and S particle velocities.

(4) The 1-dimensional case (t) is the most common one, and is represented by any record versus time taken at a given point, e.g. a seismogram, a thermogram, etc. In space (z), the vertical distribution of physical properties in the earth is an example, assuming no lateral inhomogeneity (no variation with x and y) and stationarity (no variation with t). Although it would be perfectly sound mathematically to express, for instance, the vertical velocity variation in the earth as a wavenumber spectrum, $V(k_z)$, it would probably be of little use.

(5) The 0-dimensional case is represented by constants, e.g. with time (t) at a given point, or spatially (x,y,z) by a plane, stationary surface.

Even though these examples are taken from geophysical disciplines, it is possible to take examples from almost every branch of science as well as from every-day life. In fact, everything could be expressed as a spectrum, and thus appear as transformed into another "inverted world". It is good training for the student to try to think in the two "worlds", although we feel that further exposition of a total and simultaneous transformation has to be left over to science-fiction writers.

2.5.4 Geophysical applications of spherical harmonics

Any phenomenon of global extent invites the application of spherical harmonics, and we shall here give some examples from geophysics:

(1) Gravity (Heiskanen and Vening Meinesz, 1958; Kaula, 1959, 1966a and b, 1967), and all kinds of tides.

(2) Topography (Heiskanen and Vening Meinesz, 1958; Lee and Kaula, 1967), distribution of continents and oceans (Walzer, 1972b), tectonic features (Coode, 1966, 1967), and large-scale crustal deformation.

(3) Geomagnetism (Chapman and Bartels, 1940, 1951; Fanselau and Kautzleben, 1958; Mauersberger, 1959; Yukutake and Tachinaka, 1968, the latter for the magnetic field in past epochs; and Malin and Pocock, 1969, who demonstrate that an expansion to the sixth order for a spherical earth combines compactness with sufficient accuracy).

(4) Terrestrial heat flow (Lee and MacDonald, 1963; Lee, 1963; Horai and Simmons, 1969; and others) and convection currents in the earth.

(5) Free oscillation of the earth.

(6) Global studies in meteorology (Ellsaesser, 1966, developed and applied the method to hemispheric meteorological data, assuming asymmetry between northern and southern hemispheres; Eliasen and Machenhauer, 1965, used the method in a study of wind patterns over the northern hemisphere).

(7) Correlation studies between various types of global distributions. This has in fact been attempted by Arkani-Hamed and Toksöz (1968) and Toksöz et al. (1969), who correlated coefficients in spherical harmonic expansions of heat flow, seismic P-wave crustal velocities and teleseismic travel-time residuals, crustal thickness, topography, gravimetric potential and the magnetic field. Detailed analysis is still to a great extent hampered by paucity of data. Moreover, the numerous interwoven effects suggest use of partial correlation techniques.

2.5.5 Application of the Fourier transform to arrays

The Fourier transform and its inverse can easily be written down in any special case, using the general equations [62]. Let us just take one example with application to two-dimensional array stations:

$$F(k_x,k_y) = \int_{-\infty}^{\infty} \int_{-\infty}^{\infty} f(x,y)\, \mathrm{e}^{-i(k_x x + k_y y)}\, \mathrm{d}x\, \mathrm{d}y \qquad [67]$$

or in vector form:

$$F(\mathbf{k}) = \int_{-\infty}^{\infty} f(\mathbf{r})\, \mathrm{e}^{-i\mathbf{k}\cdot\mathbf{r}}\, \mathrm{d}\mathbf{r} \qquad [68]$$

Identifying $f(x,y) = f(\mathbf{r})$ with a weight factor w_{js} for the array-element combination $\mathbf{r} = \mathbf{r}_j - \mathbf{r}_s$, we have:

$$F(\mathbf{k}) = \int_{-\infty}^{\infty} w_{js}\, \mathrm{e}^{-i\mathbf{k}\cdot(\mathbf{r}_j - \mathbf{r}_s)}\, \mathrm{d}\mathbf{r} \qquad [69]$$

or in case of discrete summation over a finite number N of elements:

$$F(\mathbf{k}) = \sum_{j=1}^{N} \sum_{s=1}^{N} w_{js}\, \mathrm{e}^{-i\mathbf{k}\cdot(\mathbf{r}_j - \mathbf{r}_s)} \qquad [70]$$

This is the two-dimensional *spectrum window* for the array (Haubrich, 1968). Cf. also Barber (1966) and Haubrich and McCamy (1969). In particular, for a linear, uniform array, the spectrum window becomes:

$$F(k) = \sum_{j=1}^{N} w_{jj} + 2 \sum_{j,s=1}^{N} w_{js} \cos k(j - s) \qquad [71]$$

which is found immediately by applying [70]. In the latter sum, $j \neq s$ and $w_{js} = w_{sj}$. Equation [71] can be simplified to the following:

$$F(k) = 2w_0 + 2 \sum_{m=1}^{M} w_m \cos km = 2 \sum_{m=0}^{M} w_m \cos km \qquad [72]$$

where:

$$\left. \begin{aligned} 2w_0 &= \sum_{j=1}^{N} w_{jj} = \text{constant} \\ m &= (j-1)\,N - \tfrac{1}{2}j\,(j+1) + s \\ M &= N(N-1)/2 \end{aligned} \right| \qquad [73]$$

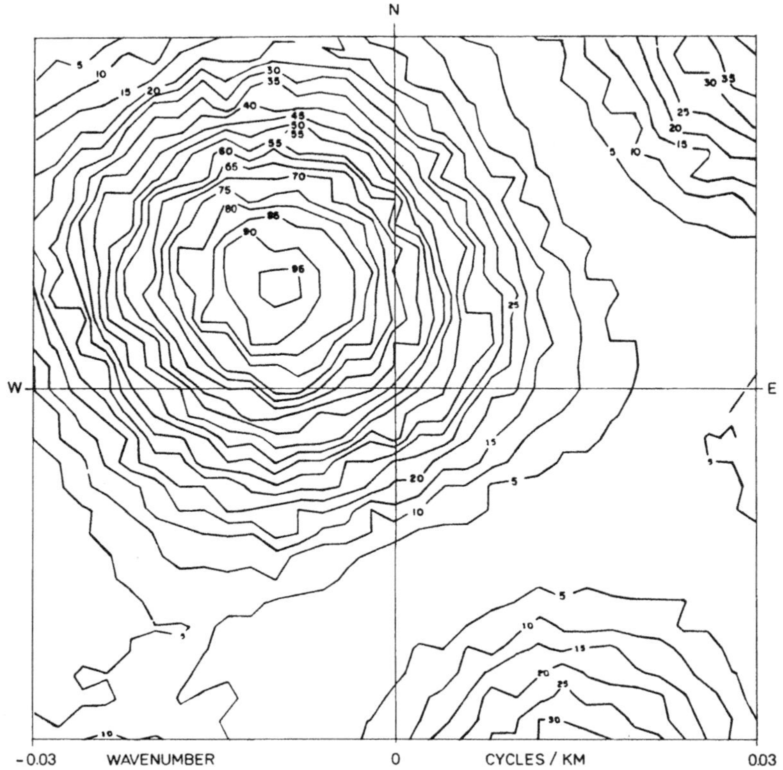

Fig.18. Wavenumber spectrum of Rayleigh-wave power at a frequency of 0.05 cps from an earthquake in Kurile Islands, November 12, 1966, as recorded at the LASA array in Montana, USA. After Capon et al. (1969).

M is the number of elements in the so-called *coarray*. The correctness of [73] can most easily be proved by induction. For a brief but clear exposition of linear arrays with all weight factors equal, the reader is referred to Officer (1958, pp.274–278). Further discussion with application to two-dimensional spatial filtering will follow in section 6.4.2 and 10.2.2.

A case is shown in Fig.18, displaying isolines of Rayleigh-wave power spectral density $E(\mathbf{k}) = |F(\mathbf{k})|^2$ for a frequency of 0.05 cps (= 20-sec period) with k_x and k_y as coordinates. The vector from the origin to any point in such graphs gives (for the chosen frequency) the direction of arrival:

$$\left.\begin{aligned} &\text{Wavenumber} = k = (k_x^2 + k_y^2)^{\frac{1}{2}} \\ &\text{Direction of arrival} = \tan^{-1}(k_y/k_x) \end{aligned}\right| \quad [74]$$

This technique, with application to the LASA array in Montana, USA, has been considerably improved by Capon (1969a). By using an optimum wavenumber window (section 4.4), different for different wavenumbers, he has reached a higher resolution and thus more accurate determination of arrival direction.

This method has made it possible to discriminate between multi-path-propagated surface waves from a given event as well as between superimposed surface waves of different origins (Capon and Evernden, 1971, Capon, 1972). The same method is naturally applicable to any other type of array data, e.g. of micro-barographs (Mack and Smart, 1972) or of ocean-wave recorders. A related array technique, suitable for detection of weak signals in noise, has recently been devised by Posmentier and Herrmann (1971). See also Barber (1958, 1963), Green et al. (1966), Haubrich (1968), Iyer (1968).

Frequently, the same type of graph is used for phase-velocity determination: $c = \omega/k$, corresponding to the maximum power E. However, as pointed out by Smart (1971), this may lead to errors. The maximum of the power $E(k_x,k_y,\omega)$ corresponds to:

$$\partial E/\partial k_x = \partial E/\partial k_y = \partial E/\partial \omega = 0 \qquad [75]$$

In a graph with k_x,k_y as axes for a constant ω (Fig.18), the first two may be nullified but this does not guarantee that also $\partial E/\partial \omega = 0$, which is necessary for correct determination of phase velocity. Therefore, a more correct procedure is the following:

(1) Use a plot of $E(k_x,k_y)$ for given ω to determine azimuth of arrival.

(2) Use a plot of $E(\omega,\mathbf{k})$, where $\mathbf{k}$ is total wavenumber (along direction of arrival), for determination of phase velocity.

Smart (1971) has also given a modified, simpler procedure for the phase-velocity determination.

A related, though different method of analyzing array data is the *Vespa process*, in which power is contoured in graphs with slowness $\mathrm{d}t/\mathrm{d}\Delta$ and time as coordinates, so-called *vespagrams*. The word vespa is derived from *velocity spectral analysis*, as power is displayed versus velocity at each moment (taking "spectral" in a wide sense). The method, as applied to seismic arrays, has proved to be efficient in distinguishing between multiple, closely arriving body-wave phases of different apparent velocities (Davies et al., 1971).

Array theory and practice have by this time filled a very large literature, to a great extent contained in institute reports, which would well deserve to be summarized in a book. In the present book, arrays will be mentioned only in connection with spectral considerations. Array treatment and spectral analysis are two parallel, partially overlapping fields, but generally with different goals. Spectral analysis is one method out of several others which enter the treatment of array data, and vice versa. Even though arrays have application to practically every field of geophysics, the permanent seismograph array stations dominate the recent literature. It is hoped that this book will make it easier for the reader to penetrate the array literature. Most of this literature deals with specially built array stations. But in addition it has been demonstrated that some ordinary station networks permit the application of array-data processing techniques, e.g. in Fennoscandia (Husebye and Jansson, 1966; Jansson and Husebye, 1968; Whitcomb, 1969). In other areas with more variable geological structure, especially variations in sedimentary layering, such station combinations are not possible.

Chapter 3

POWER SPECTRA AND FUNDAMENTALS OF OBSERVATIONAL SPECTRA

In this chapter we shall gradually take the step from spectral analysis of analytical functions to empirically observed data. The latter constitute the main concern in geophysical applications of spectral analysis. No doubt, the book by Blackman and Tukey (1959) is a real milestone in this development. By giving methods for calculation of power spectra at a time when large computers came more and more into general use, it provided for an almost explosive development in the application of power spectral methods in geophysics. These methods dominated the geophysical spectroscopy for nearly a decade after the appearance of that book. Beginning in the middle of the 1960's and during the following years, they gradually gave way to computationally more efficient methods, covered by the name Fast Fourier Transform (FFT). In this chapter we shall deal with correlation functions and power spectra, which are still of very great geophysical significance, while the FFT will be given in section 4.6.3. In the latter part of the present chapter we shall discuss the general types of signals encountered in geophysics, and we shall also study some alternatives and extensions of the spectral concepts beyond those described hitherto, also with geophysical significance.

3.1 CORRELATION FUNCTIONS

3.1.1 Autocorrelation

In this chapter we shall essentially restrict ourselves to real time functions $f(t)$ as being the only ones of geophysical interest. The autocorrelation function is defined as follows:

$$C_{11}(\tau) = \int_{-\infty}^{\infty} f_1(t) f_1(t + \tau)\, dt = \int_{-\infty}^{\infty} f_1(t) f_1(t - \tau)\, dt \qquad [1]$$

By a simple substitution we find that the autocorrelation is an even function: $C_{11}(\tau) = C_{11}(-\tau)$. It has its maximum for $\tau = 0$.

As the integral introduced above extends from $-\infty$ to $+\infty$, the question may be raised under what conditions this integral will have finite values. Without attempting an exhaustive, theoretical discussion of this matter, we may state the following practical rules (the functions $f(t)$ always assumed finite):

(1) For theoretically defined functions, the integration limits are finite, only when the functions have non-zero values in a limited range. If not, the integrals will still converge, if the functions taper off sufficiently towards $\pm\infty$ (in other words, if Dirichlet's conditions are fulfilled).

(2) For empirically observed values, the range is always limited, as we have only limited ranges of observations.

In this connection, it should be noted that alternative definitions exist, in terms of limits of finite integrals, e.g. for the autocorrelation function:

$$C_{11}(\tau) = \lim_{T\to\infty} \frac{1}{T} \int_{-T/2}^{T/2} f_1(t) f_1(t + \tau)\, \mathrm{d}t \qquad [2]$$

Corresponding formulas can be written down for the following quantities as well. For a stationary, i.e. statistically time-independent $f_1(t)$, as is generally assumed, averages like the one expressed in [2] will clearly be independent of the record length T. From a purely mathematical point of view, the infinite integral in [1] would correspond to the limit in [2] without the factor $1/T$ ahead of the integral. The expression [2] would approach zero in the limit for $f_1(t)$ fulfilling Dirichlet conditions. However, in practical applications, the integration has always to be restricted to a limited range T, and then [2] represents the average of $f_1(t)$ $f_1(t + \tau)$ over the record length T. Inclusion of the factor $1/T$ then serves as a certain normalization procedure, especially necessary in comparing series of different lengths T. Then [2] has a definite value, independent of T, just like [1]. And, after all, we are only interested in relative measures, but calculated in a way that makes comparisons possible.

The autocorrelation function, here defined for the one-dimensional case, can be immediately extended to multi-dimensional conditions. For instance, the two-dimensional spatial autocorrelation can be immediately written down as follows:

$$C_{11}(\xi,\eta) = \int_{-\infty}^{\infty} \int_{-\infty}^{\infty} f_1(x,y) f_1(x + \xi, y + \eta)\, \mathrm{d}x\, \mathrm{d}y \qquad [3]$$

This expression has found application to analyses of geophysical maps, e.g. of geomagnetism (Chapter 10). For ocean-wave studies use has been made of the corresponding distance–time correlation function (Medwin et al., 1970):

$$C_{11}(\xi,\tau) = \int_{-\infty}^{\infty} \int_{-\infty}^{\infty} f_1(x,t) f_1(x + \xi, t + \tau)\, \mathrm{d}x\, \mathrm{d}t \qquad [4]$$

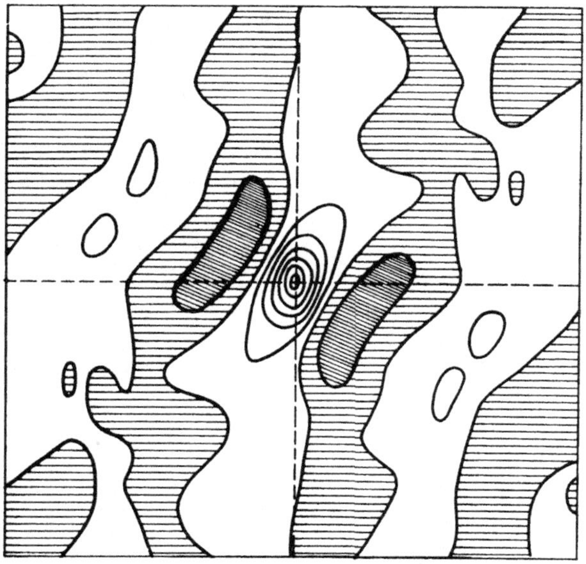

Fig.19. A display of a two-dimensional autocorrelation, derived for ocean surface waves. Note especially the symmetry of the contours. Dashed areas correspond to negative correlations. After Neumann and Pierson (1966, p.353); originally in *Meteor. Papers*, N.Y. Univ., 2:6, 1960, 88 pp.

Corresponding to the symmetry condition $C_{11}(\tau) = C_{11}(-\tau)$ in the one-dimensional case, we find by appropriate substitutions the corresponding conditions in the two-dimensional case from [3]:

$$C_{11}(\xi, \pm\eta) = C_{11}(-\xi, \mp\eta) \qquad [5]$$

where the upper signs belong together, and so do the lower signs. Equation [5] implies symmetry with respect to the origin in the $\xi\eta$-plane (which is equivalent to skew-symmetry with respect to any straight line through the origin). As a consequence it is sufficient to know the values of $C_{11}(\xi,\eta)$ in any of the ξ,η-half-planes. See Fig.19.

3.1.2 Cross-correlation

The cross-correlation function is defined as follows:

$$C_{12}(\tau) = \int_{-\infty}^{\infty} f_1(t) f_2(t + \tau)\, \mathrm{d}t = \int_{-\infty}^{\infty} f_1(t - \tau) f_2(t)\, \mathrm{d}t \qquad [6]$$

That the two expressions in [6] agree is found by a simple substitution in one of them. This means that $C_{12}(\tau) = C_{21}(-\tau)$.

Generalizations of the cross-correlation function [6] can be made in the following ways:

(1) To more than one dimension. This development is analogous to the one given in 3.1.1 for the autocorrelation function.

(2) To more than two functions f_1 and f_2. Cross-correlations involving more than two functions are of importance in turbulence studies (Hinze, 1959). Cf. section 3.6.4.

Auto- and cross-correlations are of use in calculating power spectra (section 3.3) and for detection of weak signals in noise.

3.2 CONVOLUTION

3.2.1 Definition of convolution

The concept of convolution will be of considerable use in following chapters, especially to express various filtering processes (Chapter 6). It is introduced here, partly because of its formal similarity with the cross-correlation. Convolution is defined as follows:

$$f_1(t)\star f_2(t) = \int_{-\infty}^{\infty} f_1(\tau)f_2(t-\tau)\,\mathrm{d}\tau = \int_{-\infty}^{\infty} f_1(t-\tau)f_2(\tau)\,\mathrm{d}\tau \qquad [7]$$

or, equivalently:

$$f_1(\tau)\star f_2(\tau) = \int_{-\infty}^{\infty} f_1(t)f_2(\tau-t)\,\mathrm{d}t = \int_{-\infty}^{\infty} f_1(\tau-t)f_2(t)\,\mathrm{d}t \qquad [8]$$

The right-hand sides, obtained by substituting $t-\tau$ for τ in the first expression, and substituting $\tau - t$ for t in the second, demonstrate the commutative law. It is also easily seen that the associative and distributive laws hold for convolution. The integral expressing convolution corresponds to the *Duhamel integral* (cf. Båth, 1968, p.276). Although this is the most usual definition of convolution, it sometimes appears in a slightly different form (e.g. Lanczos, 1966, p.247).

Combining the formulas for correlations [1] and [6] and for convolution [8], we find (e = even, o = odd, respectively):

$$\left.\begin{aligned}
C_{11}(\tau) &= C_{11}(-\tau) = f_1(\tau)\star f_1(-\tau);\\
C_{12}(\tau) &= C_{21}(-\tau) = f_1(-\tau)\star f_2(\tau) = [f_{1e}(\tau) - f_{1o}(\tau)]\star f_2(\tau);\\
C_{12}(-\tau) &= C_{21}(\tau) = f_1(\tau)\star f_2(-\tau) = f_1(\tau)\star [f_{2e}(\tau) - f_{2o}(\tau)]
\end{aligned}\right| \qquad [9]$$

Just as for the correlation functions, the convolution could also be extended to the multi-dimensional case.

3.2.2 The convolution theorem

If $f_1(t) \leftrightarrow F_1(\omega)$ and $f_2(t) \leftrightarrow F_2(\omega)$, then $f_1(t) \cdot f_2(t) \leftrightarrow 1/2\pi \cdot F_1(\omega) \star F_2(\omega)$ and $f_1(t) \star f_2(t) \leftrightarrow F_1(\omega) \cdot F_2(\omega)$.

Proof. The Fourier transform of $f_1(t) \cdot f_2(t)$ is by equation [20] in section 2.2.2:

$$\int_{-\infty}^{\infty} f_1(t) f_2(t)\, \mathrm{e}^{-i\omega t}\, \mathrm{d}t = \int_{-\infty}^{\infty} f_2(t)\, \mathrm{e}^{-i\omega t} \left[\frac{1}{2\pi} \int_{-\infty}^{\infty} F_1(\lambda)\, \mathrm{e}^{i\lambda t}\, \mathrm{d}\lambda \right] \mathrm{d}t$$

$$= \frac{1}{2\pi} \int_{-\infty}^{\infty} F_1(\lambda)\, \mathrm{d}\lambda \int_{-\infty}^{\infty} f_2(t)\, \mathrm{e}^{-i(\omega - \lambda)t}\, \mathrm{d}t = \frac{1}{2\pi} \int_{-\infty}^{\infty} F_1(\lambda)\, F_2(\omega - \lambda)\, \mathrm{d}\lambda$$

$$= \frac{1}{2\pi} F_1(\omega) \star F_2(\omega)$$

where λ is just an integration variable. Thus, we find:

$$f_1(t) \cdot f_2(t) \leftrightarrow 1/2\pi \cdot F_1(\omega) \star F_2(\omega) \qquad [10]$$

For the second part of the theorem, we have for the inverse Fourier transform of $F_1(\omega) \cdot F_2(\omega)$ from [23] in section 2.2.2:

$$\frac{1}{2\pi} \int_{-\infty}^{\infty} F_1(\omega)\, F_2(\omega)\, \mathrm{e}^{i\omega t}\, \mathrm{d}\omega = \frac{1}{2\pi} \int_{-\infty}^{\infty} F_2(\omega)\, \mathrm{e}^{i\omega t}\, \mathrm{d}\omega \int_{-\infty}^{\infty} f_1(\tau)\, \mathrm{e}^{-i\omega\tau}\, \mathrm{d}\tau$$

$$= \int_{-\infty}^{\infty} f_1(\tau)\, \mathrm{d}\tau \frac{1}{2\pi} \int_{-\infty}^{\infty} F_2(\omega)\, \mathrm{e}^{i\omega(t-\tau)}\, \mathrm{d}\omega = \int_{-\infty}^{\infty} f_1(\tau) f_2(t - \tau)\, \mathrm{d}\tau$$

$$= f_1(t) \star f_2(t)$$

that is:

$$F_1(\omega) \cdot F_2(\omega) \leftrightarrow f_1(t) \star f_2(t) \qquad [11]$$

Interchange of the order of integration, as done above, is permitted if the signals have finite energy. This is fulfilled by all physical signals. Thus, the spectrum of the product of two functions is the convolution of their individual spectra, and the signal corresponding to the product of two spectra is the convolution of their individual signals. The theorem can be extended to any finite number n of signals:

$$f_1(t) \cdot f_2(t) \cdot f_3(t) \ldots f_n(t) \leftrightarrow (1/2\pi)^{n-1} F_1(\omega) \star F_2(\omega) \star F_3(\omega) \star \ldots \star F_n(\omega) \quad [12]$$

and:

$$F_1(\omega) \cdot F_2(\omega) \cdot F_3(\omega) \ldots F_n(\omega) \leftrightarrow f_1(t) \star f_2(t) \star f_3(t) \star \ldots \star f_n(t) \quad [13]$$

In several examples in section 2.4, there are products of two time functions, for which the separate transforms are also given. Then, it is a good exercise to test such cases with the convolution theorem [10], according to which the transform of a product of time functions is obtained from the convolution of the individual transforms.

3.2.3 Derivative of convolution

If $f_1(t) \leftrightarrow F_1(\omega)$ and $f_2(t) \leftrightarrow F_2(\omega)$, then:

$$[f_1(t) \star f_2(t)]' = f_1'(t) \star f_2(t) = f_1(t) \star f_2'(t) \leftrightarrow i\omega\, F_1(\omega) \cdot F_2(\omega)$$

Proof. From the derivation theorem (section 2.3.6) we have $f_1'(t) \leftrightarrow i\omega\, F_1(\omega)$ and $f_2'(t) \leftrightarrow i\omega\, F_2(\omega)$. Then we find from [11]:

$$f_1'(t) \star f_2(t) \leftrightarrow i\omega\, F_1(\omega)\, F_2(\omega)$$

and:

$$f_1(t) \star f_2'(t) \leftrightarrow F_1(\omega)\, i\omega\, F_2(\omega) \quad [14]$$

Also it follows from [11] and the derivation theorem (section 2.3.6):

$$[f_1(t) \star f_2(t)]' \leftrightarrow i\omega\, F_1(\omega)\, F_2(\omega) \quad [15]$$

From these relations we readily deduce a useful result:

$$[f_1(t) \star f_2(t)]' = f_1'(t) \star f_2(t) = f_1(t) \star f_2'(t) \quad [16]$$

The difference between the derivative of a convolution and the derivative of a product of functions is to be particularly noticed.

3.3 POWER SPECTRA

3.3.1 The concept of power

Let us take a specific example, i.e., that the time function $f(t)$ represents the displacement in a seismic record. This is the general case for seismograph records. Then the corresponding frequency function $F(\omega)$, related to $f(t)$ through [20] and [23] in Chapter 2, represents the displacement spectrum. In many cases, however, it is more appropriate to consider the power of a signal instead of its amplitude or displacement. In a general sense, power is taken as proportional to the square of the amplitude. The corresponding spectra are termed power spectra.

In the time domain, the *average power* of any real function $f(t)$ is defined by the expression:

$$\lim_{T\to\infty} \frac{1}{T} \int_{-T/2}^{T/2} |f(t)|^2 \, dt \qquad [17]$$

when this limit exists. In any practical case, T has to be finite. Cf. discussion in section 3.1.1. $|f(t)|^2$ is called the *instantaneous power* of $f(t)$, and the integral

$$\int_{-\infty}^{\infty} |f(t)|^2 \, dt$$

is the *total energy* of $f(t)$, when this integral converges.

In the general case, $f(t)$ can stand for any time function, such as displacement, particle velocity, acceleration, temperature, rainfall, wind velocity, geomagnetic field intensity, electric current, voltage, etc. Therefore we understand that the term power has a much wider usage in geophysical spectroscopy than in physics.

3.3.2 Parseval's theorem

We shall now derive the relation between the power of a signal $f(t)$ and its spectrum $F(\omega)$. Let $f_1(t)$ and $f_2(t)$ be two signals whose spectra are $F_1(\omega)$ and $F_2(\omega)$ and consider the integral:

$$\int_{-\infty}^{\infty} f_1(t) f_2(t) \, dt = \int_{-\infty}^{\infty} f_1(t) \left[\frac{1}{2\pi} \int_{-\infty}^{\infty} F_2(\omega) \, e^{i\omega t} \, d\omega \right] dt$$

$$= \frac{1}{2\pi} \int_{-\infty}^{\infty} F_2(\omega) \, d\omega \int_{-\infty}^{\infty} f_1(t) \, e^{i\omega t} \, dt = \frac{1}{2\pi} \int_{-\infty}^{\infty} F_2(\omega) F_1(-\omega) \, d\omega$$

$$= \frac{1}{2\pi} \int_{-\infty}^{\infty} F_2(\omega) F_1^*(\omega) \, d\omega \qquad [18]$$

since $F(-\omega) = F^*(\omega)$ when $f(t)$ is real, according to 2.2.3. Equation [18] expresses the *power theorem*.

Now if $f_1(t) = f_2(t) = f(t)$, then $F_1(\omega) = F_2(\omega) = F(\omega)$ and the following relation follows at once:

$$\int_{-\infty}^{\infty} |f(t)|^2 \, dt = \frac{1}{2\pi} \int_{-\infty}^{\infty} F(\omega) F^*(\omega) \, d\omega = \frac{1}{2\pi} \int_{-\infty}^{\infty} |F(\omega)|^2 \, d\omega = \frac{1}{\pi} \int_{0}^{\infty} |F(\omega)|^2 \, d\omega \qquad [19]$$

since $|F(\omega)|^2$ is an even function, when $f(t)$ is real; $F(\omega)$ is Hermitian, i.e., real even and imaginary odd (section 2.2.3).

This relation is usually referred to as *Parseval's theorem* or *Rayleigh's theorem*, sometimes as *Plancherel's theorem* or the *completeness theorem*. The real quantity $|F(\omega)|^2$ is commonly referred to as *power spectrum* or *energy spectrum* or, more precisely, *power spectral density* or *energy spectral density*, i.e. power or energy per unit interval on the frequency scale. As [19] contains the square of the absolute value of the amplitude spectrum, no information is conveyed about the phase spectrum of the time function. This means that it is impossible to recover the original signal $f(t)$ if only its power is given, as distinct from the case when $F(\omega)$ is given. It also means that signals with identical amplitude spectra but different phase spectra will have identical power spectra.

On the other hand, Parseval's equation [19] permits us to derive influences on the power spectrum from any modification made on $f(t)$. For example, a simple time shift will leave the power spectrum unaffected, as seen by using the time-shifting theorem (2.3.4):

$$\int_{-\infty}^{\infty} |f(t \pm a)|^2 \, dt = \frac{1}{2\pi} \int_{-\infty}^{\infty} |e^{\pm ia\omega} F(\omega)|^2 \, d\omega = \frac{1}{2\pi} \int_{-\infty}^{\infty} |F(\omega)|^2 \, d\omega \qquad [20]$$

But a change of the time scale $t \to at$, will reduce the power by a factor of $1/a$, as can be deduced from the time-scaling theorem (2.3.3). This is important to consider when comparing spectra derived from records with different time scales.

The discussion so far is based on the assumption of finite total energy and finite average power of $f(t)$. This is also the case with transient signals. However, for signals of infinite duration, e.g. microseisms, ocean surface waves, atmospheric turbulence, the total energy of $f(t)$ may be infinite, even though its average power is finite. This is circumvented by considering only a finite portion T of $f(t)$, and assuming $f(t) = 0$ outside these limits, a procedure which is anyway necessitated in practice. Then, the infinite integral in Parseval's formula [19] is replaced by the following:

$$\lim_{T\to\infty} \frac{1}{2\pi T} \int_{-T/2}^{T/2} |F(\omega)|^2 \, \mathrm{d}\omega \qquad [21]$$

As a consequence, the quantity $|F(\omega)|^2/T$ also plays a great role in spectral analysis, sometimes termed *power spectral density*, sometimes and more often *periodogram* (section 4.6.2).

3.3.3 *Autocorrelation and power spectrum*

Applying the definition [1] of the autocorrelation function and the time-shifting theorem (section 2.3.4) to [18], we get:

$$C_{11}(\tau) = \int_{-\infty}^{\infty} f_1(t) f_1(t+\tau) \, \mathrm{d}t = \frac{1}{2\pi} \int_{-\infty}^{\infty} |F_1(\omega)|^2 \, \mathrm{e}^{i\omega\tau} \, \mathrm{d}\omega$$

$$= \frac{1}{2\pi} \int_{-\infty}^{\infty} E_{11}(\omega) \, e^{i\omega\tau} \, \mathrm{d}\omega \qquad [22]$$

and conversely:

$$E_{11}(\omega) = \int_{-\infty}^{\infty} C_{11}(\tau) \, \mathrm{e}^{-i\omega\tau} \, \mathrm{d}\tau \qquad [23]$$

where $E_{11}(\omega) = |F_1(\omega)|^2 = F_1^*(\omega) \, F_1(\omega)$ is the power spectral density, and

$$F_1(\omega) = \int_{-\infty}^{\infty} f_1(t) \, \mathrm{e}^{-i\omega t} \, \mathrm{d}t.$$

Thus the autocorrelation and the power spectrum form a Fourier pair:

$$C_{11}(\tau) \leftrightarrow E_{11}(\omega) \qquad [24]$$

As $C_{11}(\tau)$ is real and even, the formula [23] for $E_{11}(\omega)$ simplifies to the following:

$$E_{11}(\omega) = 2 \int_{0}^{\infty} C_{11}(\tau) \cos \omega\tau \, \mathrm{d}\tau \qquad [25]$$

From the property of $C_{11}(\tau)$, it follows that $E_{11}(\omega) = E_{11}(-\omega)$. Moreover, we immediately find the following relation from [22]:

$$C_{11}(0) = \int_{-\infty}^{\infty} [f_1(t)]^2 \, dt = \frac{1}{2\pi} \int_{-\infty}^{\infty} E_{11}(\omega) \, d\omega = \frac{1}{2\pi} \int_{-\infty}^{\infty} |F_1(\omega)|^2 \, d\omega \qquad [26]$$

in agreement with Parseval's formula [19]. The relation [24] between $E_{11}(\omega)$ and $C_{11}(\tau)$ is usually referred to as the *Wiener-Khintchine relation* (Goodman, 1961; Jennison, 1961). A generalization to non-stationary processes has been given by Lampard (1954).

Power spectra play a great role in geophysics and are used frequently. As we have seen we can calculate the power spectrum of a signal in two ways:

(1) Calculating first the autocorrelation and then its Fourier transform.

(2) Calculating first the Fourier transform of the given signal and then taking the square of its absolute value.

The two methods lead to identical results, as demonstrated above. Just as powers cannot convey complete information on the original signals, as the phase information is lost, the same holds also for the correlation functions.

3.3.4. Cross-correlation and cross-power spectrum

In the same way as for the autocorrelation (section 3.3.3) it is proved immediately that the cross-correlation (section 3.1.2) forms a Fourier pair with the cross-power spectrum $E_{12}(\omega) = F_1^*(\omega) \, F_2(\omega)$:

$$\left.\begin{aligned} & C_{12}(\tau) \leftrightarrow E_{12}(\omega) \\ & C_{12}(\tau) = \frac{1}{2\pi} \int_{-\infty}^{\infty} E_{12}(\omega) \, e^{i\omega\tau} \, d\omega \\ & E_{12}(\omega) = \int_{-\infty}^{\infty} C_{12}(\tau) \, e^{-i\omega\tau} \, d\tau \end{aligned}\right| \qquad [27]$$

From the property of $C_{12}(\tau)$ it follows:

$$\left.\begin{aligned} & E_{12}(\omega) = E_{12}^*(-\omega) = E_{21}(-\omega) = E_{21}^*(\omega) \\ & E_{21}(\omega) = E_{21}^*(-\omega) = E_{12}(-\omega) = E_{12}^*(\omega) \end{aligned}\right| \qquad [28]$$

In correspondence to [26], we immediately find the following relation:

$$C_{12}(0) = \int_{-\infty}^{\infty} f_1(t) f_2(t)\, dt = \frac{1}{2\pi} \int_{-\infty}^{\infty} E_{12}(\omega)\, d\omega = \frac{1}{2\pi} \int_{-\infty}^{\infty} F_1^*(\omega) F_2(\omega)\, d\omega \qquad [29]$$

Unlike the power spectrum E_{11}, which is always real and positive, the cross-power E_{12} is in general complex. Then only $|E_{12}(\omega)|$ can be used as a measure of the cross-power. In some applications, it has been found useful to split $E_{12}(\omega)$ into its *co-spectrum* (real part) and *quadrature spectrum* (imaginary part). See, for example, Panofsky and Wolff (1957):

$$E_{12}(\omega) = P_{12}(\omega) - i\, Q_{12}(\omega) = F_1^*(\omega) F_2(\omega) \qquad [30]$$

where $P_{12}(\omega)$ is the co-spectrum and $Q_{12}(\omega)$ is the quadrature spectrum (frequently abbreviated to *quad-spectrum*). Using the last expression in [30] (inserting $F_1(\omega) = a_1(\omega) - i\, b_1(\omega)$, etc), it is demonstrated that:

$$\left.\begin{aligned} P_{12}(\omega) &= a_1(\omega)\, a_2(\omega) + b_1(\omega)\, b_2(\omega) \\ Q_{12}(\omega) &= a_1(\omega)\, b_2(\omega) - a_2(\omega)\, b_1(\omega) \\ P_{12}(\omega) &= P_{21}(\omega) \\ Q_{12}(\omega) &= -Q_{21}(\omega) \end{aligned}\right| \qquad [31]$$

It is easily shown that the two parts of $E_{12}(\omega)$ can also be written as follows:

Co-spectrum:

$$P_{12}(\omega) = \frac{1}{2} \int_{-\infty}^{\infty} [C_{12}(\tau) + C_{21}(\tau)] \cos \omega\tau\, d\tau$$

Quadrature spectrum:

$$Q_{12}(\omega) = \frac{1}{2} \int_{-\infty}^{\infty} [C_{12}(\tau) - C_{21}(\tau)] \sin \omega\tau\, d\tau \qquad [32]$$

The phase lag φ of F_1 with respect to F_2 is obtained from:

$$\tan \varphi = -Q_{12}(\omega)/P_{12}(\omega) \qquad [33]$$

By analogy, we have in case of autopower: $E_{11}(\omega) = P_{11}(\omega) = a_1^2(\omega) + b_1^2(\omega)$, $E_{22}(\omega) = P_{22}(\omega) = a_2^2(\omega) + b_2^2(\omega)$, $Q_{11}(\omega) = Q_{22}(\omega) = 0$.

As a summary, we report for clarity the time functions and their corresponding transforms which have been introduced so far in this chapter:

Time function		*Fourier transform*
Autocorrelation	$C_{11}(\tau)$	$\|F_1(\omega)\|^2 = E_{11}(\omega)$
Cross-correlation	$C_{12}(\tau)$	$F_1^*(\omega)\, F_2(\omega) = E_{12}(\omega)$
Convolution	$f_1(t) \star f_2(t)$	$F_1(\omega)\, F_2(\omega)$

In reading different books, the student will be faced with a bewildering

TABLE VII

Commonly used nomenclature for correlation functions and power spectra

Function	System 1	System 2
$C_{11}(\tau)$	autocorrelation	autocovariance
$\dfrac{C_{11}(\tau)}{C_{11}(0)}$	normalized autocorrelation	autocorrelation
$C_{12}(\tau)$	cross-correlation	cross-covariance
$\dfrac{C_{12}(\tau)}{[C_{11}(0)\ C_{22}(0)]^{1/2}}$	normalized cross-correlation	cross-correlation
$E_{11}(\omega)$	power spectrum	autocovariance spectrum or autospectrum
$E_{12}(\omega)$	cross-power spectrum	cross-covariance spectrum or cross-spectrum

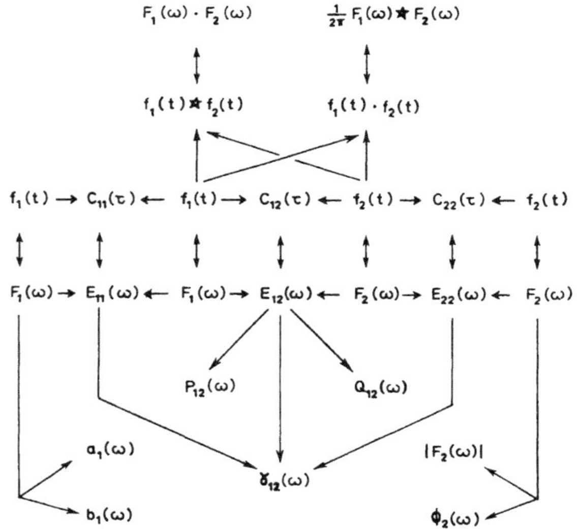

Fig.20. Schematic diagram showing the correspondence between different functions in the time domain and the frequency domain.

multitude of names for correlation and power spectral functions. By and large, these can be grouped into two parallel systems, which for clarity are listed in Table VII. In this book we shall essentially adhere to system 1. Fig.20 gives a schematic presentation of the relations between the different functions.

In the literature, different opinions are reflected as to the relative use of correlation functions and the corresponding power spectra. From a statistical point of view, both are considered important and interesting, whereas from some engineering sides, the correlation functions are only a means to calculate power spectra, and the interest is concentrated wholly on the latter.

3.3.5 Coherence

Coherence is defined in terms of power and cross-power spectra as follows (Foster and Guinzy, 1967; Hinich and Clay, 1968):

$$|\gamma_{12}(\omega)| = \frac{|E_{12}(\omega)|}{[E_{11}(\omega)\,E_{22}(\omega)]^{\frac{1}{2}}} = \left[\frac{P_{12}^2(\omega) + Q_{12}^2(\omega)}{P_{11}(\omega)\,P_{22}(\omega)}\right]^{\frac{1}{2}} \qquad [34]$$

for $E_{11}(\omega)\,E_{22}(\omega) > 0$ and $|\gamma_{12}(\omega)| = 0$ for $E_{11}(\omega)\,E_{22}(\omega) = 0$. From the definition it follows that $0 \leq |\gamma_{12}(\omega)| \leq 1$.

The quantity [34] without the absolute sign for $E_{12}(\omega)$, i.e. with its real and imaginary parts, has also been applied, under the name of *cross-correlation spectrum* (Davenport, 1961; Rudnick, 1969) or *coefficient of coherency* (Robinson, 1967b) or just *coherency*. With this definition, this quantity has naturally both a magnitude and a phase. The term *coherence spectrum* is also used (e.g. by Siedler, 1971). Sometimes, coherence is defined as the square of the quantity [34]. Theoretically, the coherence function [34] would be equal to 1, independent of frequency, while this is not so in practical application due to windowing and smoothing effects (Bendat and Piersol, 1971, p.195).

3.4 ANALYTICAL EXAMPLE OF CORRELATIONS AND POWER SPECTRA

3.4.1 Calculation procedure

The best method to become familiar with all these functions, both their computation and their interpretation, is to calculate them for given functions $f(t)$. Then it is preferable to begin with simple functions, where it is easy to grasp the general shape of the results immediately by inspection of the graphical representations.

As an exercise, we have selected two simple functions, $f_1(t)$ being a square function and $f_2(t)$ a triangular function (Fig.21 and Table VIII). In Table VIII,

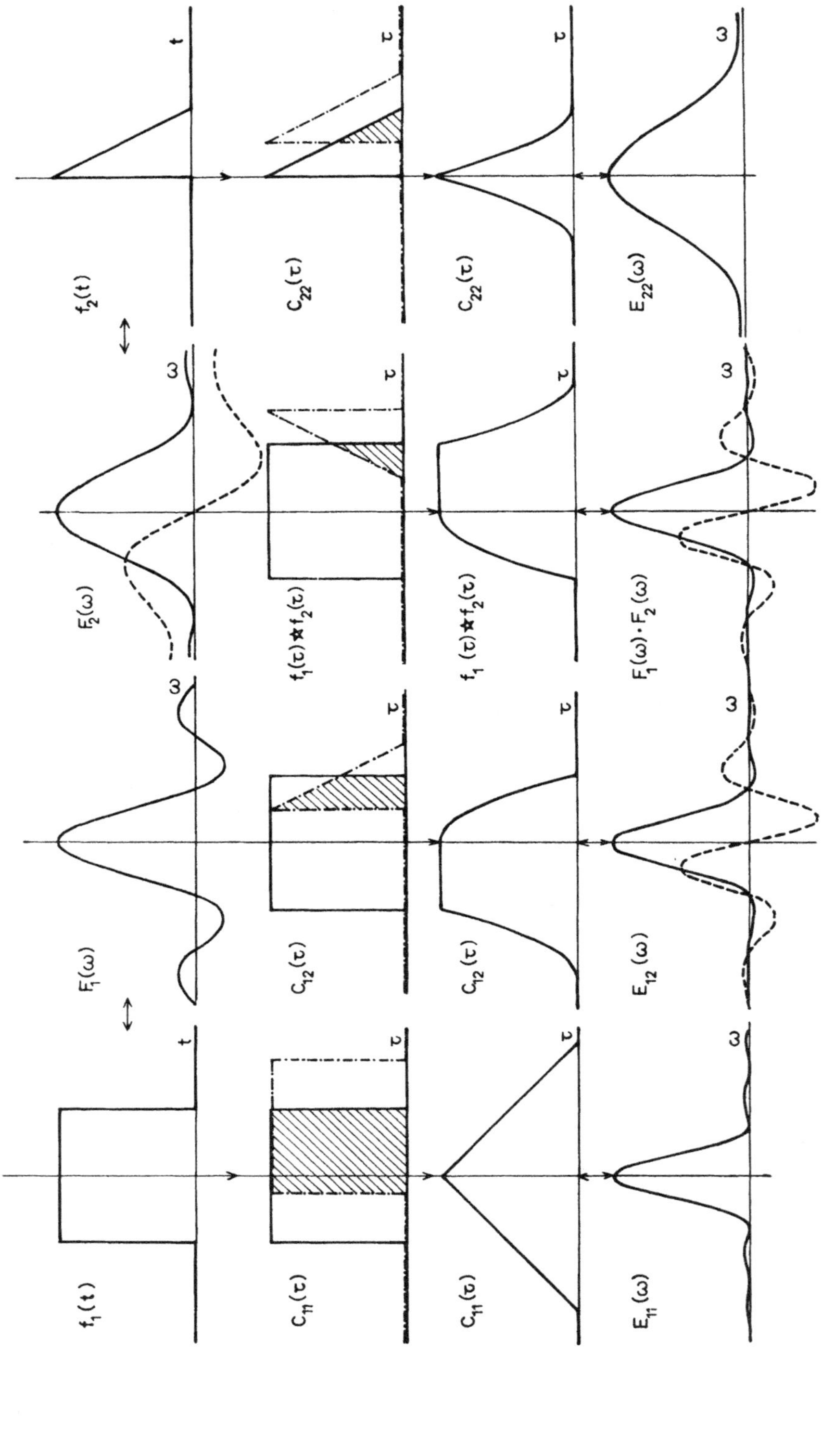

Fig.21. Graph showing correlation functions and power spectra for two simple analytic functions. Full lines: real parts, dashed lines: imaginary parts, dashed-dotted lines: displaced time functions.

TABLE VIII

Mathematical expressions corresponding to the graphs in Fig.21

$$f_1(t) = \Pi(t) = 1 \quad \text{for} \quad -\frac{1}{2} \le t \le \frac{1}{2}$$

$$F_1(\omega) = \frac{\sin(\omega/2)}{\omega/2} = \text{sinc}\,\frac{\omega}{2\pi}$$

(example 1, Table V)

$$f_2(t) = u(t)\,\Lambda(2t) = 1 - 2t \quad \text{for} \quad 0 \le t \le 1/2$$

$$F_2(\omega) = \frac{1}{4}\left[\frac{\sin(\omega/4)}{\omega/4}\right]^2 - \frac{i}{\omega}\left[1 - \frac{\sin(\omega/2)}{\omega/2}\right] = \frac{1}{4}\,\text{sinc}^2\,\frac{\omega}{4\pi} - \frac{i}{\omega}\left(1 - \text{sinc}\,\frac{\omega}{2\pi}\right)$$

(example 20, Table V)

$$C_{11}(\tau) = 1 - |\tau| \quad \text{for} \quad -1 \le \tau \le 1$$

$$C_{22}(\tau) = \frac{2}{3}|\tau|^3 - \frac{|\tau|}{2} + \frac{1}{6} \quad \text{for} \quad -\frac{1}{2} \le \tau \le \frac{1}{2}$$

$$C_{12}(\tau) = \begin{cases} (1+\tau)^2 & \text{for} \quad -1 \le \tau \le -\frac{1}{2} \\ \frac{1}{4} & \quad -\frac{1}{2} \le \tau \le 0 \\ \frac{1}{4} - \tau^2 & \quad 0 \le \tau \le \frac{1}{2} \end{cases}$$

$$f_1(\tau) \bigstar f_2(\tau) = \begin{cases} \frac{1}{4} - \tau^2 & \text{for} \quad -\frac{1}{2} \le \tau \le 0 \\ \frac{1}{4} & \quad 0 \le \tau \le \frac{1}{2} \\ (1-\tau)^2 & \quad \frac{1}{2} \le \tau \le 1 \end{cases}$$

$$E_{11}(\omega) = F_1^*(\omega)\,F_1(\omega) = \left[\frac{\sin(\omega/2)}{\omega/2}\right]^2 = \text{sinc}^2\,\frac{\omega}{2\pi}$$

$$E_{22}(\omega) = F_2^*(\omega)\,F_2(\omega) = \frac{1}{\omega^2}\left\{1 - 2\,\frac{\sin(\omega/2)}{\omega/2} + \left[\frac{\sin(\omega/4)}{\omega/4}\right]^2\right\}$$

$$= \frac{1}{\omega^2}\left(1 - 2\,\text{sinc}\,\frac{\omega}{2\pi} + \text{sinc}^2\,\frac{\omega}{4\pi}\right)$$

$$E_{12}(\omega) = F_1^*(\omega)\,F_2(\omega) = \text{sinc}\,\frac{\omega}{2\pi}\left[\frac{1}{4}\,\text{sinc}^2\,\frac{\omega}{4\pi} - \frac{i}{\omega}\left(1 - \text{sinc}\,\frac{\omega}{2\pi}\right)\right]$$

the functions are zero outside the limits given. In Fig.21 all curves have been normalized to maximum values of unity. In both the correlation formulas and in the convolution formula, the procedure is in fact to integrate over one of the functions, using the same or another function as weight. The hatched areas in Fig.21 mark the ranges for which contributions are obtained to the respective integrals. Note that the cross-correlation and the convolution only differ by the reversal of the triangle, in agreement with [9].

The autocorrelation $C_{11}(0)$ for the square attains the value 1 for $\tau = 0$, i.e., when the two squares coincide. This seems a natural value in that case. However, for the triangle, the corresponding value is only 1/6. To get this also equal to 1, use is sometimes made of an autocorrelation normalized to its value for $\tau = 0$, i.e. *normalized autocorrelation:*

$$\bar{C}_{11}(\tau) = \frac{\int_{-\infty}^{\infty} f(t) f(t+\tau)\, \mathrm{d}t}{\int_{-\infty}^{\infty} [f(t)]^2\, \mathrm{d}t} \qquad [35]$$

and its value (maximum) is always 1 for $\tau = 0$, irrespective of the shape of $f(t)$. A similar normalization can be applied to the cross-correlation and the convolution, such that these attain a maximum value of +1 (instead of +1/4 in this example), as well as to the power spectra. For instance, the normalized cross-correlation would read as follows:

$$\bar{C}_{12}(\tau) = \frac{\int_{-\infty}^{\infty} f_1(t) f_2(t+\tau)\, \mathrm{d}t}{\int_{-\infty}^{\infty} f_1(t) f_2(t)\, \mathrm{d}t} = 1 \quad \text{for} \quad \tau = 0 \qquad [36]$$

but is often written in the following form, which in general does not differ appreciably numerically from the equation just given, but has the advantage of simple relation to the variances of $f_1(t)$ and $f_2(t)$:

$$\bar{C}_{12}(\tau) = \frac{\int_{-\infty}^{\infty} f_1(t) f_2(t+\tau)\, \mathrm{d}t}{\left[\int_{-\infty}^{\infty} \{f_1(t)\}^2\, \mathrm{d}t \int_{-\infty}^{\infty} \{f_2(t)\}^2\, \mathrm{d}t\right]^{\frac{1}{2}}} \qquad [37]$$

Also other normalizations do exist.

For the power spectra, it is suggested that these be calculated in the two ways indicated above, i.e. both by transforming the autocorrelation, and by transforming $f(t)$ and taking the square of the absolute value of the transform $F(\omega)$. The two results should naturally agree. Power spectra are always even functions of ω.

3.4.2 *Interpretation of the various functions*

Let us then, with the help of this example, learn something about the meaning of these functions. The *correlation functions* measure the degree of parallelism between two series of numbers, just as in statistics. These two series can either consist of a given time series $f(t)$ and the same series displaced along the t-axis an amount τ relative to itself (autocorrelation), or they may consist of two different time series $f(t)$ with arbitrary relative time shifts (cross-correlation). For each shift τ, there is a corresponding value of the correlation. The expressions for the correlation coefficients above agree with the expression for the correlation coefficient r found in textbooks in statistics:

$$r = \frac{\Sigma(x-\bar{x})(y-\bar{y})}{[\Sigma(x-\bar{x})^2\,\Sigma(y-\bar{y})^2]^{\frac{1}{2}}} \rightarrow \frac{\Sigma f_1(t) f_2(t)}{[\Sigma\{f_1(t)\}^2\,\Sigma\{f_2(t)\}^2]^{\frac{1}{2}}} \rightarrow$$

$$\rightarrow \frac{\int_{-\infty}^{\infty} f_1(t) f_2(t)\,\mathrm{d}t}{\left[\int_{-\infty}^{\infty}\{f_1(t)\}^2\,\mathrm{d}t \int_{-\infty}^{\infty}\{f_2(t)\}^2\,\mathrm{d}t\right]^{\frac{1}{2}}} = \bar{C}_{12}(0) = \frac{C_{12}(0)}{[C_{11}(0)C_{22}(0)]^{\frac{1}{2}}} \qquad [38]$$

This means that the usual correlation coefficient r in statistics agrees with the normalized cross-correlation coefficient for zero time shift. The advantage of auto- and cross-correlations is that they entail frequency dependence. This finds its expression in the power and cross-power spectra and the coherence, and the frequency dependence is the property which makes all these functions so useful in analyses of observations.

It is clear that in the formulas above, $f(t)$ should be deviations from their zero-lines. However, measurements from some other parallel base-line can also be used, as they only entail a vertical displacement of the resulting curve, without altering its shape. Formulas for correlation functions and convolution, taking parallel base-line shifts into account, are given in section 4.5.4.

As mentioned, correlation functions have significance, besides for calculation of power spectra, also for decision about degree of parallelism between two time series. This has a great practical significance in any geophysical observation of propagating waves by means of a triangle or an array of stations. If phase identification is not immediately obvious between the stations in the array, then cross-correlation of the time series may help. The time shift, for which the cross-correlation has its maximum, would correspond to the most likely phase shift between the stations correlated. See further section 7.1.1.

For several geophysical phenomena, e.g. in meteorology and oceanography, the autocorrelation often decreases exponentially with increasing time lag:

$$C_{11}(\tau) \sim e^{-a\tau}; \qquad a > 0$$

Such processes are called *Markov processes*. From example 24, section 2.4.3, we see that the corresponding power spectrum:

$$2a/(a^2 + \omega^2)$$

has its maximum for $\omega = 0$, i.e. constitutes "red noise". For one meteorological application, see Dyer (1970). On the other hand, from example 3, section 2.4.3, we find that an autocorrelation in the shape of a sinc-function corresponds to a constant power over a certain frequency band (sometimes referred to as "white noise", even though it refers only to a limited frequency band). Fig.22 shows some fundamental types of autocorrelation functions, of which a) and b) are to be considered as basic extremes. Note that in case a) white (uncorrelated) noise

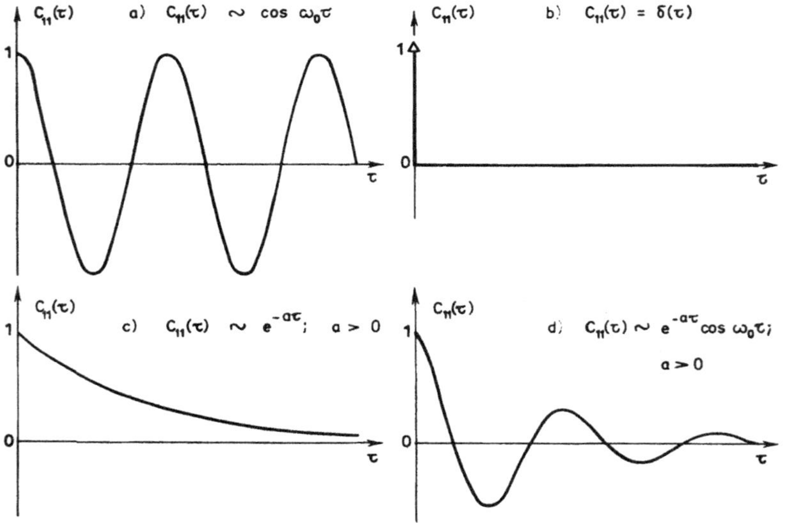

Fig.22. Some fundamental types of autocorrelation functions. a) Cosinusoidal signal. b) White noise (independent random variables). c) Red noise (Markov process). d) Red noise.

may be added, without changing $C_{11}(\tau)$. The intermediate cases c) and d) are those most frequently encountered in geophysics. By substitution in [23] or [25] the corresponding expressions for $E_{11}(\omega)$ are easily found. For example, in case b) $E_{11}(\omega)$ becomes a constant, representing *white noise* or the *base spectrum*.

Our definitions of autocorrelation and cross-correlation correspond closest to the concepts of *variance* and *covariance*, to be found in textbooks in statistics. As demonstrated in section 5.1.1, the variance is a measure of power, and in applications it has often been used for that purpose. By the same method, it can be easily demonstrated that the covariance is a measure of the cross-power.

Convolution obviously is the same as cross-correlation, only that one of the time series used should be taken in reverse order, i.e., it should be folded. The greatest significance of this operation lies in the fact that convolution in one domain corresponds to multiplication in the transform domain (section 3.2.2). Convolution in the time domain corresponds to passing a signal $f(t)$ through any kind of filter (section 6.1.2). Very clear pictorial, non-mathematical explanations of correlation functions and convolution have been published by Anstey (1964, 1966) and Anstey and Newman (1966). The latters' term retro-correlation is synonymous with convolution. Anstey (1964) gives also an extensive list of references.

The *coherence*, apart from its normalizing denominator (geometric normalization), is in fact the spectrum of the cross-correlation. Thus, if the cross-correlation $C_{12}(\tau)$ (time domain) exhibits large values at certain regularly spaced time shifts τ, this will correspond to a maximum of the coherence $\gamma_{12}(\omega)$ at the corresponding frequencies.

Power spectra and coherences are useful in specifying the direction of arrival as well as the beam-width of wave radiation, especially as they permit specifications according to frequency of the waves. With components E, N, Z (subscripts), the coherence (squared):

$$E_{NZ}^2/E_{ZZ}\,E_{NN} \qquad [39]$$

would be $= 1$ for a pure Rayleigh wave arriving from N (or S). This is easily generalized to any arrival direction θ counted from N over E:

$$\frac{E_{NZ}^2\,(1+\tan^2\theta)}{E_{ZZ}\,E_{NN}\,(1+\tan^2\theta)} = \frac{E_{NZ}^2 + E_{EZ}^2}{E_{ZZ}(E_{NN}+E_{EE})} \qquad [40]$$

For a pure Rayleigh wave, with certain fixed ratios between the three component displacements, arriving from a direction θ, this ratio (beam-width) is $= 1$, whereas for isotropic radiation it is $= 0$. This method of displaying arrival direction and beam-width has been used by Haubrich et al. (1963) for microseisms, but it has naturally general applicability to characterize directional properties of propagating waves of any kind.

In order to acquire still more familiarity with all these functions, it is advisable to try with other simple functions $f(t)$ as well. Sequences of numbers can well be used, and may serve as an introduction as to how digital computers handle such calculations (then $f(t)$ is replaced by a sequence of numbers).

Even though we have been using time t as independent parameter, it should be understood that all functions have validity for any kind of independent parameter. In geophysical applications it is often the case that the independent parameter is a linear coordinate x instead of time t.

3.5 SPECTRA OF OBSERVATIONAL DATA—FUNDAMENTAL PROBLEMS

3.5.1 Basic considerations of spectral calculation for observational data

Hitherto, our spectral calculations have been restricted to analytical functions. When we now turn to spectral calculations of observational data a number of difficulties appear, which are not involved in application to analytical functions. In this section we shall deal with the main reasons for these difficulties. A thorough understanding of these problems is necessary for the development in following chapters and for the whole field of geophysical spectroscopy.

Whereas the Dirichlet conditions may present problems for analytic functions, they are no problem for empirical data, which as a rule automatically fulfill these conditions. On the other hand, the infinite integration interval may be no problem in the analytical case, but is obviously an impossibility in the observational case. These facts, reviewed in Table IX, are the most significant distinctions between the analytical and the observational case.

TABLE IX

Review of problems connected with Fourier transform calculation

Condition	Function $f(t)$	
	analytically given function	empirically observed curve
Dirichlet conditions	fulfillment has to be checked in every case	automatically fulfilled in practically every case[1]
Infinite integration interval	presents generally no difficulties	impossible to fulfill in observational series

[1] The condition for Fourier integral that $\int_{-\infty}^{\infty} |f(t)|\, dt$ is finite (section 2.1) is fulfilled for observational curves within limited intervals.

The consequence of a limited data window is in fact that a correct spectrum is impossible to obtain. We can see this in the following manner:

(1) Take first the hypothetical case of an infinitely long time window, i.e., that the whole signal enters the operation. In this case, the time window is $w(t) = 1$ with infinite extent, and we get:

$$f(t) \cdot w(t) = f(t) \cdot 1 \leftrightarrow 1/2\pi \cdot F(\omega) \star 2\pi\, \delta(\omega) =$$

(section 3.2.2 and example 10, section 2.4.3)

$$= F(\omega) \star \delta(\omega) = \int_{-\infty}^{\infty} F(\varpi)\, \delta(\omega - \varpi)\, d\varpi = F(\omega) \qquad [41]$$

(section 3.2.1)

which means that we get the true spectrum $F(\omega)$.

(2) Take a time window of finite length T of the simplest type, i.e. the rectangular or box-car window. Then, by the same reasoning as above only using example 1 in section 2.4.3 instead, we get:

$$f(t) \cdot w(t) = f(t) \cdot \Pi\left(\frac{t}{T}\right) \leftrightarrow \frac{T}{2\pi} F(\omega) \star \frac{\sin \omega T/2}{\omega T/2}$$

$$= \frac{T}{2\pi} \int_{-\infty}^{\infty} F(\omega) \frac{\sin(\omega - \varpi)T/2}{(\omega - \varpi)T/2}\, d\omega \neq F(\omega) \qquad [42]$$

In this case we do not get the true spectrum $F(\omega)$ but instead the convolution integral, which represents a certain smoothing of the correct $F(\omega)$. The degree of smoothing depends on the window length T, such that the shorter T is, the stronger is the smoothing effect (Fig.23). The spectrum thus calculated is therefore called the *average or weighted spectrum* (Blackman and Tukey, 1959). As $F(\omega) = |F(\omega)|e^{i\Phi(\omega)}$, both amplitude and phase spectrum will be distorted.

In addition to the smoothing effect, we have to consider that the side-lobes of the frequency window $W(\omega)$, see Fig.23, lead to undesirable effects. Negative side-lobes lead to spectral leakage. The ideal is to have a narrow central lobe and insignificant side-lobes. In the limit this ideal is represented by the $\delta(\omega)$-function, which is unattainable, as we have seen. But by modifying the shape of the data window $w(t)$, we can improve the qualities of the spectral window $W(\omega)$. Much effort has gone into this business ("window carpentry"), as we shall see in the next chapter. Either different window shapes can be applied in the time domain, or else, by certain smoothing procedures in the frequency domain, it is possible to convert the effect of the chosen window into that of some other window (section 4.4), but to completely eliminate the windowing effect is not possible.

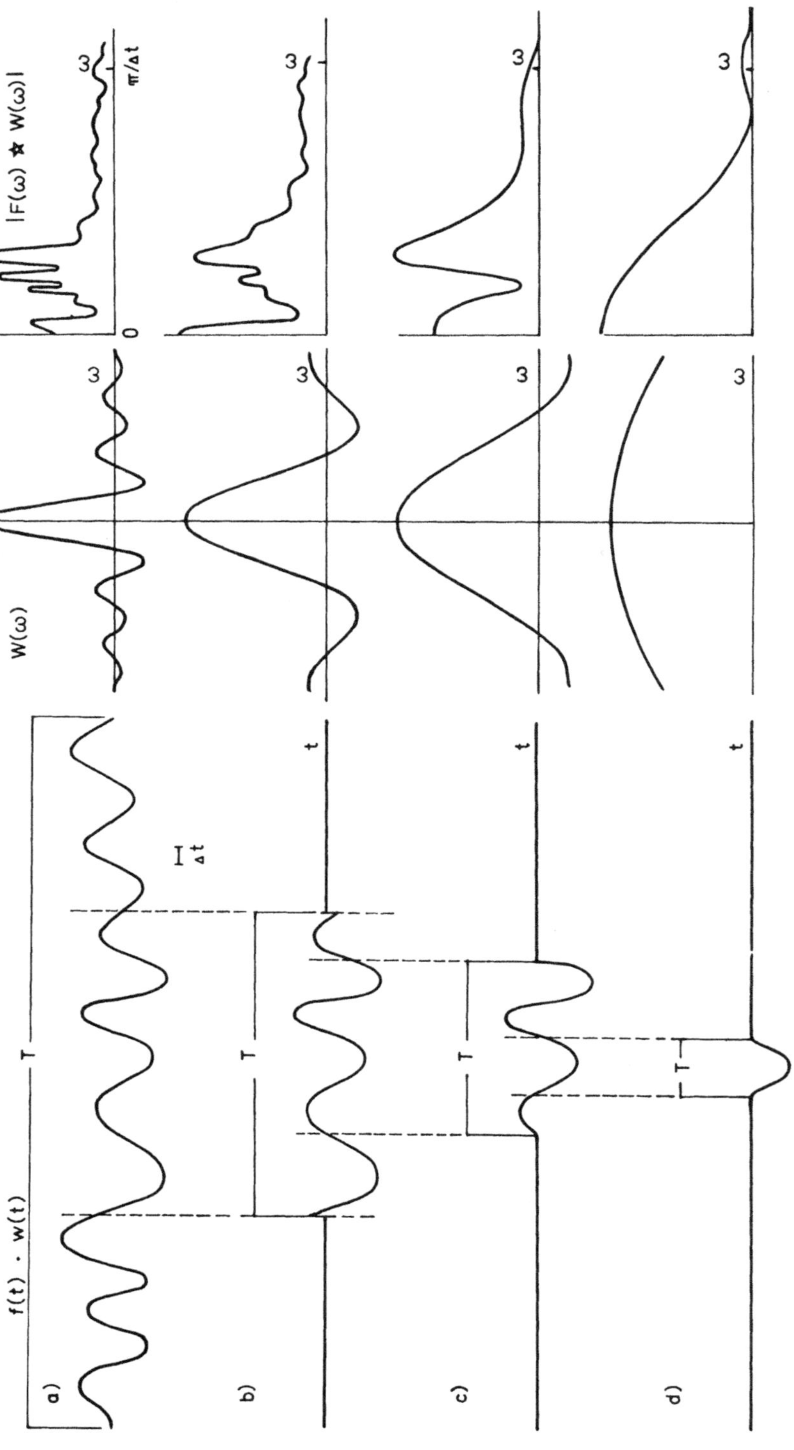

Fig.23. Spectral resolution as affected by varying the length *T* of the data window. The figure is partly adopted from Holzmann (1960b).

Recently, also other methods have been tried to retrieve the true spectrum. Obviously, [42] is an integral equation in $F(\omega)$. Methods have been devised for its solution in case of a box-car window and frequency-limited $F(\omega)$, i.e. $F(\omega) = 0$ for $|\omega| > \Omega$. This is naturally an approximation, because as we have seen in section 2.4.4, a time-limited function $f(t)$ cannot have a frequency-limited spectrum. Then, [42] becomes a Fredholm integral equation of the first kind (Båth, 1968, p.315):

$$\bar{F}(\omega) = \int_{-\Omega}^{+\Omega} F(\omega) \frac{\sin(\omega - \bar{\omega})T/2}{\pi(\omega - \bar{\omega})} \mathrm{d}\omega \qquad [43]$$

where $\bar{F}(\omega)$ is the observed spectrum, $F(\omega)$ is the true spectrum, and $\sin[(\omega - \bar{\omega})T/2]/\pi(\omega - \bar{\omega})$ is the kernel. The solution is effectuated by applying methods to solve integral equations, in this case by expanding the integral into a series of eigenfunctions. The method has been tested on numerical examples by Abramovici (1973), which clearly demonstrates its efficiency. The spectrum calculated in this way shows a much better approximation to the true spectrum than $\bar{F}(\omega)$ does. However, this methodical improvement has as yet hardly found any geophysical application.

Besides the shape of the time window $w(t)$, also its length T has an important bearing on the characteristics of the computed spectrum. This we can understand from Fig.23 and in a more quantitative way from the corresponding mathematical expressions. The main rules are the following:

(1) A large T will lead to more detail, i.e. to better resolution, in the computed spectrum than a small T.

(2) A small T will lead to better stability or better reliability of the computed spectral estimates than a large T, because in case of a smaller T the spectral smoothing extends over a larger frequency interval.

(3) Limitations on the choice of T can be imposed by the type of record used. For example, in seismic records with a sequence of different waves arriving after each other, time windows should be chosen such that only the wave under study is included and no others.

One extreme, the infinite T, unattainable in practice, will lead to a correct spectrum. The other extreme, a very small T approaching zero, will lead to a "white spectrum", i.e. no resolution at all. In any actual case, a compromise has to be made between various considerations (of resolution, stability, computer-time economy, etc.) in the choice of an appropriate window length T. If the frequency convolver $W(\omega)$ is much wider than spectral details, these will be smoothed out too much. Therefore, a general rule is that $W(\omega)$ should be so narrow that the spectrum does not exhibit wide fluctuations within the range of the frequency convolver. In Fig.23 we have been considering an artificial case, but later, in section 4.4.8, we shall give actual examples from geophysical observations.

We have here discussed the influence of window length T from the viewpoint of the Fourier transform. But identical results are obtained if instead we consider a Fourier series expansion. Then the window length T defines the fundamental period and its higher harmonics, of cyclic frequencies $1/T$, $2/T$, $3/T$, ... A window of length $2T$ will similarly give the frequencies $1/2T$, $2/2T$, $3/2T$, ..., i.e. a two-fold better resolution, and so on. A specific example may make this even more obvious. Say that we are interested in periods down to 1 hour and start with a $T = 24$ hours. Then the fundamental period is 24 hours and the periods ($T/2$, $T/3$, ...) of the higher harmonics (the spectral lines) occur at 12, 8, 6, 4.8, 4, 3.4, 3, 2.7, 2.4, 2.2, 2.0, 1.8, 1.7, 1.6, 1.5, 1.4, 1.3, 1.26, 1.20, 1.14, 1.09, 1.04, and 1.0 hour, whereas if $T = 12$ hours, the periods in the same range will be 12, 6, 4, 3, 2.4, 2, 1.7, 1.5,

TABLE X

Review of spectra calculations and their limitations: $F(\omega) = S(\omega) + N(\omega)$, where S denotes signal and N noise

Type of time function $f(t)$	Theoretical spectra $N(\omega) = 0$	Empirical spectra $N(\omega) \neq 0$
Infinite signal duration	$\int_{-\infty}^{\infty}$ possible exactly, provided Dirichlet's conditions fulfilled: $F(\omega) = S(\omega)$	$\int_{-\infty}^{\infty}$ possible only approximately, as no empirical series extends from $-\infty$ to $+\infty$ (always a window, in addition to noise): $F(\omega) \star W(\omega) = [S(\omega) + N(\omega)] \star W(\omega)$. Exactly correct $S(\omega)$ not obtainable.
Bandlimited signal duration, zero elsewhere (a pulse)	$\int_{-T/2}^{T/2}$ possible exactly, provided Dirichlet's conditions fulfilled: $F(\omega) = S(\omega)$	$\int_{-T/2}^{T/2}$ possible approximately, because $\int_{-T/2}^{T/2} = \int_{-\infty}^{\infty}$ with no change for the signal: $S(\omega) \star W(\omega) = S(\omega) \star \delta(\omega) = S(\omega)$. But $N(\omega) \neq 0$, thus: $F(\omega) \star W(\omega) = S(\omega) + N(\omega) \star W(\omega)$. Taking the difference of two equal intervals (during and before the signal): $[F_2(\omega) - F_1(\omega)] \star W(\omega) = S(\omega) + [N_2(\omega) - N_1(\omega)] \star W(\omega) = S(\omega) + \Delta N(\omega) \star W(\omega)$. Correct $S(\omega)$ obtainable: (1) if $N(\omega) = 0$; or (2) if $\Delta N(\omega) = 0$, i.e. perfect noise stationarity (almost never fulfilled).

1.3, 1.2, 1.1, 1.0 hours, and for $T = 6$ hours, they will only be 6, 3, 2, 1.5, 1.2, 1.0 hours.

Our discussion has immediate application to continuous signals, such as microseisms, ocean surface waves, meteorological turbulence, gravity, geomagnetism. In such cases the signals never die out completely, but exhibit only energy variations, which are slow in relation to involved periods. In such cases a representative record length has to be chosen for analysis, and the degree of representativeness has to be judged from comparison of spectra from records of varying lengths; anyway, they should include many wavelengths even of the longest wave studied.

For transient earthquake waves, the situation appears different, as such waves have as a rule only a limited duration and the analysis is extended over the duration of the wave under study, excluding the rest. This would be a satisfactory procedure in the complete absence of noise (microseisms). But as noise is always present, in varying degree, the situation is not essentially different from the one discussed in the preceding paragraph. A review of the case where a seismic record is the sum of an earthquake wave and background noise: $f(t) = s(t) + n(t)$, is given in Table X.

The effects of data windows (in time and/or space) can be summarized as follows: (1) to broaden spectral lines; (2) to introduce spurious spectral lines (side-bands) by the side-lobes.

If the limited record length constitutes one major difference between observed and analytical signals, the other main difference is the way in which the signal is represented. Analytical signals, for which we have had numerous examples in this and the preceding chapter, are represented by exact mathematical formulas. This is not the case for an observed record. Then, there are essentially two different ways to proceed:

(1) To use the record directly as given, i.e. in analogue form. The spectral analysis then requires special apparatuses, which were briefly reviewed in section 1.6.

(2) To reformulate the record into mathematical form, which can then be used for calculations just as for an analytical function.

Because method 2 does not require any special apparatus for the construction of spectra, it is the one most generally used. It only requires the availability of a digital electronic computer.

To reformulate the record into a form, equivalent to an analytic function, requires determination of its numerical values at equidistant points along the time axis. These values are the digits of the record, and the process of their determination is called digitization or to digitize the record.

The procedure could then be envisaged as follows. The record is put into analytical form by fitting a polynomial, a Fourier series or any other appropriate function to the measured digits. By including a sufficiently large number of terms in the analytical expansion, it would be possible to achieve any desired approximation to the given record. Having the record then in analytical form, the rest of the work would just proceed as for any other analytical function, as we have done above.

However, the calculation of an analytical expression for the given record is an unnecessary intermediate step. Instead, it is possible to use the digits immediately in the calculation of spectra or any other related function. We only need to have the formulas not in integral form as they have been given so far in this book, but instead in discrete summation form. This is the procedure always followed in digital spectral calculations.

Digitizing of records entails further problems, partly connected with methods to execute this operation and their inherent errors, partly connected with the digital presentation as such. In the next chapter we shall investigate such effects in more detail.

Summarizing the present section, we have seen that spectral treatment of observational records differs from that of analytical functions in two very important aspects: (1) use of a time window of finite length; (2) need to have the record in digital form. It is important that we have a clear understanding of these facts, as they are of influence in spectral calculations of geophysical data.

The order 1 and 2 of these aspects also reflects their relative significance: item 1 cannot be dispensed with in any spectral analysis of observational data, while item 2 can be dispensed with provided the necessary analogue machinery is available.

3.5.2 Truncation of analytical signals

From the preceding section, we have learnt that truncation of signals may have significant effects on the resulting spectrum. As observational series do not provide an absolutely true spectrum for comparison, they do not lend themselves to any absolute evaluation of truncation effects, but only to relative evaluation of such effects. On the other hand, we know the absolutely true spectrum for analytically given functions, and then truncating such signals can give valuable information on the corresponding deformation of the spectrum.

Again it should be emphasized that analytical signals with known spectra are very useful also in empirical work, especially to test various procedures, the reliability of methods, including computer programming.

We shall illustrate the effects on spectra arising from truncation of analytic signals in the time domain and consider the following examples:

(1) *Truncated Heaviside unit-step function* (Fig.24). As expected, when the truncation point t_0 approaches infinity, the spectrum approaches the one of $u(t)$. Cf. example 16, section 2.4.3.

(2) *Truncated exponentially decreasing function:* $f(t) = u(t)\,e^{-at}$ (Fig.24). This corresponds better to such an impulsive onset as we encounter in the seismic

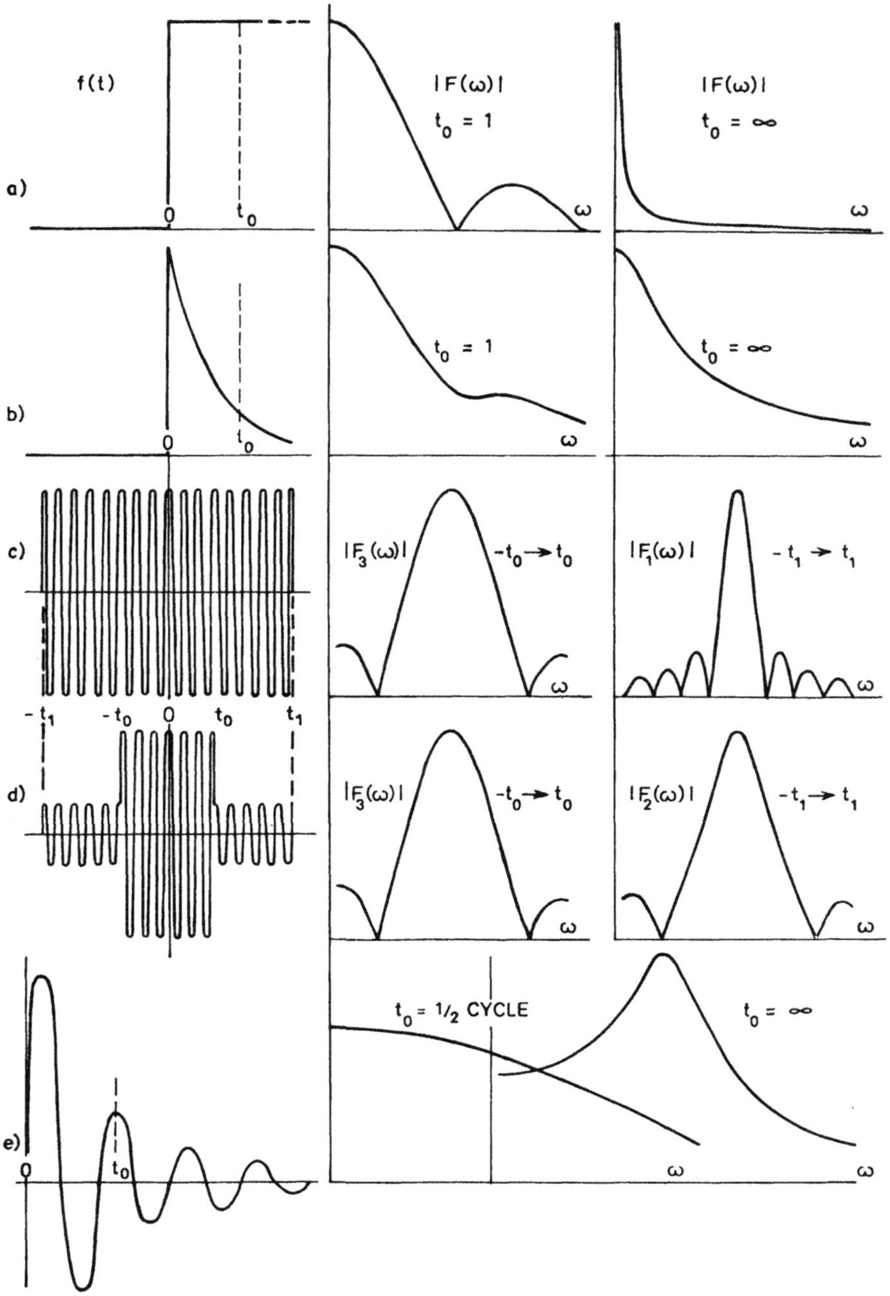

Fig.24. Examples of truncation of analytically given functions. Case e) is after Molotova (1966).

P-phases. In practice, it may often be difficult to settle upon the truncation point t_0, because the gradually lower signal amplitudes may be hidden by noise. Cf. example 23, section 2.4.3.

(3) *Truncation of an infinite cosine-curve*, whose maximum amplitudes are concentrated in a limited time interval. In such a case, it is customary to extend the time interval over the portion occupied by the maximum amplitudes and neglect the rest. This is illustrated in Fig.24, where three spectra have been calculated:

F_1 corresponding to the total time interval from $-t_1$ to $+t_1$ with constant amplitude ($= 1$);

F_2 corresponding to the same time interval but with reduced amplitude in the ranges from $-t_1$ to $-t_0$ and from $+t_0$ to $+t_1$;

F_3 corresponding to the range from $-t_0$ to $+t_0$ with constant amplitude ($= 1$).

We see that F_3 affords a better approximation to F_2 than to F_1. This means that under such circumstances it may be permitted to limit the spectral analysis to the time interval occupied by the maximum amplitudes. The error committed will be the smaller, the less significant amplitudes that the neglected portions have. This situation is frequently encountered in seismology, especially in the records of surface waves.

(4) *Truncated attenuating sine-wave:* $f(t) = e^{-at} \sin \omega_0 t$; $a = \omega_0/4$. Amplitude spectra, corresponding to different truncations of this function, are shown in Fig.24, redrafted from Molotova (1966). As is evident from this figure, truncation needs careful consideration when spectra are used for conclusions about the earth's structure (amplitude spectra for attenuation and reflection and transmission coefficients, phase spectra for phase-velocity and dispersion studies as well as for phase changes at discontinuity surfaces). When the area covered by the included part of the signal amounts to about 80% or more of the total signal, spectra may be considered trustworthy. Moreover, it helps if only relative use is made of spectra (*ratio* of amplitudes, *difference* between phases) instead of absolute determinations.

(5) *Truncated sinusoid* (constant amplitude). The truncated sinusoid offers another example of geophysically important truncated functions. With regard to the length of a record we can conveniently differ between three cases:

(a) An infinite extent: spectra are line spectra, located at $\pm\omega_0$, where ω_0 is the frequency of the sinusoid (examples 13 and 14 in section 2.4.3).

(b) Semi-infinite sinusoids (examples 18 and 19, section 2.4.3) and finite sinusoids (example 5, section 2.4.3), the latter with signal length many times the period of the sinusoid, both produce band spectra, centered around $\pm\omega_0$, though of different shape. The spectrum gets more blurred, the shorter the time interval is. This phenomenon is expressed in the time-scaling theorem (2.3.3), and we shall return to it later in discussing the uncertainty principle (sections 4.3.4 and 5.3.3).

(c) The situation when the signal extent is comparable to or less than the

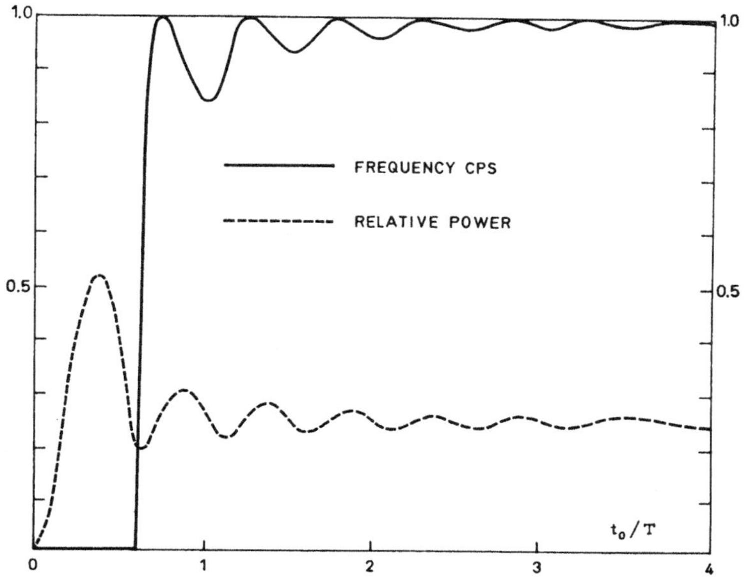

Fig.25. Truncated sinusoid ($T = 1$ sec, $\varphi = 0$): Frequency position of the spectral peak and its relative power as function of truncation length. After Toman (1965).

period of the sinusoid is in fact also covered by example 5, section 2.4.3, but deserves a more detailed investigation.

This case can happen in many geophysical studies, where the record interval covers just a fraction of a significant periodic phenomenon. If such a phenomenon is interesting in the analysis, and therefore has not been eliminated by detrending, the question appears to what reliability we can determine power and frequencies, when observations are available only for a fraction of one period.

Writing the function $f(t)$ as:

$$f(t) = \begin{cases} \sin(\omega_0 t + \varphi) & \text{for } 0 \leq t \leq t_0 \\ 0 & \text{otherwise} \end{cases}$$

its Fourier transform [44]

$$F(\omega) = \int_0^{t_0} \sin(\omega_0 t + \varphi)\, e^{-i\omega t}\, dt$$

can be evaluated after lengthy but straight-forward arithmetic. This is left as an exercise to the student. The result for $\varphi = 0$ is shown in Fig.25. Note especially how the frequency of the spectral power peak varies with:

$$t_0/T = \omega_0 t_0/2\pi$$

The effect of severely truncated sinusoids (or cosinusoids) may be illustrated also by the following case, which is practically identical with the preceding one, but is computationally simpler and also has a direct relation to example 5 in section 2.4.3, i.e.:

$$f(t) = \Pi(t/2t_0) \cos \omega_0 t$$

whose Fourier transform is:

$$F(\omega) = \frac{\sin (\omega_0 - \omega)t_0}{\omega_0 - \omega} + \frac{\sin (\omega_0 + \omega)t_0}{\omega_0 + \omega} \qquad [45]$$

Results analogous to those shown in Fig.25 are obtained, considering that the frequency of the spectral power peak is the solution of the equation $\mathrm{d}F(\omega)/\mathrm{d}\omega = 0$.

Truncated sinusoids have been discussed in the literature by Kharkevich (1960), Gratsinskii (1962), Kats (1963), Toman (1965, 1966), Jackson (1967), Ulrych (1972). See also Spetner (1954). It is suggested that the correct period T of a geophysical phenomenon, severely truncated, can be obtained from a knowledge of t_0/T and φ, but this naturally presupposes at least approximate knowledge of the period T, which is sought. Referring to Fig.25, this implies that one should try to get t_0/T as big as possible, at least well above the steep rise in the frequency curve.

The explanation for the frequency shifts (Fig.25) is to be found in interference between the sinc-functions which constitute $F(\omega)$ and $|F(\omega)|^2$. Ulrych (1972) demonstrates that these frequency shifts can be eliminated by the use of the so-called *maximum entropy spectral method*, which in addition shows much higher resolution than earlier methods and nothing of the undesirable sinc-function properties. This new method has been developed by J. P. Burg, and the reader is referred to Lacoss (1971) for details.

Gratsinsky (1962) demonstrates truncation by rectangular windows on signals with interfering reflections. The main trouble is that such truncation leads to additional high-frequency false minima in the spectra (due to the rectangular windows used), but also that the primary peak alters: its amplitude decreases, its width increases, its frequency position alters.

As a general conclusion, it can be stated that for any given truncated time function an infinite number of spectra can be constructed, depending on the width and the shape of the time window applied.

In some cases, especially for non-stationary processes, truncation of signals is made on purpose for the study of so-called *running spectra*. These will be discussed in more detail in section 3.6.5.

Of the cases studied above, the convergence is especially poor when the record does not show any central concentration of larger amplitudes. The spectral analysis of such a wave train can be looked upon from the viewpoints of the Fourier series and of the Fourier transform methods:

(1) Fourier series: the portion of the record chosen for analysis is taken as the fundamental period and the record is assumed to be repetitive of this portion beyond both ends of the interval analyzed. The resulting spectrum is a line spectrum, corresponding to the terms in the Fourier series expansion. Thus, a continuous time function may correspond to a line spectrum.

(2) Fourier transform: the record is assumed to have zero amplitude outside the range chosen for analysis. The resulting spectrum is continuous, but with a shape corresponding to the Fourier series expansion. Thus, a transient time function (a *wavelet*) corresponds to a continuous spectrum.

It is clear that none of the assumptions (repetitive amplitudes or zero amplitudes, respectively, beyond the analyzed record portion) is strictly correct. In both cases, the assumptions lead to greater error, the smaller the analyzed portion is.

In this section we have investigated these effects for analytical signals, which are well under control. But the same problems are encountered in geophysical series of measurements. For example, in the construction of microseismic spectra, it is advisable to test with record portions of increasing lengths, until the spectra exhibit sufficient degree of convergence on the aspects under study. As another example, Webster (1969b) finds that deep-sea ocean current measurements, less than two months in length, do not give results of more general or representative value. In the next section we shall proceed to a more detailed discussion of geophysical signals.

Our discussion has been limited to truncation in the time domain, but it is equally applicable to truncation in the space domain. The latter problem happens when an array of stations is used to observe a propagating wave, and then the array covers only a part of the whole space extent of the wave motion. Then it is often assumed that space-stationarity prevails. For a discussion of such problems, see Linville and Laster (1966).

3.5.3 Different types of signals in geophysics

Under the general term of signal we summarize all phenomena which we may subject to harmonic or spectral analysis, and this is no doubt a generalization of the common meaning of signal. The types of signals encountered in geophysics as well as in many other fields, e.g. communication engineering, are of three basic types, as summarized in Table XI. The different types of signals differ both in duration and in their general appearance. The reasons for these differences are to be found in the generating mechanisms. A periodic generator will give rise to a *periodic signal*, of which the atmospheric temperature is a familiar example. On the other hand, a transient or aperiodic phenomenon will give rise to *transient or aperiodic signals*. Earthquakes are examples of this. Signals which are neither

TABLE XI

Review of different types of signals

Type of signal	Examples from geophysics	Fourier series (line spectrum): periodic signal repetition outside measured interval	Fourier transform (continuous spectrum): zero signal outside measured interval
Periodic (stationary)	any periodic phenomenon, such as atmospheric temperature variation over 1 year or 1 day	yes	no
Transient (non-stationary)	seismic waves	no	yes
Random (stationary, non-stationary)	microseisms, turbulence, ocean surface waves	(no)	(no)

periodic, nor transient, are called *random signals*. Due to lack of periodicity, they do not permit precise prediction, and statistics is needed for their treatment. A very lucid account of random signals or random processes with special reference to ocean surface waves has been given by Kinsman (1965, p.325 ff). Microseisms and atmospheric turbulence are other examples of random signals.

In addition to the distinction between different signals made in Table XI, from the statistical point of view we differ between *stationary* and *non-stationary signals*, depending upon whether statistical quantities are time-independent or not. Thus, a periodic signal would be stationary as long as the source action is constant. A transient signal is to be considered as non-stationary, whereas random signals may be stationary or non-stationary, again depending on variations in source conditions or medium properties.

In case of transient signals in geophysics, the source action is limited in space and time, and there is a complete distinction between source and propagation path. On the other hand, for stationary signals in geophysics, the source is extended both in space and time, and there is only partial separation between source and path. This is important to bear in mind in any effort to make deductions about source and path properties from a given record. Even for transient signals, it may often be difficult to make a clear distinction between source and path properties. Then, it is to be expected that such distinction will be considerably more difficult for stationary signals.

Even though seismic waves are generally considered as transient, it has proved advantageous in dealing with accelerograms of strong-motion earthquakes to consider such records as a non-stationary random process or random superposition of a large number of pulses (Housner, 1955; Thomson, 1959; Iyengar and Iyengar, 1969; Liu, 1969).

The signal terminology is applicable not only to the case when the independent variable is time. In many cases, the independent variable may be a space coordinate. Moreover, the terminology is applicable to functions of several independent parameters.

The *Fourier series expansion* implies integration over a whole (fundamental) period, or an integer number of such periods, and leads to a line spectrum. Clearly, it presupposes a knowledge of the fundamental period. Moreover, it assumes repetition of the signal over an infinite range outside the measured interval. Such conditions are fulfilled only for periodic signals, and only approximately for random signals.

Frequently, the fundamental period is not known. Methods for Fourier series expansion in such a case have been discussed by Jeffreys (1964). His method is briefly as follows. Identifying the Fourier series expansion [3.4] in Chapter 2, for a finite record length T, with a "true" expression for a record:

$$f(t) = a \cos \lambda t + b \sin \lambda t \qquad [46]$$

of unknown period $2\pi/\lambda$ and unknown coefficients a, b, we get expressions for a_0, a_n, b_n in terms of a, b, λ, n. From these formulas it is obvious that the largest Fourier coefficients are obtained for $2\pi n \simeq \lambda T$, especially for $2\pi n < \lambda T < 2\pi(n+1)$. Taking values of $n = \lambda T/2\pi$ and $n + 1 = \lambda T/2\pi$, we get four equations (a_n, b_n) with three unknowns (a, b, λ). We could proceed with $n - 1$ and $n + 2$, etc, which yields eight equations in the same three unknowns. These equations could be solved by a least-squares procedure, including uncertainty estimates. For further extension of these ideas, see Shimshoni (1967, 1968 a and b).

There is another source of uncertainty to consider in harmonic analysis, i.e., when there is not one fundamental period but two or more closely spaced fundamental periods. A straight-forward harmonic analysis, assuming only one of these as fundamental period, will lead to incorrect results. A method to deal with the case of two close fundamental periods has been given by Yampolsky (1960).

The *Fourier transform* leads to a continuous spectrum and the method assumes zero amplitude outside the measured interval. This may be exactly fulfilled by transient signals, but may be considered only as an approximation in case of random signals.

Both approaches (Fourier series expansion and Fourier transform method) can be applied in any case, but in individual cases there may be reasons for preferring one or the other. In this connection, it is of interest to note that Nowroozi (1965) in analysis of free vibrations of the earth found that power and cross-power spectral analyses in general lack the precision of the harmonic analysis (Fourier series expansion). Also, Saltzman and Fleisher (1960) used line spectra in a study of atmospheric circulation. See also Barsenkov (1967a and b) who found good agreement between harmonic and spectral analysis for earth's tidal data. In an analysis of lunar daily geomagnetic variations, Gupta and Chapman (1969) found much larger amplitudes by spectral than by harmonic analysis, but the same frequencies by the two methods.

In line spectra, obtained by harmonic analysis, each line represents a single wave, whereas in continuous spectra, a peak does not represent a single wave but a group of waves with close frequencies. Harmonic analysis is generally used only when frequencies (at least the fundamental frequency) is known or suspected before-hand, whereas spectral analysis works without preconceived opinions. From the theoretical point, it is suggested as an exercise to the student to get both the Fourier series and the Fourier transform for given functions (using for example those of section 2.4.3) and to compare the results. A thorough discussion of different types of signals and their treatment, with special regard to communication engineering, has been given by Panter (1965). See also Kinsman (1965, especially pp.437–438). An advantage of power-spectrum analysis over harmonic analysis is that it also includes rules for calculation of cross-power spectra. This provides efficient means to compare different series of data in the spectral domain.

Kinsman's (1965, p.438) remark is especially worth mentioning, that the amplitude spectra of samples of a stationary process will not be stable, but instead the *variability* of the amplitudes can be considered stable. And as a measure of the variability we use the variance, hence the significance of variance or power spectra. Even so, stationarity is no doubt always an idealization of real processes. Theoretical correlation and spectral studies of space- and time-stationary seismic waves were made by Aki (1957), with application to traffic noise (called "microtremors"), and statistical tests for stationarity, with special reference to microseisms, are discussed by Haubrich (1965).

A point of philosophy may be justified here. A process, that is exactly stationary, does not exist in nature. All processes are only more or less non-stationary. The question if a process is "stationary" or non-stationary is essentially a matter of time scale. In geophysics, we may consider phenomena with "slow" variation, i.e. with durations extending over several days, as stationary or quasi-stationary. Examples are ocean surface waves and microseisms as well as weather. On the other hand, we are accustomed to consider a "rapidly varying" phenomenon, such as a P-wave, as transient and non-stationary.

One approach would probably be to consider a process as stationary if the ratio of its duration (A) to the interval measured (B) exceeds some assigned limit, with B including many periods. For example, for a microseismic storm, let $A = 2$ days, $B = 1$ min, then $A/B \simeq 3000$. For a P-wave, on the other hand, any normal A/B will be appreciably smaller, i.e. below the "stationary limit".

3.5.4 *Spectral terminology*

A variety of terms appear in the geophysical literature for defining spectra. Without clear specification it is not always easy to know which particular spectrum an author is referring to. There does not seem to exist any fixed rules for identifying spectra. Part of the confusion derives from the fact that reference is made to different functions or variables entering the equation for a spectrum.

Starting from a general form $X(x)$ for a spectrum function, of which for example $F(\omega)$ is just one special case, we can list the following three specifications which are necessary for an unambiguous identification of a geophysical spectrum: (1) the physical quantity that X represents; (2) the dimension of X; and (3) the nature of the independent variable x.

Table XII gives a summary. For full specification one item from each of the three columns is needed, and any combination of the three columns is permitted. This means for example, that we could talk about power–frequency spectrum of rainfall, amplitude–wavenumber spectrum of gravity, phase–frequency spectrum of acceleration, etc. Terms like temporal spectrum and spatial spectrum, sometimes used instead of frequency spectrum and wavenumber spectrum, should rather be avoided, as being dimensionally incorrect.

In addition to the general-type spectra mentioned, more special types of spectra may be used for particular studies, e.g. response spectra in earthquake engineering (section 7.1.7).

The use of the term amplitude deserves a special comment. Any periodic

TABLE XII

Terminology necessary to specify any spectrum $X(x)$

Spectral function X		Independent parameter x
physical quantity	dimension	
Displacement	amplitude (density)	frequency
Velocity	power (density)	wavenumber
Acceleration	phase	
Temperature		
Wind	co-spectrum < amplitude / power	
Rainfall		
Geomagnetic elements	quad-spectrum < amplitude / power	
Gravity		
Etc.		

curve, whether this represents displacement, velocity, acceleration or any other property, has a certain amplitude. This is the general meaning of this word. But in seismology, also a more restricted use of amplitude exists, that is, referring amplitude just to displacement, as distinct from velocity, acceleration, etc. Then we do not talk about amplitude of displacement, but just amplitude; similarly, we do not talk about amplitude of velocity, but just about velocity, etc.

The spectra we have dealt with so far are of the simplest kind. In extensions to higher-order spectra of different kinds, it will be necessary to amplify the nomenclature correspondingly. This will be described in the next section.

In some geophysical applications, the term spectrum is used without any direct relationship to the spectra listed above. For instance, Taner and Koehler (1969) and Cook and Taner (1969) describe *velocity spectra*, which have proved useful to identify reflecting horizons in seismic prospecting work. They use as ordinate the time-weighted root-mean-square average velocity over a number of layers, and as abscissa the two-way normal incidence travel time for reflected waves. A sequence of curves, one for each abscissa value, is plotted which show the power as the third coordinate.

3.6 ALTERNATIVES AND EXTENSIONS OF SPECTRAL METHODS

In this section we shall investigate various alternatives, extensions and generalizations of the methods presented hitherto in this book. Even though we find it most appropriate to include this section at this point, mainly for its close relation to the theoretical developments made so far, the reader could proceed directly to Chapter 4, without loss of continuity, if he so desires. The present section could then be read at any later stage, for example after Chapter 6.

3.6.1 Alternatives to the Fourier series expansion

Unquestionably, we are justified to speculate on the mathematical development made thus far. We have seen in Chapter 2 that the expression for the spectrum $F(\omega)$ in [20] and [23] has been derived from an expression of the corresponding time function $f(t)$ in terms of an infinite sum of sine and cosine functions, [1] in Chapter 2. This procedure is nowadays so commonly used in spectral analysis that only very few reflect on the possibility of doing the spectral analysis in some other, and hopefully better way. Without entering into any detailed exploration of this interesting aspect, we may state a few things, at least intuitively. The development as given here, and which is the traditional one, has the advantage of being applicable to almost any time function $f(t)$. The Dirichlet conditions for the application of the method are not severe, considering physical phenomena, and the method can without special consideration be employed to any curve encountered in geophysics.

However, it may be envisaged that special curve forms could be approximated more accurately with fewer terms than the expansion in [1] of Chapter 2. For instance, it is expected that a square-shaped curve $f(t)$ would be easier to represent by a summation over a number of square waves, than by using sine and cosine functions. Of course, any given curve can be approximated by a *step-curve*, the approximation to the given curve being the better, the smaller the steps are chosen. This is in fact the basic principle underlying the Walsh spectrum, to be discussed in more detail later (section 3.6.2). Already Tomoda (1954) devised a method for expansion of a given curve in square-wave functions, and then used this expansion to derive the usual Fourier coefficients.

Seismic body waves are transient wavelet pulses, often of short duration. In such cases also other methods may be preferred. I quote from my book (Båth, 1968, pp.93–94) the following: "In spectral analyses of transient seismic pulses, the Fourier series expansion is usually used." "But this method is not efficient, since we are then expanding transient phenomena in a series of steady-state functions." "A more efficient approach is to expand such transients in a series of functions with properties similar to the transients themselves." "*Laguerre functions* have this property" (see fig.38, p.94, in Båth, 1968). "The advantage of using Laguerre functions in such cases over the Fourier series expansions is that the number of required terms is much less (just due to the fact that the Laguerre functions approximate the transient pulses so much better)." "Whereas a few hundred terms of a Fourier series may be required, the same approximation to a given signal can be achieved by only a few dozen Laguerre functions." For more details, see Dean (1964). This remark concerns the series expansion of $f(t)$ in the time domain, whereas the usual notion of frequency has been lost.

For special purposes, expansion also in other functions may be of value. For instance, an arbitrary function $f(t)$ can be expanded in a series of Bessel functions, which due to its similarity with the Fourier series, is called the *Fourier-Bessel series* (Båth, 1968, p.127). Expansion in Fourier-Bessel series has found geophysical application, especially for the spatial variation of gravity (Tsuboi, 1954). Alternatively, expansion into sums of terms of $(\sin x)/x$, where x is a spatial coordinate, has been used for the same purpose (Tomoda and Aki, 1955; Tsuboi and Tomoda, 1958). See further section 10.2.1. These expansions have the advantage over the ordinary Fourier series that they produce vanishing amplitudes at some distance from the measured interval, contrary to undiminished amplitudes in the Fourier series case. On the other hand, also for these expansions the frequency concept is lost.

Similarly, a function $f(t)$ can be expanded into a *Legendre series* for $|t| \leq 1$ (Båth, 1968, p.161). Expansion into Legendre polynomials, as one type of generalized Fourier analysis, is given by Harmuth (1972, pp.54–57). Expansion of functions into *spherical harmonics* is of particular value to geophysics, as mentioned already in section 2.5.1. Moreover, we have to observe that the complex Fourier transform discussed above is only one case out of quite a number of different integral transforms (Båth, 1968, p.211).

Still another expansion is in the form of *polynomials*, especially cubic polynomials, which have found application for two-dimensional spatial expansion of potential functions (gravity, geomagnetism). Piece-wise expansion in bicubic spline functions has been found to yield higher accuracy for large wavelengths than Fourier series for spectral estimation. See further section 10.2.1.

From mathematics, several other series expansions (Taylor, Maclaurin, Laurent) are familiar. Even though different expansions may be appropriate in different cases, it is only in the case of the Fourier series and transform and the spherical harmonic analysis that we are justified in talking about frequency and spectra in the usual sense of these terms.

Several forms of *modified Fourier series expansions* exist, for example by using time-dependent amplitudes, as being better signal-adaptive than expansions with constant amplitudes. One such form is the following:

$$f(t) = \sum_{i=1}^{\infty} A_i(t)\, e^{-a_i t} \sin(\omega_i t + \varphi_i) \qquad [47]$$

with special cases, such as:

$$\text{attenuated sinusoidal function:}\ f(t) = \sum_{i=1}^{k} A_{0i}\, e^{-a_i t} \sin \omega_i t \qquad [48]$$

$$\text{Berlage function:}\ f(t) = \sum_{i=1}^{k} A_{0i}\, t^{n_i}\, e^{-a_i t} \sin \omega_i t \qquad [49]$$

Among applications, we note that Seyduzova (1970) applied the attenuated sinusoidal function to approximate S-waves in near recordings of Tashkent aftershocks in 1966. The basic idea behind these formulas is the same as above, i.e. by using functions which comply more closely to the given records, it would be possible to represent these by a smaller number of terms than in a regular Fourier series expansion in sine and cosine functions. For time-limited functions, it is natural to expect better representation by using time-dependent amplitudes than constant amplitudes. This principle also underlies a method described by Spitznogle and Quazi (1970). Another modified Fourier expansion is obtained by applying weights p_n to the different harmonics in such a way that a limited number N of terms will represent the function $f(t)$ equally well as a large number of ordinary Fourier terms:

$$f(t) = \sum_{n=0}^{\infty} (a_n \cos nt + b_n \sin nt) = \sum_{n=0}^{N} p_n (a_n \cos nt + b_n \sin nt) \qquad [50]$$

An example of this expansion in oceanography can be found in Longuet-Higgins et al. (1963). See also Neumann and Pierson (1966, pp.354–356).

Still other signal-adaptive forms could be envisaged, e.g. time–dependent frequency for dispersive waves. However, any such effort has to comply to the purpose of the analysis, i.e.:

(1) If only a mathematical form of a record is wanted, then signal-adaptive forms with as few terms as possible are naturally advantageous.

(2) If, on the other hand, the purpose of the analysis is to gain information about the constituents of the record, then more terms, each of a simple form, are usually more informative and easier to interpret (e.g. in terms of amplitude and phase).

In addition, we easily understand that using *different* approximating functions for different functions $f(t)$, according to their general shape, would entail greater complications, both in theory and in numerical practice. Therefore, equation [1], Chapter 2, no doubt represents the greatest potentiality in representing any given curve, due to its superior flexibility, and due to the unified and simple mathematical treatment which it permits. The mathematical simplicity is especially clear from the orthogonality relations which hold for sine and cosine functions. Thanks to the orthogonality property, it is possible to calculate the coefficients of the different harmonics independently, as is obvious from [2] and [3] in Chapter 2. Other orthogonal functions would lead to a corresponding simplicity.

3.6.2 Walsh sequential spectrum

This subject has close relation to the preceding section, and therefore we consider it most appropriate to deal with Walsh spectra in this context, even though some items mentioned below have connections with later sections in this book (this concerns formulas for discrete summation, section 4.5, and filtering, Chapter 6).

The Walsh sequential spectrum represents a method to analyze given curves into a summation of square waves, and in this respect it differs basically from the harmonic analysis involved in Fourier series and spectra. It is immediately clear that any given curve can be resolved into square-wave constituents to any desired degree of accuracy, just as well as by harmonic functions. In the Walsh analysis, such a calculation has been put into a strict mathematical system.

The Walsh functions, introduced into mathematics by Walsh (1923), assume only the values $+1$ or -1. We denote them by $w_j(x)$, where the subscript j is the sequency (to be defined below) and x is running parameter (time, space). Several different descriptions of the Walsh functions appear in the literature but we follow Gubbins et al. (1971), because it is clear and brings out the analogies between Walsh and Fourier transforms. A far more extensive treatment can be found in the book by Harmuth (1972), both of fundamental properties, with frequent

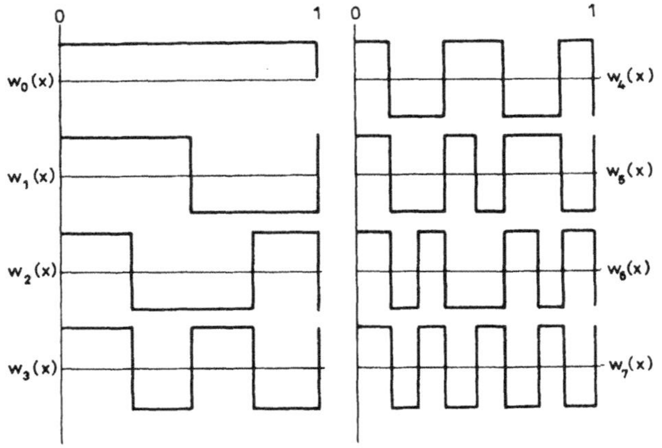

Fig.26. The first eight Walsh functions. After Gubbins et al. (1971).

comparisons with Fourier transforms, together with applications, especially to communication problems.

The complete definition of the Walsh functions is contained in the following equation set:

$$\left.\begin{aligned} &w_j(x) = 0 \qquad \text{for } x < 0 \quad \text{and} \quad x > 1 \\ &w_0(x) = 1 \qquad \text{for } 0 \leq x \leq 1 \\ &w_{2j}(x) = w_j(2x) + (-1)^j \, w_j[2(x - 1/2)] \\ &w_{2j+1}(x) = w_j(2x) + (-1)^{j+1} \, w_j[2(x - 1/2)] \\ &j = 0, 1, 2, \ldots \end{aligned}\right| \qquad [51]$$

Fig.26 shows the first eight Walsh functions, and to become familiar with their build-up it is recommended to the reader to compare these graphs with [51]. It is seen that each Walsh function is formed by its preceding order by compressing this on the first half of the x-range, followed by a repetition of this signal on the second half of the x-range, with same or opposite sign.

Several properties of the Walsh functions follow easily from their definition:

(1) The subscript j in $w_j(x)$ is equal to the number of zeros in the interval $0 < x < 1$ and is called *sequency*, which is a generalized frequency corresponding to frequency in the Fourier analysis. As j increases, the sequency increases, which is analogous to increasing frequency for Fourier series, as clearly seen in Fig.26.

(2) Walsh functions with even sequency are symmetrical about $x = 1/2$, those with odd sequency are antisymmetrical about $x = 1/2$. See Fig.26. This has led to some definitions of cal and sal functions, by analogy with cos and sin functions:

$$\begin{aligned} w_{2j}(x) &= \text{cal}\,(j,x) \\ w_{2j+1}(x) &= \text{sal}\,(j,x) \end{aligned} \qquad [52]$$

(3) The Walsh functions are orthonormal, i.e.:

$$\int_0^1 w_i(x)\, w_j(x)\, \mathrm{d}x = \delta_{ij} = \begin{cases} 1 \text{ for } i = j \\ 0 \text{ for } i \neq j \end{cases} \qquad [53]$$

This property makes them suitable for construction of transforms and spectra.

In fact, the finite Walsh transform and its inverse are as follows:

$$\begin{aligned} F(j) &= \frac{1}{N} \sum_{i=0}^{N-1} f(x_i)\, w_j(x_i) \\ f(x_i) &= \sum_{j=0}^{N-1} F(j)\, w_j(x_i) \end{aligned} \qquad [54]$$

written in digital summation form. As seen, there is complete analogy to the corresponding Fourier formulas (section 4.5.3). The spectrum $F(j)$ is the *Walsh spectrum* and the analysis is often referred to as *sequential analysis*, because of the term sequency (as distinct from frequency analysis or Fourier spectrum analysis). The analogous formulas for two dimensions can be written down immediately:

$$\begin{aligned} F(k,l) &= \frac{1}{N^2} \sum_{j=0}^{N-1} \sum_{i=0}^{N-1} f(x_i,y_j)\, w_k(x_i)\, w_l(y_j) \\ f(x_i,y_j) &= \sum_{l=0}^{N-1} \sum_{k=0}^{N-1} F(k,l)\, w_k(x_i)\, w_l(y_j) \end{aligned} \qquad [55]$$

Further evaluation of the Walsh transform, deduction of theorems corresponding to those for the Fourier transform, and comparison with the Fourier transform, especially with regard to geophysical applications, are of the greatest interest and value. The Walsh transform can be calculated by the FFT method ("Fast Walsh Transform") but faster than the Fourier transform. There is now an extensive literature on the mathematics of the Walsh spectrum, and numerous applications, especially in engineering. But so far, only few geophysical applications have been published. One example is Båth and Burman (1972) who applied this method to vertical-component Rayleigh waves from an underground nuclear explosion on Novaya Zemlya and compared the results with Fourier spectra.

Bois (1972) emphasizes the use of the Walsh spectrum for transmission of information, e.g. seismic prospection recordings, in condensed form. This is achieved by first Walsh-transforming the given record, then decimating the spectrum according to some principle, and finally retrieving the original record by inverse transformation of the decimated spectrum. The decimated spectrum represents the condensed version suitable for transmission.

Gubbins et al. (1971) demonstrate the use of Walsh and some other square functions for two-dimensional digital filtering. The square functions are superior to the Fourier methods because of the absence of the Gibbs phenomenon (section 2.1.2). Windows used with Fourier methods to reduce such undesirable effects are not needed with Walsh functions because these are adapted to discontinuities in the curve to be analyzed. A detailed account of sequency filtering with Walsh functions is given by Harmuth (1972). It is tempting to expect that the Walsh analysis would be well suited to spectralize telegraphic signals (having square-peaked waveform), used for instance to model geomagnetic field reversals (Naidu, 1971). Fourier series analysis of square-peaked waveforms has naturally been made already long ago (see for example Manley, 1945).

In conclusion we can summarize some of the relative merits of Walsh and Fourier transforms as follows:

(1) Walsh transform is easier and quicker to calculate, because only additions and subtractions are involved and no multiplications or calculations of exponential functions.

(2) Walsh transform is superior in having no Gibbs phenomenon and not requiring any special windowing in the time domain.

(3) On the other hand, complications arise in calculation of correlations and convolutions by the Walsh transform, which in the seismological case implies problems in applying seismograph response functions.

3.6.3 Cepstrum analysis

An obtained spectrum $F(\omega)$ can naturally be subjected to spectral analysis itself, which yields a new spectrum. Such a "second" spectrum of $f(t)$ has found application in certain geophysical studies in revealing periodicities of $F(\omega)$. Naturally the procedure can be continued practically without limit, but higher-order spectra do not lend themselves to any easy and useful interpretation. The extension of the spectral concept can be formulated as follows:

$$f(t) \leftrightarrow F(\omega) \leftrightarrow \bar{f}(\bar{t}) \qquad [56]$$

where the last term denotes the second spectrum of the initially given function $f(t)$. This extension of the spectrum concept has also necessitated a corresponding extension of the terminology. The second spectrum $\bar{f}(\bar{t})$ will here be called *cepstrum* (for the moment in a more general sense), and its argument $\bar{t}$, which has the same dimension as the argument of $f(t)$, is called *quefrency*. Both are artificial names, formed by paraphrasing the words spectrum and frequency, respectively. See Bogert et al. (1963), who used cepstrum analysis for echo studies, including pP, defining cepstrum in a special way (see below).

It is left as an exercise to the reader to form successive spectra from examples given in Table V, section 2.4.3. For instance, continuing the pair:

$$\Pi(t) \leftrightarrow \text{sinc}\,(\omega/2\pi) \qquad [57]$$

we find that they will be alternatively Π- and sinc-functions, indefinitely, but with increasing ordinates. The same holds for the pair:

$$\Lambda(t) \leftrightarrow \text{sinc}^2\,(\omega/2\pi) \qquad [58]$$

The pair $\delta(t) \leftrightarrow 1$ will be alternatively 1 and δ, also with increasing ordinates. A Gaussian function will be preserved even by as many successive operations as we like, but also here, ordinates increase, and the curve will show two alternating forms, varying from step to step.

We may state these results such that the functions $\Pi(t)$, $\Lambda(t)$, $\delta(t)$ are of period 2 with regard to repeated transforms, by which we mean that after two successive transformations the original function is retrieved. Then, obviously the Gaussian function is of period 1. Generalizing to any function $f(t)$ we find by repeated application of the Fourier transform formula that this is of period 4, apart from a constant factor (cf. section 2.3.2):

$$f(t) \leftrightarrow F(\omega) \leftrightarrow 2\pi f(-t) \leftrightarrow 2\pi\, F(-\omega) \leftrightarrow (2\pi)^2 f(t) \qquad [59]$$

It is only in certain special cases of $f(t)$ that the period might be less than 4, as shown by the examples above. For instance, if $f(t)$ is even, i.e. $f(t) = f(-t)$, then the period will obviously be 2, at most. Iterated diffraction (see, for example, Lansraux and Delisle, 1962) constitutes the optical correspondence to the repeated formation of cepstra.

As is evident from the explanation just given, the definition of cepstrum as above will hardly give anything new about a given function $f(t)$ and its spectral composition. Still we made this approach to the cepstrum concept, mainly for its pedagogical value. Instead, cepstrum exists usually with a more special definition, more useful in applications, i.e. as the square of the Fourier transform of the logarithm of power (or amplitude) spectrum, or in formula (Bogert et al., 1963; Noll, 1964):

$$\bar{f}(\bar{t}) = \left| \int_{-\infty}^{\infty} \log |F(\omega)|^2\, e^{-i\bar{t}\omega}\, d\omega \right|^2 \qquad [60]$$

The advantage of using the logarithm is that if $F(\omega)$ is a product of two or several factors, like $F(\omega) = G(\omega)\ H(\omega)$, the cepstrum will be a sum of separate terms rather than a product, thus exhibiting the various contributions with a better separation. Howcver, Cohen (1970) in addition uses a definition equivalent to the one just given, only without the log.

Considering that $|F(\omega)|^2$ is an even function, being the transform of the even function $C_{11}(\tau)$, then also log $|F(\omega)|^2$ is even, by which [60] can be written as follows:

$$\bar{f}(\bar{t}) = 4\left[\int_0^\infty \log|F(\omega)|^2 \cos \bar{t}\omega \, d\omega\right]^2 \qquad [61]$$

We also see that a cepstrum defined as in [61] but without the log, will give us $4\pi^2\ C_{11}^2(\tau)$, i.e. proportional to the square of the autocorrelation function.

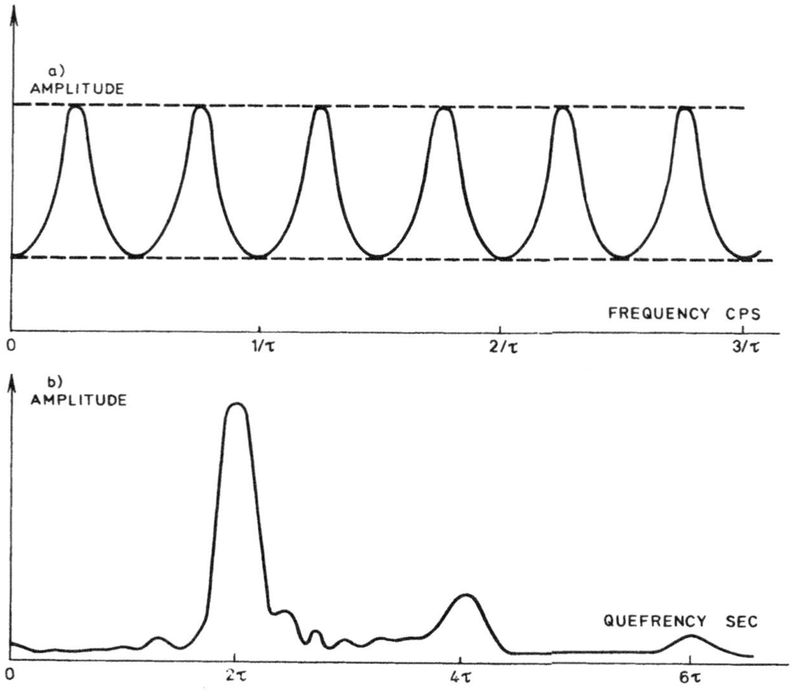

Fig.27. Example of cepstrum analysis. a) Modulus of the vertical response of a layer over a half-space to a P-wave at normal incidence. b) Cepstrum, here defined as the modulus of the Fourier transform of the amplitude in a). τ is the one-way vertical travel time of P in the layer. After Bakun (1971).

Still another definition of cepstrum exists, namely as the inverse transform of log $F(\omega)$, used among others by Tsai (1972) for filtering multiple-path and multiple-source effects from surface-wave records.

Cepstrum analysis is a natural procedure to undertake in any case that a spectrum has a significant oscillatory character. This is in fact very often the case. Examples from seismology include earthquake source parameters (oscillations due to finite sources and to multiple events, section 8.1), reverberations and other reasons for constructive and destructive interference (section 7.1; Fig.27 shows an example of this after Bakun, 1971), echo analysis being a special case of the latter (section 8.4.3), as well as response curves for array stations. These are all examples of one and the same phenomenon: interference (see section on "interference analysis", 8.4.3), constructive interference leading to spectral maxima, destructive interference leading to spectral minima. Usually the maxima and minima appear at regular frequency intervals, which bear genetic relations to the underlying cause. Therefore, readings of frequencies of spectral maxima or minima may give valuable information about source and path properties, otherwise not easily retrievable from a seismic record. In some cases, the spectral minima are sharper, corresponding to more well-defined frequencies, while the maxima are rounded off and less well defined. In such cases, it is preferable to read frequencies of minima. In other cases the reverse happens, i.e., maxima are sharper, and then *their* frequencies should be read. Similar circumstances apply to records of other geophysical properties, as soon as a regular interference comes into the picture.

In a paper containing much general information on cepstrum analysis, Noll (1967) applies the technique to analysis of speech, using short-time cepstra. These

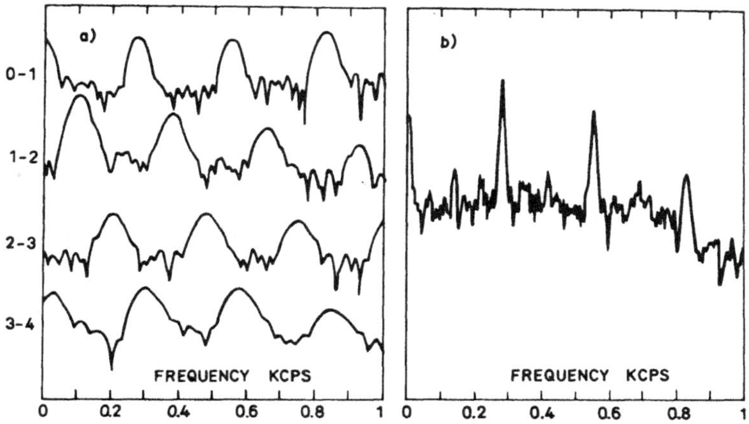

Fig.28. Example of harmonic product spectrum. a) Spectrum of a 40-msec segment of female utterance (the spectrum comprises the range 0–4 kcps, splitted in four parts, with integer frequencies at the left-hand side and tenths along the bottom). b) Harmonic product spectrum of a). After Schroeder (1968).

are formed by applying moving but partially overlapping Hamming windows (section 4.4.4) to the given function $f(t)$, and deriving a cepstrum for each position of the window, which thus exhibits the spectral variation with time.

An alternative method to the cepstrum analysis to determine fundamental frequencies (ω) is offered by the *harmonic product spectrum* (Schroeder, 1968), defined as (A = amplitude):

$$\sim \log \sum_{n=1}^{N} |A(n\omega)| = \log \left[|A(\omega)| + |A(2\omega)| + \ldots + |A(N\omega)|\right] \qquad [62]$$

and which does not require an additional Fourier transformation. An example is shown in Fig.28.

Still another method of calculating spectra has found geophysical application, especially to earth-tide observations (Pil'nik, 1970), for the purpose of increasing the signal/noise ratio, and for periodicity research in climatology (Stringer, 1972, p.92). This method consists in forming successive correlation functions and their corresponding spectra, which can be simply illustrated as follows:

Usual or first autocorrelation:

$$\left.\begin{aligned} C_{11}(\tau) &= \int_{-\infty}^{\infty} f(t) f(t+\tau)\,\mathrm{d}t \\ C_{11}(\tau) &\leftrightarrow E_{11}(\omega) \end{aligned}\right| \qquad [63]$$

Second autocorrelation, i.e. autocorrelation of the first autocorrelation:

$$\left.\begin{aligned} C_{11}^{(2)}(\bar{\tau}) &= \int_{-\infty}^{\infty} C_{11}(\tau)\, C_{11}(\tau+\bar{\tau})\,\mathrm{d}\tau \\ C_{11}^{(2)}(\bar{\tau}) &\leftrightarrow E_{11}^{(2)}(\omega) \end{aligned}\right| \qquad [64]$$

Similarly, second cross-correlation can be formed:

$$\left.\begin{aligned} C_{12}^{(2)}(\bar{\tau}) &= \int_{-\infty}^{\infty} C_{11}(\tau)\, C_{22}(\tau+\bar{\tau})\,\mathrm{d}\tau \\ C_{12}^{(2)}(\bar{\tau}) &\leftrightarrow E_{12}^{(2)}(\omega) \end{aligned}\right| \qquad [65]$$

3.6.4 Non-linear phenomena—bispectrum

Another generalization concerns extension to higher orders of the development given hitherto. If the autocorrelation defined above in [1] is termed the one-dimensional correlation, we may define a corresponding two-dimensional autocorrelation function, termed the *bicorrelation:*

$$C(\tau_1,\tau_2) = \int_{-\infty}^{\infty} f(t)f(t + \tau_1)f(t + \tau_2)\,\mathrm{d}t \qquad [66]$$

and its corresponding power spectrum, called the *bispectrum:*

$$B(\omega_1,\omega_2) = \int_{-\infty}^{\infty}\int_{-\infty}^{\infty} C(\tau_1,\tau_2)\,\mathrm{e}^{-i\omega_1\tau_1 - i\omega_2\tau_2}\,\mathrm{d}\tau_1\,\mathrm{d}\tau_2 \qquad [67]$$

In general, $B(\omega_1,\omega_2)$ has both real and imaginary components, like an ordinary Fourier spectrum. Moreover, the inverse to $B(\omega_1,\omega_2)$ can be written down immediately in full analogy to the inverse of the ordinary Fourier transform. From the formula for $B(\omega_1,\omega_2)$, it is immediately clear that $B(-\omega_1,-\omega_2) = B^*(\omega_1,\omega_2)$. For symmetry reasons, it is seen that $C(\tau_1,\tau_2)$ and $B(\omega_1,\omega_2)$ are fully defined if they are given in the octants $\tau_1 \geq \tau_2 \geq 0$ and $\omega_1 \geq \omega_2 \geq 0$, respectively. Therefore, in graphical displays it is customary to show only one octant (see, e.g., MacDonald, 1965; and Chang, 1969). In case of discrete sampling, the upper limits of the frequencies are set at the Nyquist frequency (section 4.3.2).

These concepts may yield useful information for non-Gaussian time series, whereas for a Gaussian (normal) process, the third moments are zero, as well as the bispectrum (Hinich and Clay, 1968). Trispectra and higher-order spectra can be formed by continuing this process.

Bispectra have been used in wave analyses to study non-linear effects (i.e. interactions between different wave modes), such as of ocean waves in shallow water (Hasselmann et al., 1963; see also Kenyon, 1968), between sea-level oscillations of different frequencies (Munk et al., 1965; Roden, 1966b), interaction between atmospheric pressure variations of different periods, especially the annual period and its higher harmonics (MacDonald, 1965), mass transport by ocean waves (Chang, 1969), etc. For ocean surface waves, non-linearity decreases with depth, and therefore bottom-pressure records may show much less of this effect than surface elevation observations of the same wave motion. Moreover, the bottom-pressure records show much less of high frequencies, the water layer serving as a low-pass filter.

Zadro and Caputo (1968) and Zadro (1971) applied bispectral analysis to the free oscillations of the earth to investigate spectral peaks found observationally but not contained in the theory. These peaks have frequencies equal to the sum or difference between theoretically confirmed peaks, and are due to non-linear

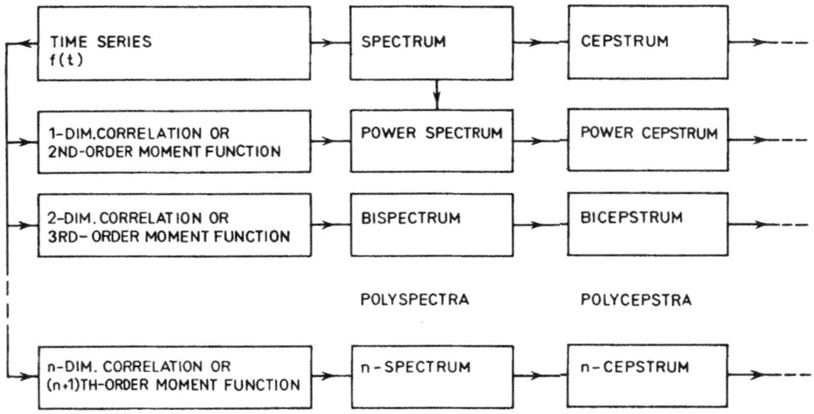

Fig.29. A pedigree of various spectral functions.

effects in the earth's elastic behaviour. Application of bispectrum analysis to gravimetry has been outlined by Kaula (1966b), who uses the term *degree skewance* as a synonym for bispectrum. However, much greater complications will be encountered in this application than for example in ocean-wave studies. For further discussion and application of the bispectrum, see Rosenblatt and Van Ness (1965), Haubrich (1965), Kinsman (1965), and Godfrey (1965). Also *bicoherence* and *biphase* have been introduced (Roden, 1966b) in analogy with the definitions for simple spectra, as well as *cross-bispectrum*.

Our presentation has been limited to single (univariate) or double (bivariate) time series. For the extension to combined study of more than two time series (multi-variate analysis), the reader will find thorough treatments, using matrix theory, in Robinson (1967b) and Jenkins and Watts (1968). Multi-variate analysis may find seismological application to the parallel recordings of seismometers within an array. Combined with multi-dimensional Fourier analysis, this would represent the most general case.

The pedigree in Fig.29 demonstrates more clearly the relations between the different kinds of higher-order spectra and the cepstra. The scheme is given for a single one-dimensional time series $f(t)$. Generalization of this scheme to $f_n(x_1, x_2, x_3, \ldots, x_m)$, i.e., multi-variate ($n$), multi-dimensional ($m$) series would represent the most general case. As a rule only the lowest-order spectra lend themselves to any clear physical interpretation. Among polyspectra, there are only the first-order polyspectrum = the usual power spectrum (or cross-power spectrum) and the second-order polyspectrum = the bispectrum that find useful physical applications, the latter for investigation of non-linear effects. Polyspectra have to be kept clear from cepstrum, as seen in Fig.29. Alternative names for *power spectral density*, such as *autocorrelation spectrum* and *second-moment spectrum*, find an immediate explanation from Fig.29. Higher-order cepstra or bicepstra have not yet found any geophysical application.

The generalizations thus introduced permit construction of spectra for any given function $y = f(x)$ and conversely, construction of the inverse transform to any given spectrum $y = f(x)$, provided the integrals exist. In the scheme in Fig.29 we could envisage an infinite extension from the time series also in the other direction, by successively forming inverse transforms. The conclusion is that the question, whether any given function $y = f(x)$ is a spectrum or not, has no general significance, as $y = f(x)$ can both be a spectrum of a function as well as the inverse of a spectrum. Naturally, this practically infinite extension in both directions is nothing unique to Fourier transforms; it applies to every mathematical operation, with certain restrictions. Clearly, among Fourier transforms it is only the function itself and its first Fourier transform that are of dominating significance.

3.6.5 Spectra of non-stationary processes

Extensions of correlations and spectra to non-stationary functions (section 3.5.3), essentially from the mathematical-statistical side, have been published by Kharkevich (1960), Kampé de Fériet and Frenkiel (1962) and Goodman and Dubman (1969). A corresponding extension to observational series would be most welcome. One important problem to consider is that we are not entitled simply to take over concepts from the stationary case and apply them immediately to non-stationary cases. In fact, concepts as spectrum, frequency, etc., which have well-defined meanings in the stationary case, need to be *defined* also for non-stationary cases. A critical examination of such definitions has been made by Loynes (1968), who also reviews various proposals to measure non-stationary spectra.

Frequency has a well-defined meaning for stationary processes, but for non-stationary processes there is a continuous change of frequency. We understand that there is a certain dilemma between the need for a finite sampling interval to determine a frequency and a spectrum, and the fact that frequency is continuously varying, the latter fact requiring an infinitely small sampling interval. A certain compromise in the length of the sampling interval seems unavoidable. A well-known case with frequency varying along the trace is offered by dispersive surface waves in seismology. Methods for their treatment will be presented in section 7.2.

The influence of various types of non-stationarities in the wind field on calculated spectra is demonstrated by Baer and Withee (1971). Non-stationarities in the form of spikes (wind gusts) produce enhanced high-frequency energy and those in the form of shifts (change of wind direction) entail increased low-frequency energy. They conclude that special observational techniques are required for non-stationary time series for any efficient sampling system.

The concept of running spectra (a special kind of truncated spectra, cf.

section 3.5.2) provides for a certain progress in this direction. The procedures employed use the usual expression for the Fourier transform, [23] in Chapter 2, only with the difference that the limits of the integral are modified so as to take account of non-stationarity. There are two main ways to do this:

(1) By integrating from $-\infty$ to some instant of time t_0 and differentiating the result.

(2) By integrating from t_0 to $t_0 + \delta t_0$, or similar, i.e. over a small limited portion of time.

We shall now discuss these two methods in the order mentioned.

Method 1. The so-called *running transform*, *running spectrum*, or *moving spectrum*, is defined as:

$$F(t_0,\omega) = \int_{-\infty}^{t_0} f(t)\,\mathrm{e}^{-i\omega t}\,\mathrm{d}t \qquad [68]$$

assuming $f(t) = 0$ for $t > t_0$. See, for example, Page (1952), Turner (1954), Schroeder and Atal (1962), Loomis (1966), Seyduzova (1970). Applying Parseval's theorem (section 3.3.2) to the running transform [68], we find:

$$\frac{1}{2\pi}\int_{-\infty}^{\infty} |F(t_0,\omega)|^2\,\mathrm{d}\omega = \int_{-\infty}^{t_0} |f(t)|^2\,\mathrm{d}t$$

Differentiation of this expression with regard to t_0 gives:

$$\frac{1}{2\pi}\int_{-\infty}^{\infty} \frac{\partial}{\partial t_0}|F(t_0,\omega)|^2\,\mathrm{d}\omega = |f(t_0)|^2 \qquad [69]$$

using the rule for differentiation with regard to the upper limit of an integral. Here, the right-hand side is the *instantaneous power*, i.e. the power at time t_0, as defined in section 3.3.2. As a consequence of [69] the expression:

$$\frac{\partial}{\partial t_0}|F(t_0,\omega)|^2 \qquad [70]$$

is the *instantaneous power (or energy) spectral density*, again in conformity with the definitions in section 3.3.2. It can be calculated from $f(t)$ by carrying out the differentiation of $|F(t_0,\omega)|^2$, using [68]. However, as pointed out by Turner (1954), [70] does not give a unique representation of the instantaneous power spectral density.

Method 2. An alternative definition of instantaneous spectrum was discussed by Priestley (1965b, 1967). Instead of using the derivative as above, he performs integration over a short time interval, say from t_0 to $t_0 + \delta t_0$, and calls this the *evolutionary spectrum* at time t_0. See also Priestley (1971). A related method is to calculate spectra for successive, overlapping time segments. A presentation of running and instantaneous spectra can be found in the book by Kharkevich (1960).

The *moving-window Fourier spectrum* is an equivalent, time-dependent spectrum defined as:

$$F(t_0,\omega,\Delta t_0) = \int_{t_0-\Delta t_0/2}^{t_0+\Delta t_0/2} f(t)\,\mathrm{e}^{-i\omega t}\,\mathrm{d}t \qquad [71]$$

where t_0 is the center of the window, of length Δt_0. F may be displayed as isolines in graphs with t_0 and ω as axes. This method has been applied by Trifunac and Brune (1970) for analysis of a complex (multiple) seismic event. Similar techniques are applied in some recent studies of seismic surface-wave dispersion (section 7.2). Another equivalent method for computation of the complex running frequency spectrum was developed and applied to short-period geomagnetic variations by Dobeš et al. (1971):

$$F(t_0,\omega,\Delta t_0) = \int_{t_0-\Delta t_0}^{t_0} w(t - t_0)f(t)\,\mathrm{e}^{-i\omega t}\,\mathrm{d}t \qquad [72]$$

where $w(t - t_0)$ is a time window of length Δt_0, and t_0 is the endpoint of the window.

A useful computational method for obtaining the running spectrum of discrete series $f(t)$ is the *complex demodulation*, which can be briefly explained by giving the steps involved:

$$f(t) \rightarrow f(t)\,\mathrm{e}^{-i\omega t} \rightarrow \text{smooth} \rightarrow F(t,\omega) \rightarrow |F(t,\omega)|^2 \qquad [73]$$

The last item represents the running power spectrum.

A method for calculation of time-dependent power spectral density for accelerograms of strong motion has been developed by Liu (1970) under the name of *evolutionary power spectral density.* Like other time-dependent spectra, this is apt to demonstrate any non-stationarity in the process.

On the experimental side, it should be noticed that running spectra are produced by certain spectrographs, which display power in plots with time and frequency as axes (sections 1.6.2 and 1.6.3). Cf. also Noll (1964), Wood (1964) and Kato et al. (1966). This mode of presentation is applicable to any kind of phenomenon and moreover, it is able to give a much more complete information than simple frequency spectra, in which the development with time is lost.

For simplicity, the discussion in this section has been referred to the time and frequency domains. But the same discussion applies to the space–wavenumber domains. Only the terms stationarity and non-stationarity, used for time series, are substituted by homogeneity and heterogeneity, respectively, in case of space series.

Also, it should be mentioned that the concepts developed above are applicable to correlation functions as well. For instance, a running or local autocorrelation function has found some application to spatial gravity studies, especially field zoning (Beryland, 1971).

Chapter 4

COMPUTATION OF SPECTRA OF OBSERVATIONAL DATA

In Chapter 3 we learnt about the basic principles by which spectral analysis of observational data differs from spectral analysis of theoretically given functions. We saw that use of finite record lengths is indispensable and that digitization of records is the dominant procedure, even if not indispensable. If Chapter 3 aimed at answering the important question *Why*? with regard to observational procedures, we shall in the present chapter aim at answering the equally important question *How*? In other words, we are going to describe the operations of observational spectral analysis, methods used, their limitations, errors and other effects, and will end up with computational schemes for the spectral calculation. We shall follow the general order in which the various operations are undertaken in any practical case.

4.1 DIGITAL AND ANALOGUE DATA

4.1.1 Digital versus analogue methods

A main distinction both in data production and in data handling exists between the digital and analogue methods.

(1) Digital methods start from a series of numerical values of $f(t)$. In pre-computer time, harmonic analysis was made by means of special computational schemes, often adapted to the use of a simple desk calculator. In recent time, this method has been replaced by calculation on large electronic computers, using special computer programs.

(2) Analogue methods start from an analogue representation of $f(t)$, in graphical, electrical or optical form. Such methods were briefly reviewed in section 1.6. Among these, practically only those which use the record in electrical analogue form (variable band-pass filtering or sound spectrographs) or in optical analogue form (optical spectrographs) are still of significance in geophysical research, but only of relatively limited application. Other analogue methods are by now more or less only of historical interest.

The vast application of spectral methods in geophysics is mainly a result of the general availability of large electronic computers during the last fifteen years. Their high speed of computation, with suitable programming, and their general

TABLE XIII

Combinations of recording and analysis methods in seismology

Analysis	Recording	
	analogue	digital
Analogue	1a: FM recording on magnetic tape	1b: does not exist
Digital	2a: most common analysis of usual seismic records or analogue magnetic tape records	2b: digital seismograph records (magnetic tape or otherwise)

applicability to any type of computational problem have opened the way for geophysical spectroscopy. Therefore, in the following we shall concentrate our attention to the use of modern computers in the execution of the analysis, which have by now almost completely replaced the earlier used methods. A summary review of the present situation of recording and analysis in geophysics is given in Table XIII.

Observational data are mostly an intermediate product, i.e. the output from a recorder and at the same time the input to a computer or other operation. As output, its form depends on the instrument. On the other hand, as input to a computer or other operation, its form has to comply to the needs set by the computer etc. If the two forms are not identical, another stage enters the operation, i.e., to modify the given output into such a form that it will be accepted by the computer or other device used. In many geophysical branches, this is the common state of affairs, corresponding to item 2a in Table XIII, i.e., the instrumental output (record) is in analogue form, but the computer needs the data in digital form. Hence the problem to digitize records.

4.1.2 Methods of digitization

No doubt, the digitizing process can be considered as a real bottle-neck to be forced in many efforts to construct spectra. Therefore, effort has been put into the problem of digitizing records in a simple and accurate way. This phase of the work has in fact been partially lagging behind other phases of the work. Restricting ourselves to an outline of the main principles and avoiding detailed instrumental descriptions, we can on the whole distinguish between four methods for digitization or analogue-to-digital conversion of a given record. We shall deal with them in order of increasing sophistication.

(1) *Hand digitization*, including the reading of scales or similar. This is the most elementary method, which has been practiced in various forms. Generally

an enlarged record (five to ten times the original) has to be used. In one method two perpendicular scales (for time and amplitude, respectively) are used for direct reading on the enlarged photographic record. In another method, the original record is projected onto a square-millimeter paper on a screen, a trace is drawn and from this the readings are made directly. Several variations of this method exist. The read data have then generally to be transferred into punched cards or punched tape to be used in a digital computer. Hand digitizing may be considered as the least efficient way of performing the conversion, even though its accuracy, when carefully done, compares favourably with other more sophisticated methods.

(2) *Semi-automatic digitization*, using special, mechanical digitizers. To facilitate the digitizing, especially to save time and hopefully also to increase the accuracy, a number of different digitizers have been constructed in various laboratories. By and large, they agree in basic principles but may differ in constructional details. Quite a common type is the drum digitizer, in which the record is placed on an ordinary seismograph recorder drum and is read by means of a microscope. The readings are automatically converted into electrical analogue form and connected to a digital voltmeter for punching. Digitizers of this general type are described by Belotelov and Rykunov (1963), Matumoto (1960), Wickens and Kollar (1967). Most such equipment can be built within limited observatory budgets, like the one constructed by Howell (1966a). As far as accuracy and speed are concerned, they all compare very well with commercially available, more expensive equipment, usually constructed on the same basic principles. The most practical solution, considering the extensive and intensive need for digitizing, is probably that computer centers are also equipped with general-purpose digitizers. Photogrammetric instrument systems, frequently used in physical geography for map readings etc, offer an advantageous method for digitization of all kinds of records, including seismograms. This represents the most recent (1972) efficient solution of digitizing problems at Uppsala, by means of manual curve following (on an enlarged trace on a TV screen) and automatic digitizing and tape-recording with preset sampling intervals. This method needs no record magnification.

(3) *Fully automatic digitizers*, using electronic equipment. In equipment of this type, the manual work entering into items 1 and 2 above has been eliminated and the procedure is fully automatic. We will mention three examples of this method from the literature. Adams and Allen (1961a) describe a method for automatic digitization of seismic records, originally developed for digitizing weather maps. The seismogram, recorded as a transparent trace on opaque background, is scanned by a light-ray in a programmed manner. Each time the light-ray crosses the seismogram trace, this is sensed by a cathode-ray tube and the coordinates of that point are stored. Also multiple-trace records can be digitized in this way, provided the traces do not overlap. The method has proved to be both accurate and fast.

Bogert (1961a) performs automatic digitization of seismic records with electronic equipment, earlier developed for processing speech and visual data. An automatic digitizer, specially developed for records of electromagnetic micro-variations, has been reported by An (1965), who also discusses some principle errors committed in digitizing continuous curves.

(4) *Digitized outputs from recording instruments.* The digitizing work would be eliminated completely if the instrumental output could be produced in digital form, so as to be directly usable in a computer. A few such seismograph types have been developed but they have found only limited application.

In a study of seismic noise, Haubrich and Iyer (1962) use a digitally recording equipment, where the output is stored on punched paper tape. Spectral analysis and other operations are then performed directly by a digital computer. Digital recording of seismic waves and of ocean waves were also made by Haubrich et al. (1963) for the purpose of spectral calculations.

In the direct digitizing seismograph of De Bremaecker et al. (1962), the output of a displacement capacitance transducer is frequency-modulated and a frequency counter gives the equivalent digital form. Frequency modulation is preferred over amplitude modulation, because frequency counters are accurate and less expensive than digital voltmeters of the same accuracy. The digital record can be obtained in any desired form: printed, punched, or magnetic tape. Another development of this kind is the Pasadena digital seismometer (Miller, 1963; Phinney and Smith, 1963; S. W. Smith, 1965). It uses digital voltmeters and magnetic-tape recording. See also Burke et al. (1970).

In the digital seismograph constructed by Aki et al. (1965), direct digitizing of seismometer output is not used, but instead the output of a galvanometer connected to the seismometer is digitized. In this way, the galvanometer serves as a low-pass filter, practically eliminating aliasing (section 4.3). Digital seismographs have made it possible to accumulate amplitude and phase spectra on a routine basis, which is certainly useful in providing a library for research. A seismograph system for field work with digital tape recording has been devised by Allsopp et al. (1972).

In discussing various techniques of digitization in this section, our attention has been mostly focussed on seismograph records. The reason is that from the literature this branch appears to be the one which has received most attention and where most developments have been made. But, naturally, the methods listed above have general applicability to any kind of geophysical records, with minor modifications as the case may be.

4.1.3 Magnetic-tape recording

Analogue analyses, including filtering, spectral analysis, rotation of axes, identification of phases, using analogue magnetic-tape records, have been fully described by Sutton and Pomeroy (1963).

Magnetic-tape recording, especially in digital form, has several advantages over the conventional analogue photographic recording, which can be summarized in the following points:

(1) More information stored, because of large dynamic range and recording in a broad frequency band.

(2) Reproducibility of records in various ways, e.g. band-pass filtering, application of various corrections, combination of different traces, etc.

(3) Convenience of operations, especially by automatic data handling, and good precision, both in recording and computations, such as of spectra, correlation functions etc. Variable narrow-band filtering may well be used for construction of spectra from tape records (see, for example, Willis and Johnson, 1959; Grossling, 1959; Milne, 1959).

Even though digital seismographs (see, for example, Miller, 1963; De Bremaecker et al., 1963) and magnetic-tape recording, especially in digital form, are gaining an increasing application, we consider it still very important to study in detail the derivation of spectra from the common analogue, generally photographic, seismograph records. This is even more important, considering that a full understanding of this procedure is necessary for a complete appreciation of the numerous precautions to be taken in calculating spectra from seismic records of any kind, including every more or less automatic procedure. Moreover, the ordinary paper records are still in such a vast majority, that for any research of wider coverage, recourse has to be taken to these records. These remarks, here made with reference to seismic records, can be immediately extended to cover all types of geophysical records to be dealt with in this book.

4.2 ERRORS OF DIGITIZATION

As in every physical measurement, there are a number of errors involved in the digitizing process, which can affect the computed spectrum. In this section we shall consider the major sources of error in the practical procedure and the possibilities to remedy their effect. Limitations on the digitizing accuracy are set by the quality of the record used, by the digitizing apparatus as well as by the measuring capability of the operator himself. In addition to measuring inaccuracies, we have also to consider that there are limitations inherent in the nature of the digitization process as such.

To assess errors introduced as a result of imperfect digitization, comparisons have to be made between procedures, where only one parameter is varied at a time. As parameter in this context we consider every factor which is of influence on the digitizing process, such as digitizing apparatus or technique used, the person who executed the digitizing (comparison made both between different persons and the same person under different situations), the quality of record used, etc. Comparisons can be made both with the use of geophysical records as well as of theoretical curves. A recent detailed and useful examination of errors in digitization, with special reference to strong-motion accelerograms, has been given by Trifunac et al. (1973).

4.2.1 Errors in fitting given curves by digital data

Digitizing a given continuous curve implies its replacement by discrete data points, equally spaced along the abscissa. By means of the given digits, the original continuous curve may be approximated in various ways, of which Fig.30 illustrates the three most common. That digitizing involves certain errors, i.e. deviations from the given, true curve, is immediately obvious. In order to assess such errors let us consider the case when the given curve is approximated by sloping straight lines through the measuring points, with slopes equal to those of the given curve.

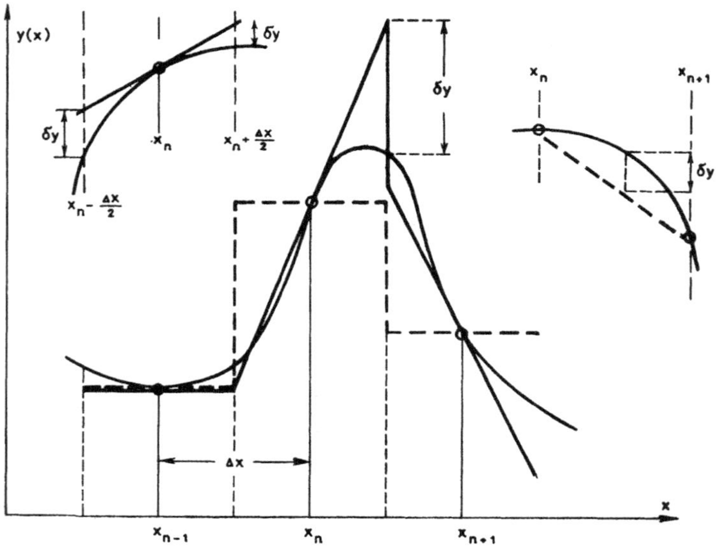

GIVEN CURVE

APPROXIMATION BY STEP CURVE (ZERO SLOPE)

APPROXIMATION BY TANGENTIAL STRAIGHT LINES

APPROXIMATION BY STRAIGHT-LINE CONNECTIONS

Fig.30. Various approximations to a continuous curve by digitized data.

With reference to Fig.30 (especially the inset upper left-hand figure) we then have the following equations:

(1) For the straight line:

$$y(x) = y(x_n) + (x - x_n)\left(\frac{\mathrm{d}y}{\mathrm{d}x}\right)_n$$

(2) For the given curve, including second-order derivative:

$$y(x) = y(x_n) + (x - x_n)\left(\frac{\mathrm{d}y}{\mathrm{d}x}\right)_n + \frac{(x - x_n)^2}{2}\left(\frac{\mathrm{d}^2y}{\mathrm{d}x^2}\right)_n$$

The ordinates in the point $x_n + \Delta x/2$ become:

(1) For the straight line:

$$y\left(x_n + \frac{\Delta x}{2}\right) = y(x_n) + \frac{\Delta x}{2}\left(\frac{\mathrm{d}y}{\mathrm{d}x}\right)_n$$

(2) For the given curve:

$$y\left(x_n + \frac{\Delta x}{2}\right) = y(x_n) + \frac{\Delta x}{2}\left(\frac{\mathrm{d}y}{\mathrm{d}x}\right)_n + \frac{1}{2}\left(\frac{\Delta x}{2}\right)^2\left(\frac{\mathrm{d}^2y}{\mathrm{d}x^2}\right)_n$$

The difference between (1) and (2) yields the sought expression for δy:

$$|\delta y|_{\max} \simeq \frac{(\Delta x)^2}{8}\left|\frac{\mathrm{d}^2y}{\mathrm{d}x^2}\right|_{\max} \qquad [1]$$

Clearly, the larger the curvature, i.e. $y''(x)$, the larger the error, which is also clearly shown in Fig.30. The same expression holds with good approximation for the case of straight-line connections between the discrete points (inset upper right-hand figure in Fig.30). For other methods of reproducing the given curve from the discrete data, similar relations can be derived.

Of the three methods envisaged in Fig.30 to reproduce a given curve from digitized data, the straight-line connection method may be preferred. It has generally higher accuracy than the step curve, which emanates from our discussion above, and it does not require any knowledge of slopes as the tangential straight-line method does.

The errors dealt with so far derive from lacking coincidence between the curve constructed from the discrete data points and the continuous curve. But in addition we have to consider errors due to limited accuracy in the digitizing method used. These can be split as follows:

(1) Errors in timing, i.e. in Δx, by which the measured Δx can be taken as the sum of the true Δx, denoted $\Delta_0 x$, and an error δx:

$$\Delta x = \Delta_0 x + \delta x \qquad [2]$$

(2) Errors or quantization in amplitude readings; if A is the quantization interval, then this error leads to an uncertainty of:

$$|\delta y|_{max} = A/2 \qquad [3]$$

Both errors due to quantization and to limited accuracy are equivalent to random noise and can be treated statistically as such (Schiff and Bogdanoff, 1967; Manzoni, 1967; Gold and Rader, 1969, chapter 4).

Combining [1] and [3], we get an instruction how to choose our Δx and A, in order to get errors δy at most equal to some preassigned value Δy:

$$|\delta y|_{max} \simeq \frac{(\Delta x)^2}{8} \left| \frac{d^2 y}{dx^2} \right|_{max} + \frac{A}{2} \leq \Delta y$$

Rewriting this with reference to our time function $f(t)$, we have:

$$|\delta f|_{max} \simeq \frac{(\Delta t)^2}{8} |f''(t)|_{max} + \frac{A}{2} \leq \Delta f \qquad [4]$$

If the reading accuracy A is set by the digitizing equipment, this equation tells us the maximum Δt we could use for a given curve in order to have no errors exceeding Δf. This will have to be enough as regards analytical expressions for digitizing errors. We shall proceed by investigating the effects for given functions $f(t)$, both theoretical and observational.

4.2.2 Tests with theoretically given curves

For a curve whose mathematical form is given the corresponding spectrum can be analytically calculated. Therefore, the spectrum obtained by digitization and Fourier-transformation can be compared with the theoretically computed spectrum. This therefore yields an absolute test on the digitizing process.

Even though this provides for an immediate check on digitization accuracy, Kulhánek and Klíma (1970) preferred to let the signal pass through a seismograph to get a trace record and better simulate seismological reality. Among several processes they used the following two:

(1) a) Assume ground motion in analytical form $f(t)$.

b) Calculate analytically ground amplitude spectrum $F(\omega)$.

c) Apply instrumental response $I(\omega)$ to get analytical trace amplitude spectrum $G(\omega)$.

In short: $f(t) \rightarrow F(\omega) \rightarrow F(\omega) \cdot I(\omega) = G(\omega)$.

(2) a) Assume again ground motion in the form $f(t)$.

b) Calculate analytically record trace $g(t)$ from the equation:

$$g(t) = \frac{1}{2\pi} \int_{-\infty}^{\infty} G(\omega)\, e^{i\omega t}\, d\omega = \frac{1}{2\pi} \int_{-\infty}^{\infty} F(\omega)\, I(\omega)\, e^{i\omega t}\, d\omega \qquad [5]$$

c) Digitize trace $g(t)$ manually.

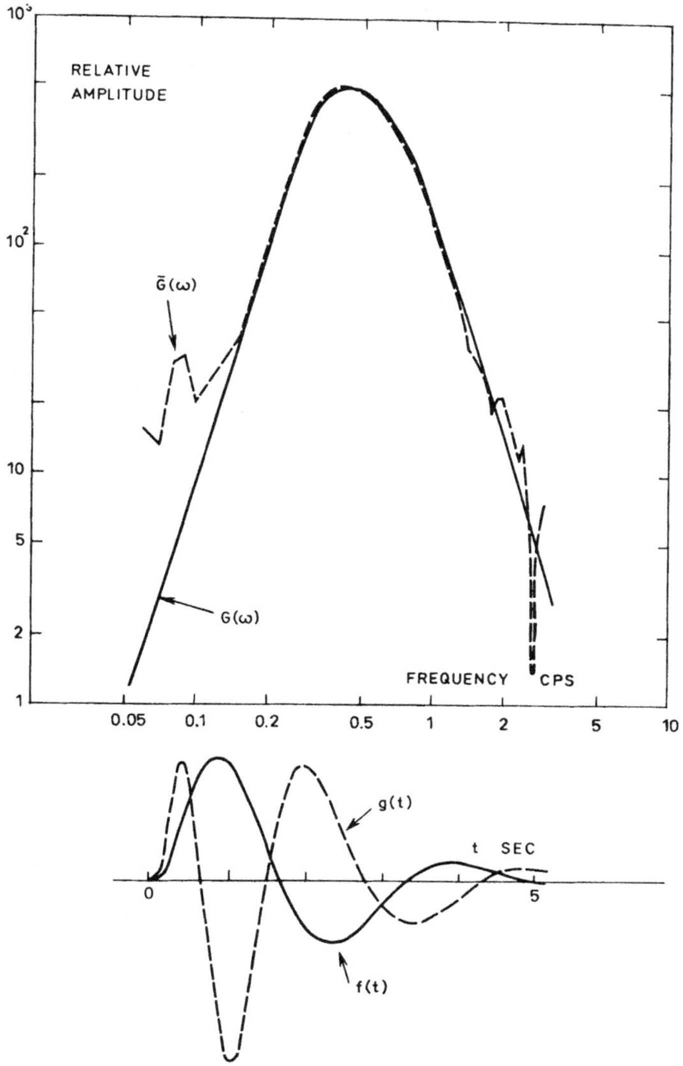

Fig.31. Comparison of spectra for a Berlage pulse, calculated analytically $G(\omega)$ and determined from digitized data $\bar{G}(\omega)$. After Kulhánek and Klíma (1970).

d) Calculate trace spectrum $\bar{G}(\omega)$.

In short: $f(t) \rightarrow g(t) \rightarrow$ digitize $\rightarrow \bar{G}(\omega)$.

The two spectra $G(\omega)$ and $\bar{G}(\omega)$ differ in the respect that digitizing enters into the calculation of $\bar{G}(\omega)$ but not of $G(\omega)$. Therefore, differences between $G(\omega)$ and $\bar{G}(\omega)$ are expected to give an idea of digitizing errors.

Kulhánek and Klíma (1970) used the Berlage function (Fig.31):

$$f(t) = t\,e^{-at} \sin \omega_0 t \qquad [6]$$

whose spectrum is (section 2.4.3, example 27, Table V):

$$F(\omega) = 2\omega_0 z/(z^2 + \omega_0^2)^2 \qquad [7]$$

where a, ω_0 are constants and $z = a + i\omega$.

For the case studied by Kulhánek and Klíma (1970), it is seen that deviations occur both at high and low frequencies relative to the maximum, and the most reliable band extends from about 0.15 cps to 1.5 cps, i.e. approximately symmetrical around the maximum on the logarithmic frequency scale. At least qualitatively, this can be understood as follows. Around the maximum recorded amplitudes, percentage errors in digitized samples are at their minimum. But for considerably smaller amplitudes (in this case about 6% of the maximum and lower), percentage digitizing errors become considerable. In addition, comparison of traces can be used to assess digitization errors, i.e., on the one hand $g(t)$ obtained analytically from $f(t)$ by applying instrumental response, on the other hand $\bar{g}(t)$ obtained by applying instrumental response to digitized $f(t)$.

4.2.3 Tests with observational curves

With an observational curve, e.g. a seismogram, comparison can be made between spectra obtained by repeated digitization of the same record, by different techniques or different persons, etc. The comparisons thus rendered possible will only yield a relative measure of the precision, as the absolutely true spectrum is not known in this case. Tests with repeated machine-digitization of short-period P-wave records exhibit differences essentially only for frequencies below 0.5 cps and above 5 cps. An absolute measure of accuracy can be provided by reproducing the original record from the digital data and then comparing the result with the originally given seismogram (Fig.32).

One field where much attention has been focussed on digitization errors is engineering seismology, where often large amounts of accelerograms have to be digitized for further analysis. See for example, Schiff and Bogdanoff (1967), Nigam and Jennings (1969). Lessons from their investigations have certainly much wider application, and precautions studied have to be taken into account in

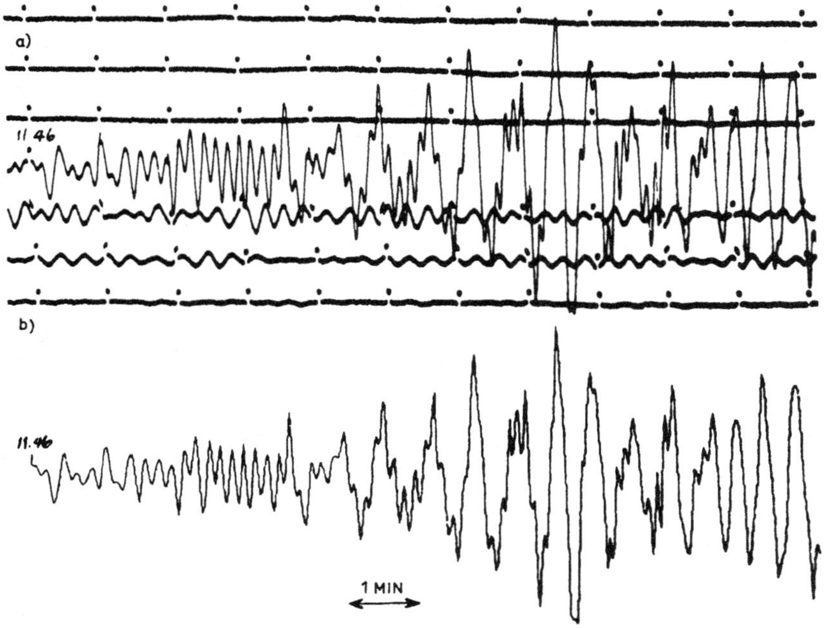

Fig.32. Comparison of a) an original record and b) a record reproduced from digitized data, showing good agreement. After Crampin and Båth (1965).

digitization of any kind of records. We shall first consider two factors, depending upon the character of the records to be analyzed.

(1) *Record distortion.* In measurements from geophysical records we assume that these keep the assigned minute lengths exactly, for instance, that on short-period seismic records minute lengths are exactly 60 mm, or 1 mm = 1 sec. Often the records do not keep this rate perfectly exact, due to recording drum speeds and quality of photographic paper, etc. But even extreme deviations from this rate, i.e. minute lengths of 58.5 to 61.5 mm, do not produce deviations exceeding 0.15 cps in the frequency range of 1–5 cps. These deviations are negligible. Moreover, if desired, the deviations mentioned can be wholly eliminated if duly accounted for in the magnification of the original record, which is often done before digitizing. On the other hand, the photographic enlargement (five to ten times) can in itself produce distortions of the original record. The quality of this enlargement can be tested by making a corresponding enlargement of a millimeter scale and measuring it. Schiff and Bogdanoff (1967) mention a case with accelerograms where paper distortion may have caused serious errors.

Special care is required in digitizing records written by a pen on an arm, which produces circular arcs instead of deflections perpendicular to the time axis. Examples are old-time direct-recording instruments (mostly on smoked paper) as well as some modern types of recorders. For each amplitude read, appropriate time corrections have then to be applied. If, for smaller deflections, such corrections are omitted, spurious frequencies will be introduced into the calculated spectrum.

(2) *Trace characteristics.* Photographic traces necessarily exhibit some line thickness, further augmented by the enlargement made for the digitizing process. To estimate the center line of such a trace may be rather difficult, especially at sharp turning points (maxima and minima) on short-period records, where the thickness is further enhanced. For such reasons it is always advisable to digitize both the upper and the lower edge of the trace, and to take the average of the two sets of measurements, before computation is made. Usually, all spectra (both from the top trace, the bottom trace and the average trace) agree well. When digitization errors become larger (i.e. outside the range 0.5–5 cps), the three spectra begin to diverge, but then the one corresponding to the average trace is generally also an average of the other two spectra.

4.2.4 Zero-line characteristics

In trying to further explore the reasons for digitizing errors, both absolute and relative, we have to consider that the measurements in principle involve readings along two perpendicular scales: a time axis and an amplitude axis. Errors in either of these axes will lead to digitizing errors; for instance, the amplitude measurement depends both on the zero- or base-line used and on the character of the trace measured.

We distinguish between several types of zero-line characteristics, each of which will falsify our results if precautions for their elimination are not taken. One type is due to a constant displacement of the zero-line parallel to itself (item 1 below). Another type is due to misalignment of the digitizer base-line in relation to the record zero-line, introducing effects of relative slope (item 2). The third type is due to the character of the record itself, showing a disturbed zero-line of natural origin (item 3). We shall discuss the three types of zero-line characteristics in this order.

(1) *Displacement of zero-line by a constant amount.* For convenience of measurement, this procedure is a rule in digitizing operations, i.e., the digits of $f(t)$ are not measured from the true zero-line but from a parallel line, chosen as arbitrary zero-line and located such that all digits will have the same sign. We shall now investigate the effect of such a constant displacement superposed on the measurements. In the time domain it may be said to represent a d.c. component. This is also true for the corresponding Fourier series expansion. The constant shift will only affect the constant (d.c.) term in this expansion. Cf. section 2.1.1. We have to envisage the adopted zero-line as well as the record extending to infinity in both directions.

In case of the Fourier transform, however, the situation will be different

(Buchbinder, 1968). Then, both the record and the adopted zero-line are zero outside the measured record length. This means that we are in fact not measuring just the record, but the record plus a rectangular (box-car) function of the same length as the record. Fourier transforming such measurements we get the calculated spectrum superposed by a sinc-function (cf. example 1, Table V, section 2.4.3), or in formula:

$$[f(t) + a_0] \cdot w(t) = f(t) \cdot w(t) + a_0 \cdot w(t) \leftrightarrow 1/2\pi\, F(\omega)\star W(\omega) + a_0\, W(\omega) \qquad [8]$$

where $f(t)$ is measured from the true zero-line, a_0 is the constant shift of the displaced zero-line, and $w(t)$ is the rectangular time window. In the Fourier transform case, we cannot consider this shift as a d.c. component, as it will affect the spectrum. As the sinc-function is broader, the shorter the time window is, we understand that for shorter time windows, a larger portion of the calculated spectrum will be affected.

Correction for the displaced zero-line can be made either in the time or the frequency domain, but as a rule it is most convenient to make it in the time domain, before further calculations are done. This rule also holds for the following cases of zero-lines which need adjustment.

(2) *Misalignment of scales.* If the zero-line used in digitizing a record is sloping or deformed in some way, this will affect the calculated spectrum. Obviously, the function analyzed is not just the true record $f(t)$ but in addition the time-varying zero-line $f_1(t)$, i.e., the function analyzed is $f(t) + f_1(t)$. The corresponding spectrum will then be $F(\omega) + F_1(\omega)$, which deviates from the true spectrum $F(\omega)$. For given functions $f_1(t)$ it is an easy exercise to calculate what the spectral modification $F_1(\omega)$ will be. Such examples have been given by Gratsinsky (1962). In practice, the problem may be that $f_1(t)$ is not known beforehand but needs to be found out.

The effect of a straight but sloping base-line can be seen as follows, first in the time domain. A slope a implies that instead of true readings $f(t)$ we obtain false readings $f(t) - at$. As explained in section 2.1.2 in discussing the saw-tooth curve, the term at can be expanded in a Fourier series as follows:

$$at = 2a\left[\frac{\sin t}{1} - \frac{\sin 2t}{2} + \frac{\sin 3t}{3} - \ldots\right] \qquad [9]$$

This means that the slope term at will cause contamination of the frequencies involved in $f(t)$ and thus lead to erroneous results. As expected, we see from [9], that the contamination from at will be the smaller, the smaller the slope a is, and will vanish completely for $a = 0$.

Correspondingly, in the frequency domain the signal spectrum $F(\omega)$ will be contaminated by a spectrum $F_1(\omega)$ due to the sloping base-line. The resulting spectrum $F(\omega) + F_1(\omega)$ will be incorrect, the bigger the slope, i.e. the bigger $F_1(\omega)$ is. Considering example 21, Table V, section 2.4.3, it is clear that low frequencies will be most affected by such slope effects.

A useful exercise to estimate the effect of slope in any given case is to repeat the digitization and spectral calculation process with a series of artificially introduced, known slopes. Cf. Kasahara (1957). In the digitizing of seismic records, it may frequently be difficult to assess the exact zero-line. Especially for signals of short duration, it is easy to introduce an incorrect slope of the zero-line, because the record interval is too short to define it well. Then, it is advisable to determine the exact zero-line from a longer record, extended on both sides of the signal to be analyzed.

Irregularity of the zero-line in accelerograms, mostly used for calculation of response spectra in earthquake engineering, has been noted by several observers. Part of the trouble is due to the fact that most accelerometers are triggered by the earthquake, and that a zero-line prior to the recording is missing. Wiggins (1962)

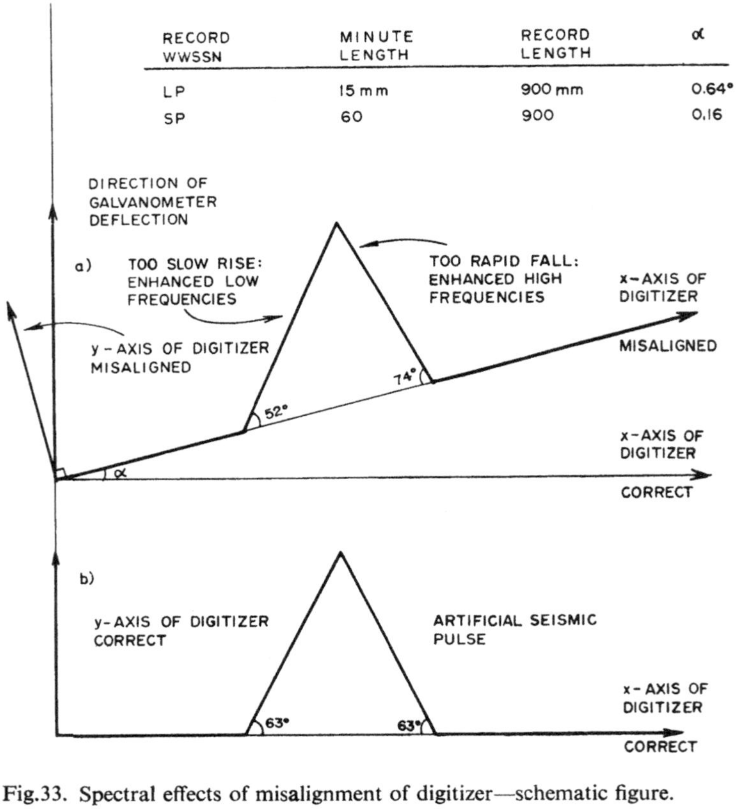

Fig.33. Spectral effects of misalignment of digitizer—schematic figure.

estimated such effects, due to false impulses and false superposed long-period oscillations, on the computed spectra and found their greatest influence to be in the long-period range. Similar results are reported by Irikura et al. (1971), who corrected the accelerograms by a second-degree polynomial with constants determined under various different assumptions.

Special care is required in the case of drum recording, common in seismology, where the record is truly a helix and thus not strictly perpendicular to the galvanometer deflection. In an endeavour to avoid trends and misalignments of the digitizer, this is usually aligned with its x-axis parallel to the trace. This will then lead to inaccuracies in the spectra (enhanced low and high frequencies, as illustrated in Fig.33), and the true spectrum would be obtained if instead the y-axis were aligned in the direction of galvanometer deflection. With drum recording it is case a) in Fig.33 which happens, not the ideal case b), and the procedure is to (1) align digitizer correctly as indicated (guided by special galvanometer test swings), or (2) apply linear detrending. This error source is often thought to be negligible, but as demonstrated by James and Linde (1971) the spectral errors may be appreciable, especially for long-period (LP) instruments (even if their α is only about 0.3°); see Fig.34. As readings are always taken perpendicular to the digitizer x-axis, there are also second-order angular effects entering in the case of a sloping zero-line.

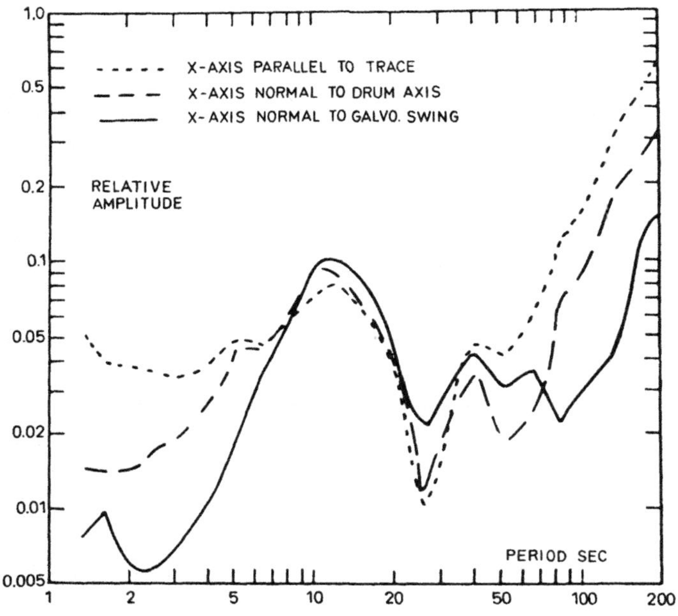

Fig.34. Spectral effects of misalignment of digitizer—actual cases demonstrating the effects shown in Fig.33. After James and Linde (1971).

The situation is slightly different if the base-line itself exhibits some undulations. Without calculation it is clear that such undulations will falsify a spectrum by contamination from the spectrum of the base-line (cf. Manley, 1945, p.249). Such effects may be due to faults in the digitizing equipment and can then only be removed by checking this equipment against a straight line. Instead of correcting measurements, the best procedure then is to discard or adjust the digitizing apparatus. For example, if the base-line shows a superposed displacement of rectangular shape of limited duration, this will naturally lead to a superposition of its corresponding sinc-function in the computed spectrum. Cf. item 1 above.

(3) *Unwanted fluctuations in the record.* In addition to poor base alignment or poor digitizing apparatus, similar effects may be introduced by the nature of the record itself. This happens when the signal to be analyzed is superimposed on a wave motion of much longer period. Such cases are common in meteorology, e.g. in analyzing daily temperature variations over a interval of a couple of months or so, when the daily variation is superimposed on part of the annual variation. It may also happen in seismic records, for example, when a short-period P-wave is riding on long-period surface waves from a preceding earthquake, or if there has been any other kind of slow zero-line drift of the seismograph trace where the P-wave arrives. Other examples are provided by disturbance from earth tides in records of the earth's free oscillations (Bolt and Marussi, 1962; Satô et al., 1963b) or disturbance from instrumental drift in earth-tide records (Pan, 1971). In any case, the long-period trend should be removed before spectral analysis is made or otherwise duly corrected for.

Due to the greater variety of base-line distortions under this item than generally found under item 1, methods for their removal, called *detrending*, will also have to be of greater variability (Robinson, 1967b, p.22; Bendat and Piersol, 1971, p.288). We give here below some examples of detrending methods in geophysics, in a general order of increasing sophistication.

Linear-trend removal is applied by Mizoue (1967) to observations of secular crustal movements, and by Márton (1970) to archeomagnetic data. Winch (1965) developed a similar method for elimination of non-cyclic magnetic variations.

Polynomials have been developed for the elimination of unknown trends in harmonic analysis of sea level, ocean tides and crustal deformations. In this case, a polynomial, preferably an orthogonal Chebyshev polynomial, is added to the harmonic expansion (Munkelt, 1959; Horn, 1960; Popov and Chernyavkina, 1960; Rossiter, 1967). This method may have a more general applicability. See also Lomnicki and Zaremba (1957), Rinner (1960), Zetler (1960) and Rossiter (1962). Centerline adjustment by a second-degree polynomial was developed by Schiff and Bogdanoff (1967) with application to accelerograms of strong-motion earthquakes. In this case, parabolic base-line correction has been customary, but has recently been improved, especially for longer periods, by a method based on successive filtering (Trifunac, 1971b).

Fourier-series elimination is useful when data consist partly of periodic components, partly of non-periodic components, i.e.: (1) one part which has clear physical relations, with known periods, to certain generating factors, e.g. gravitational effects; (2) another part, which is much less predictable and which depends on various modifying factors to the effect 1.

Examples of such phenomena are the tides, both atmospheric, oceanic and solid earth. Other geophysical examples can be envisaged, including all cases where at least part of the observed phenomenon is due to a well-defined source, with well known periodical properties.

The observed quantities $f(t)$ thus consist of two parts:

$$f(t) = f_1(t) + f_2(t) \qquad [10]$$

$f(t)$ is expanded into a Fourier series and this series is subtracted from the given data $f(t)$. The Fourier expansion essentially corresponds to $f_1(t)$, and the difference will yield $f_2(t)$; in short:

$$f_2(t) = f(t) - f_1(t) \qquad [11]$$

The residual $f_2(t)$ is then subjected to spectral analysis. The advantage with this procedure is that minor variations in $f_2(t)$ can be subject to more accurate analysis than when they are hidden by the simultaneous $f_1(t)$. Such a method was applied by Dobson (1970) in evaluating stereo-photographic observations of the ocean surface for ocean-wave spectral studies, using a Fourier expansion with ten harmonics for the elimination of $f_1(t)$.

Two-dimensional detrending by least-square methods, especially for an ocean surface, is described by Kinsman (1965, pp.462–463), and for interpretation of gravity and magnetic anomalies by Bulakh (1970). Application of spherical harmonic expansions for detrending of global magnetic fields is discussed by Bullard (1967).

4.2.5 Unequal digital spacing

Sometimes, unequal time spacing is used on purpose, with denser points near sharp bends of the curve. This can lead to a more accurate reproduction of the given curve than equal spacing. The question of equal or unequal spacing has several aspects which can be summarized as follows:

(1) Depending on digitizing method used: Equal spacing is practically a necessity when dealing with magnetic-tape records or in automatic digitization, whereas in manual digitization it is optional.

(2) Depending on frequencies present in the record: In case of equal spacing the digitizing interval should be $\Delta t \leq 1/2v_N$ where v_N is the highest frequency

present (section 4.3). For unequal spacing there is no such rule, except that Δt can vary along the trace according to variation of v_N. This is just another way of saying that Δt should be smaller at sharp bends, which include high frequencies, and that it can be larger along more straight portions, corresponding to lower v_N.

(3) Depending upon accuracy of reproduction: Unequal spacing can lead to more exact reproduction, provided a skilfull choice of reading points is applied, than readings at equal intervals can do. That is, the unequal spacing can lead to more accurate reproduction with a much smaller amount of readings. This follows naturally from point (2) above.

(4) Depending upon subsequent calculations: For most calculations, it may be simplest to have the readings at constant, equal intervals. However, the inconvenience of unequal intervals has to be weighed against the smaller number of necessary digits.

The methods for spectral calculation we shall give later (sections 4.5 and 4.6) assume constant digitizing intervals. But with a given curve $f(t)$ digitized at unequal intervals, the procedure to calculate its spectrum $F(\omega)$ can proceed along any of the following two alternative ways:

(1) Produce equispaced digits from the given unequally spaced digits, before spectra calculation. The equalization of digital intervals is usually effectuated in any of the following two ways:

(a) By linear interpolation: applicable when digits are so close that straight-line segments can be used to connect the points with enough accuracy.

(b) By sinusoidal interpolation: sometimes used, e.g. on accelerograms, when only maxima and minima in the records have been read. Then, the sinusoidal curve gives a better approximation to the true curve than straight-line interpolation. See Beaudet and Wolfson (1970). Alternatively, polynomial interpolation can be applied.

(2) Calculate the Fourier spectrum directly from the unequally spaced digits.

An example of the latter procedure is when $f(t)$ is approximated with straight-line segments of unequal lengths. Applying the derivation theorem (section 2.3.6):

$$f''(t) \leftrightarrow -\omega^2 F(\omega) \qquad [12]$$

we get the following expression for $F(\omega)$:

$$F(\omega) = -\frac{1}{\omega^2}\int_{-\infty}^{\infty} \frac{\mathrm{d}f'}{\mathrm{d}t} e^{-i\omega t}\,\mathrm{d}t = -\frac{1}{\omega^2}\sum_{n=0}^{N} (f'_{n+1} - f'_n)\,\mathrm{e}^{-i\omega t_n} \qquad [13]$$

where f'_n is the slope of the nth segment with endpoint at t_n. This method, which eliminates the need for equispaced points with prescribed digitizing interval, has been applied to spectrum analysis of transient signals by Pease (1967).

By analogy, using the derivation theorem for the first derivative $f'(t)$:

$$f'(t) \leftrightarrow i\omega F(\omega) \qquad [14]$$

we envisage the following formula for calculating $F(\omega)$, also from unequally spaced digits:

$$F(\omega) = \frac{1}{\omega} \sum_{n=0}^{N} (f_{n+1} - f_n) \, e^{-i/2[\omega(t_{n+1}+t_n)+\pi]} \qquad [15]$$

In [13] slopes have to be calculated, which is not needed in [15], whereas the exponential factor is slightly more complicated in [15] than in [13]. The relative merits of [13] and [15] can be tested on samples, theoretical or observational, and this is left as an exercise to the reader.

Methods for spectral calculation have been developed for special kinds of unequal spacing, as for example when the readings (of time) are taken at the zero-points of the given function $f(t)$ (Brillinger, 1968). With regard to evaluation of accelerograms of strong-motion earthquakes, problems of unequal spacing have been investigated by Nigam and Jennings (1969).

4.3 EFFECTS OF DIGITIZATION—ALIASING

4.3.1 Sampling theorem

In section 3.5.1 we have seen that the record length T defines the fundamental

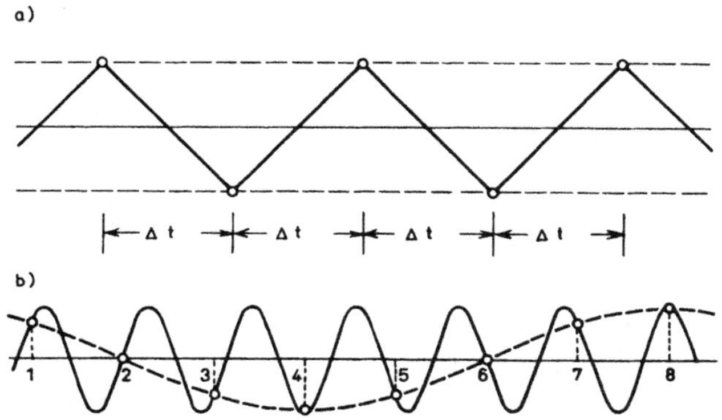

Fig.35. Digital sampling. a) Sketch demonstrating intuitively the need for at least two points per period. b) A well-known sketch demonstrating the effect of aliasing. By varying the sampling interval and the starting point, it is easily seen that different results will be obtained, regarding both amplitude, frequency and phase. Cf. also Brier (1968).

period or the lowest frequency in the spectral analysis. At the other end, the digitization sampling interval Δt defines the highest resolvable frequency in the spectral calculation. Intuitively, we understand that at least three points (three digits), i.e. two time intervals $= 2\,\Delta\, t$, are the minimum of information needed to define one period (Fig.35). This means that the shortest period which can be detected is $2\,\Delta\, t$, or that the highest resolvable frequency is $1/(2\,\Delta t)$. This frequency limit for the calculated spectrum is referred to as the *folding frequency* or the *Nyquist frequency* (after Nyquist, 1928), sometimes as the *Shannon limit*. In other words, the folding frequency is equal to one-half the *sampling frequency*, $1/\Delta t$.

In a more exact way, the *sampling theorem* can be seen as follows. Assume that $F(\omega)$ has a cut-off frequency ω_N such that $F(\omega) = 0$ for $|\omega| \geq \omega_N$, in other words ω_N is the highest frequency of the spectrum $F(\omega)$. Then the spectrum can be represented by multiplication of $F(\omega)$ with a rectangular function, of which the inverse transform can be formed (see example 2, Table V, section 2.4.3):

$$F(\omega) = F(\omega) \cdot \Pi\left(\frac{\omega}{2\omega_N}\right) \leftrightarrow f(t) \star \frac{\sin \omega_N t}{\pi t} = f(t) \star \frac{\omega_N}{\pi} \operatorname{sinc} \frac{\omega_N t}{\pi}$$

$$= f(t) \star \frac{1}{\Delta t} \operatorname{sinc} \frac{t}{\Delta t} = \sum_n f(t_n) \frac{\sin \frac{\pi}{\Delta t}(t - t_n)}{\frac{\pi}{\Delta t}(t - t_n)} = f(t) \qquad [16]$$

This expresses the sampling theorem: For $F(\omega)$ with a cut-off at ω_N, the function $f(t)$ is fully specified by its values at equal spacing Δt, where:

$$\Delta t = \pi/\omega_N = 1/2v_N \qquad [17]$$

For a fuller discussion of the sampling theorem, the reader is referred to standard textbooks, e.g. Papoulis (1962), Panter (1965), who produce proofs similar to each other. Monroe (1962, pp.68–72), Bracewell (1965, pp.192–193) and Kinsman (1965, pp.443–444) give other very clear expositions of the proof, including instructive illustrations.

4.3.2 Frequency limits of spectra

The rules for the frequency limits as caused by limited record length T and finite digitizing interval Δt, can be summarized in the following rules:

(1) The record length T defines the lower frequency limit in the spectrum: $v_1 = 1/T$.

(2) The digitizing interval or sampling interval Δt (also called the *Nyquist interval*) defines the upper frequency limit in the spectrum: $v_N = 1/(2\Delta t) = N/2T$, where $N + 1$ is the number of digits. Contamination of computed spectra by

TABLE XIV

Resolution in dependence on record length

Function	Range	Resolution (fundamental frequency)
$f(t) \leftrightarrow F(\omega)$	$-T/2 \leq t \leq T/2$	$1/T$
$C_{11}(\tau) \leftrightarrow E_{11}(\omega)$	$-T \leq \tau \leq T$	$1/2T$

frequencies higher than v_N is termed *aliasing*, sometimes *spectrum folding* (section 4.3.3).

(3) The spectral resolution power, i.e. the difference between successive frequencies analyzed, is $\Delta v = 1/T$. This is called a *Nyquist co-interval.*

The rules (1) and (2) can be summarized by stating the limits of frequency:

$$1/T \leq v \leq N/2T \qquad [18]$$

or that:

$$v = m/2T$$

where $m = 2, 3, \ldots, N$.

The lowest (fundamental) frequency, which defines the spectral resolution (i.e. interval between successive frequencies), is equal to 1/record length. This has some consequence for a Fourier transform as distinct from its power spectrum, obtained by transforming the autocorrelation function. For a record length of T, the range of the variable τ in $C_{11}(\tau)$ is obviously $2T$. Hence we get the results summarized in Table XIV.

Before proceeding we shall give the following comments on these rules. The two frequency limits are related by the equation $(N/2)v_1 = v_N$. For simplicity, let us consider the expansion of $f(t)$ into a Fourier series from its $N + 1$ digits; see [1] in Chapter 2. We can then write down $N + 1$ linear equations with $N + 1$ unknowns, i.e. a_0 and a_n and b_n. We thus see that the given readings of $f(t)$ permit determination of a_n and b_n up to the $N/2$ harmonic. If N is even, the highest harmonic is of order $N/2$ and if N is odd it is of order $(N + 1)/2$. This result, here derived for $f(t)$, is equally applicable to $F(\omega)$, by virtue of the Fourier transform. The result substantiates the inference about the Nyquist or folding frequency, above obtained more intuitively in a graphical way.

With reference to a figure showing the digital representation of a curve (e.g. Fig.42 in section 4.5.1 or perhaps even better, by tracing the digits equally spaced around a circle), we can summarize the discussion in the last paragraph as follows:

Number of intervals $= N$
Number of digits $= N + 1$, i.e. 0, 1, 2, ..., N
Number of significant harmonics $n = N/2$ for N even
$= (N + 1)/2$ for N odd.

This is the system generally followed in this book, and care has to be observed in comparison with slightly other presentations, e.g. counting digits as 1, 2, 3, ..., N, which leads to $N - 1$ intervals.

4.3.3 *Aliasing—its explanation and its avoidance*

By aliasing we mean the phenomenon that spectral components with frequencies higher than v_N present in a record, will contaminate spectral components with frequencies lower than v_N. And the contamination (aliasing) works that way that components with frequencies symmetrical to the Nyquist frequency v_N will affect each other, hence the name folding frequency for v_N. Aliasing implies naturally a falsification of the spectrum of affected frequencies below v_N. Aliasing is thus a special kind of spectral contamination. In section 4.2 we have studied some other kinds of spectral contamination, due to other causes. While the effects discussed in 4.2 are essentially of a low-frequency nature, aliasing is due to the presence of high-frequency components in a record.

The phenomenon of aliasing requires special explanation, and for this we shall use a simple example (for more detailed derivation, see Manley, 1945, p.187). In a Fourier series expansion, harmonics up to order $N/2$ will be obtained correctly only if there are no higher harmonics present. If higher harmonics in fact exist, they will influence also the values of harmonics of order lower than $N/2$. Consider the Fourier series expansion:

$$f(t) = a_0 + \sum_n a_n \cos nt + \sum_n b_n \sin nt \qquad [19]$$

We assume four readings (digits) of $f(t)$ be given:

$$\left.\begin{aligned} f(0) &= a_0 + \sum_n a_n \\ f\left(\frac{\pi}{2}\right) &= a_0 + \sum_n a_n \cos\frac{n\pi}{2} + \sum_n b_n \sin\frac{n\pi}{2} \\ f(\pi) &= a_0 + \sum_n a_n \cos n\pi \\ f\left(\frac{3\pi}{2}\right) &= a_0 + \sum_n a_n \cos\frac{3n\pi}{2} + \sum_n b_n \sin\frac{3n\pi}{2} \end{aligned}\right| \qquad [20]$$

From these we form the following four expressions:

$$
\begin{aligned}
&\sum_{k=0}^{3} f\left(\frac{k\pi}{2}\right) = 4a_0 + 4a_4 + 4a_8 + \ldots \\
&\sum_{k=0}^{3} f\left(\frac{k\pi}{2}\right) \cos\frac{k\pi}{2} = 2a_1 + 2a_3 + 2a_5 + 2a_7 + \ldots \\
&\sum_{k=0}^{3} f\left(\frac{k\pi}{2}\right) \sin\frac{k\pi}{2} = 2b_1 - 2b_3 + 2b_5 - 2b_7 + \ldots \\
&\sum_{k=0}^{3} f\left(\frac{k\pi}{2}\right) \cos k\pi = 4a_2 + 4a_6 + 4a_{10} + \ldots
\end{aligned} \qquad [21]
$$

If $f(t)$ does contain or is assumed to contain harmonics only up to $n = 2$, then the constants can be determined exactly from the last four equations: the left-hand members are given from the given digits and of the right-hand members only the first terms remain. If, however, $f(t)$ also contains harmonics higher than the second, the system [21] is insufficient for determination of any coefficients. For instance, if it contained also the 3rd harmonic, the existence of a_3 and b_3 would influence the values of a_1 and b_1, as seen from [21]. This result can easily be generalized and it is found that the harmonics higher than v_N will influence the corresponding ones which are symmetrical around v_N. In the case just mentioned $v_N = 2$, and then the presence of a_3, b_3 will influence the values of a_1, b_1. As we can see from [21], the even part of the spectrum (represented by coefficients a) will be larger and the odd part (coefficients b) will be smaller, when higher harmonics are neglected. This is a general property of aliasing.

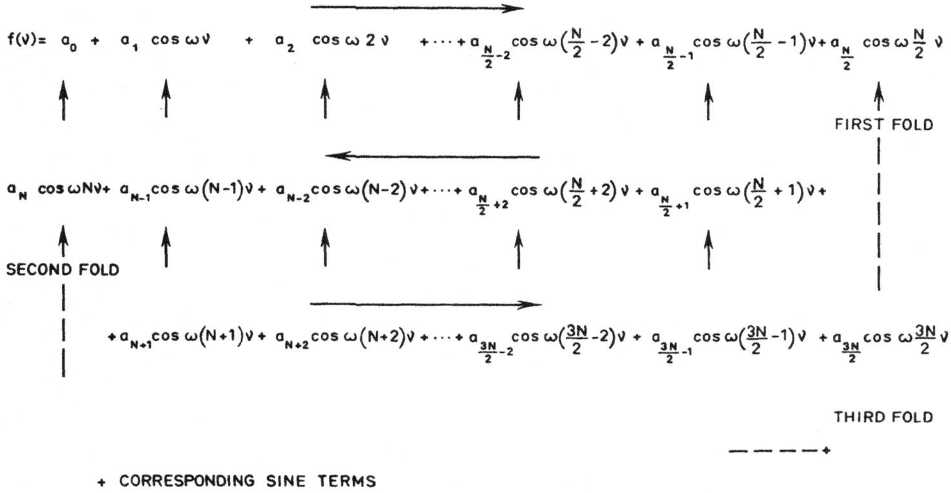

Fig.36. Calculation scheme illustrating the effect of aliasing.

Let us take another example, which can easily be deduced by an extension of the reasoning above and which illustrates the effect of aliasing. Assume 24 digits are given, i.e. the Nyquist frequency = 12. We distinguish two cases:

(1) $f(t)$ contains only the harmonics 1, 3, 7, 10 and 11. Then these will be exactly determined from the given digits.

(2) $f(t)$ contains as well the harmonics 13, 14 and 20, i.e. over 12. Then, only the harmonics 1, 3, and 7 will be accurate, whereas 10 will be confused with 14 ($= 12 \pm 2$) and 11 confused with 13 ($= 12 \pm 1$), and in addition a 4th harmonic will be introduced as a consequence of the 20th ($= 12 \pm 8$). These are symmetrical with respect to the Nyquist or folding frequency.

In summary we can illustrate aliasing by a general Fourier series expansion, as shown in equation Fig.36. We have $N + 1$ discrete observations of $f(t)$, v is an integer $= 0, 1, 2, \ldots, N$ and we have written $\omega = 2\pi/N$ for brevity. The horizontal arrows show the direction in which the series should be read and the vertical arrows indicate which terms will combine, as their cosine factors are identical. The folding frequency is $N/2$ and this, as well as its multiples, i.e. $N/2$, N, $3N/2$, ... are the points at which the series is folded. It is quite clear that truncating the series after the term $N/2$ will lead to correct results only if the coefficients of all following terms vanish, i.e. only if the corresponding frequencies are missing in the record analyzed.

As we have seen, aliasing is a result of using discrete formulas instead of the continuous ones. The aliasing discussed so far, and the one which is almost exclusively considered in the literature, depends on the finite sampling interval Δt. However, there is also another reason for aliasing, with different folding frequencies, depending upon the method by which discrete formulas are evaluated (section 4.5). The trapezoidal rule, with equal weights to the digits, is the safest. If other formulas are used for the evaluation, with unequal weights to different digits, additional aliasing is introduced, although with reduced amplitudes. This additional aliasing could be expected intuitively, as unequal weights imply another sampling; its effect has been demonstrated by Bakun and Eisenberg (1970).

The phenomenon of aliasing, as a direct consequence of the digitization of the records, is an undesirable effect. Several different measures have been explored to eliminate this spectral contamination, which we can list under the following points:

(1) A zero digitizing interval is naturally an impossibility in geophysical practice, but it could be approached by choosing Δt as small as possible. However, decreasing Δt means increasing the folding frequency and this also increases the frequency range occupied by the aliases. But in general, the higher the folding frequency, the smaller is the contribution of the aliases. Careful visual examination is usually sufficient to tell where the limit should be set. The sharp bends in a given record set the upper limit of the frequencies involved. This defines the Nyquist frequency and the necessary digitizing interval to be applied right through the

whole record. The strongest effect of aliasing is to be expected near the folding frequency. Therefore, if computed spectra exhibit rises close to this end of the frequency range, it is advisable to exclude this part from further discussion. An inconvenience by having small Δt is the increased amount of digitizing work and the increased requirement for computer storage and computer time.

(2) From the schematic figure illustrating aliasing (Fig.35b) it is seen intuitively that an alternative method to avoid aliasing, instead of taking close readings, would be to sample the record at unequally spaced points, chosen randomly. In fact, this can be proved mathematically and conditions can be stated for a random sampling to be alias-free (Shapiro and Silverman, 1960). For instance, a Poisson sampling is alias-free. A completely arbitrary sampling may still exhibit aliasing, though with modified properties. For a more complete discussion of this point, see R. H. Jones (1972). In connection both with this point and item (1) above, actual tests using the saw-tooth curve (section 2.1.2) would be instructive. This is left as an exercise to the reader.

(3) A third method to avoid aliasing effects, often used, is to filter the record before further analysis. A well-defined low-pass filter would do. However, if a filter is not carefully selected, then unwanted effects may arise: on the one hand, only partial elimination of aliasing, on the other hand, undesirable effects on the wanted part of the spectrum (Baer and Withee, 1971). Filtering methods will be discussed in Chapter 6.

In geophysics, aliasing effects are encountered in digital data sampling in time as well as in space. However, often not much is said about aliasing and its possible effects on spectra and on conclusions based on them. A notable exception is a paper by O'Neill and Ferguson (1971), who carefully consider aliasing effects upon their observations, which include wind speed and humidity for deduction of horizontal moisture flux. We quote from their paper the following passage: "The power spectra presented in this paper confirm the presence of considerable power at the highest frequencies." "These considerations must weigh heavily in the selection of an observing interval". "In effect, they imply that the shortest practical sampling interval be used in an attempt to provide both a capability of 'seeing' physically significant shorter period fluctuations and an opportunity to use filtering techniques to reduce physically extraneous noise contributed by the highest frequencies." "Additionally, more frequent observations have the benefit of increasing the sample size (and hence the statistical significance of the moments) for a given duration of experiment." In fact, these words could be taken as generally valid for any kind of geophysical observations. As example, O'Neill and Ferguson (1971) obtained a 50% error reduction in horizontal moisture flux by decreasing the sampling interval from 12 to 2 hours. Already earlier, Muller (1966) gave an extensive account of aliasing effects, with special reference to meteorological observations.

4.3.4 Practical procedures

On the basis of the rules in section 4.3.2, we can formulate some practical rules for obtaining the best spectrum, i.e. with wide frequency range and high resolution. The record length T should be large for a correct spectrum (by comparison with the requirement of the Fourier theory). Then the lower limit v_1 is always small. To get the upper limit v_N large, we need in addition that N/T is large. As T is assumed large, this can only be achieved by also choosing N large or Δt small. The latter condition will also contribute to remove the effect of aliasing. The resolution will be large if T is large. Thus, in practice choose T as large as possible (avoiding contamination from adjacent phases) and then choose N such that N/T is large or Δt small. In terms of the two frequency expressions v and ω, the lower and upper limits of the spectrum and the frequency interval between successive determinations become as follows:

$$\left.\begin{array}{lll} v_1 = 1/T; & v_N = 1/2\Delta t = N/2T; & \Delta v = 1/T \\ \omega_1 = 2\pi/T; & \omega_N = \pi/\Delta t = N\pi/T; & \Delta\omega = 2\pi/T \end{array}\right| \qquad [22]$$

The true spectrum can never be recovered unless the doubly infinitely long record ($|T| \rightarrow \infty$) is digitized at zero digitizing interval ($\Delta t \rightarrow 0$). In [22] $\Delta v = 1/T$ expresses the resolution as the frequency difference between consecutive spectral estimates. To clarify concepts, a *small* frequency interval (Δv, $\Delta\omega$) means *high* resolution. An alternative but equivalent way to express resolution is to say that there are T spectral estimates per cps.

For any given record, the record length T can usually be fixed relatively easily. The maximum frequency that we want to cover (v_{max}) can also be fixed relatively well from a visual examination of the record. Then, to avoid aliasing as much as possible, we choose a Nyquist frequency v_N higher than v_{max}, say about 1.5 v_{max}. Thus, having T and v_N given, these will fix the required digitizing interval Δt by the relation $\Delta t = 1/3v_{max}$. The number of intervals is $N = T/\Delta t$ and the number of digits is $N + 1$.

The formulas for resolution in [22], which we can write as follows:

$$\Delta v \cdot T = 1 \quad \text{or} \quad \Delta\omega \cdot T = 2\pi \qquad [23]$$

express the *principle of reciprocal spreading*. This says that the longer the time interval T, the more finely the spectral structure will become. See Fig.23. No doubt, [23] expresses one of the fundamental limitations to spectral analysis. It is common to all methods of frequency analysis. We derived a mathematical formulation for this principle in Chapter 2, equation [34]. We shall return to this question in section 5.3.3 in connection with a discussion of spectral widths.

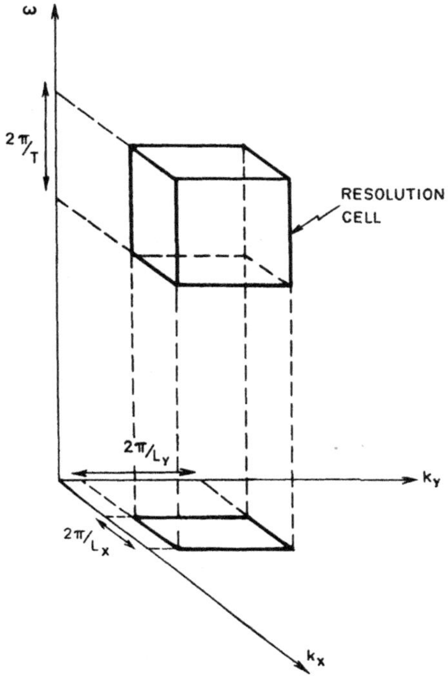

Fig.37. Three-dimensional resolution cell in the frequency–wavenumber domain.

These considerations have here been developed for frequency, but they are equally applicable to wavenumber (k). Corresponding to the frequency resolution $\Delta\omega = 2\pi/T$ in [22] and [23], we have wavenumber resolution $\Delta k = 2\pi/L$ with the components:

$$\Delta k_x = 2\pi/L_x, \qquad \Delta k_y = 2\pi/L_y, \qquad \Delta k_z = 2\pi/L_z \qquad [24]$$

The combined consideration of resolution power both for frequency and wavenumber is of significance in the treatment of array-station data. Treating such data ensemble, we are faced not only with a time coordinate (as in single station records) but also with space coordinates (corresponding to the extent of the array). See Fig.37 and Green et al. (1966) for more detail.

4.3.5 Summarizing expression for the effects of T and Δt

In rounding off this section, it is appropriate to see that several of the effects discussed can be summarized in more general formulas. This is especially true for the combined effect of limited record length (T) and of certain averaging intervals (Δt). Providing the variance with two subscripts $\sigma^2_{T,\Delta t}$, indicating the values of T and Δt which it refers to, we can write down the following three expressions:

From [5] in Chapter 5:

$$\sigma^2_{\infty,0} = \frac{1}{2\pi} \int_{-\infty}^{\infty} |F(\omega)|^2 \, d\omega \qquad [25]$$

From the formula for time-domain smoothing [31] in section 6.3.1, we have:

$$\sigma^2_{\infty,\Delta t} = \frac{1}{2\pi} \int_{-\infty}^{\infty} |F(\omega)|^2 \frac{\sin^2 \omega\Delta t/2}{(\omega\Delta t/2)^2} \, d\omega \qquad [26]$$

By analogy with the last equation:

$$\sigma^2_{\infty,T} = \frac{1}{2\pi} \int_{-\infty}^{\infty} |F(\omega)|^2 \frac{\sin^2 \omega T/2}{(\omega T/2)^2} \, d\omega \qquad [27]$$

By the definition of variance we have the following identity:

$$\sigma^2_{\infty,\Delta t} = \sigma^2_{\infty,T} + \sigma^2_{T,\Delta t} \qquad [28]$$

where the last term is an average over all intervals T. Solving for $\sigma^2_{T,\Delta t}$ and using [26] and [27], we get:

$$\sigma^2_{T,\Delta t} = \frac{1}{2\pi} \int_{-\infty}^{\infty} |F(\omega)|^2 \left[\frac{\sin^2 \omega\Delta t/2}{(\omega\Delta t/2)^2} - \frac{\sin^2 \omega T/2}{(\omega T/2)^2}\right] d\omega$$

$$= \frac{1}{2\pi} \int_{-\infty}^{\infty} |F(\omega)|^2 \frac{\sin^2 \omega\Delta t/2}{(\omega\Delta t/2)^2} \left[1 - \frac{\sin^2 \omega T/2}{(\omega T/2)^2}\right] d\omega$$

(assuming $T/\Delta t$ large)

$$= \int_{-\infty}^{\infty} |F(v)|^2 \, \text{sinc}^2 (v\Delta t) \, [1 - \text{sinc}^2 (vT)] \, dv \qquad [29]$$

introducing the cyclic frequency $v = \omega/2\pi$ and the sinc-function. Especially the expression next to the last one occurs quite frequently in connection with spectral studies of meteorological turbulence. As a check, it is seen from this equation that as $\Delta t \to 0$ and $T \to \infty$, $\sigma^2_{T,\Delta t} \to \sigma^2_{\infty,0}$ as given by [25].

The factor of F^2 defines the power spectral window which results from finite sampling length T and finite averaging interval Δt. In fact, referring to section 6.3.3, we see that $\mathrm{sinc}^2\,(v\Delta t)$ implies low-pass filtering and $1 - \mathrm{sinc}^2\,(vT)$ implies high-pass filtering, and therefore the product is equivalent to band-pass filtering. Obviously, from [29], the pass-band is defined by $2\pi/\Delta t > |\omega| > 2\pi/T$. For further discussion, see Ogura (1957b, 1959), F. B. Smith (1962), Pasquill (1962a).

4.4 WINDOW FUNCTIONS

4.4.1 General considerations

In section 3.5.1 we learnt that the use of limited record lengths is inevitable in spectral analysis of observational series, and as a consequence that use of a data window is equally inevitable. Using no particular window is equivalent to applying the rectangular (box-car) window. We also saw that application of windows will lead to distorted (smoothed) spectra and that the true spectrum is impossible to recover.

To keep the spectral distortion close to a minimum, we request the following properties from the spectral window, which corresponds to the applied time window:

(1) A high concentration to the central (main) lobe: requires a broad time window (reciprocal spreading, section 2.4.4).

(2) Small or insignificant side-lobes: requires smooth time window without sharp corners.

The rectangular window leads to a spectral window (sinc-function), which is quite good on point 1, but not good on point 2, having high-frequency oscillations, also with negative side-lobes. Even though the rectangular window leaves the time function undistorted, it may lead to severe distortion in the frequency domain.

The undesirable effect 2 is due to the sharp corners and vertical sides of the rectangular window, which introduce high frequencies. Therefore, some compromise has to be sought. With some other window, which tapers off gradually towards both ends of the record interval under investigation, we certainly make some distortion of the signal but at the same time we can avoid the high-frequency oscillations of its frequency window. A window which leaves both domains undistorted is obviously an impossibility. Table XV summarizes the situation in general terms.

There is no straight-forward procedure to derive the shape of the best window. Instead, as the approach is based on a compromise between different factors, the procedure has rather been by trial and error. This explains the voluminous efforts

TABLE XV

The problem to select a suitable window with minimum distortion in both the time and the frequency domains

Domain	Rectangular time window	Tapering time window
Time	no distortion within measured range	tapering towards both ends of measured range
Frequency	strong distortion, by convolution with sinc-function	milder distortion, mainly because of smaller side-lobes than for sinc-function

made in this field, of which the more significant results will be outlined below. Kurita (1969a) gave a most complete discussion of various window functions and their relative merits.

A time or lag window $w(t)$ is applied in the time domain, i.e. either directly to the given observations $f(t)$ or to auto- or cross-correlation functions before these are transformed into power spectra.

A spectral window $W(\omega)$ is the Fourier transform of the corresponding time window (data window) $w(t)$, and therefore calculation of $W(\omega)$ for any given $w(t)$ is just the same problem as dealt with in section 2.4. However, there is one important difference. In the examples in Chapter 2, most time functions are of infinite extent and then integration can be performed in closed form. Dealing

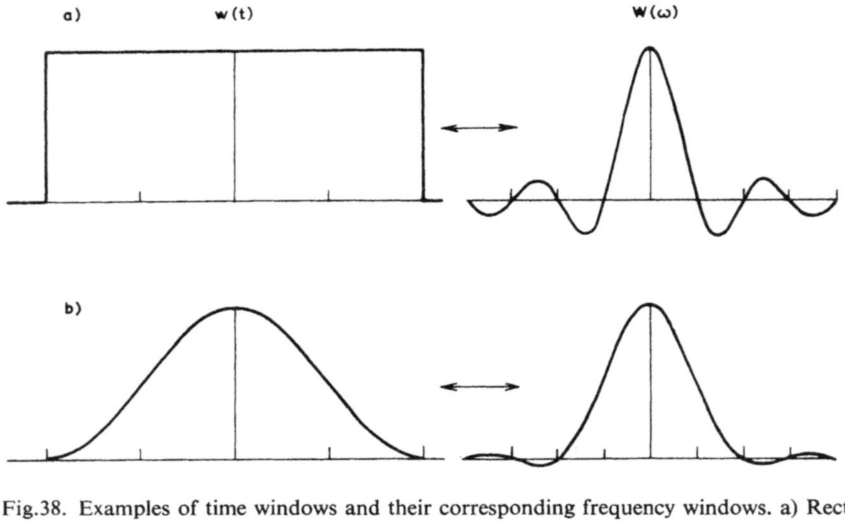

Fig.38. Examples of time windows and their corresponding frequency windows. a) Rectangular window. b) Hanning window.

with windows $w(t)$, on the other hand, these are for obvious reasons always of a limited time duration, i.e. transients. This entails certain complications in calculating their transforms, which are continuous, and for some spectral windows only a series expansion can be given. Most windows $w(t)$ are real and even, and then also $W(\omega)$ is real and even. We shall be giving the formulas for $w(t)$ and $W(\omega)$ in analytic form, but in any application to digital data, they have to be used in their digital form. This will not be given explicitly here. We shall discuss the most common window types and give examples from the literature where respective windows have found application. Some of the windows discussed below are shown graphically in Fig.38.

On the whole, any symmetric time function $w(t)$ which tapers off towards both ends could be investigated for its spectral properties and thus for its suitability as a window. From this viewpoint, it is advantageous to discuss different window types according to their shape, and we differ between the following main types:

(1) Trigonometric windows, which use trigonometric functions of time, i.e. sinc-functions $\sin x/x$ or cosine-functions.

(2) Power windows, using time raised to some power.

(3) Exponential windows, using exponential time functions.

In all the following cases, the factor $1/T$ appears on the right-hand side of the expression for $w(t)$. This makes the dimension of $w(t)$ as 1/time and consequently $W(\omega)$ becomes dimensionless, and mathematically more homogeneous, as a rule having T only in the combination ωT. In practical calculations, however, it is customary to omit the factor $1/T$ in $w(t)$, thus making $w(0) = 1$ always, leading to a frequency window $= T\, W(\omega)$.

4.4.2 *Rectangular window*

(also called box-car window or Bartlett window):

$$w(t) = \begin{cases} 1/T & 0 \leq |t| \leq T \\ 0 & |t| > T \end{cases} \qquad [30]$$

Its transform was derived in example 1, Table V:

$$W(\omega) = 2\,(\sin \omega T)/\omega T = 2\,\text{sinc}\,(\omega T/\pi) \qquad [31]$$

Not using a special window is in fact equivalent to passing the function through the box-car window. The purpose of special windows is to get a rounded or tapered signal that avoids the high frequencies introduced by the sharp rise and fall of the rectangular window. The sinc-function in $W(\omega)$ has large side-lobes and is therefore not very suitable for a true representation of a spectrum. Cf. Gibbs' phenomenon, section 2.1.2. Nevertheless, the rectangular window is among the most commonly used windows, and the undesirable spectral effects are compensated for by smoothing in the frequency domain (section 4.6.1).

The way in which the length T of a rectangular window influences the spectral resolution and stability was demonstrated already in section 3.5.1 and illustrated in Fig.23 in connection with our discussion of fundamental principles of observational spectra.

A modified version of the rectangular window is the *trapezoidal window*, discussed and used among others by Casten et al. (1969). Due to its sloping sides, i.e. more general tapering off of the signal, its properties are superior to those of the rectangular window.

4.4.3 Sinc-windows

Sinc-windows are based on the sinc-function and both its first and second power have been tried.

(1) *Fourier kernel window or Daniell window*:

$$w(t) = \begin{cases} \dfrac{1}{T}\dfrac{\sin \pi t/T}{\pi t/T} = \dfrac{1}{T}\operatorname{sinc}\dfrac{t}{T} & 0 \leq |t| \leq T \\ 0 & |t| > T \end{cases} \qquad [32]$$

The transform was derived in examples 2 and 3, Table V, but then for an infinite time duration. In the present case, with a truncated time window $-T \leq t \leq T$, the transform cannot be expressed in closed form, but can be written in the form of an infinite, convergent series ($n = 0, 1, 2, 3, \ldots$):

$$W(\omega) = \frac{1}{\pi}\sum_{n=0}^{\infty}\frac{(-1)^n}{(2n+1)\cdot(2n+1)!}\left[(\pi - \omega T)^{2n+1} + (\pi + \omega T)^{2n+1}\right] =$$

$$= \frac{1}{\pi}\left[\operatorname{Si}(\pi - \omega T) + \operatorname{Si}(\pi + \omega T)\right] \qquad [33]$$

where Si denotes the sine integral. In deriving this formula, the sine-function in [32] is replaced by exponential functions (Euler's formula) and then the integration is carried out. Equation [33] approximates to the rectangular function, valid in case of infinite time duration. Therefore, this window represents the closest inverse to the rectangular window.

(2) *Fejér kernel window* (also called Cesàro window):

$$w(t) = \begin{cases} \dfrac{1}{T}\left[\dfrac{\sin \pi t/T}{\pi t/T}\right]^2 = \dfrac{1}{T}\operatorname{sinc}^2\dfrac{t}{T} & 0 \leq |t| \leq T \\ 0 & |t| > T \end{cases} \qquad [34]$$

This case is related to example 7, Table V, but as we are now concerned with a limited time interval, the Fourier transform $W(\omega)$ leads to an infinite, convergent series (just corresponding to [33] above), which approximates to the triangular function ($n = 0, 1, 2, \ldots$):

$$W(\omega) = \frac{1}{2\pi^2} \sum_{n=0}^{\infty} \frac{(-1)^n}{(2n+1)\cdot(2n+1)!} \cdot \left[(2\pi + \omega T)^{2n+2} + (2\pi - \omega T)^{2n+2} - 2(\omega T)^{2n+2}\right] =$$

$$= \frac{1}{2\pi^2}\left[(2\pi + \omega T)\,\mathrm{Si}(2\pi + \omega T) + (2\pi - \omega T)\,\mathrm{Si}(2\pi - \omega T) - 2\omega T\,\mathrm{Si}(\omega T)\right] \qquad [35]$$

The Fejér window was applied among others by Kurita (1969b).

4.4.4 Cosine-windows

The cosine-windows are all based on the tapering properties of the cosine-function. These are among the most used windows and appear in several different forms.

(1) *Hanning window* (also called Tukey window or simply the cosine-window):

$$w(t) = \begin{cases} 1/2T \cdot (1 + \cos \pi t/T) & 0 \leq |t| \leq T \\ 0 & |t| > T \end{cases} \qquad [36]$$

The transform can easily be expressed in closed form:

$$W(\omega) = \frac{\sin \omega T}{\omega T} + \frac{1}{2}\left[\frac{\sin(\omega T + \pi)}{\omega T + \pi} + \frac{\sin(\omega T - \pi)}{\omega T - \pi}\right] =$$

$$= \mathrm{sinc}\,\frac{\omega T}{\pi} + \frac{1}{2}\left[\mathrm{sinc}\left(\frac{\omega T}{\pi} + 1\right) + \mathrm{sinc}\left(\frac{\omega T}{\pi} - 1\right)\right] \qquad [37]$$

As distinct from [30] (rectangular time window), this $W(\omega)$ is a sum of three sinc-functions, displaced relative to each other. This entails that the side-lobes from the three sinc-functions to a large extent cancel each other, which is favourable.

From [37] it is obvious that as T becomes very small, $W(\omega)$ will approach 1, i.e., it becomes constant independent of ω. This case corresponds to low resolution and high stability of the spectrum. Conversely, a large T leads to high resolution and low stability. Of course, this is in fact nothing but another example of the influence of the data length T on spectral resolution and stability, which we encountered in section 3.5.1. See further section 4.4.8.

The Hanning window has found wide-spread application, e.g. by Utsu (1966), Kurita (1968), Rossiter and Lennon (1968), Terashima (1968), Gupta and Chapman (1969), Hasegawa (1971b), and Stuart et al. (1971). It was applied as a moving time window by Pfeffer and Zarichny (1963) for analysis of atmospheric wave propagation, and by McGarr et al. (1964) for analysis of P-waves amidst noise. The advantage of a moving window it to display the spectral development with time. As immediately obvious, the Hanning window is identical with the cosine-squared window:

$$w(t) = \begin{cases} 1/T \cdot \cos^2 \pi t/2T & 0 \le |t| \le T \\ 0 & |t| > T \end{cases} \qquad [38]$$

In this form it has been used among others by Maruyama (1968), Bloch and Hales (1968), Landisman et al. (1969). Also the first-power cosine-taper is used quite often (e.g. Landisman et al., 1969).

Sometimes a half cosine-window has been used, a so-called "modified Hanning window", e.g. by Bakun (1971), who places the maximum of the window over the P onset.

(2) *Hamming window* (same alternative names as for [36]):

$$w(t) = \begin{cases} 1/T \cdot (0.54 + 0.46 \cos \pi t/T) & 0 \le |t| \le T \\ 0 & |t| > T \end{cases} \qquad [39]$$

This is closely related to [36], and its transform $W(\omega)$ is given immediately:

$$W(\omega) = 1.08 \frac{\sin \omega T}{\omega T} + 0.46 \left[\frac{\sin(\omega T + \pi)}{\omega T + \pi} + \frac{\sin(\omega T - \pi)}{\omega T - \pi} \right] =$$

$$= 1.08 \operatorname{sinc} \frac{\omega T}{\pi} + 0.46 \left[\operatorname{sinc}\left(\frac{\omega T}{\pi} + 1 \right) + \operatorname{sinc}\left(\frac{\omega T}{\pi} - 1 \right) \right] \qquad [40]$$

The windows [36] and [39] can be written in the following common form:

$$w(t) = 1/T \cdot (1 - 2a + 2a \cos \pi t/T) \qquad [41]$$

where $a = 0.25$ for [36] and $a = 0.23$ for [39].

Like [37], this $W(\omega)$ is also a sum of three sinc-functions. By the slight modification in the weight factors, a still better elimination of side-lobe effects is achieved. Obviously, the Hanning and Hamming spectral windows correspond to an averaging (with unequal weights) over three consecutive values of the spectral function of the rectangular window [30]. This averaging or smoothing has the consequence that the spectral windows [37] and [40] have much smaller side-lobes (of the order of one-tenth of [31], an advantage), whereas the central lobe (main lobe) is much broader for [37] and [40] than for [31] (a certain disadvantage). The comparison is clear from Fig.38. A comparison of the Hamming window with the rectangular and the cosine-windows has been published by Gudmundsson (1966) for application to geomagnetic space functions. The Hamming window is considered as one of the well-behaved windows and for that reason it is often used, e.g. by Noll (1964), Fernandez and Careaga (1968), Kanamori (1967b), and others.

(3) *Cosine-tapered rectangular window* (combination of [30] and [36]). Often a data window is wanted which is fairly flat over most of the signal duration (like the rectangular window), but tapers off near the two ends of the signal (for instance, like a cosine-window). This can be easily achieved by combining the windows [30] and [36] in such a way that the window begins to the left with the left-hand side of the cosine-window, is followed by a rectangular window over most of the signal, and ends to the right with the right-hand side of the cosine-window, or in formulas (cf. Bingham et al., 1967):

$$\left.\begin{aligned} w_1(t) &= \frac{1}{2T}\left(1 + \cos\frac{5\pi t}{T}\right) && \text{for } -T \le t \le -\frac{4T}{5} \\ w_2(t) &= \frac{1}{T} && \text{for } -\frac{4T}{5} \le t \le \frac{4T}{5} \\ w_3(t) &= \frac{1}{2T}\left(1 + \cos\frac{5\pi t}{T}\right) && \text{for } \frac{4T}{5} \le t \le T \end{aligned}\right| \qquad [42]$$

The corresponding spectral window $W(\omega)$ can be easily deduced and the following expression is obtained:

$$W(\omega) = \frac{\sin \omega T + \sin 4\omega T/5}{\omega T[1 - (\omega T/5\pi)^2]} \qquad [43]$$

4.4.5 Power windows

Power windows include some power function of t. We can again differ among several different, but related types.

(1) *Parzen window* (also simply called power window):

$$w(t) = \begin{cases} \dfrac{1}{T}\left[1 - \left(\dfrac{|t|}{T}\right)^m\right] & 0 \le |t| \le T \\ 0 & |t| > T \end{cases} \qquad [44]$$

where m is a positive integer. The number of terms in its transform depends on the exponent m:

$$W(\omega) = \frac{2 \sin \omega T}{\omega T} - \frac{2}{(\omega T)^{m+1}} \int_0^{\omega T} t^m \cos t \, dt \qquad [45]$$

For the last integral in [45] a simple formula can be given, if we carry out the successive integrations:

$$\int_0^{\omega T} t^m \cos t \, dt = [f(t) \sin t + f'(t) \cos t]_0^{\omega T} =$$

$$= f(\omega T) \sin \omega T + f'(\omega T) \cos \omega T - f'(0) \qquad [46]$$

where:

$$f(t) = m! \sum_{n=0}^{N} (-1)^n \frac{t^{m-2n}}{(m-2n)!}$$

$$n = 0, 1, 2, \ldots, N$$

$$N = \begin{cases} m/2 & m \text{ even} \\ (m-1)/2 & m \text{ odd} \end{cases}$$

These formulas permit an easy evaluation of the transform of the power window, i.e. of $W(\omega)$ in [45], for any value of m. The power window has been much used by Kurita (1966, 1969b and c, 1970), Mikumo and Kurita (1968), and others.

The Parzen window appears frequently in the literature under the following more special form of a power window:

$$w(t) = \begin{cases} \frac{1}{T}\left[1 - 6\left(\frac{t}{T}\right)^2 + 6\left(\frac{|t|}{T}\right)^3\right] & 0 \le |t| \le \frac{T}{2} \\ \frac{2}{T}\left[1 - \frac{|t|}{T}\right]^3 & \frac{T}{2} \le |t| \le T \\ 0 & |t| > T \end{cases} \qquad [47]$$

It is left as an exercise to prove that the corresponding spectral window is of the form $(\sin x/x)^4$, thus with no negative side-lobes. Another related window is the following (Arsac, 1966; Blum and Gaulon, 1971):

$$w(t) = 1/T \cdot [1 - (t/T)^2]^2 \qquad [48]$$

(2) *Triangular window* (also called Bartlett window, cf. [30], or Fejér window, cf. [34]) is obtained from [44] as a special case, putting $m = 1$:

$$w(t) = \begin{cases} \dfrac{1}{T}\left(1 - \dfrac{|t|}{T}\right) & 0 \leq |t| \leq T \\ 0 & |t| > T \end{cases} \qquad [49]$$

The transform is derived in example 6, Table V (comes also out as a special case of [45] and [46]):

$$W(\omega) = \left(\frac{\sin \omega T/2}{\omega T/2}\right)^2 = \text{sinc}^2 \frac{\omega T}{2\pi} \qquad [50]$$

This spectral window has no negative side-lobes, as distinct from most other windows.

Examples of the use of this window can be found in Crampin and Båth (1965), Husebye and Jansson (1966), Capon (1970), Treitel et al. (1971), and others. Another special case of the power window [44] is obtained by putting $m = 2$, which yields a parabolic window.

4.4.6 Exponential windows

An exponential window contains the time in the exponent. The most important of these is the *Gaussian window*:

$$w(t) = \begin{cases} 1/T \cdot e^{-at^2} & 0 \leq |t| \leq T \\ 0 & |t| > T \end{cases} \qquad [51]$$

where a is a positive constant. In case of infinite time interval, the Gaussian curve is self-reciprocal (example 25, Table V). In the present time-limited case, the transform is:

$$W(\omega) = \frac{1}{T}\sqrt{\frac{\pi}{a}}\, e^{-\omega^2/4a} \,\text{erf}\,(T\sqrt{a}) \qquad [52]$$

where the error function erf T is defined as:

$$\text{erf}\, T = \frac{2}{\sqrt{\pi}} \int_0^T e^{-u^2}\, du$$

Equation [52] is obtained by writing the transform of [51] in the shape of the error function, and then performing a contour integration in the complex u-plane. As $T \to \infty$, erf $T \to 1$, and the formula derived in section 2.4.3 is obtained. Obviously, this spectral window has no negative side-lobes and in fact no oscillations which could lead to spurious maxima and minima in the calculated spectra.

Phinney (1964) used a Gaussian window as a time-lag window, applied to the autocorrelation function. Kasahara (1957), Hirasawa and Stauder (1964) and Kishimoto (1964) used a time-shifted Gaussian function:

$$w(t) = 1/T \cdot \mathrm{e}^{-a(t-t_0)^2} \qquad [53]$$

where t_0 is the point around which the signal is to be analyzed.

A related window is the *exponential window*:

$$w(t) = 1/T \cdot \mathrm{e}^{-a|t|} \qquad [54]$$

A one-sided exponential window was used by Dobeš et al. (1971) for calculation of running spectra (section 3.6.5).

4.4.7 Methods of windowing

In Chapter 2 we presented a table of Fourier transforms for a number of given functions $f(t)$, Table V. And in the present chapter, we have now given a corresponding list of a number of window functions and their transforms. Having these functions given, we are free to make any combination of the two sets according to [42] in Chapter 3. This will illustrate how various functions $f(t)$ will appear in their Fourier-transformed shape $F(\omega)$, with the application of various window types. Tests on analytical functions are usually very informative and are recommended to acquire close familiarity with windowing operations and their results.

Applying a certain window function $w(t)$ to a time series $f(t)$ is nothing but applying certain weights to the values of $f(t)$. Table XVI lists the relative weights for $t = 0$, $T/2$ and T for the different windows studied here. The list provides a quick, rough comparison of the different windows. It may be remarked that window 36 (Hanning) corresponds to the weighting procedure 1, 2, 1 applied to three consecutive values, and is named after J. von Hann. For the points chosen in Table XVI, same weights are obtained for no. 49. Also, it is seen that no. 44 (power) approaches to no. 30 (rectangular) for increasing values of m. The spectral weights in Table XVI demonstrate that the windows 32, 34, 36, 39 and 49 exhibit only minor differences. The window operation is one of multiplication in the time domain and one of convolution in the frequency domain. This means a certain smoothing of the spectrum. Different windows will cause somewhat different smoothing of the spectra.

TABLE XVI

Relative weights corresponding to different time windows

Reference equation in Chapter 4	Window function $w(t)$	Weight in time domain			Weight in frequency domain[1]		
		$t = 0$	$t = T/2$	$t = T$	$\omega T = 0$	$\omega T = \pi$	$\omega T = 2\pi$
30	rectangular	1	1	1[2]	1	0	0
32	Fourier kernel	1	0.64	0	0.56	0.22	(0)[3]
34	Fejér kernel	1	0.41	0	0.48	0.26	(0)
36	Hanning	1	0.50	0	0.50	0.25	0
39	Hamming	1	0.54	0.08	0.54	0.23	0
42	cosine-tapered rectangular	1	1	0	1	0.10	−0.10
44	power $m = 2$	1	0.75	0	0.62	0.19	(0)
	$m = 5$	1	0.97	0			
	$m = 20$	1	1.00	0			
49	triangular	1	0.50	0	0.54	0.23	0
51	Gaussian $a = 2/T^2$	1	0.61	0.13	0.62	0.19	(0)
	$a = 4/T^2$	1	0.37	0.02			

[1] These weights are normalized so that their sum (over three consecutive values) is equal to 1.
[2] This is sometimes given as 0.50 instead, as this corresponds to the value obtained from the Fourier transform.
[3] Zero in brackets means that the value is not exactly zero, but for practical purposes it is zero.

The practical procedures in applying windows may be summarized in the following two alternative methods:

(1) a. The given observational series $f(t)$ is passed through the rectangular window (i.e., the function $f(t)$ is taken unchanged within a limited interval of time).

b. $f(t)$ is Fourier-transformed into $F(\omega)$.

c. $F(\omega)$ is smoothed usually over three consecutive points, by applying weights, frequently 0.23, 0.54, 0.23, corresponding to the Hamming spectral window (this smoothing is done to counteract the high frequencies introduced because of the rectangular time window).

(2) a. $f(t)$ is multiplied by any window, for example from our list above.

b. The resulting product is Fourier-transformed into $F(\omega)$. This time no smoothing is effected on $F(\omega)$, as this was done by the application of a suitable time window.

The two procedures (1) and (2) are equivalent, only that steps are taken in different order. Denoting the true spectrum by $F(\omega)$, the spectral windows of the rectangular function and the Hamming function by $W_1(\omega)$ and $W_5(\omega)$, respectively, and the spectral smoothing function by $\bar{S}(\omega)$, we have the following expressions for the operations outlined above. Method 1, i.e. first convolving with $W_1(\omega)$, then smoothing by $\bar{S}(\omega)$ is:

$$[F(\omega)\star W_1(\omega)]\star \bar{S}(\omega)$$

which by the associative law for convolution is equal to:

$$F(\omega)\star [W_1(\omega)\star \bar{S}(\omega)]$$

which finally equals:

$$F(\omega)\star W_5(\omega) \qquad [55]$$

i.e. method 2. This demonstrates the equivalence between methods 1 and 2. By suitable selection of smoother $\bar{S}(\omega)$, agreement with any wanted window can be achieved. However, this procedure leads only to a conversion from one window $W_1(\omega)$ to another $W_5(\omega)$, and the true spectrum $F(\omega)$, is still not recovered. In order to achieve this, the last expression above should have been:

$$F(\omega)\star \delta(\omega) = F(\omega) \qquad [56]$$

But $\delta(\omega)$ corresponds to a time window of infinite length (example 10, Table V), which is impossible in practical application. Nor is it possible to find a smoother $\bar{S}(\omega)$ which will execute a conversion to $\delta(\omega)$, i.e.:

$$W_1(\omega)\star \bar{S}(\omega) = \delta(\omega) \qquad [57]$$

or in the time domain: $w_1(t) \cdot \bar{s}(t) = 1$, i.e.:

$$\bar{s}(t) = \frac{1}{w_1(t)} = \begin{cases} 1 \\ \infty \end{cases} \text{depending upon whether } w_1(t) = \begin{cases} 1 \\ 0 \end{cases}$$

Thus, Dirichlet's conditions (section 2.1) are not fulfilled, and $\bar{S}(\omega)$ does not exist. Therefore, the true spectrum $F(\omega)$ cannot be recovered, only an approximation to it. An analogous discussion was carried through in section 3.5.1.

4.4.8 Window type and window length

In the choice of window to be applied, an ideal solution is impossible, and a compromise has to be made between various factors, the most significant of which are: (1) bandwidth of spectral window, and (2) side-lobes of spectral window.

We want to have a spectral window that is concentrated in its main lobe

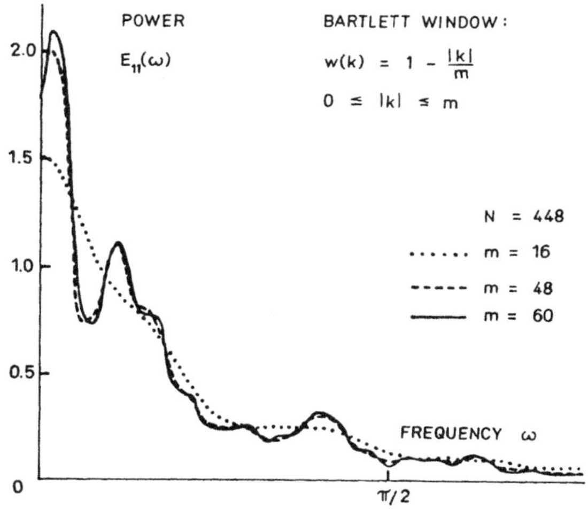

Fig.39. Effect of increasing the data length m in calculation of power spectrum from autocorrelation, using the triangular window. N = number of sampled points. After Jenkins (1965).

around $\omega = 0$ and has only small side-lobes. Negative side-lobes should preferably be as small as possible. These demands are clear when we consider that the spectral window by its convolution with $F(\omega)$ plays the role of a weighting function. From this viewpoint, we see in Table XXIII (section 5.3.2) that the rectangular window has the narrowest width. But the rectangular shape of this time window is undesirable, because of the high frequencies which are thus erroneously introduced (Gibbs' phenomenon).

Empirical tests show that the bandwidth of the spectral window is a more significant parameter than the type of window used. The bandwidth should be of the same order as the narrowest detail that one wants to investigate in a spectrum. But the smaller the bandwidth, the smaller will the stability of the spectral estimates be. The influence of bandwidth using a Bartlett window [49] is illustrated in Fig.39 from Jenkins (1965). In the formulation of Fig.39, increasing m means increasing the bandwidth in the time domain, thus decreasing the bandwidth in the frequency domain. According to Jenkins (1965), it is advisable to start with a small T and then increase this until the spectrum is no longer affected. The three spectra in Fig.39 clearly demonstrate the result of such a procedure.

Some striking geophysical illustrations of spectral resolution and stability as dependent upon window length are shown in Fig.40. Trembly and Berg (1968) gave amplitude spectra of short-period P-waves for three different window lengths, R. Y. Anderson and Koopmans (1963) gave power spectra for varve series, and Caner and Auld (1968) produced magnetotelluric power spectra for two window lengths, one being nearly 15 times the other. From such considerations of spectral resolution versus stability, it is clear that in any comparison between spectra of

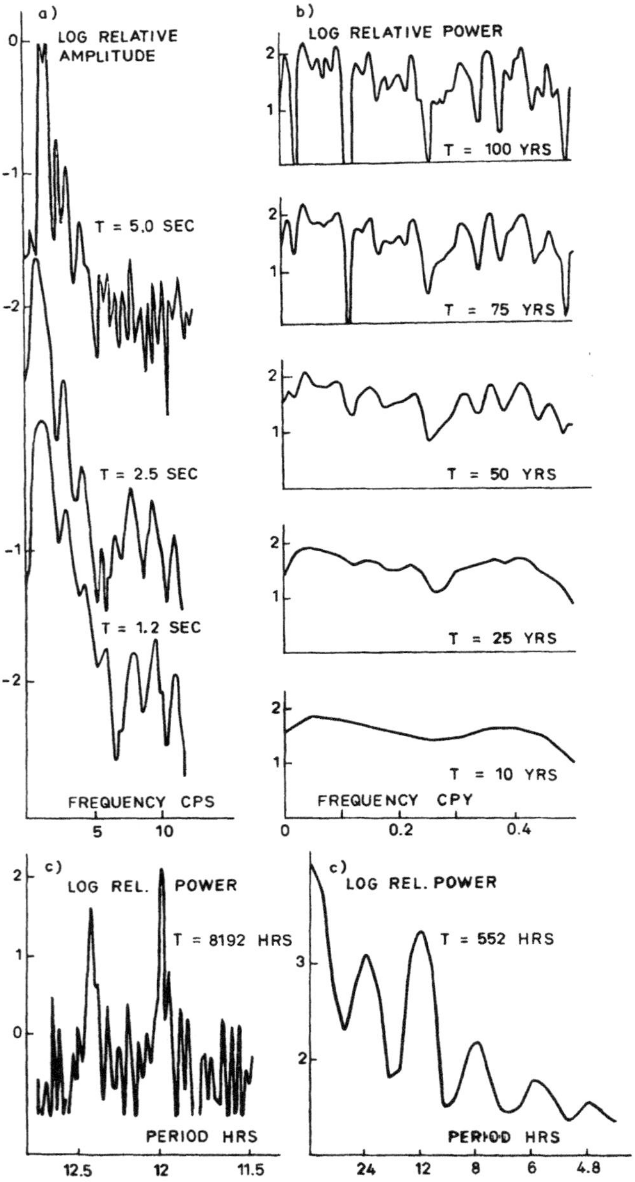

Fig.40. Geophysical examples of spectral resolution and stability as influenced by data length. a) Seismic P-wave amplitude spectra after Trembly and Berg (1968). b) Power spectra for varve series after Anderson and Koopmans (1963). c) Magnetotelluric power spectra after Caner and Auld (1968). Note that in a) a sequence of spectral minima in the range 5–11 cps for $T = 1.2$ sec has largely disappeared for $T = 5.0$ sec, while resolution has increased.

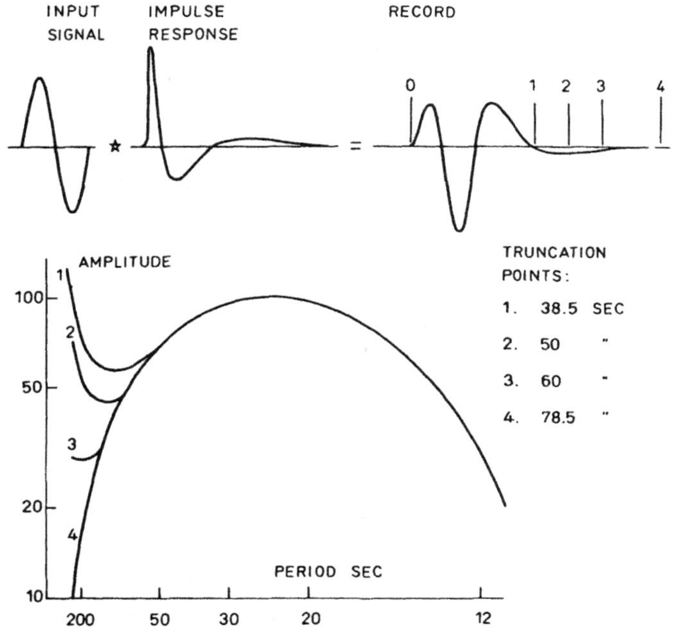

Fig.41. Spectral effect of truncation in the time domain when a rectangular window (4.4.2) is applied to the record of a 20-sec period single-cycle sine wave as input, written by a seismograph with 30 sec seismometer period and 100 sec galvanometer period. After Linde and Sacks (1971).

the same parameter under different conditions, it is of great significance to keep window lengths constant. See also McGarr et al. (1964), Priestley (1965a) and Jenkins and Watts (1968).

In efforts to reach a sufficiently good spectral resolution, there are obviously two alternatives: either to keep T constant and to search among different window types, or to let T vary and keep the window type constant. No doubt, the latter procedure is the most efficient one.

Truncation has been discussed in section 3.5.2, for a number of theoretical curves. As demonstrated by Linde and Sacks (1971) truncation may have great effect on the long-period portion of the spectrum of body waves (see Fig.41 for an input signal of one dominant period, with spectra corrected for instrumental response in the frequency domain). The truncation leads incorrectly to a too flat spectrum for long periods. In summary, they conclude that the truncation incorrectly increases long-period amplitudes and this effect is enhanced for shorter windows, for added noise, for Hamming window as compared to the rectangular window, and for increasing signal period. It is recommended to use a window length at least equal to twice the dominant wave period, but still excluding later unwanted arrivals. Sometimes it may be impossible to fulfill both requirements at the same time. Moreover, it is recommended always to analyze an equally long record portion before the signal arrival, to see the influence of the noise spectrum.

In a comparative study of analogue and digital methods for spectral calculations of seismic records, McIvor (1964) presented a detailed discussion on window requirements. Relative merits of different windows are also extensively discussed in the literature. See, for example, Ku et al. (1971), who in connection with filtering of gravity data, found the Hanning and the Hamming windows to provide the best compromise between resolution and stability.

4.4.9 Various generalizations

Several other window types with some modifications from those given here can be found in the literature. See, for instance, Robinson (1967a, p.61), Priestley (1965a), and others. Other methods have also been used to taper records towards the ends. As one example, Teng and Ben-Menahem (1965) taper the signal by hand, which is justified if the amplitude of the signal near the point of truncation is small. This procedure is analogous to application of the window [42] above.

As already emphasized, the data windows $w(t)$ are time-limited, and as a consequence, their transforms $W(\omega)$ are continuous with infinite extent. They consist of a main lobe and an infinite sequence of gradually smaller side-lobes. To suppress the influence of the side-lobes, an additional spectral window may be applied, which enhances the central frequencies and suppresses the higher and lower ones. This process is applied by multiplication in the frequency domain. For more details, see Blackman and Tukey (1959). Dziewonski et al. (1969) apply a Gaussian window in the frequency domain, which is equivalent to convolution by its inverse transform in the time domain.

In some cases, it has been suggested to use two consecutive time windows, so as to further suppress the influence of the side-lobes. The idea is to combine one window with essentially positive side-lobes with another with essentially negative side-lobes, so that the side-lobe effects of the two windows will largely cancel each other (see Wonnacott, 1961).

We have restricted the discussion of windows to time functions $f(t)$ and their corresponding spectra. However, appropriate windows can be applied to any function for which we want to construct a spectrum from a limited data set. As we saw in section 3.3.3, the autocorrelation function and the power spectrum form a Fourier pair. Then, windows can be applied to the autocorrelation, before making the transformation. As the autocorrelation is a function of the time-lag τ, the applied window is also a function of τ, and then termed a *lag window*. Cf. Blackman and Tukey (1959, p.12).

Similarly, we could envisage the extension of window types to multi-dimensional Fourier transforms, including the space domain, which is of significance when dealing with array data. Space windows can be shaped exactly as the time windows, we have studied above, only substituting space (x) for time (t). Discussion on suitability of different windows and of their side-lobes will be identical, the only difference is the set of data they are applied to. For example, Linville and Laster (1966) discuss a space window of the Bartlett type. Different windows in time and space could be used simultaneously.

4.5 DISCRETE FORMULAS

4.5.1 Basic principles

The formulas, integrals etc, in Chapters 2 and 3 were developed for continuous functions, and these have to be modified so as to fit the discrete samples. Quite generally we may state that to express a continuous integral in discrete form, just means to express the area under the curve by means of the discrete ordinates, or in formula:

$$\int_a^b y(x)\,\mathrm{d}x \rightarrow \sum_{n=0}^{N} y(n\Delta x)\,\Delta x \qquad [58]$$

$$a = 0,\ b = N\Delta x$$

where $y(x)$ represents the given curve, Δx is the digitizing interval, and $N + 1$ is the

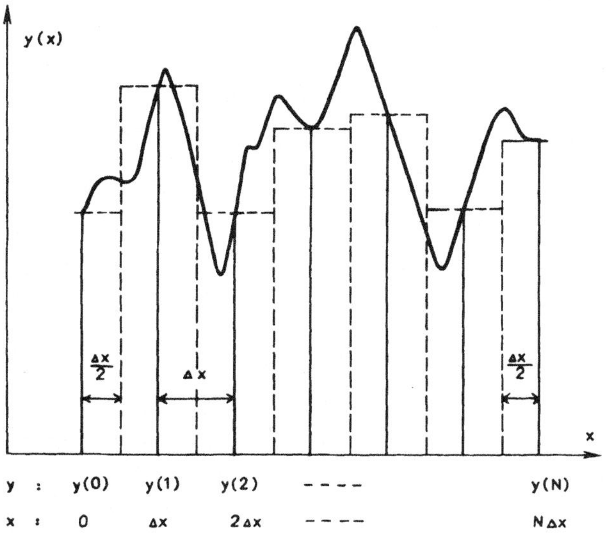

Fig.42. Sketch demonstrating the development of discrete summation formulas.

total number of sample points (Fig.42). For a finite integration interval, the corresponding summation extends over this interval.

For an infinite integration interval, we consider the following alternative expressions for finite intervals (cf. [3] and [16] in Chapter 2 and [1] and [2] in Chapter 3):

$$\int_{-\infty}^{\infty} y(x)\,\mathrm{d}x \rightarrow \lim_{X\rightarrow\infty} \frac{1}{X} \int_{-X/2}^{X/2} y(x)\,\mathrm{d}x \rightarrow \frac{1}{N\Delta x} \sum_{n=0}^{N} y(n\Delta x)\,\Delta x$$

$$= \frac{1}{N} \sum_{n=0}^{N} y(n\Delta x) = \frac{1}{N}\left[\frac{1}{2}y(0) + y(1) + y(2) + \ldots + y(N-1) + \frac{1}{2}y(N)\right] \qquad [59]$$

using the trapezoidal rule for expressing the area under the curve. With the two extreme values equal, i.e. $y(0) = y(N)$, which is mostly the case in our applications, the sum reduces to the following form:

$$\frac{1}{N}[y(1) + y(2) + \ldots + y(N)] = \frac{1}{N}\sum_{n=1}^{N} y(n\Delta x)$$

or to:

$$\frac{1}{N}[y(0) + y(1) + y(2) + \ldots + y(N-1)] = \frac{1}{N}\sum_{n=0}^{N-1} y(n\Delta x) \qquad [60]$$

and these are the forms we will mostly be using in the following. Here $X = N\Delta x$ is the finite range measured, and $y(0)$, $y(1)$, $y(2)$, ..., $y(N)$ are the given discrete, equidistant values. The factor $1/N$ forms a kind of normalization factor, to give the same result irrespective of the length of the sample, a stationary process provided.

In summations from 1 to N or from 0 to $N - 1$, all terms have the same weight (= 1), and in summations from 0 to N, the first and last terms have half weight (= 1/2), and the others weight 1. Among different quadrature rules for evaluating discrete formulas, the trapezoidal one used here is recommended, i.e. with weights as indicated, before those with unequal weights. The reason is that using unequal weights is equivalent to superposing a certain oscillation of the zero-line, which introduces spurious spectral components, i.e. spectral contamination (cf. section 4.2.4).

4.5.2 Fourier series coefficients

By digitizing process, the continuous curve $y(x)$ is replaced by a polygon, the more accurately the shorter Δx is. Methods for Fourier series expansion, which take advantage of this representation of $y(x)$, have been developed (Haase, 1961).

For a series of $N + 1$ discrete observations at equal spacing and with $\nu = 1, 2, \ldots, N$, the discrete summation formulas for the Fourier coefficients are as follows (for more details, see for example Stumpff, 1937, pp.35–38, or Manley, 1945, p.186), as obtained from [3] in Chapter 2:

$$\left.\begin{aligned} a_0 &= \frac{1}{N}\sum_{\nu=1}^{N} f(\nu) \\ a_n &= \frac{2}{N}\sum_{\nu=1}^{N} f(\nu)\cos\frac{2n\pi}{N}\nu \\ b_n &= \frac{2}{N}\sum_{\nu=1}^{N} f(\nu)\sin\frac{2n\pi}{N}\nu \end{aligned}\right| \qquad [61]$$

and the discrete Fourier series expansion becomes:

$$f(\nu) = a_0 + \sum_{n=1}^{N/2} a_n \cos\frac{2n\pi}{N}\nu + \sum_{n=1}^{N/2} b_n \sin\frac{2n\pi}{N}\nu \qquad [62]$$

The symmetry properties of sine- and cosine-functions permit us to rewrite the discrete formulas [61] for a_n and b_n in the following form, which has the advantage of computer-time saving:

$$\begin{bmatrix} a_n \\ \\ b_n \end{bmatrix} = \frac{2}{N}\sum_{\nu=1}^{N/2}\left[f(\nu) + (-1)^n f\left(\frac{N}{2}+\nu\right)\right]\begin{bmatrix}\cos \\ \\ \sin\end{bmatrix}\frac{2n\pi}{N}\nu \qquad [63]$$

Evidently, the point $\nu = N/2$ represents a folding point along the sequence of ν-values. Relations between the coefficients a_n, b_n for the continuous case (section 2.1.1) and the discrete case [63] can be derived and which show the effect of aliasing (see Hamming, 1962, pp.278–280).

4.5.3 Fourier transform

Equation [59] can be immediately applied to the Fourier transform formula [20] in Chapter 2, which becomes:

$$F(\omega) = \int_{-\infty}^{\infty} f(t)\,\mathrm{e}^{-i\omega t}\,\mathrm{d}t \rightarrow \frac{1}{N}\sum_{n=1}^{N} f(t_n)\,\mathrm{e}^{-i\omega t_n} \qquad [64]$$

Increased accuracy can be achieved by allowing for the variation of $f(t)$ and of $e^{-i\omega t}$ within each digitizing interval Δt. Then, we write $F(\omega)$ as:

$$F(\omega) = \frac{1}{N\Delta t} \sum_n \int_{t_n - \Delta t/2}^{t_n + \Delta t/2} f(t)\, e^{-i\omega t}\, dt \qquad [65]$$

We admit a linear variation of $f(t)$ within each compartment, i.e. we put:

$$f(t) = f(t_n) + (t - t_n)\left(\frac{df}{dt}\right)_n$$

Inserting this into the expression for $F(\omega)$, we arrive at the following formula after the integrations have been carried out:

$$F(\omega) = \frac{1}{N} \frac{\sin \omega\Delta t/2}{\omega\Delta t/2} \sum_n f(t_n)\, e^{-i\omega t_n} - \frac{1}{N} \frac{i}{\omega} \left[\frac{\sin \omega\Delta t/2}{\omega\Delta t/2} - \cos \frac{\omega\Delta t}{2}\right] \sum_n \left(\frac{df}{dt}\right)_n e^{-i\omega t_n}$$

Expanding the sinc- and cos-functions into power series and including terms up to $(\Delta t)^2$, we get:

$$F(\omega) = \frac{1}{N}\left[1 - \frac{\omega^2 (\Delta t)^2}{24}\right] \sum_n f(t_n)\, e^{-i\omega t_n} - \frac{1}{N}\, i\omega \frac{(\Delta t)^2}{12} \sum_n \left(\frac{df}{dt}\right)_n e^{-i\omega t_n} \qquad [66]$$

Taking the digitizing interval Δt sufficiently small, we may neglect the square terms in Δt and we obtain:

$$F(\omega) = \frac{1}{N} \sum_n f(t_n)\, e^{-i\omega t_n} \qquad [67]$$

This formula is rewritten in the following form, which is better adapted to the discrete case:

$$F(n) = |F(n)|\, e^{i\Phi(n)} = \frac{1}{N} \sum_{\nu=1}^{N} f(\nu)\, e^{-i(2\pi n/N)\nu} \qquad [68]$$

This is the discrete Fourier transform, often abbreviated DFT, where the given time series is $f(1), f(2), \ldots, f(N)$, i.e. N data points, all at equal spacing [$f(0) = f(N)$ for $N + 1$ data points]. Equation [68] is also immediately obtained from [61]:

$$F(n) = 1/2\,(a_n - ib_n) \qquad [69]$$

where a_n and b_n are the Fourier coefficients in discrete form.

The Nyquist or folding frequency is $N/2$ for N even and $(N + 1)/2$ for N odd (section 4.3). Therefore, $F(n)$ need only be calculated for $n = 1, 2, 3, \ldots$ up to the Nyquist frequency, i.e. $N/2$ or $(N + 1)/2$, respectively. This is also easily seen as N given quantities $f(v)$ can only determine N other quantities a_n and b_n, and there are two unknowns in each $F(n)$. The two unknowns in each $F(n)$ constitute the amplitude and phase of each $F(n)$.

We could rewrite the formula [68] as follows:

$$F(n) = \frac{1}{N} \sum_{v=1}^{N/2} \left[f(v) + (-1)^n f\left(\frac{N}{2} + v\right) \right] e^{-i(2\pi n/N)v} \qquad [70]$$

The inverse transform of $F(n)$ can also be easily given in discrete form:

$$f(v) = \sum_{n=1}^{N} F(n)\, e^{i(2\pi v/N)n} = 2 \sum_{n=1}^{N/2} F(n)\, e^{i(2\pi v/N)n} \qquad [71]$$

This can be checked by inserting expression [68] for $F(n)$, whereby the right-hand side is found to be equal to $f(v)$. This formula could also be written in a form corresponding to [59]. Equation [71] is the inverse discrete Fourier transform, sometimes abbreviated IDFT. The second expression [71] for $f(v)$ might be more attractive considering the folding property. This expression is easily seen as for N even we have, for example:

$$F\left(\frac{N}{2} - 1\right) \exp\left[i \frac{2\pi v}{N} \left(\frac{N}{2} - 1\right)\right] = F\left(\frac{N}{2} + 1\right) \exp\left[i \frac{2\pi v}{N} \left(\frac{N}{2} + 1\right)\right] \qquad [72]$$

using expressions for a_n and b_n. For N odd, the upper limit is $(N + 1)/2$ instead.

More accurate discrete formulas for $f(v)$ can be deduced fully analogously to the case for $F(\omega)$, given in [66]. This is left as an exercise to the reader. See also Harkrider (1964) and Kogeus (1968).

The discrete summation formulas for $F(n)$ and $f(v)$, [68] and [71], can also be written in the form of polynomials:

$$F(n) = \frac{1}{N} \sum_{v=1}^{N} f(v)\, z^v \qquad [73]$$

where $z = e^{-i(2\pi n/N)}$

and:

$$f(v) = \sum_{n=1}^{N} \frac{F(n)}{z^n} \qquad [74]$$

where $z = e^{-i(2\pi v/N)}$

The expressions [73] and [74], which are met with rather frequently in literature, are called *z-transforms.*

The rules of calculation, derived in section 2.3 for the continuous Fourier transform, can be shown to hold also for the discrete transform (Gold and Rader, 1969). A well-written discussion of discrete spectral analysis has been given by Huang (1966).

4.5.4 Correlation functions and convolution

Applied to discrete, digital readings, the formula [1] in Chapter 3 for the autocorrelation function becomes:

$$C_{11}(\tau) = \frac{1}{N - \tau} \sum_{t=1}^{N-\tau} f(t) f(t + \tau) \qquad [75]$$

where $\tau = 0, 1, 2, \ldots, m$ and the given time series is $f(1), f(2), \ldots, f(N)$.

In forming the averages, expressed as $C_{11}(\tau)$, see [75], we see that fewer and fewer terms will be included with increasing m, and this leads to fluctuations in the averages $C_{11}(\tau)$, in addition to those due to the fact that the functions run out of correlation. Therefore, m is limited to a small fraction of N, and it is generally recommended that m should not be more than 10–15% of N for power-spectrum computations. On the other hand, the smaller m is, the shorter time-lag interval is covered by the corresponding correlation function, and hence the lower is the spectral resolution. Therefore, a certain compromise between correlation stability and spectral resolution is necessary in deciding upon m. See further sections 3.5.1 and 4.4.8. In some other texts, the normalizing factor $1/(N - \tau)$ for $C_{11}(\tau)$ is replaced by $1/N$, which leads to more stable estimates (smaller mean square error).

In order to give a definite meaning to the correlation coefficients and the power, it is necessary that $f(t)$ is counted from its mean value. If not, then an appropriate correction for the mean has to be made, in order to avoid disturbing components in the spectrum which mask wanted components (section 4.2.4, point 1). Let $x(t)$ denote the directly measured values, such that $f(t) = x(t) - \bar{x}$. By substitution into [75] for $C_{11}(\tau)$, it is then found, assuming stationarity, that:

$$C_{11}(\tau) = \frac{1}{N - \tau} \sum_{t=1}^{N-\tau} x(t)\, x(t + \tau) - \left[\frac{1}{N - \tau} \sum_{t=1}^{N-\tau} x(t)\right]^2 \qquad [76]$$

where the last term on the right-hand side is the square of the average $\bar{x}$.

Corresponding to [75], we can immediately form the discrete summation formula for cross-correlation:

$$C_{12}(\tau) = \frac{1}{N - \tau} \sum_{t=1}^{N-\tau} f_1(t) f_2(t + \tau) \qquad [77]$$

where $\tau = 0, 1, 2, \ldots, m$ and the given time series are:

$$f_1(1), f_1(2), \ldots, f_1(N) \qquad \text{and} \qquad f_2(1), f_2(2), \ldots, f_2(N)$$

When measurements are taken from an arbitrary zero-line, we obtain the following formula for the cross-correlation coefficient $C_{12}(\tau)$, putting $f_1(t) = x(t) - \bar{x}$ and $f_2(t) = y(t) - \bar{y}$ with stationarity assumed:

$$C_{12}(\tau) = \frac{1}{N - \tau} \sum_{t=1}^{N-\tau} x(t)\, y(t + \tau) - \frac{1}{(N - \tau)^2} \sum_{t=1}^{N-\tau} x(t) \sum_{t=1}^{N-\tau} y(t) \qquad [78]$$

This is the counterpart of [76] for the autocorrelation. Both [76] and [78] provide corrections for the use of an arbitrary, parallel zero-line, before further calculations are made.

Similarly, we form a discrete summation formula for convolution:

$$f_1(t) \star f_2(t) = \frac{1}{N - t} \sum_{\tau=1}^{N-t} f_1(\tau) f_2(t - \tau) \qquad [79]$$

where $t = 0, 1, 2, \ldots, m$
and the given time series are:

$$f_1(1), f_1(2), \ldots, f_1(N) \qquad \text{and} \qquad f_2(1), f_2(2), \ldots, f_2(N)$$

For an arbitrary, parallel zero-line, [79] changes into the following:

$$f_1(t) \star f_2(t) = \frac{1}{N - t} \sum_{\tau=1}^{N-t} x(\tau)\, y(t - \tau) - \frac{1}{(N - t)^2} \sum_{\tau=1}^{N-t} x(\tau) \sum_{\tau=1}^{N-t} y(\tau) \qquad [80]$$

Discrete summation formulas for power and cross-power will be given in the following section.

4.6 METHODS FOR SPECTRAL CALCULATION

The role of modern electronic computers for the development of geophysical spectroscopy has already been emphasized. In fact, it is the availability of large computers that has made the great advances in this field during the last decade possible. In this section we shall give the dominating methods for calculation of spectra to the extent that the student should be able to apply them. Computer programs are not given here but covered by references at appropriate points. Useful collections were given by Robinson (1966, 1967b). An examination of the relative merits of different methods for spectra calculation, with particular reference to geophysical data, has been published by Hinich and Clay (1968).

In our efforts to give a complete outline of the practical procedures, it was also found necessary to mention filtering methods. Even though these will not be dealt with more in detail until in Chapter 6, we do not anticipate that they will cause difficulties to the reader at this stage.

4.6.1 Indirect method or correlation-transform method

The basic principle of this method is first to calculate the autocorrelation of the given time series $f(t)$ and then by Fourier-transforming the autocorrelation to get the power spectrum. Likewise, by calculating the cross-correlation of two different time series and then Fourier-transforming this, we get the cross-power spectrum. The fundamental formulas were derived in Chapter 3.

We shall first limit the discussion to autocorrelation and power calculation, the development being analogous for cross-correlation and cross-power. As we have to deal with digitized measurements, we have to use formulas adapted to this case. The calculation of autocorrelation for discrete measurements was given in [75], section 4.5.4. For the calculation of power spectrum from autocorrelation, we develop the corresponding formula [25] in section 3.3.3, as follows:

$$E_{11}(\omega) = 2 \int_0^\infty C_{11}(\tau) \cos \omega\tau \, d\tau$$

becomes in digital form:

$$E_{11}(n) = \frac{2}{m} \sum_{\tau=0}^{m} C_{11}(\tau) \cos \frac{2n\pi}{m} \tau =$$

$$= \frac{1}{m} \left[C_{11}(0) + 2 \sum_{\tau=1}^{m-1} C_{11}(\tau) \cos \frac{l\pi}{m} \tau + C_{11}(m) \cos l\pi \right] = E(l) \qquad [81]$$

putting $l = 2n$ and noting that the two extreme values enter with half weight (only half interval along τ-axis included in the two end segments; cf. section 4.5.1). $E(l)$ is the power spectrum at frequency l, and:

$$0 \leq l \leq m \qquad \text{or} \qquad 0 \leq n \leq m/2$$

The upper limits correspond to the Nyquist frequency (section 4.3.1). The formula [81] is frequently quoted in the literature, from Blackman and Tukey (1959), and appears in slightly different forms, but is seldom derived.

Now we have the necessary tools for calculating a power spectrum by the indirect method. We can summarize the operations in the following steps:

(1) From the given time series $f(t)$, the average value is subtracted. In this way, we have henceforth to deal only with deviations from the average and formulas simplify. This is data preparation, which has always to be done.

(2) Apply filtering, if any part of the spectrum is unwanted (section 6.2.3), pre-whitening to eliminate side-lobe effects of strong peaks if any (section 6.3.2), and detrending to eliminate trends in the data if any (section 4.2.4). Like (1), these are also data preparation, but only to be done when conditions require this.

(3) Calculate the autocorrelation by the discrete formula:

$$C_{11}(\tau) = \frac{1}{N - \tau} \sum_{t=1}^{N-\tau} f(t) f(t + \tau)$$

where $\tau = 0, 1, 2, \ldots, m$
and the given time series is $f(1), f(2), \ldots, f(N)$. See section 4.5.4.

(4) Get the power spectrum by the formula [81], derived above:

$$E(l) = \frac{1}{m}\left[C_{11}(0) + 2\sum_{\tau=1}^{m-1} C_{11}(\tau)\cos\frac{l\pi}{m}\tau + C_{11}(m)\cos l\pi\right]$$

$E(l)$ are termed the *raw spectral density estimates.*

(5) If the record has been written by an instrument with frequency-dependent response, e.g. a seismograph, then the spectral estimates should be corrected for this:

$$E(l)\,|I(l)|^2 = E_i(l)$$

where $|I(l)|$ is the instrumental amplitude response (in case of power, we are not concerned with phase response). This step may be omitted if we have an instrument with $|I(l)| = 1$ or if we are only concerned with comparative measurements from the same instrument or from instruments with identical characteristics.

(6) The raw spectral estimates $E_i(l)$ are smoothed in order to yield *refined spectral density estimates.* Smoothing can be done according to several, practically identical formulas (section 4.4). We give here the Hamming-Tukey and the Hanning smoothing coefficients (upper and lower numerals, respectively), which are those most commonly used:

$$
\left.
\begin{aligned}
\bar{E}_i(0) &= \left.\begin{matrix}0.54\\ \\ 0.5\end{matrix}\right\} E_i(0) + \left.\begin{matrix}0.46\\ \\ 0.5\end{matrix}\right\} E_i(1) \\
\bar{E}_i(l) &= \left.\begin{matrix}0.23\\ \\ 0.25\end{matrix}\right\} E_i(l-1) + \left.\begin{matrix}0.54\\ \\ 0.5\end{matrix}\right\} E_i(l) + \left.\begin{matrix}0.23\\ \\ 0.25\end{matrix}\right\} E_i(l+1) \\
&\quad \text{for } 1 \leq l \leq m-1 \\
\bar{E}_i(m) &= \left.\begin{matrix}0.46\\ \\ 0.5\end{matrix}\right\} E_i(m-1) + \left.\begin{matrix}0.54\\ \\ 0.5\end{matrix}\right\} E_i(m)
\end{aligned}
\right| \qquad [82]
$$

denoting the smoothed values by an overbar. The reason for doing this or similar kind of smoothing is partly to eliminate the effect of noise, partly to eliminate the effect of the rectangular time window (equivalent to using no special window). Then, the smoothing operation in the frequency domain tends to eliminate side-lobe effects introduced by the rectangular window.

From this viewpoint, it is important that the smoothing in the frequency domain matches the time window in such a way that side-lobe effects are effectively removed, but that no additional, unnecessary smoothing is done. The latter could remove significant spectral features, which could be important in the interpretation of the spectra. In fact, if in the expression for $\bar{E}_i(l)$ above, we substitute on the right-hand side the expressions for $E_i(l-1)$, $E_i(l)$ and $E_i(l+1)$ from [81] we obtain, using the Hanning smoothing coefficients:

$$
\bar{E}_i(l) = \frac{1}{m}\left[C_{11}(0) + \sum_{\tau=1}^{m-1} C_{11}(\tau)\left(1 + \cos\frac{\pi\tau}{m}\right)\cos\frac{l\pi\tau}{m}\right] \qquad [83]
$$

This is equivalent to applying the Hanning lag window in the time domain.

Thus, we have two equivalent procedures: either to apply some lag window in the time domain (in this case applied to $C_{11}(\tau)$), or to smooth by corresponding factors in the frequency domain. This is just another example of the equivalence between muliiplication in the time domain and convolution in the frequency domain. Cf. also section 4.4.7. In a comparison of these alternative methods, Kurita (1969a) finds that the longer the time window, the smaller will the difference between different procedures be.

(7) If pre-whitening has been done (item 2), then its effect is removed by "post-colouring", which is the frequency-domain inverse to the time-domain pre-whitening process applied (section 6.3.2). Thus, pre-whitening is done to remove side-lobe effects from large peaks in the spectrum of the given trace,

whereas windowing and/or smoothing is done to smooth off effects of a rectangular window.

(8) As a final step, it is desirable to calculate confidence limits of the obtained spectrum (section 5.1.3).

The indirect method as outlined here, has been the dominating method in all spectral calculations during most of the 1960's, especially after the publication of the book by Blackman and Tukey (1959). Instructions how to make digital computer programs for these calculations have been published by Southworth (1960).

In a very useful and instructive paper, Stuart et al. (1971) have investigated various factors, such as sampling, digitizing methods, etc., for different types of records, with special reference to geomagnetic micropulsation records, and they demonstrate the capabilities of the power spectral density method. The formula for $E(l)$ provides for some useful rules for sampling methods in order to achieve any desired degree of resolution and stability. The argument of the cosine-function is:

$$\omega t = 2\pi t/T_l = l\pi\tau/m$$

i.e., the period of the lth component is:

$$T_l = 2mt/l\tau = 2m\Delta t/l \qquad [84]$$

The period resolution (defined as the period difference between consecutive spectral estimates) is given by:

$$\Delta T_l = T_l - T_{l+1} \simeq 2m\Delta t/l^2 \qquad [85]$$

Eliminating l between [84] and [85] we get:

$$m\Delta t = T_l^2/2\Delta T_l \qquad [86]$$

With given values of T_l and ΔT_l (i.e. a requested resolution), we have $m\Delta t$ determined, and if one of them, e.g. Δt is specified, then the maximum lag m is given. If in addition we want a certain number of degrees of freedom (section 5.1.3) $\mu = 2N/m$, then also the total number of data points N is determined.

Let us take one example from Stuart et al. (1971). We want a resolution of $\Delta T_l = 2$ sec at a period of $T_l = 40$ sec. If Δt is given as 10 sec, we get $m = 40$. Moreover, requesting 20 degrees of freedom, i.e. $2N/m = 20$, this gives finally $N = 400$. Corresponding considerations are applicable also to the following spectral methods.

We shall terminate this section by giving the corresponding discrete formulas for the cross-power, derivations being analogous to the preceding case. First, we calculate the cross-correlation from its discrete summation formula (section 4.5.4):

$$C_{12}(\tau) = \frac{1}{N-\tau}\sum_{t=1}^{N-\tau} f_1(t) f_2(t+\tau)$$

where $\tau = 0, 1, 2, \ldots, m$
and the given time series are:

$$f_1(1), f_1(2), \ldots, f_1(N) \qquad \text{and} \qquad f_2(1), f_2(2), \ldots, f_2(N)$$

The cross-power is conveniently split into its real part (co-spectrum) and its imaginary part (quad-spectrum), from [27] and [30] in Chapter 3:

$$E_{12}(\omega) = \int_{-\infty}^{\infty} C_{12}(\tau)\, e^{-i\omega\tau}\, d\tau = P_{12}(\omega) - iQ_{12}(\omega) \qquad [87]$$

where:

$$P_{12}(\omega) = \int_{-\infty}^{\infty} C_{12}(\tau) \cos \omega\tau\, d\tau$$

$$Q_{12}(\omega) = \int_{-\infty}^{\infty} C_{12}(\tau) \sin \omega\tau\, d\tau$$

As $C_{12}(\tau) = C_{21}(-\tau)$, we get (cf. [32] in Chapter 3):

$$\left.\begin{aligned} P_{12}(\omega) &= \int_{0}^{\infty} [C_{12}(\tau) + C_{21}(\tau)] \cos \omega\tau\, d\tau \\ Q_{12}(\omega) &= \int_{0}^{\infty} [C_{12}(\tau) - C_{21}(\tau)] \sin \omega\tau\, d\tau \end{aligned}\right| \qquad [88]$$

Writing these formulas in discrete summation form, considering the half weights for the two extreme segments, we have:

$$P_{12}(l) = \frac{1}{m}\left[C_{12}(0) + \sum_{\tau=1}^{m-1} \{C_{12}(\tau) + C_{21}(\tau)\} \cos \frac{l\pi}{m}\tau + C_{12}(m) \cos l\pi\right] \qquad [89]$$

noting that $C_{12}(m) = C_{21}(m)$, the only contribution to the cross-correlation then being $f_1(m)\, f_2(m)$, and:

$$Q_{12}(l) = \frac{1}{m} \sum_{\tau=1}^{m-1} [C_{12}(\tau) - C_{21}(\tau)] \sin \frac{l\pi}{m} \tau \qquad [90]$$

Quite clearly, the autocorrelation and power come out as a special case of these formulas, i.e. $E(l) = P_{11}(l)$ is immediately obtained from $P_{12}(l)$, while $Q_{11}(l) = 0$. Otherwise, the calculation procedure is the same as above. Lag windows can be applied by multiplying the correlation coefficients by discrete factors corresponding to the chosen window.

4.6.2 Direct method or periodogram method

The direct method is based on a straight-forward calculation of the Fourier transform of the given time series. In digitized form the formulas are as follows (for details, see section 4.5):

Cosine transform of $f(\nu)$:

$$a(\omega) \rightarrow a(n) = \frac{1}{N} \sum_{\nu=1}^{N} f(\nu) \cos \frac{2n\pi}{N} \nu$$

Sine transform of $f(\nu)$: [91]

$$b(\omega) \rightarrow b(n) = \frac{1}{N} \sum_{\nu=1}^{N} f(\nu) \sin \frac{2n\pi}{N} \nu$$

Complex Fourier transform of $f(\nu)$:

$$F(\omega) = a(\omega) - ib(\omega)$$

Amplitude spectrum:

$$|F(\omega)| = [a^2(\omega) + b^2(\omega)]^{\frac{1}{2}} \qquad [92]$$

Power spectrum:

$$|F(\omega)|^2 = a^2(\omega) + b^2(\omega)$$

Phase spectrum $\Phi(\omega)$:

$$\tan \Phi(\omega) = - b(\omega)/a(\omega)$$

These formulas provide us with the necessary tools for application of the direct method. The procedure is as follows, using for easier reference the same numbers as above for the indirect method:

(1) Same as above.
(2) Same as above.
(3) Not applicable.
(4) Spectral calculations by formulas given for the direct method.
(5) Instrumental correction both of amplitude and phase spectrum.
(6) Same as above.
(7) Same as above.
(8) Same as above.

Evidently, this method does not include the autocorrelation coefficient, but if wanted this can be calculated as the inverse transform of the power $|F(\omega)|^2$, as explained in section 3.3.3. The direct method yields the phase spectrum $\Phi(\omega)$, which is not included in the indirect method. The direct method may not be very convenient in practical applications and is not very much used.

As an alternative, it is possible to work instead with the Fourier coefficients a_n and b_n, for which the digital formulas read as follows (section 4.5.2):

$$\left.\begin{aligned} a_n &= \frac{2}{N} \sum_{v=1}^{N} f(v) \cos \frac{2n\pi}{N} v \\ b_n &= \frac{2}{N} \sum_{v=1}^{N} f(v) \sin \frac{2n\pi}{N} v \end{aligned}\right| \qquad [93]$$

after which the *average* power for each frequency n is obtained from:

$$\frac{1}{2}(a_n^2 + b_n^2)$$

Otherwise, procedures are the same as for the direct method.

Instructions how to make digital computer programs for Fourier series calculations have been given by Goertzel (1960) and Robinson (1967b). See also Davenport and Root (1958, p.107) and R. H. Jones (1965), where the periodogram is defined as (cf. section 3.3.2):

$$\frac{|F(\omega)|^2}{T} = \frac{a^2(\omega) + b^2(\omega)}{T} \qquad [94]$$

where T is the record interval ($= N$).

The instrumental correction has been introduced as step 5 above, which is most convenient, i.e. as a multiplication in the frequency domain. If a "ground" record, i.e. of the incoming signal, is wanted in the time domain, this could best be obtained by inverse transformation from the frequency domain, or by deconvolution in the time domain (considering that the recorded trace is the con-

TABLE XVII

Alternative procedures to calculate modified periodograms

Step	Procedure		
	I (direct)	II (direct)	III (indirect)
1	$f(t)$	$f(t)$	$f(t)$
2	$f(t) \cdot w(t)$	$F(\omega)$	$C_{11}(\tau)$
3	$F(\omega)$	$\cdot I(\omega)$	$\lvert F(\omega)\rvert^2$
4	$\cdot I(\omega)$	smooth	$\cdot \lvert I(\omega)\rvert^2$
5	square	square	smooth
6	$\bar{P}(\omega)$	$\bar{P}(\omega)$	$[\bar{P}(\omega)]$

$w(t)$ = special data window, $I(\omega)$ = seismograph response.

volution of the incoming signal and instrumental impulse response, cf. section 6.5.4).

In most analyses of records, a series of operations have to be carried out. It is then of importance to consider the order in which the various steps are done, as incorrect order of operations could lead to incorrect results. As an example, we take the calculation of the Fourier periodogram, i.e. $|F(\omega)|^2/T$. By modified periodogram $\bar{P}(\omega)$ we mean the one obtained when smoothing is executed before squaring $|F(\omega)|$. We can list two alternative ways to calculate $\bar{P}(\omega)$, which give essentially the same result (procedures I and II in Table XVII). In fact, the only difference between procedures I and II is the application of a data window in I as compared to spectrum smoothing in II. Whereas I and II lead to practically the same result, a different result would be obtained by interchanging the order of 4 and 5 in procedure II. This is in fact what is done in procedure III (Table XVII), where we first calculate the autocorrelation coefficient. Cf. Bingham et al. (1967).

4.6.3 Fast Fourier Transform method (FFT)

A computationally far more efficient method than those described hitherto was developed around the middle of the 1960's. The principle of the FFT method can be explained as follows, essentially based on Cochran et al. (1967). The given observational, discrete series is denoted x_t with $t = 1, 2, \ldots, N$, where N is an even number. This series is split into two parts:

$y_t = x_{2t-1} = x_1, x_3, \ldots, x_{N-1}$ comprising all odd-numbered digits

and:

$z_t = x_{2t} = x_2, x_4, \ldots, x_N$ comprising all even-numbered digits [95]

and with:

$t = 1, 2, \ldots, N/2.$

Note that the two series y_t and z_t overlap each other.

By means of the expression for the Fourier transform in digital form (equation [64], section 4.5.3), we can then immediately form the discrete Fourier transforms (DFT) of the three series (the total series and the two partial series):

$$X_n^{(N)} = \frac{1}{N} \sum_{t=1}^{N} x_t \, e^{-i(2\pi n/N)t}$$

$$Y_n^{(N/2)} = \frac{2}{N} \sum_{t=1}^{N/2} y_t \, e^{-i(4\pi n/N)t} \qquad [96]$$

$$Z_n^{(N/2)} = \frac{2}{N} \sum_{t=1}^{N/2} z_t \, e^{-i(4\pi n/N)t}$$

noting that the number of terms in $X_n^{(N)}$ is N, but in $Y_n^{(N/2)}$ and $Z_n^{(N/2)}$ only half as much, $N/2$. The three time series are related, and so are also the transforms, as results from the following expansion of $X_n^{(N)}$:

$$\begin{aligned} X_n^{(N)} &= \frac{1}{N} \sum_{t=1}^{N/2} \left[y_t \, e^{-i(2\pi n/N)(2t-1)} + z_t \, e^{-i(2\pi n/N)2t} \right] \\ &= e^{i(2\pi n/N)} \frac{1}{N} \sum_{t=1}^{N/2} y_t \, e^{-i(4\pi n/N)t} + \frac{1}{N} \sum_{t=1}^{N/2} z_t \, e^{-i(4\pi n/N)t} \\ &= \frac{1}{2} e^{i(2\pi n/N)} \, Y_n^{(N/2)} + \frac{1}{2} Z_n^{(N/2)} \end{aligned} \qquad [97]$$

Replacing n by $n + N/2$ we also find:

$$Y_{n+N/2}^{(N/2)} = Y_n^{(N/2)} \quad \text{and} \quad Z_{n+N/2}^{(N/2)} = Z_n^{(N/2)} \qquad [98]$$

as $$\frac{4\pi n}{N} t \rightarrow \frac{4\pi}{N}\left(n + \frac{N}{2}\right) t = \frac{4\pi n}{N} t + 2\pi t$$

which leaves the exponential function unchanged, t being an integer. As a consequence we get:

$$\begin{aligned} X_{n+N/2}^{(N)} &= \frac{1}{2} e^{i(2\pi/N)(n+N/2)} \, Y_{n+N/2}^{(N/2)} + \frac{1}{2} Z_{n+N/2}^{(N/2)} \\ &= -\frac{1}{2} e^{i(2\pi n/N)} \, Y_n^{(N/2)} + \frac{1}{2} Z_n^{(N/2)} \end{aligned} \qquad [99]$$

For easier review, let us rewrite [97] and [99], which are of basic importance in understanding the FFT method:

$$
\left.
\begin{aligned}
X_n^{(N)} &= \frac{1}{2}\,e^{i(2\pi n/N)}\,Y_n^{(N/2)} + \frac{1}{2}\,Z_n^{(N/2)} \\
&\qquad\qquad \text{for } n = 0, 1, 2, \ldots, (N/2) - 1 \\
X_{n+N/2}^{(N)} &= -\frac{1}{2}\,e^{i(2\pi n/N)}\,Y_n^{(N/2)} + \frac{1}{2}\,Z_n^{(N/2)} \\
&\qquad\qquad \text{for } n + N/2 = N/2, N/2 + 1, \ldots, N - 1
\end{aligned}
\right| \qquad [100]
$$

Equations [100] permit an evaluation of the transform of the given series x_t from the transforms of the two partial series y_t and z_t. If then $N/2$ is also an even number, the process can be continued, i.e. the series y_t and z_t can each be split into two partial series, and the transforms formed for each one of these, and so on as long as we have an even number of digits. The latter condition can be fulfilled by choosing a number of digits such that $N = 2^k$ where k is any positive integer. Then, the process above can be continued until we are left with only one term in each of the partial series. In that case, the Fourier transform equals the reading itself (as seen from the expression for $X_n^{(N)}$ in [100]). A simple application of the method can be found in Jenkins and Watts (1968, p.315).

Therefore, in applying the FFT method we should choose a number of observations $= 2^k$, i.e. as follows:

for $k =$	1,	2,	3,	4,	5,	6,	7,	8,	9,	10,	11,	12,
we have $N =$	2,	4,	8,	16,	32,	64,	128,	256,	512,	1024,	2048,	4096,

respectively.

In applications, N or k could be determined as follows:

(1) Select the total interval to be included: a representative sample for a random process, the whole signal in case of transients.

(2) Select the digitizing interval, trying to eliminate possible effects of aliasing.

(3) In general, the ratio of 1 and 2 would not be exactly 2^k. Then, to achieve this, prolong the measured interval to the next higher 2^k in case of random processes, or fill out with zeros to the next higher 2^k in case of transients.

Like the direct method, FFT gets the Fourier transform directly of the given record, in digitized form. This means that power spectra have to be obtained by squaring the spectral estimates. Also, like the direct method, the autocorrelation coefficient is not involved, but if wanted it can be obtained by inverse transformation of the power spectrum (Bingham et al., 1967). The calculation procedure to be followed is the same as for the direct method, only that under item 4 the FFT method is substituted. Similarly, in the direct method and FFT, the cross-power has to be calculated from the Fourier transforms of the given time series, as $|F_1^*(\omega)\ F_2(\omega)|$, following fundamental relations in Chapter 3. Then, the cross-correlation can be obtained by inverse transformation of the cross-power. Tests on observed data show that power spectra calculated by FFT are practically identical with those obtained by the correlation-transform method (see for example, Schule et al., 1971). Likewise, convolution of two time functions is easiest to calculate by first forming the two Fourier transforms by the FFT method, then multiply these and finally take the inverse transform of the product.

The main advantage of the Fast Fourier Transform method compared to other methods for spectral analysis is the increased speed, because of a much lower number of operations, while the accuracy is the same as with other methods. The main advantage is thus saving of computer time and storage. It can be shown that the number of arithmetic operations is about $2kN = 2N \log_2 N$ for FFT against about N^2 for other methods. For $N = 4096$, $k = 12$ we find that N^2 is about 171 times as large as $2N \log_2 N$. This ratio is about equal to the ratio between the computer times for the two methods for a given computer.

The method was in fact presented many years ago and was rediscovered by Cooley and Tukey (1965) and is sometimes called the Cooley-Tukey algorithm. A historical review is given by Cooley et al. (1967a). Since the latter part of the 1960's, the method has got an ever increasing application in all fields where spectra are calculated, and it has more and more replaced the correlation-transform method, outlined above.

Corresponding to the increasing range of applications of the FFT method, there is also an increasing amount of literature on various aspects of this subject. Among the FFT literature, the following may be mentioned: Gentleman and Sande (1966), Welch (1967), Cooley et al. (1967b), Bingham et al. (1967), Gold and Rader (1969), Glassman (1970), Bendat and Piersol (1971, pp.300 and 322), Otnes and Enochson (1972). Especially publications of the IEEE (Institute of Electrical and Electronics Engineering) contain extensive accounts of various aspects of the FFT method, and special issues have been devoted to this subject, such as Proc. IEEE, vol. 55, Oct. 1967; IEEE Trans. Audio and Electroacoustics, vol. AU-15, June 1967; and vol. AU-17, June 1969. Computer programs for the FFT can be found in Robinson (1967b, pp.62–64), Glassman (1970), and others.

In all methods explained here for spectral estimation, we have referred the treatment to a time series. Naturally, the methods are equally applicable to a space series. Moreover, the methods can be easily extended to two or more variables, by the use of appropriate formulas from Chapter 2. For example two-dimensional FFT, which can be developed analogously to the one-dimensional case above, can be found in Naidu (1969), Black and Scollar (1969), Lewis and Dorman (1970), and others. Transforms for a two-dimensional array may conveniently be calculated by first taking the transforms of the rows, i.e. with respect to one of the variables. Then the resulting column is transformed with respect to the second variable. A modified FFT, particularly useful for saving storage space and computer time in transforming two-dimensional data fields (e.g. aeromagnetic data), has been devised by Naidu (1970a). A new fast algorithm for computation of frequency–wavenumber spectra has been given by Smart and Flinn (1971).

There is still another method for spectral calculation, the so-called *direct segment method*. This is related to the FFT, with a corresponding saving in the number of operations. In the direct segment method, the total set of data is divided into non-overlapping segments, and spectra are calculated for each segment. The same discrete formulas as in FFT can be used for this purpose. Then, averages are formed over all segments. This method has proved especially useful in dealing with seismograph array-station records (Capon et al., 1967; Lacoss et al., 1969). See also Jenkins and Watts (1968). This method is in fact one version of the FFT method, which is termed *decimation in frequency*, as distinct from the description we gave above for the FFT, which is called *decimation in time* (Cochran et al., 1967).

4.6.4 Newer developments

The methods described are all based on the same basic principles, and yield therefore practically the same results. The FFT method meant an important step forward, but only on the computational side (considerable saving of computer time and storage). It did not imply any improvement in computed spectra, as it is based on the same fundamental principles.

Recently, a radically different approach to the computation of power spectra, called the *maximum entropy spectral method*, has been developed by J. P. Burg. It has been described by Lacoss (1971) and others. Like the indirect method, it is based on transformation of correlation functions, but, on the contrary, it is data-adaptive. Its advantage over the older methods both in correct frequency information and above all in a much higher resolution is well demonstrated by the power spectra for truncated sinusoids. A very informative case is illustrated in

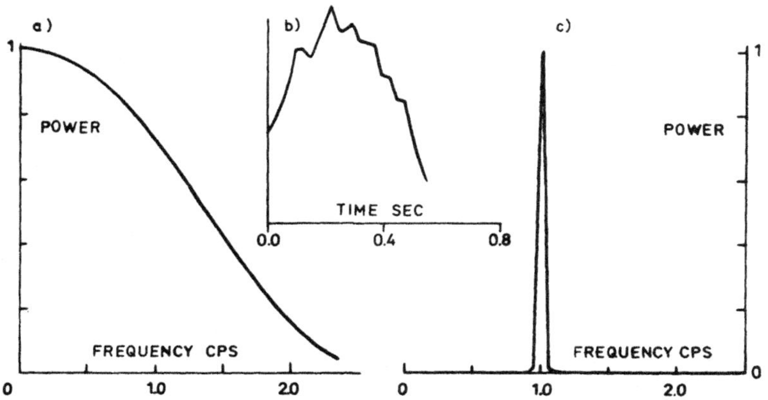

Fig.43. Maximum entropy power spectrum (c) as compared with square of the modulus of the Fourier transform (a), in both cases starting from an incomplete sinousoid (b). After Ulrych (1972).

Fig.43. In fact, the maximum entropy method is able to yield such a spectrum as one would expect intuitively, before one has learnt about all the fallacies connected with window functions and their Fourier transforms. The new method reaches an exactness and sharpness that matches the optical spectra (cf. section 1.7.). No doubt, the new method will be very important in future geophysical applications, especially when the measured interval is relatively short in comparison with periods involved and when the spectrum contains narrow spikes. A related, though apparently less efficient method, is the *maximum likelihood method.*

A new method for computation of high-resolution spectra has also been presented by Capon (1969a), further developed by Pisarenko (1972). Another promising method, especially for time-limited signals, is the *complex exponential representation* of the given time function (Spitznogle and Quazi, 1970). In this method use is made of an expansion of the given digital data in terms of a series, resembling a Fourier series expansion, but with the difference that the exponent of the e-function is not purely imaginary as in the Fourier series, but complex (i.e. analogous to a Laplace transform with complex exponent; cf. Båth, 1968, pp.233–235). This means that the spectral amplitudes are not constant but varying exponentially with time (cf. section 3.6.1). This method has generally higher spectral resolution than the conventional Fourier transform method.

Among newer developments, we also mention the *Walsh sequential spectrum* (section 3.6.2), which (like the other newer methods) is still in its infancy as far as geophysical applications are concerned but which promises to open up new pathways for spectral research.

Considering that new methods are coming up, with increased precision and efficiency as compared to the more traditional Fourier methods, it is tempting to see, like a vision, the time when new "revolutionary" methods dominate geophysical research and when Fourier methods belong to the historical past.

4.6.5 Brief summary of some practical considerations

Besides the computational schemes outlined above, a number of precautions have to be taken in any spectral analysis, the most important of which we summarize here.

(1) Windowing has always to be applied, but the type of window used is generally of less significance than its length. The length should be at least twice the dominant period, but also exclude unwanted signals. Truncation effects are discussed in sections 3.5.2 and 4.4.8. In uncertain cases, try with a series of windows of different lengths.

(2) The sampling interval should be carefully selected: on the one hand, it should be small enough to avoid aliasing errors (section 4.3), on the other hand, it should not be unnecessarily small which may lead to inconveniences considering computer time and storage. The best sampling interval can be decided

upon by inspection of given records or of test spectra with trial sampling intervals.

(3) The exact zero-line should be determined from a sufficiently long record or else some kind of detrending should be applied before spectral analysis (section 4.2.4).

(4) To evaluate possible spectral effects of noise, it is advisable always to analyze an equally long record portion before the signal.

(5) In general, geophysical spectra are only used for comparisons, e.g. of amplitudes at different frequencies or of amplitudes at a given frequency in different sets of data. In other words, they are used for relative studies, and in general not for absolute information. This means that relative spectra are sufficient and that absolute spectra are not needed. Then it is important to note that the relative spectra (used for comparisons) should be comparable, i.e., they must have been obtained under similar or equivalent conditions, concerning both data recording and data handling. The restriction to relative spectra permits the application of arbitrary, but equivalent definitions, as for example of the correlation coefficients (section 3.1.1).

(6) While the preceding rules mainly concern handling of data, it is in every case at least equally important to have a good control on the quality of observed data, records etc. which are analyzed, such as their amplitude and phase characteristics. Broad-band responses are to be preferred at least in preliminary attacks on phenomena of unknown spectral composition. It is important to remember that conclusions from spectra can only be made for the spectral range actually observed, and quite different properties may prevail outside this range.

Chapter 5

RELIABILITY AND PRESENTATION OF SPECTRA

The development so far has led us up to the point where we have obtained the spectral estimates corresponding to any given observational curve or set of data. Before being able to use these estimates for conclusions of geophysical significance, it is necessary to make a closer examination of the calculated spectra. The first task is to try to evaluate the reliability of the spectral estimates.

As the numerical evaluation of spectral parameters will generally constitute the foundation for conclusions to be made, the question of accuracy is significant. In Chapter 4, we have learnt about various limitations to the accuracy of a spectrum and how to minimize errors. In measuring various parameters from observed spectra, additional errors of measurement are bound to be introduced. In any special study, these deserve careful attention.

Another task is to try to express the spectra in some way that is useful for their further study. On the latter point we may distinguish between: (1) quantitative expressions in terms of certain spectral parameters, and (2) graphical presentation in some suitably chosen coordinate system. There is no unique way to describe items 1 and 2, but the choice depends in each case on the particular type of problem studied. Therefore, great variations in the procedures are to be expected.

For a seismogram trace, with one time axis and one amplitude axis, there is not much hesitation which parameters should be used. They are arrival times, amplitudes and corresponding periods, sometimes supplemented by direction of first swing (especially of P-waves). For a seismic spectrum, however, with one frequency axis and an amplitude or power axis, there is no a priori given parameter to be used. And this holds for any kind of geophysical spectra. To find a correct parameter for any spectral study is an important task. In the literature we find quite a number of different parameters for different studies, and the multitude of characteristics used may be quite bewildering to the student. In this chapter we shall aim at presenting a consistent collection of spectral parameters. Some of these may be of more use for some studies, others for other studies.

5.1 RELIABILITY OF HARMONIC ANALYSIS AND SPECTRAL ESTIMATION

5.1.1 Elements of variance analysis

We use the following notation:

$f(t)$ = given function;
$f_k(t)$ = Fourier expansion of $f(t)$ into k harmonics;
σ = standard deviation of $f(t)$;
σ_k = standard deviation of $f(t) - f_k(t)$.
Variance is defined as the square of the standard deviation, hence we have:

$$\left.\begin{aligned}\sigma^2 &= \frac{1}{2\pi}\int_0^{2\pi} [f(t) - a_0]^2\,\mathrm{d}t \\ \sigma_k^2 &= \frac{1}{2\pi}\int_0^{2\pi} [f(t) - f_k(t)]^2\,\mathrm{d}t\end{aligned}\right| \qquad [1]$$

By means of the orthogonality relations for trigonometric functions and the Fourier series expansion (section 2.1.1), cf. Manley (1945, p.174), we then easily find the following equation:

$$\sigma_k^2 = \sigma^2 - \frac{1}{2}\sum_{n=1}^{k}(a_n^2 + b_n^2) \qquad [2]$$

σ_k gives a measure of the accuracy of representing $f(t)$ by a finite number of Fourier terms. As the number of terms increases, this error will decrease. The representation of $f(t)$ can be considered good when σ_k/σ is small. Near discontinuities of $f(t)$, the convergence is particularly poor or absent (Gibbs' phenomenon, section 2.1.2).

It should be noted that the variance has a simple relation to power. If in the first equation of [1] we introduce the expression for $f(t) - a_0$ from the Fourier series expansion and then carry out the integrations, we easily find:

$$\sigma^2 = \frac{1}{2}\sum_{n=1}^{\infty}(a_n^2 + b_n^2) \qquad [3]$$

i.e., the variance is equal to the average a.c. power. For this reason, *variance spectrum* is sometimes used as synonymous with *power spectrum*. Similarly, we find:

$$\frac{1}{2\pi}\int_0^{2\pi} [f(t)]^2\,\mathrm{d}t = a_0^2 + \frac{1}{2}\sum_{n=1}^{\infty}(a_n^2 + b_n^2) \qquad [4]$$

i.e., the average power (left-hand side) is equal to the d.c. power plus the total a.c. power, averaged over one cycle (right-hand side).

The expression for σ^2 refers to the average total a.c. power, but the sum can also be split into terms corresponding to different frequencies or frequency bands. In [4], the sum is over discrete frequencies n. We could rewrite [4] in terms of an integral over a continuous frequency range. The total variance thus gives a measure of the total power, using Parseval's theorem, section 3.3.2:

$$\sigma^2 = \frac{1}{2\pi} \int_{-\infty}^{\infty} |F(\omega)|^2 \, d\omega, \qquad a_0 = 0 \tag{5}$$

and variances over limited frequency ranges are proportional to the corresponding powers. Introducing the cyclic frequency $v = \omega/2\pi$ in [5] we have:

$$\int_{-\infty}^{\infty} \frac{|F(v)|^2}{\sigma^2} \, dv = 1 \tag{6}$$

Therefore, $(|F(v)|^2/\sigma^2)dv$ is the fraction of the total power per frequency interval dv centered at v. Referred to unit frequency interval, this fraction is $|F(v)|^2/\sigma^2$, called *power spectral density*. With reference to section 2.2.2, it is clear that power spectral density can be used in two slightly different forms:

(1) Defined as power per unit frequency interval (absolute spectral density).

(2) Defined as fraction of total power per unit frequency interval (relative spectral density).

Instead of power, we could insert any other spectralized quantity, as amplitude, etc.

5.1.2 Reliability tests in harmonic analysis

Essentially on the basis laid down in section 5.1.1, a number of reliability or significance tests have been developed for the terms in a Fourier series expansion.

Let us assume that we have readings of atmospheric temperature at some point for each minute for a whole year, and that we make a Fourier analysis of these readings. To cover the whole range of frequencies in the same way as outlined in section 4.3.2, we should need an expansion into 19 terms, i.e. from the fundamental period of 1 year down to a period of 2 minutes, corresponding to the Nyquist frequency. The first term, with a period of 1 year, would no doubt be most significant, possibly some of the following terms would also. But at some point, the terms would no doubt cease to have physical significance, and they would just reflect high-frequency "noise", i.e. random fluctuations and errors, in the given trace. Then, the Fourier coefficients would not decrease any more with increasing order, but they would rather scatter around some constant values. This is an indication that the significant series could be truncated at the point where the Fourier coefficients cease to decrease.

In the temperature case, the record length (for example one year) corresponds to the most significant period. This is a general rule for observational series to which one uses to apply harmonic analysis, which presupposes a knowledge of the fundamental period. Procedures in case of lack of such knowledge were discussed in section 3.5.3. In analyses of seismic records, on the other hand, often periods, considerably shorter than the record length, may carry the largest amplitudes ("hidden periodicities").

Stumpff (1937, pp.188–195) and (1940, pp.60–62) devised a method to test the statistical significance of the terms in a harmonic analysis. This method may in brief be described as follows. The whole observation interval is subdivided into a number N of smaller intervals, which are all equal in length and do not overlap (in a way, corresponding to the direct segment method, section 4.6.3). The Fourier coefficients a_n and b_n (for the term to be tested) are calculated for each of the N sub-intervals. For easier apprehension, these are suitably plotted as *periodogram vectors* in the complex $F(\omega)$-plane, whereby their endpoints form a point cloud. Then the following statistics is calculated:

(1) The *expectance* Ex or root-mean-square radius of the point cloud, defined as:

$$Ex = [\sum_N (a_n^2 + b_n^2)]^{1/2}/N \qquad [7]$$

(2) The distance h_c from the origin to the center of gravity of the point cloud, which is obtained from:

$$h_c = [(\sum_N a_n)^2 + (\sum_N b_n)^2]^{1/2}/N \qquad [8]$$

Then, the numerical values of Ex and h_c are compared and the Fourier term n is considered statistically significant if $h_c \geq 3\ Ex$, otherwise not. Such a test has to be made for each term in a Fourier expansion and the series is truncated where the terms begin to lose statistical significance.

Geophysical time series do not consist of completely random numbers, but they show conservation or quasi-persistence, at least for some interval of time. The influence of this property on significance tests has been discussed by Chapman and Bartels (1940, 1951, pp.582–593) and by Conrad and Pollak (1950, pp.396–410). The effect is that in the relation $h_c \geq 3\ Ex$, the factor 3 will be increased, easily by two to three times.

Another significance test of harmonic analysis is due to Fisher (1929). This gives the probability that the ratio:

$$(a_n^2 + b_n^2)_{max} / \underset{N}{\Sigma}(a_n^2 + b_n^2) \qquad [9]$$

exceeds a certain parameter, by which the confidence limits can be assigned. In applications one starts with the maximum $(a_n^2 + b_n^2)$, then proceeds with the next largest and so on, until a certain significance level is lost. The ratio used here can be compared to $(h_c/Ex)^2$ in the earlier test, and we see that the numerators differ, and so do the corresponding significance parameters. Fisher's test has been put into convenient tabular form by Nowroozi (1967), improved by Shimshoni (1971).

Alternatively, it is always possible to recalculate the time function $f(t)$, either from its Fourier series expansion or from inversion of its Fourier transform, and then to compare this $f(t)$ with the originally given curve. This procedure was applied to geomagnetic records by Střeštík (1970).

It could be envisaged that tests, here developed for a Fourier series expansion, properly modified, could be applied to Fourier transforms of geophysical records and other time or space series. The analogous procedure would be to subdivide the total interval and to calculate the transforms for each interval and compare them. A procedure similar to this one has sometimes been applied to analyses of microseismic records, in order to decide the total length of a record necessary to get a reliable spectrum.

5.1.3 Reliability tests of spectral estimates

A full discussion of this item would require a considerable amount of statistics, and only a very brief outline can be given here. We define a quantity, chi-square = χ^2, as a measure of the dispersion of the estimates x_i:

$$\left.\begin{aligned} \chi^2 &= \sum_i (x_i - \bar{x}_i)^2/\bar{x}_i \qquad \text{for } \bar{x}_i \neq 0 \\ \chi^2 &= \sum_i x_i^2 \qquad \text{for } \bar{x}_i = 0 \end{aligned}\right| \qquad [10]$$

Provided that x_i is distributed according to a Gaussian curve, χ^2 is distributed according to a so-called χ^2-distribution, which we denote by $f(\chi^2)$ for simplicity. The function $f(\chi^2)$ contains the number of degrees of freedom μ as a parameter, and it has the property that:

$$\int_0^\infty f(\chi^2)\, d\chi^2 = 1 \qquad [11]$$

Then the probability Pr that χ^2 is located between a and b is as follows:

$$Pr\,(a < \chi^2 < b) = \int_a^b f(\chi^2)\,\mathrm{d}\chi^2 \qquad [12]$$

If the integral from 0 to a equals 0.05 and the integral from b to ∞ also equals 0.05, then $Pr = 0.90$, i.e. there is a 90-% probability (9 chances in 10) that χ^2 is located between a and b. Other limits can be chosen. From [12] it is at least formally clear that the values of a and b depend both on μ and on Pr.

Table XVIII lists 95-% confidence limits from Munk et al. (1959). A corresponding graphical presentation can be found among others in Jenkins and Watts (1968, p.82). The limits a,b have to be multiplied by the respective values of power in each case. These limits mean that there is 95% probability that the true value lies within these limits. Deviations which go beyond these limits can be considered as significant.

The assignment of the number of degrees of freedom therefore enters this discussion. For a Fourier series containing n harmonics, the number of degrees of freedom is $\mu = 2n$, as there are two degrees of freedom associated with each Fourier component. For a Fourier spectrum $F(\omega)$, Blackman and Tukey (1959) give the following formula for the number of degrees of freedom:

$$\mu = 2T\left[\int_0^\infty F(\omega)\,\mathrm{d}\omega\right]^2 \Big/ \int_0^\infty F^2(\omega)\,\mathrm{d}\omega \qquad [13]$$

where T is the total data length and $F(\omega)$ represents the spectrum whose stability is investigated.

TABLE XVIII

Confidence limits of power spectra assuming chi-square distribution (after Munk et al., 1959)

Degrees of freedom (μ)	Confidence limits	Degrees of freedom (μ)	Confidence limits
1	0.2–1000	15	0.55–2.4
2	0.21–40	20	0.59–2.1
3	0.32–14	50	0.69–1.55
4	0.36–8.3	100	0.78–1.35
5	0.39–6.0	150	0.81–1.27
6	0.42–4.8	200	0.83–1.23
8	0.46–3.8	300	0.86–1.18
10	0.49–3.1		

Example: Degrees of freedom $\mu = 100$. The computed value is 10. There is a 95% probability that the true value lies between 7.8 and 13.5.

When power spectra are obtained as Fourier transforms of autocorrelation functions, the number of degrees of freedom of the spectra is $\mu = 2N/m$, where N = the sample size (number of readings) and m = the maximum number of lag values used in the calculation of the autocorrelation. The expression $\mu = 2N/m$ is only approximate but very well sufficient. It can be seen as follows: N = number of data points = total number of elementary frequencies, i.e. $1/N$, $2/N$, ..., $N/N = 1$. $1/m$ = *spectral* width of window used, i.e. its fraction of the total frequency range. Thus, $N \cdot (1/m) = N/m$ = number of elementary frequencies within spectral window. Each frequency corresponds to 2 degrees of freedom, as by virtue of the Fourier theorem there are two constants needed to characterize each frequency component in a waveform. Hence, the total number of constants to be determined, in other words, the total number of degrees of freedom is $= 2N/m$, as given above. For degrees of freedom in case of two coordinates, see for example Kinsman (1965, p.464). On the other hand, when spectra are calculated directly by the FFT method, each raw estimate has two degrees of freedom. Averaging over n estimates yields new estimates with $2n$ degrees of freedom (see section 4.6.3).

In interpretation of spectra, frequently special attention is focussed on

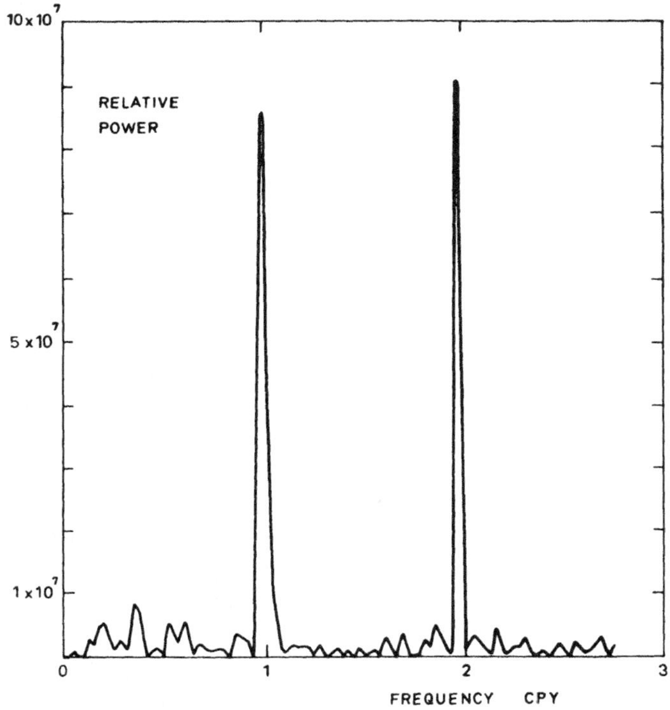

Fig.44. Exceptionally pronounced peaks in power–frequency spectrum of river runoff. After Reed (1971).

maxima and minima in the spectra. It is then of importance to be sure that any maxima or minima, chosen for interpretation, are significant, e.g., that the difference between a maximum and adjacent parts of the spectrum exceeds chosen confidence limits. It is only very rarely that spectra show such pronounced peaks as in Fig.44 and which would make confidence tests superfluous. The rule is that maxima stand out far less conspicuously from their background, and then significance tests are important.

The significance test outlined here assumes some properties of the given time series, e.g. a Gaussian distribution. An alternative, more reliable and purely empirical test of significance, free from any such assumptions, would be to produce spectra for several different samples from the same population, if possible, and then to compare the spectra. Such methods are also substantiated by Tukey (1961, p.210), who says that "there is no substitute for some sort of repetition as a basis for assessing stability of estimates and establishing confidence limits". On the other hand, it can be proved (see especially Jenkins and Watts, 1968) that correlation functions and power spectra for random (stochastic) observational series, as of microseisms, do not converge in any statistical sense by increasing the observational range T. In general, influence from sampling fluctuations may be large, and therefore only really outstanding peaks or troughs in a spectrum could warrant any efforts at physical interpretation. A simple and useful "rule of thumb" is expressed by Stuart et al. (1971), who say that "a spectral line is significant if it contains at least three computed points which deviate from the noise and has a maximum two or three times greater than the surrounding noise level". "It is stressed that independent tests, by repeating the computation with different parameters or redigitized data, is the only sure method of eliminating doubts."

5.1.4 Reliability tests of spectral curves

In section 5.1.3 we were concerned with reliability of digital (point) values of spectral estimates, i.e. the output from a digital computer. In constructing a continuous spectral curve from these digits, additional uncertainty creeps in.

As parameters evaluated from empirically obtained spectral curves are usually made the basis for conclusions regarding the underlying phenomena, it is of great significance to be clear about the degree of accuracy in each case. The question of accuracy concerns partly the accuracy of an obtained spectrum, partly the accuracy of parameters determined from it. Whereas the latter can be determined to any degree of accuracy for any given spectrum, the main limitation in accuracy lies with the determination of the spectral curve itself. Various sources of error and methods to remedy or assess their effects have been discussed above (especially sections 4.2 and 4.3). At this point, we like to illustrate one source of inevitable inaccuracy, i.e. the one arising from connecting discrete points in a given spectrum

TABLE XIX

Example illustrating inaccuracy in straight-line interpolation

Abscissa (frequency) x:	1 read	2 interpolated	3 read
Ordinate (amplitude) y:	1 read	8 interpolated	15 read
Ordinate (power) y^2:	1	64 square of y	225
Ordinate (power) y^2:	1	113 interpolated	225

by straight lines. If we have determined both an amplitude and a power spectrum for exactly the same phenomenon, it is clear that the straight-line approximation cannot hold in both cases. Let us take an example, shown in Table XIX. A straight line in the amplitude spectrum, defined by the points $(x,y) = (1,1)$ and $(3,15)$, converted into a straight line in the power spectrum, defined by the points $(x,y^2) =$ $(1,1)$ and $(3,225)$, gives for the middle point, by interpolation $y^2 = 113$. This should be compared with $y^2 = 64$, which would be obtained if a middle point $(x,y) = (2,8)$ were included in the amplitude spectrum. The difference between 64 and 113 is quite large.

The result can easily be put into a more general form. If the given, read points are (x_1,y_1) and (x_2,y_2), the difference of power between: (1) linear interpolation between squared y-values, and (2) the square of interpolation between given y-values, is easily seen to be $\frac{1}{4}(y_1 - y_2)^2$. In the example above the difference amounts to 49.

Thus, differences for this reason may be quite large, and there is no way to tell offhand which is correct. They are probably both incorrect, because the straight-line interpolation does not hold, not in the amplitude spectrum, nor in the power spectrum. Or, as Tukey (1961, p.207) states it, "drawing smooth freehand curves through spectral estimates is often much more useful than connecting them by segments of straight lines." The result of our discussion above would be the same if we choose to plot spectra as histograms instead of using straight-line connections, even though it could be argued that histograms are to be preferred, considering that spectral estimates are averages over certain frequency ranges and not just point values.

5.1.5 General comments on spectral reliability

It has sometimes been maintained that one should really know the character of the spectrum before it is computed, just to be able to choose measuring techniques and data handling methods in the right way. This has been emphasized especially with regard to meteorological turbulence spectra (Panofsky and Deland, 1959), but has more general validity. However, we seem by this statement to be lumping two problem steps together: (1) measuring techniques to get a record, and (2) spectral analysis of the obtained record.

If all precautions have been taken in item 2, the spectrum will be as true as possible in relation to the given record, but not necessarily in relation to the underlying phenomenon to be analyzed. Even though the two items are closely interrelated, it is preferable to consider them separately, especially as this may assist in deciding where in the operations improvements have to be put in.

In spectral studies it is of importance to differ between absolute accuracy and relative accuracy. In most studies, we are not dealing with just a single spectrum to derive various properties from it. If so, we are certainly concerned with the absolute accuracy of this spectrum. Generally a study consists of comparisons of properties among a lot of spectra, all derived under similar conditions (similar instruments and similar analysis methods). In such cases we are only concerned with the relative accuracy of these spectra, in other words, if a *variation* from one spectrum to another is significant or not. Let us take an example. We may have a single spectrum where a maximum and a minimum fall within the adopted significance range and are thus deemed as insignificant. However, if we have ten spectra, obtained under similar conditions, all of them showing the corresponding maxima and minima, then we are permitted to put some significance into them, even though each individual spectrum would be insignificant. Alternatively, a significance test comprising the whole set of spectra could be formulated. This is achieved from the consideration that a certain confidence interval will decrease proportionally to $1/\sqrt{\mu}$ for an increase of the number μ of degrees of freedom. Most spectral studies in fact work with such comparisons, trying to establish how a certain feature changes from spectrum to spectrum with corresponding modification in experimental conditions.

Moreover, we have to consider the different character of different kinds of data, especially regarding their duration. Many series, such as in meteorology, oceanography, geomagnetism, etc., have in fact an infinite duration in time. The seismological counterpart consists of microseismic records. In such cases, it is important that the portion selected for analysis is long enough to be representative. It is in this connection that the request for stationarity comes in. The series, to be spectralized, should be stationary in time and/or homogeneous in space so that its statistical properties, such as averages over the selected portion, will have a meaning. It is different with transient phenomena, such as seismic signals, where it is generally possible to include the whole duration of the signal in a spectral analysis, and the problem of stationarity does not enter in general. Power spectra of non-stationary records will show a superposition of stationary frequencies and of peaks due to non-stationarity.

Considering the discussion in the last two paragraphs, especially the need for homogeneous measurements and for representative window lengths in the time domain, we understand that great attention has to be paid to the choice of window length. In any set of data whose spectra shall be compared, this length should be kept constant. Moreover, the window should be long enough to include the whole of a transient signal or a representative portion of a stationary signal. The use of a constant window length is particularly important in the latter case. Truncation may severely distort a spectrum, introducing spurious minima and maxima, etc., and this has to be considered in any interpretation of spectra (sections 3.5.2 and 4.4.8).

At least, some general rules to assess the reliability of a spectrum can be given. The same record can be spectralized under varying conditions, e.g. varying record length, varying sampling intervals, etc., and the results compared. If available, different records representing the same statistical population, e.g. different portions of records of microseisms, could also be analyzed and results compared, assuming record stationarity to hold. Spectral analysis of theoretical curves, for which spectra are known analytically, also affords a good check on the methods (see section 2.4.3). If methods of unknown reliability are used for spectral analysis, then it is advisable to check them against other known methods (cf. Parks, 1960) or with results for theoretical functions.

5.2 SPECTRAL PARAMETERS—THEORETICAL APPROACH

5.2.1 Integrals and derivatives

Given, quite generally, a dependent quantity y and an independent variable x, i.e. $y(x)$, there are naturally a number of ways to describe this relation. If $y(x)$ is given in analytical form (in some cases together with limiting conditions), this provides for all necessary and sufficient information of the relation. Any wanted property can be derived analytically from the given relation $y(x)$. Likewise, if $y(x)$ represents an empirical equation, obtained for example by least-square fitting to observational data, this equation also provides for all needed information. In this case, the information will be loaded with some errors depending upon the accuracy with which the empirical equation $y(x)$ represents the observed data.

Applying these considerations to a spectrum $F(\omega)$, we can state that the analytical representation is possible in case the original function $f(t)$ is given in analytical form. We have given examples of this in Table V. The equations for $F(\omega)$, there given, provide for complete information on the spectra. In the observational case, being the most important case for us, however, there is generally no corresponding procedure. It is not customary to reproduce the spectrum as an empirically derived equation, which is then used for calculating properties of $F(\omega)$. We could envisage expansion of $F(\omega)$ into a Fourier series or alternatively, to calculate the Fourier transform of $F(\omega)$. The latter procedure or some modifi-

TABLE XX

Specification of certain integrals and derivatives of $y(x)$

n	$\int x^n\, y(x)\, \mathrm{d}x$	m	$y^{(m)}(x)$
0	area under the curve	0	level (height) of the curve
1	first moment with respect to the y-axis	1	slope of the curve
2	second moment or moment of inertia with respect to the y-axis	2	curvature of the curve

cation of it, is in fact materialized in the cepstrum analysis (section 3.6.3). Beyond that, there is hardly any such procedure.

But what are the alternatives in representing a function? Instead of giving it in mathematical form, or giving numerical values of constants for some specified functional form, we could give other parameters which in some way are related to these constants. Of course, the most complete and most original way to do this, is just to report the calculated values of $F(\omega)$, i.e. the output from the computer. Even though these contain all available information, they have from our present viewpoint to be considered only as raw material, and we have to reproduce them in some form with a clear intelligible meaning.

Looking again at our general expression $y(x)$ we could envisage a number of ways to do this, by means of integrals and derivatives:

$$\int x^n\, y(x)\, \mathrm{d}x, \qquad y^{(m)}(x), \qquad n, m, = 0, 1, 2, \ldots \tag{14}$$

Normally, in specifying properties of a function, only a few terms of these are considered, as summarized in Table XX. The names given to the two latter integrals in Table XX derive from the fact that the differential $y\mathrm{d}x$ is proportional to the mass of a plate of the shape of $y(x)$ and of uniform thickness and density. The parameters listed are essentially a characterization of y. But in addition to these parameters, also the corresponding values of x are of significance, especially the values of x where $y(x)$ exhibits maxima, minima, cusps or steep slopes, etc.

5.2.2 Moments of distributions

For a description of spectral curves, a technique developed already in statistics for application to probability distributions is immediately applicable. We give here a summary of such parameters and refer the reader to statistical textbooks for more details. The book by Aitken (1962) describes the various measures in a simple way. Several definitions refer to specific values of the frequency ω; this is the case for parameters 1, 3–5 below. Moments provide useful characteristic measures; they are used in nearly all parameters below. The definitions have here been given only for $F(\omega)$, but can equally well be applied to power spectrum $|F(\omega)|^2$ or to corresponding wavenumber spectra, also in several components. Moreover, integrations are extended to infinite frequency, although in the practical case this has naturally to be replaced by a finite upper limit.

(1) *Mode* = the frequency ω_x for which the ordinate is maximum:

$$|F(\omega_x)| = \max \qquad [15]$$

Spectra with more than one maximum are termed *dimodal* for two maxima, etc., and *multimodal* for several maxima.

(2) *Moment about* $\omega = 0$:

$$M_n = \int_0^\infty \omega^n \, F(\omega) \, d\omega \qquad [16]$$

n is the order of the moment, and $\omega \geq 0$.

(3) *Median* = the frequency ω_m by which the area under the spectrum is divided into two equal halves:

$$\int_0^{\omega_m} F(\omega) \, d\omega = \int_{\omega_m}^\infty F(\omega) \, d\omega = \frac{1}{2} \int_0^\infty F(\omega) \, d\omega$$

or:

$$[M_0]_0^{\omega_m} = [M_0]_{\omega_m}^\infty = \frac{1}{2} M_0 \qquad [17]$$

(4) *Quartiles* = the frequencies ω_{q1} and ω_{q2} by which the lower 1/4 and the upper 1/4 of the area under the spectrum are separated:

$$\int_0^{\omega_{q1}} F(\omega) \, d\omega = \int_{\omega_{q2}}^\infty F(\omega) \, d\omega = \frac{1}{4} \int_0^\infty F(\omega) \, d\omega$$

or:

$$[M_0]_0^{\omega_{q1}} = [M_0]_{\omega_{q2}}^\infty = \frac{1}{4} M_0 \qquad [18]$$

(5) *Arithmetic mean* = the frequency ω_c of the centroid (center of gravity):

$$\omega_c = \int_0^\infty \omega\, F(\omega)\, \mathrm{d}\omega \Big/ \int_0^\infty F(\omega)\, \mathrm{d}\omega = M_1/M_0 \qquad [19]$$

(6) *Moments about the mean*:

$$m_n = \int_0^\infty (\omega - \omega_c)^n\, F(\omega)\, \mathrm{d}\omega \qquad [20]$$

n is the order of the moment, and $\omega \geq 0$.

(7) *Second moment*, also referred to as variance:

$$m_2 = \int_0^\infty (\omega - \omega_c)^2\, F(\omega)\, \mathrm{d}\omega \qquad [21]$$

(8) *Width, dispersion or spread*: there is a whole series of measures for this parameter, and we shall return to more details later, in section 5.3.2.

(9) *Skewness* = the degree of asymmetry of a spectrum, defined as:

$$m_3 = \int_0^\infty (\omega - \omega_c)^3\, F(\omega)\, \mathrm{d}\omega$$

or in absolute measure (normalized):

$$m_3/m_2^{3/2} \qquad [22]$$

(10) *Kurtosis* = the degree of flattening of the spectrum, defined as:

$$m_4 = \int_0^\infty (\omega - \omega_c)^4\, F(\omega)\, \mathrm{d}\omega$$

or normalized:

$$m_4/m_2^2 \qquad [23]$$

Moments higher than the fourth have no practical or theoretical application to the spectral characterization.

5.2.3 Relations between moments and derivatives

The following rules get their simplest forms for $\omega = 0$. As this frequency is generally not covered by observed spectra, the rules may be more of a mathematical interest than a true practical value.

We let $f(t) \leftrightarrow F(\omega)$ be a Fourier pair. Then we first prove the following general formula:

$$F^{(n)}(0) = \frac{d^n F(0)}{d\omega^n} = (-i)^n \int_{-\infty}^{\infty} t^n f(t)\, dt = (-i)^n m_n \qquad [24]$$

where m_n is the n-th moment of $f(t)$.

Proof. We get two series expansions of $F(\omega)$, first by a power series expansion of $e^{-i\omega t}$ in the Fourier transform (integration term by term is permitted for finite moments):

$$F(\omega) = \int_{-\infty}^{\infty} f(t) \sum_{n=0}^{\infty} \frac{(-i\omega t)^n}{n!} dt = \sum_{n=0}^{\infty} (-i)^n m_n \frac{\omega^n}{n!} \qquad [25]$$

secondly by a Taylor series expansion:

$$F(\omega) = \sum_{n=0}^{\infty} \frac{d^n F(0)}{d\omega^n} \frac{\omega^n}{n!} \qquad [26]$$

Identifying coefficients of equal powers of ω in [25] and [26], we find [24].

The general formula [24] permits immediate deduction of a number of important special cases:

(1) Putting $n = 0$ (for zeroth moment) in [24] we find:

$$F(0) = \int_{-\infty}^{\infty} f(t)\, dt \qquad [27]$$

i.e., $F(0)$ is equal to the area under the curve $f(t)$.

(2) $n = 1$ (for first moment) in [24] gives the following formula:

$$F'(0) = -i \int_{-\infty}^{\infty} t f(t)\, dt \qquad [28]$$

i.e., the slope of $F(\omega)$ at the origin is proportional to the first moment of $f(t)$.

(3) Taking the ratio of [28] and [27] we obtain:

$$\frac{F'(0)}{F(0)} = -i \frac{\int_{-\infty}^{\infty} t f(t)\, \mathrm{d}t}{\int_{-\infty}^{\infty} f(t)\, \mathrm{d}t} = -i \cdot t_c \qquad [29]$$

where t_c is the abscissa of the centroid (center of gravity) for $f(t)$.

(4) Putting $n = 2$ (for the second moment or moment of inertia) in [24] gives:

$$F''(0) = -\int_{-\infty}^{\infty} t^2 f(t)\, \mathrm{d}t \qquad [30]$$

i.e., the curvature $F''(\omega)$ of $F(\omega)$ at its origin is proportional to the second moment or the moment of inertia of $f(t)$. This is nothing but another expression of the reciprocal spreading in time and frequency domains. Cf. Robinson (1967a).

(5) Taking the ratio of [30] and [27], we find:

$$\frac{F''(0)}{F(0)} = -\frac{\int_{-\infty}^{\infty} t^2 f(t)\, \mathrm{d}t}{\int_{-\infty}^{\infty} f(t)\, \mathrm{d}t} = -\overline{t^2} \qquad [31]$$

where $\overline{t^2}$ represents the mean-square abscissa for $f(t)$. This quantity has a significance as the square of the radius of gyration in dynamics, and is related to the variance of a frequency distribution function $f(t)$ in statistics.

From the Fourier transform formula, we find in addition the following upper limits for $|F(\omega)|$ and $|F'(\omega)|$:

$$\left.\begin{aligned} |F(\omega)| &\leq \int_{-\infty}^{\infty} |f(t)|\, \mathrm{d}t \\ |F'(\omega)| &\leq \int_{-\infty}^{\infty} |t f(t)|\, \mathrm{d}t \end{aligned}\right| \qquad [32]$$

Proceeding in the same way as above, but instead starting from the inversion formula for $f(t)$ we derive the following formula, which is an exact counterpart to [24]:

$$f^{(n)}(0) = \frac{\mathrm{d}^n f(0)}{\mathrm{d}t^n} = \frac{i^n}{2\pi} \int_{-\infty}^{\infty} \omega^n F(\omega)\,\mathrm{d}\omega = \frac{i^n}{2\pi} M_n \qquad [33]$$

Here M_n is the n-th moment of the $F(\omega)$-curve. We can derive a number of special cases from this formula, just as we have done from [24]. This is left as an exercise to the reader, and we limit ourselves to the special case $n = 0$:

$$2\pi f(0) = \int_{-\infty}^{\infty} F(\omega)\,\mathrm{d}\omega \qquad [34]$$

Thus, the area under a spectrum curve is proportional to the value of the inverse function $f(t)$ at its origin. In analogy to [29] it is also easily found that the absicssa ω_c of the centroid of $F(\omega)$ satisfies the following equation:

$$f'(0)/f(0) = i\omega_c \qquad [35]$$

We do not give any examples to illustrate these formulas, but as an exercise the student is advised to test the formulas on the examples given in Table V. The formulas presented here may also serve as convenient checks on any derived Fourier transform.

The moment formulas are very useful in framing many expressions. As just one exercise, we find easily that the variance of acceleration is simply related to the fourth-order moment of the power spectrum $E(\omega)$ of $f(t)$:

$$\sigma^2 = \int_{-\infty}^{\infty} |f''(t)|^2\,\mathrm{d}t = \frac{1}{2\pi} \int_{-\infty}^{\infty} \omega^4\,|F(\omega)|^2\,\mathrm{d}\omega =$$

$$= \frac{1}{2\pi} \int_{-\infty}^{\infty} \omega^4\,E(\omega)\,\mathrm{d}\omega = \frac{1}{2\pi}\,[M_4]_{-\infty}^{\infty} \qquad [36]$$

This is seen by the derivation theorem (2.3.6), equation [5] above, Parseval's theorem [19] in Chapter 3, and by section 3.3.3.

The formulas above have been given for $F(\omega)$ without splitting them into formulas for the amplitude spectrum $|F(\omega)|$ and the phase spectrum $\Phi(\omega)$. As a hint to such an extension we find that the following formula is obtained by repeating the general derivation above:

$$\frac{d^n|F(0)|}{d\omega^n} = e^{-i\Phi(\omega)}(-i)^n \int_{-\infty}^{\infty} t^n f(t)\, dt \qquad [37]$$

from which special cases can be immediately obtained as above. However, in applications, $F(0)$ is usually not observed, referring as it does to an infinite period.

The moment formulas have equal validity for a space variable as for the time t and they can also be extended to more than one variable, defining the moment m_{pq} for two variables x and y as follows:

$$m_{pq} = \int_{-\infty}^{\infty}\int_{-\infty}^{\infty} x^p\, y^q f(x,y)\, dx\, dy \qquad [38]$$

Proofs proceed analogously to the case of one variable and are left as an exercise.

5.3 SPECTRAL PARAMETERS—EMPIRICAL APPROACH

In practical application to spectra, only some of the parameters discussed in the general section 5.2 are of use. We can conveniently differ between parameters based on derivatives of different orders (section 5.3.1) and those based on integrals (section 5.3.2). The comments apply to amplitude or power spectra or other spectra equally well.

5.3.1 Level, slope and curvature of a spectrum

The level (height), slope and curvature of a spectrum represent its zeroth, first and second derivative, respectively.

Due to the complexity and uncertainty of computed spectra, most interpretations are made only qualitatively or by comparison between spectra obtained under similar conditions. For example, the amplitude $y(x)$ or $F(\omega)$ or any quantity depending on it, does not have much absolute significance in a spectrum. On the other hand, the abscissas or frequencies x or ω corresponding to special features of the spectrum, such as peaks, troughs, slopes, etc., have much more of absolute significance and are therefore usually quoted. The ordinates are mostly used only in comparisons of spectra.

Some particular features in spectra have received more attention than others. Thus, maxima and minima in spectra have repeatedly been interpreted as expressions of constructive and destructive interference, respectively, between different wave trains. In such cases, most emphasis has been laid on the frequencies at which maxima and minima are found than on the actual height of the spectral curve at these points.

Maxima and minima in a curve are characterized by a vanishing first derivative, i.e., $y'(x) = 0$ with $y''(x) < 0$ for maximum, $y''(x) > 0$ for minimum and $y''(x) = 0$ for an inflection point. We have to note that spectra frequently exhibit cusps instead of true maxima and minima. At a cusp we have $y'(x) = \infty$ and $y''(x) < 0$ for an upward cusp and $y''(x) > 0$ for a downward cusp.

However, it is important to note that spectra also exhibit isolated minima, not due to any interference. This is the case when the analyzed wave motion consists of two or several components of different origin with well separated spectral peaks. Then, minima will naturally show up between these peaks, but such minima do not carry any information in themselves. The microseismic spectrum offers several exceptionally clear examples of such fortuitous minima (for more details, see section 9.3.3). Likewise, ocean wave spectra are expected to show similar minima, if the spectra cover a sufficiently broad band.

Likewise, in using spectral slope as a parameter, there may be spurious influences due to the data handling. We have seen in sections 3.5.1 and 4.4.8 that a longer data window will lead to higher resolution and thus greater local slopes in a spectrum than a shorter data window. Therefore, it is necessary in using slope as spectral parameter to consider the following:

(1) Like most other parameters, slope is useful only as a relative parameter.

(2) In comparing slopes of different spectra, these must have been calculated under the same conditions; especially the data windows must have equal length.

(3) Local slopes should not be used, due to their uncertainty, but only average slopes over some frequency range should be trusted.

In many cases, spectra exhibit very pronounced slopes or steep rises, which are not a result just of data handling but represent some true property of the studied phenomenon. For instance, in seismology, such a steep slope may correspond to the onset of a new wave mode, and the steep rise appears at its cut-off frequency. Several successive steep rises may then correspond to successive modes, each with its characteristic cut-off frequency. If such a case is at hand, it can be checked by the fact that the cut-off frequencies should bear simple relations to each other.

Slopes of spectra have clear physical significance in many geophysical problems. Then, the slope is frequently expressed as $\mathrm{d} \ln E(\omega)/\mathrm{d}\omega$ or $\mathrm{d} \ln E(k)/\mathrm{d}k$ (E = power, ω = frequency, k = wavenumber). We will come across such slopes at several places in this book, but for convenience we summarize the most important cases here:

(1) In seismology, for example to indicate depth dependence (section 8.4.1) or attenuation (section 7.4).

(2) In oceanography, meteorological turbulence, especially for the so-called inertial subrange, and in microseismic studies (Chapter 9).

(3) In geomagnetism and gravity, to give estimates of depth to certain bodies which cause magnetic or gravity anomalies, respectively (Chapter 10).

5.3.2 *Area and width of a spectrum*

Area and width of a spectrum will include integrals over the spectral curve. In practical cases, we are only dealing with positive frequencies. Integrations over frequency are then at most extended from zero or the lowest frequency involved up to the Nyquist or folding frequency (section 4.3.2). Integrations over more restricted frequency ranges also exist. The area under the spectral curve has been used mainly for calculation of spectral energy and of centroid coordinates.

(1) *Spectral energy.* In the case of power spectra where the ordinate is $|F(\omega)|^2$, the area under the curve corresponds to the energy content of the signal, as is evident from Parseval's theorem (section 3.3.2):

$$\int_{-\infty}^{\infty} |f(t)|^2 \, dt = \frac{1}{2\pi} \int_{-\infty}^{\infty} |F(\omega)|^2 \, d\omega \qquad [39]$$

(2) *Spectral centroid.* The centroid corresponds to the center of gravity for the area under the curve. For a spectrum $F(\omega)$ the coordinates of the centroid become as follows:

$$\left.\begin{aligned} \omega_c &= \int_0^{\omega_N} \omega \, F(\omega) \, d\omega \Big/ \int_0^{\omega_N} F(\omega) \, d\omega \\ F_c &= \int_0^{\omega_N} |F(\omega)|^2 \, d\omega \Big/ 2 \int_0^{\omega_N} F(\omega) \, d\omega \end{aligned}\right| \qquad [40]$$

where ω_N = the Nyquist frequency (Fig.45). In applications, the logarithms of ω_c and F_c may be better to use. One example is given by Al-Sadi (1973) who used

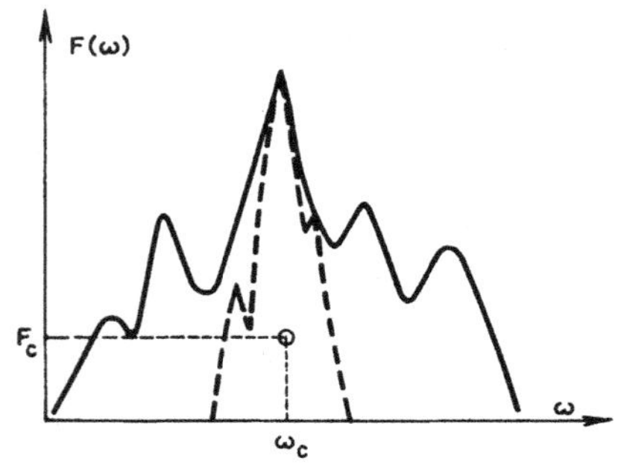

Fig.45. Sketch demonstrating the same centroid (ω_c, F_c) for two spectral curves of different shape.

the centroid coordinates of the power spectrum $E(\omega) = |F(\omega)|^2$ with advantage.

The significance of the centroid coordinates is that they indicate where in the spectrum the energy is mainly concentrated, and thus furnish additional information to just the area under the curve. On the other hand, they do not provide any information on the shape of the spectrum. For instance, the two spectra sketched in Fig.45 have the same centroid coordinates. The missing information is furnished by measuring the width of the spectrum.

(3) *Cumulative spectrum.* Another occasion when the area under the curve

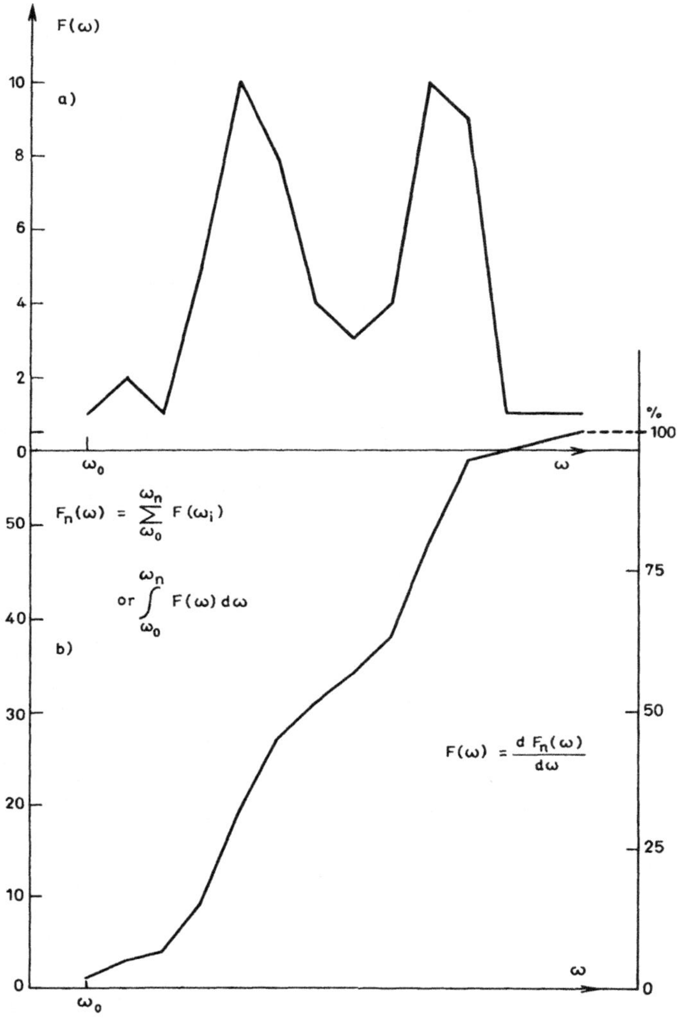

Fig.46. An artificial spectrum $F(\omega)$ in a) and its cumulative spectrum in b).

enters is the cumulative spectrum (or frequency-evolutionary spectrum), defined as the running sum of the area under a spectrum:

$$F_n(\omega) = \int_{\omega_0}^{\omega_n} F(\omega)\,\mathrm{d}\omega \rightarrow \sum_{i=0}^{n} F(\omega_i) \qquad [41]$$

Being the integral of the given spectrum $F(\omega)$, the cumulative spectrum is considerably smoother and has been recommended for obtaining mean spectra from an ensemble of measurements (Duclaux et al., 1969). A constructed spectrum $F(\omega)$ and its cumulant are shown in Fig.46.

In the definition [41] the lower limit ω_0 is kept constant, while the upper limit ω_n runs through the spectrum. Alternatively, we could have both limits running, which defines a parameter like:

$$\int_{\omega_n}^{\omega_{n+1}} F(\omega)\,\mathrm{d}\omega \qquad [42]$$

This is called the *spectral distribution function.* The integral over a spectrum between any two frequency limits, equivalent to the area under the curve, appears sometimes under the name of *integrated spectrum* (e.g., Jenkins and Watts, 1968, p.224).

(4) *Spectral width.* To investigate the energy spread on either side of the spectrum centroid, a technique analogous to the computation of the standard deviation in statistics is applied to power spectra. The *spectral-energy spread* $\Delta\omega_s$ of a power spectrum is defined as:

$$\Delta\omega_s = \left[\frac{\int_{-\infty}^{\infty} (\omega - \omega_c)^2\, |F(\omega)|^2\, \mathrm{d}\omega}{\int_{-\infty}^{\infty} |F(\omega)|^2\, \mathrm{d}\omega} \right]^{\frac{1}{2}} \qquad [43]$$

For $\omega_c = 0$, $\Delta\omega_c$ becomes the *radius of gyration* of the power spectrum.

Width is obviously a characteristic measure of a spectrum. In addition to the width measure in [43], several other widths have been defined. The *equivalent width* is here defined as:

$$\Delta\omega_e = \int_{-\infty}^{\infty} F(\omega)\,\mathrm{d}\omega \bigg/ F(0) \qquad [44]$$

and the *mean-square width* as:

$$\overline{\omega^2} = \int_{-\infty}^{\infty} \omega^2 F(\omega)\, d\omega \bigg/ \int_{-\infty}^{\infty} F(\omega)\, d\omega \qquad [45]$$

Note that equivalent width sometimes appears with a different definition (see, for example, Blackman and Tukey, 1959). We could envisage a generalization of the width formula [45] as follows:

$$\int_{-\infty}^{\infty} \omega^n F(\omega)\, d\omega \bigg/ \int_{-\infty}^{\infty} F(\omega)\, d\omega \qquad [46]$$

where $n = 1, 2, 3, \ldots$ In this form or the corresponding discrete summation form, the formula has proved useful for discriminating between earthquakes and explosions (section 8.2).

Among other ways to express the concentration in a spectrum, especially power spectrum, we indicate the range $\omega_1 - \omega_2$ (called *active band* by Kharkevich, 1960), within which a certain given fraction η of the total power falls:

$$\eta = \int_{\omega_1}^{\omega_2} |F(\omega)|^2\, d\omega \bigg/ \int_{-\infty}^{\infty} |F(\omega)|^2\, d\omega \qquad [47]$$

It gives a quantitative measure of the fraction of the total power that is contained between two specified frequency limits. Like other formulas in this section, also this quantity can be applied to any kind of curve, in any domain.

Analogous formulas can be used for measuring width of any part, e.g. a maximum, of a spectrum. Such width measures have importance in relation to resolution. For example, if a spectral maximum has a width exceeding the difference between two significant spectral components, these cannot be resolved. The resolution can then be effected by using a longer record.

Other quantities related to the spectral widths exist in the literature. For example, Neumann and Pierson (1966, p.343) define the average time interval between zero upcrosses in an ocean-wave record from the mean-square width of the corresponding power spectrum as follows:

$$\overline{T} = 2\pi/(\overline{\omega^2})^{\frac{1}{2}} \qquad [48]$$

In many cases, spectra are smoothed in the hope of getting more stable estimates, sacrificing some of the resolution. Considering the frequent use of the area under the spectral curve, it is to be recommended in such smoothing to apply an equal-area principle. That is, replacing the observed spectrum by a smooth curve or a step curve such that the area under the curve is preserved, may yield more reliable spectral levels within each frequency band. The division into frequency ranges is often a rather delicate undertaking, and has to be decided from case to case. If very wide frequency ranges are used, then significant features may be smoothed out; on the other hand, with narrow ranges, the smoothing may be too inefficient.

The formulas have in this section been written in integral form. In practical applications, these have to be replaced by discrete summation formulas. This is possible by direct application of the principles in section 4.5.1 and there is no need to give them here.

The parameters of this and the preceding section are the most common and useful ones. But in any particular study, it is possible to choose other parameters, mostly modifications of those given here but better adapted to the problem being studied. Examples of several such parameters were given by Al-Sadi (1973) in a study of P-wave spectra in relation to focal depth (section 8.4.1).

5.3.3 Reciprocal spreading

The width measures provide us with another method to demonstrate the principle of reciprocal spreading between time and frequency domains, which we encountered already in section 2.4.4.

Equivalent widths are given in Table XXI for a number of analytical functions from Table V, both for the time and the frequency domains. In addition, in Table XXII and Fig.47 we have given all the width measures listed above for one function, the Gaussian function (cf. example 25 in Table V). Table XXII includes also the products of the corresponding widths in time and frequency domains. This

TABLE XXI

Equivalent widths for a number of functions selected from Table V, section 2.4.3

Function	Equivalent width	
	time function	spectrum
$\Pi(t)$	1	2π
sinc t	1	2π
$\Pi(t)\cos\omega_0 t$	sinc $(\omega_0/2\pi)$	$2\pi/\text{sinc}\,(\omega_0/2\pi)$
$\Lambda(t)$	1	2π
$\delta(t)$	1	2π
$e^{-a\|t\|}$	$2/a$	πa
e^{-at^2}	$\sqrt{(\pi/a)}$	$2\sqrt{(\pi a)}$

TABLE XXII

Widths of the Gaussian curve in time and frequency domains (see also Fig.47)

Width measure	e^{-at^2}	$\sqrt{\frac{\pi}{a}}\, e^{-\omega^2/4a}$	Product
Energy spread	$1/2\sqrt{a}$	$\sqrt{a}$	1/2
Equivalent width	$\sqrt{\pi/a}$	$2\sqrt{\pi a}$	2π
Mean-square width	$1/2a$	$2a$	1
Root-mean-square width	$1/\sqrt{2a}$	$\sqrt{2a}$	1

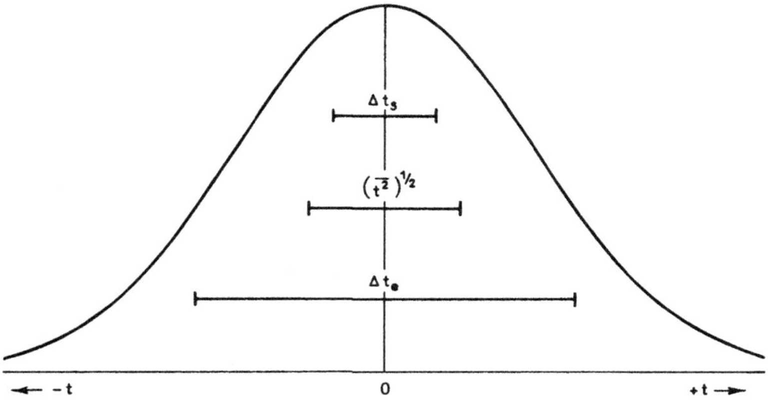

Fig.47. Different width measures illustrated for the Gaussian function in the form $e^{-\pi t^2}$. Notation: Δt_s = energy spread, $(\overline{t^2})^{1/2}$ = rms width, Δt_e = equivalent width. Cf. Table XXII.

product is always a constant. In fact, it is easy to prove from the transform formula, that the product of the equivalent widths in time and frequency domains is always = 2π, i.e.:

$$\Delta t_e \cdot \Delta\omega_e = 2\pi \qquad [49]$$

Equation (49) expresses the *uncertainty principle*, with which we became familiar in section 2.4.4. It expresses the reciprocal relation between the time and frequency domains. It is in fact the same relation that found its expression in the time-scaling theorem or similarity theorem (2.3.3). Moreover, we can state that it has every resemblance to Heisenberg's uncertainty principle (also called principle of indeterminism) in quantum mechanics.

Let us take one striking example of [49]. Suppose that we have a sine-wave of one single frequency and of infinite extent in the time domain. Then, $\Delta t_e = \infty$ and by [49], $\Delta\omega_e = 0$, i.e., we would have a sharp spectral line, corresponding to the given sine-wave frequency. But if we truncate the same sine-wave, then Δt_e will be finite and also $\Delta\omega_e$ finite, i.e., we do not get any more a sharp spectral line, but a blurred spectrum centered around the sine-wave spectrum. The smaller Δt_e is, the greater is obviously $\Delta\omega_e$ by [49], i.e., the more blurred is the spectrum.

TABLE XXIII

Equivalent widths for window functions discussed in section 4.4

Reference equation in Chapter 4	Window function	Equivalent width	
		time window (limited extent)	spectral window (unlimited extent)
30	rectangular	$2T$	$\frac{\pi}{T} = \frac{3.14}{T}$
32	Fourier kernel	$\frac{2T}{\pi}\,\mathrm{Si}(\pi) = 1.18\ T$	$\frac{\pi^2}{T\,\mathrm{Si}(\pi)} = \frac{5.33}{T}$
34	Fejér kernel	$\frac{2T}{\pi}\,\mathrm{Si}(2\pi) = 0.90\ T$	$\frac{\pi^2}{T\,\mathrm{Si}(2\pi)} = \frac{6.96}{T}$
36	Hanning	T	$\frac{2\pi}{T} = \frac{6.28}{T}$
39	Hamming	$1.08\ T$	$\frac{\pi}{0.54\ T} = \frac{5.82}{T}$
42	cosine-tapered rectangular	$1.8\ T$	$\frac{\pi}{0.9\ T} = \frac{3.49}{T}$
44	power	$2T\frac{m}{m+1} = \begin{cases} 1.33\ T\ (m=2) \\ 1.67\ T\ (m=5) \\ 1.90\ T\ (m=20) \end{cases}$	$\frac{\pi}{T}\,\frac{m+1}{m} = \begin{cases} \frac{4.71}{T}\ (m=2) \\ \frac{3.77}{T}\ (m=5) \\ \frac{3.30}{T}\ (m=20) \end{cases}$
49	triangular	T	$\frac{2\pi}{T} = \frac{6.28}{T}$
51	Gaussian	$\sqrt{\frac{\pi}{a}}\ \mathrm{erf}(T\sqrt{a})$	$2\sqrt{\pi a}$

Thus, increased accuracy in timing entails decreased accuracy in frequency determination, and vice versa. The most extreme example of this result and of the uncertainty principle consists of the unit spike in the time domain which has a "white" spectrum and vice versa (examples 9 and 10 in Table V). Similarly, the product of the energy spreads $\Delta t_s \, \Delta\omega_s$, which is 1/2 for the Gaussian curves (Table XXII), can be generalized and shown to be $\geq 1/2$ in general (for proof, see Bracewell, 1965, p.160).

For window functions (section 4.4), considerations of spectral widths are of special significance, and for that purpose we have collected in Table XXIII the expressions for the equivalent widths for a number of window functions. For any given time window of length T, the equivalent widths of the spectral windows can be immediately compared with each other.

5.4 SPECTRAL COORDINATES

5.4.1 Logarithmic versus linear scales

In determination of areas under a spectral curve, it is necessary to pay attention to the coordinates used. If both the ordinate $F(\omega)$ and the abscissa ω are in linear scales, the area under the curve between two frequencies is given simply by the integral:

$$\int_{\omega_1}^{\omega_2} F(\omega) \, \mathrm{d}\omega \qquad [50]$$

However, in many applications, for instance, in meteorological turbulence studies, the frequency range is very wide and then a logarithmic frequency scale is more practical. However, to preserve the value [50] of the area under the curve, it is then necessary to plot $\omega F(\omega)$ as ordinate and not simply $F(\omega)$. This is clear as $F(\omega)\,\mathrm{d}\omega = \omega\, F(\omega)\,\mathrm{d}\ln\omega$. Such a spectrum is referred to as a *logarithmic spectrum*. Corresponding scale modification has to be made in every case that aerial measurements are essential and modified coordinates are used, the principle being to conserve the value expressed in [50]; alternatively, recalculations of measured areas will be necessary. The rule given holds whether $F(\omega)$ is an amplitude or a power spectrum. In case of power spectra, the ordinate is given as power density, $|F(\omega)|^2$, i.e. as power per unit frequency range, while $\omega|F(\omega)|^2$ expresses power per unit time interval (section 2.2.2).

However, it is important to observe that whereas this coordinate transformation preserves the areas under the curves, it does not preserve other properties, e.g. the position of maxima or minima in the spectra. The following relation is easily shown to hold:

$$\frac{\mathrm{d}\ \omega F(\omega)}{\mathrm{d}\ \ln\omega} = \omega^2\frac{\mathrm{d}F(\omega)}{\mathrm{d}\omega} + \omega F(\omega) \qquad [51]$$

Natural logarithms are used, unless otherwise indicated. In practical applications, these are replaced by logarithms to the base 10, with no change of properties. If we consider a stationary point (maximum, minimum, inflection) in the transformed system, i.e. that $\mathrm{d}\ \omega F(\omega)/\mathrm{d}\ \ln\omega = 0$, then it follows from [51] that $\mathrm{d}F/\mathrm{d}\omega = -F/\omega$, which means a downslope of F with ω. Conversely, a stationary point in the F-ω-system, i.e. $\mathrm{d}F/\mathrm{d}\omega = 0$, entails that $\mathrm{d}\omega F(\omega)/\mathrm{d}\ \ln\omega = \omega F(\omega)$, i.e. an upslope in the transformed system. Even though maxima and minima do practically coincide in the two systems in case of especially pronounced peaks or troughs (this is easily shown for the case when $F(\omega)$ is a Gaussian curve), this phenomenon is of importance in other cases.

Still another way of representing the area under a curve is used, as seen by the expression below:

$$\int_{\omega_1}^{\omega_2} F(\omega)\,\mathrm{d}\omega = \int_{\omega_1}^{\omega_2} \omega\,F(\omega)\,\mathrm{d}\ln\omega = \ln 2\cdot\int_{\omega_1}^{\omega_2} \omega\,F(\omega)\,\mathrm{d}\log_2\omega \qquad [52]$$

This has importance in expressing, for example, power per *octave*. If $F(\omega)$ represents power, then $\ln 2\cdot\omega F(\omega)$ is the power per $\mathrm{d}\log_2\omega = 1$, i.e. per octave. One octave is a frequency range corresponding to a doubling of the frequency (log = logarithm to the base 10):

$$\mathrm{d}\log_2\omega = \mathrm{d}\log\omega/\log 2 = 1$$

$$\log(\omega_b/\omega_a) = \log 2$$

$$\omega_b = 2\omega_a$$

Then, the frequency range $\omega_a - \omega_b$ is one octave. The scale is logarithmic in ω: $\mathrm{d}\log_2\omega = 1$ or $\mathrm{d}\log\omega = 0.3$ per octave.

Frequently, power is given in decibels (dB) which is a logarithmic scale in power, often plotted versus a linear frequency scale. In this case, the areas under the curve do not represent the true values, i.e. $\int|F(\omega)|^2\,\mathrm{d}\omega$. Nor do ratios of areas under the spectral curve for different frequency ranges correspond to the true ratios. On the other hand, stationary points (maxima, minima, inflections) have the correct frequency positions, i.e. the same as those in a graph of $|F(\omega)|^2$ vs ω.

It is a common property among many geophysical power spectra to exhibit a downward slope for increasing frequency (examples are given in Table LI). This means that the power $E(\omega)$ can be written as:

$$E(\omega) \sim \omega^\gamma$$

where $\gamma < 0$. For slope determination it is better to take the logarithms and we get, letting Δ denote differences:

$$\Delta \ln E(\omega)/\Delta \ln \omega = \gamma \qquad [53]$$

Instead of giving the slope in terms of γ, it is frequently given as *decibels per octave*. The relation between the slope γ and decibels per octave is easily seen as follows: 1 dB = 10 Δ log E, 1 octave = $\Delta \log \omega = \log 2 = 0.3$. Hence it follows that 1 decibel/octave = 3γ.

Summarizing, it is very important in every operation undertaken by means of a spectrum to check that the operation does lead to correct estimates. As the spectral curve shape will in general change with changed coordinates, every property which is bound to the curve shape has to be carefully checked. In spectral comparison, frequently ratios are taken between different parts of the same spectrum or between different spectra. The coordinates chosen will in general also influence the results of such operations. For clarity, we have summarized in Table XXIV some examples, for a power spectrum, writing $|F(\omega)|^2 = E(\omega)$ for brevity. The values obtained for the spectrum with $E(\omega)$ as ordinate and ω as abscissa, both in linear scales, are considered to be correct, and this forms the norm with which we have to compare each individual case. It is left as an exercise to the student to develop correction formulas for the coordinate systems in Table XXIV, as well as for any other system.

The choice of ordinate scale in a spectrum, i.e., whether linear or logarithmic ordinate should be used, has also another aspect. Any curve, e.g. a spectrum, will show up most details in a graph where the ordinate is most spacious, and much less detail where the ordinate is less spacious (more compressed). With a linear ordinate, spacing (ΔF) is the same for all ordinate values, whereas for a logarithmic ordinate, spacing ($\Delta F/F$) will be decreased for high ordinate values and increased for low ordinate values. As a consequence, peaks will show up better with the linear scale,

TABLE XXIV

Examples of spectral modification of $E(\omega)$ due to choice of coordinates

Parameter	Ordinate $\omega E(\omega)$ Abscissa $\ln \omega$	Ordinate $\ln E(\omega)$ Abscissa ω	Ordinate $E_2(T)$ Abscissa T
Area under curve	preserved	not preserved	preserved
Location of stationary points	not preserved	preserved	not preserved
Slope of curve	not preserved	not preserved	not preserved
Curvature	not preserved	not preserved	not preserved
Location of centroid	not preserved	not preserved	not preserved
Width of curve	not preserved	not preserved	not preserved

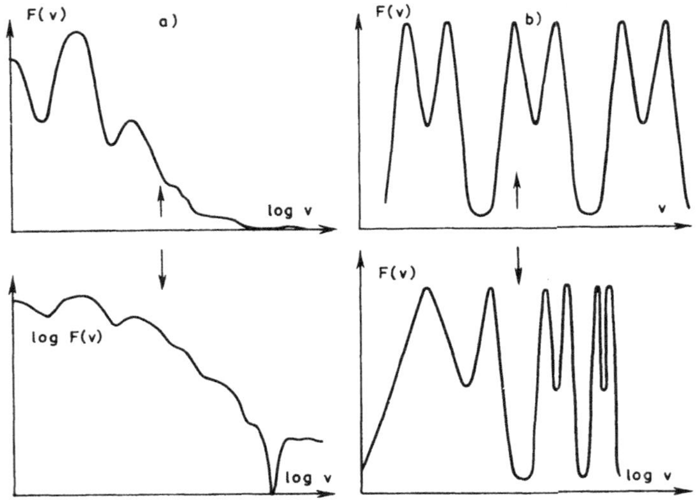

Fig.48. Linear scales versus logarithmic scales for a) the ordinate and b) the abscissa. Cf. De Bremaecker (1964).

whereas zeros are better indicated by the logarithmic scale (Fig.48). Note that the peak (near $T = 5$ sec) appears significant in the linear plot, but hardly so in the logarithmic. On the other hand, the sharp drop (near $T = 1.5$ sec) seems significant in the logarithmic plot, but not in the linear graph. Clearly, any such decisions cannot be made without knowing the confidence limits in either case. Corresponding rules hold for logarithmic versus linear frequency (or period) scales. The logarithmic scale will show higher resolution in the low-frequency range and lower resolution in the high-frequency range (Fig.48). More examples of the two alternatives can be found in Wunsch (1972), regarding power density spectra of sea level. The true frequency scale would be the linear one, as this is the scale in which observations are given (for frequencies $v = 1/T, 2/T, 3/T, \ldots$, where T is total record length). Logarithmic scales are often used because they easily cover a much wider range than linear scales, but then it is important to be aware of the modifications that the spectrum will undergo.

The use of a logarithmic scale, at least for the ordinate, e.g. power, has the advantage that the vertical bars indicating the range of reliability (section 5.1.3) will have constant length, whereas with a linear ordinate, they will increase in length with increasing ordinate. This is easily seen as follows. If the confidence limits are given by the factors a and b, it means that the true value of power $E(\omega)$ falls between the limits $aE(\omega)$ and $bE(\omega)$ with a certain confidence. The length of the vertical bar indicating the confidence will be $(b - a)\, E(\omega)$ in the linear scale, i.e. proportional to $E(\omega)$, while in the logarithmic power scale it will be $\log b - \log a$, i.e. independent of $E(\omega)$.

5.4.2 Frequency versus period

It is customary to read frequencies or periods corresponding to extrema in spectra, especially maxima or minima. However, great care is required in such readings, especially in collating those read from frequency and from period spectra. The reciprocal relation between frequency and period, $\omega = 2\pi/T$, entails the fact that a maximum read in a spectrum $E_1(\omega)$ at a certain frequency ω, does not occur at the corresponding period in the spectrum $E_2(T)$.

Let $E_1(\omega)$ and $E_2(T)$ represent the same power spectrum versus linear ω- and T-scales, respectively. Then the relation between the two presentations is that the total power is identical:

$$\int_0^\infty E_1(\omega)\,\mathrm{d}\omega = \int_0^\infty E_2(T)\,\mathrm{d}T$$

which leads to: [54]

$$E_2(T) = \frac{2\pi}{T^2} E_1\left(\frac{2\pi}{T}\right)$$

From [54] we find the following set of three equations:

$$\begin{aligned} &\mathrm{d}E_2(T)/\mathrm{d}T = 0 \\ &\mathrm{d}E_1(\omega)/\mathrm{d}\omega = -2\,E_1(\omega)/\omega \\ &\mathrm{d}\ln(E_1\omega^2)/\mathrm{d}\ln\omega = 0 \end{aligned} \qquad [55]$$

which tell us that an extremum in the $E_2(T)$-spectrum [55.1] does not correspond to an extremum in the $E_1(\omega)$-spectrum [55.2], whereas [55.3] shows that with coordinates $\ln(E_1\omega^2)$ versus $\ln\omega$ extrema would be conserved. A summary review is given in Table XXIV. Corresponding relations hold if instead we start from an extremum condition in the $E_1(\omega)$-spectrum:

$$\begin{aligned} &\mathrm{d}E_1(\omega)/\mathrm{d}\omega = 0 \\ &\mathrm{d}E_2(T)/\mathrm{d}T = -2\,E_2(T)/T \\ &\mathrm{d}\ln(E_2T^2)/\mathrm{d}\ln T = 0 \end{aligned} \qquad [56]$$

When different results (for instance, of location of stationary points, slopes,

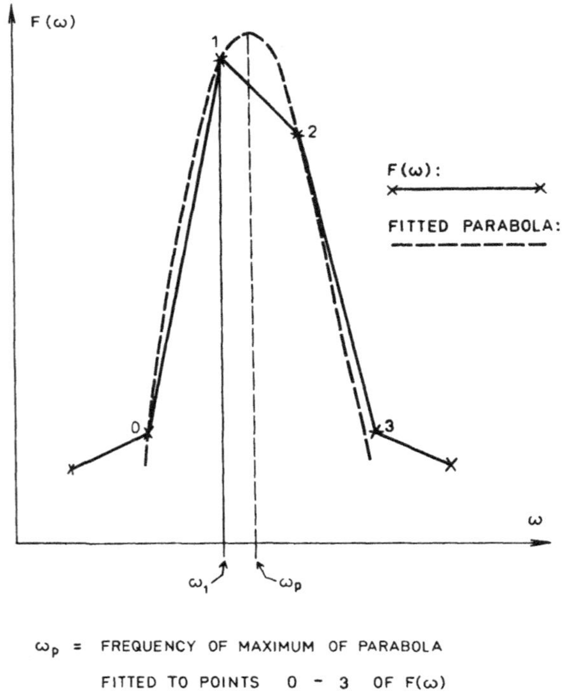

Fig.49. Parabolic interpolation versus straight-line segments.

widths, etc.) are obtained for the same spectrum but in different coordinate systems, it may be asked which spectrum yields the correct information. Of course, each representation gives correct results, but only for its own representation. For location of stationary points, however, the primary curve, i.e. linear $E(\omega)$ versus linear ω, may be considered as the standard. The most important point, however, is that in comparing different spectra, the *same* representation is used, whether as $E_1(\omega)$ or as $E_2(T)$, and that they are not mixed without due precaution. And for one and the same spectrum, information is always obtained on the relative distribution of power on different frequency bands or period bands, respectively.

This phenomenon has been discussed in detail in an important paper by Chiu (1967). Besides the mathematics given above, the reader can convince himself of the correctness of getting two different periods for the maxima of one and the same spectrum simply by taking an actual example. Starting for instance from a spectrum symmetric around its maximum in the representation $E_1(\omega)$ vs ω, and transferring this into $E_2(T)$ vs T, the latter spectrum will not be symmetrical, because of the factor $2\pi/T^2$. This leads to asymmetry, as it has different effects on the two sides of the original maximum, in addition to causing a shift of the maximum. The phenomenon can be understood as both spectra represent the

TABLE XXV

Period ranges corresponding to given frequency ranges

Period T (sec)	Period range ΔT (sec)	
	$\|\Delta\omega\| = 1$ rad/sec	$\|\Delta v\| = 1$ cycle/sec
0.1	0.0016	0.01
1	0.159	1
10	15.9	100
100	1592	10 000

total energy (same in both cases), but differently spread out over the independent parameters.

However, the reading of frequencies of spectral maxima of any curve may be open to discussion also for other reasons. As illustrated in Fig.49, ω_1 would be the frequency of the maximum if read from the $F(\omega)$-curve directly. But if there is reason to suspect a harmonic component with a frequency near this maximum, it is recommended to use instead ω_p as frequency of the maximum. Here ω_p corresponds to the parabola fitted to the $F(\omega)$-maximum, point 1, and its adjacent points.

In connection with the question of spectral reliability, it is also instructive to consider corresponding ranges in terms of period and of frequency. As a rule, spectral estimates are given for each frequency band, and all these bands have equal width all though any given spectrum. However, if instead we recalculate these estimates into their corresponding period bands, these are found to be of very unequal length through the spectrum. The relation between the respective widths can be written down immediately:

$$|\Delta T| = \frac{T^2}{2\pi}|\Delta\omega| = T^2|\Delta v| \qquad [57]$$

Table XXV gives a few numerical examples, which clearly illustrate how bad the resolution may be at longer periods, even though the frequency ranges are kept constant.

In summary, a careful consideration of the coordinates to be used for the spectra is very important in any study. A system should be chosen which is appropriate to the problem under investigation, paralleled with due account of the spectral accuracy. Also, in comparison of different investigations, one has always to check the coordinate systems used against each other.

5.4.3 Tripartite plots

For special problems, it may be advantageous to use special scales for presenta-

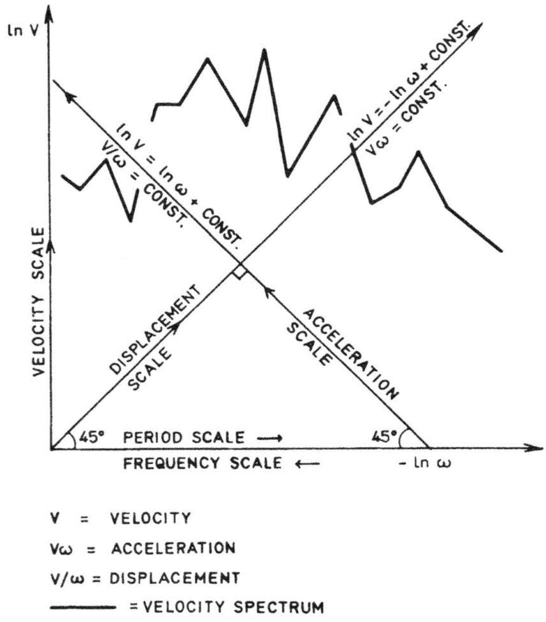

Fig.50. Tripartite plot of velocity, acceleration and displacement as used in evaluations of accelerograms in earthquake engineering. In applications, log is usually substituted for ln.

tion of spectra, naturally with due regard to the precautions mentioned above. One outstanding example is provided by the tripartite plots, frequently used in earthquake engineering for displaying velocity spectra derived from accelerograms (section 7.1.7).

If V = velocity, then $V\omega$ = acceleration and V/ω = displacement, and these simple relations permit plotting all three quantities in one and the same graph. The principle is clear from Fig.50. All scales are logarithmic. In reading such a graph, we note that to any given point, we get the values of frequency, velocity, acceleration and displacement from the respective scales. This means that one curve of a velocity–frequency spectrum also gives immediate information on corresponding acceleration and displacement spectra. For application to an actual case, see for example Krishna et al. (1969).

5.4.4 Instrumental response

In connection with readings of minima in spectra, consideration of the instrumental response is also important. This is illustrated in Fig.51 with a constructed case. The case of a steeply dipping response curve is in fact very common, especially for short-period seismographs. A minimum which may appear significant in the

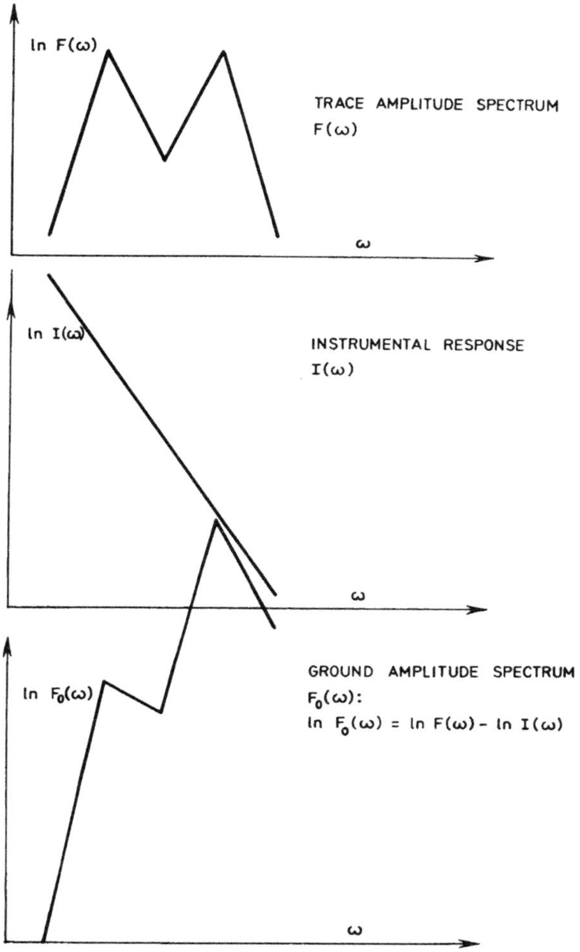

Fig.51. Schematic illustration of the influence of instrumental response on trace spectra.

trace amplitude spectrum may have almost disappeared in the (true) ground amplitude spectrum.

With $F(\omega)$ = trace spectrum, $F_0(\omega)$ = ground spectrum, and $I(\omega)$ = instrumental response, we have:

$$F(\omega) = F_0(\omega)\, I(\omega) \qquad [58]$$

We can easily convince ourselves that in addition to the effect just mentioned also other parameters will be different in $F(\omega)$ and $F_0(\omega)$. For example, the spectral slopes are different, related to each other by the equation:

$$\frac{dF(\omega)}{d\omega} = I(\omega)\frac{dF_0(\omega)}{d\omega} + F_0(\omega)\frac{dI(\omega)}{d\omega} \quad [59]$$

Further comparisons of spectral parameters of $F(\omega)$ and $F_0(\omega)$ can easily be deduced, and there is no need to give this in detail here.

It is more seldom the spectral analysis of seismograms works with only one sample, from just a one-component record. The usual procedure, which is more reliable, is to work with relative measurements, collating different components, different seismographs, different stations, etc, but then the relative response characteristics come into the picture. It is recommended to use instruments with identical response characteristics, which will greatly facilitate the work, including visual comparison of different records. If not available, it is advisable to apply the response characteristics, for amplitude and/or phase, as the case may be, as soon as a component has been Fourier-transformed and before any further operation is undertaken.

One example of the complications which can arise in case of unequal characteristics is given in Table XXVI. Calculations, which are left as an exercise to the reader, are in this example practically identical to those for Lissajou figures (see, for example, Manley, 1945, chapter 10). We can express the results of such calculations in the following formula:

$$\tan 2\psi = 2b \tan\theta \cos\varphi/(1 - b^2 \tan^2\theta) \quad [60]$$

TABLE XXVI

Review of various cases for a P-wave recorded by two horizontal components

Case	Recorded P-wave motion	Recorded azimuth
1 amplitude response identical: $b = 1$ phase response identical: $\varphi = 0$	linear	correct
2 amplitude response different: $b \neq 1$ phase response identical: $\varphi = 0$	linear	incorrect
3 amplitude response identical: $b = 1$ phase response different: $\varphi \neq 0$	elliptic	incorrect[1]
4 amplitude response different: $b \neq 1$ phase response different: $\varphi \neq 0$	elliptic	incorrect[1]

[1] If taken along the major axis of the ellipse.

where ψ = angle between E-direction and major axis of ellipse, θ = true angle of arrival of P, counted from E-direction, $b = |I_N(\omega)|/|I_E(\omega)|$, i.e. the ratio of amplitude response, and φ = phase difference between E and N.

When there is a phase difference between the two components, the incoming straight-line motion in the P-wave is modified into elliptic motion. This is to be noticed especially in particle-motion studies. At the same time, the maxima in the two horizontal records are not simultaneous. Our calculations (Table XXVI) refer to a P-wave for simplicity, but can naturally be extended to any wave with specified incoming particle motion.

5.4.5 Rotation of coordinate axes

Many observations in geophysics are made in relation to some coordinate system, e.g. a rectangular Cartesian x,y,z-system. It is then of importance to investigate to what extent measured quantities depend on the orientation of the chosen axes. We shall restrict the following discussion to a two-dimensional x,y-system.

In the preceding section we studied the combination of two rectangular components with in general unequal response characteristics, especially to see how the resulting ground particle motion was affected. In this section, we shall not combine the two components, but only investigate the effect of axes rotation, especially on various power functions.

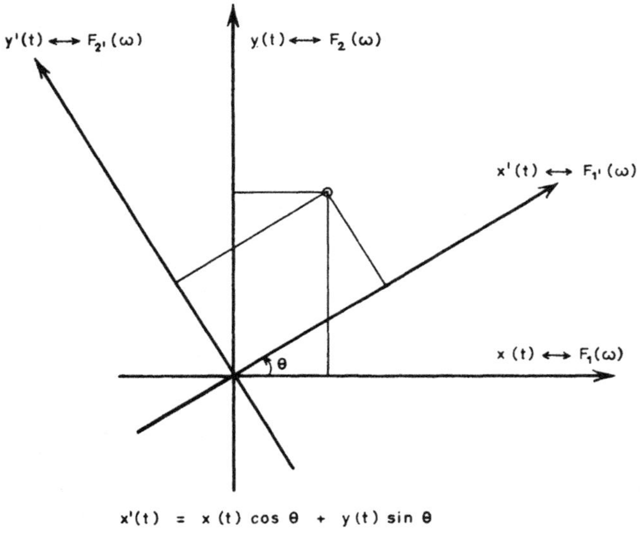

$$x'(t) = x(t)\cos\theta + y(t)\sin\theta$$

$$y'(t) = -x(t)\sin\theta + y(t)\cos\theta$$

Fig.52. Rotation of coordinate axes for investigation of its influence on spectra.

Notation is clear from Fig.52 where $F_1(\omega)$ is the Fourier transform of some quantity measured along the x-axis, etc. As the Fourier transformation is linear, the same formulas hold for the coordinate transformation of $F(\omega)$ as for x,y themselves, i.e.:

$$\left.\begin{aligned} F_{1'}(\omega) &= F_1(\omega)\cos\theta + F_2(\omega)\sin\theta \\ F_{2'}(\omega) &= -F_1(\omega)\sin\theta + F_2(\omega)\cos\theta \end{aligned}\right| \qquad [61]$$

Then, we can easily form the coordinate-transformed expressions for various spectral quantities, such as the co-spectrum:

$$\begin{aligned} P_{1'1'} = F^*_{1'} F_{1'} &= (F^*_1 \cos\theta + F^*_2 \sin\theta)(F_1 \cos\theta + F_2 \sin\theta) \\ &= P_{11}\cos^2\theta + P_{22}\sin^2\theta + P_{12}\sin 2\theta \end{aligned} \qquad [62]$$

and similarly:

$$P_{2'2'} = P_{11}\sin^2\theta + P_{22}\cos^2\theta - P_{12}\sin 2\theta \qquad [63]$$

The coordinate-transformed cross-power spectrum is obtained from:

$$\begin{aligned} E_{1'2'} = F^*_{1'} F_{2'} &= P_{1'2'} - iQ_{1'2'} \\ &= (F^*_1 \cos\theta + F^*_2 \sin\theta)(-F_1 \sin\theta + F_2 \cos\theta) \end{aligned} \qquad [64]$$

which, after separation into real and imaginary parts yields, respectively:

$$\left.\begin{aligned} P_{1'2'} &= P_{12}\cos 2\theta - 1/2\,(P_{11} - P_{22})\sin 2\theta \\ Q_{1'2'} &= Q_{12} \end{aligned}\right| \qquad [65]$$

It is of special importance to be clear about which quantities are invariant upon a coordinate rotation. We can list the following:

(1) The quad-spectrum, which was just demonstrated:

$$Q_{1'2'} = Q_{12} \qquad [66]$$

(2) The total power (it is quite natural that this should be independent of the orientation of the axes, but it can also be proved from the formulas given):

$$P_{1'1'} + P_{2'2'} = P_{11} + P_{22} \qquad [67]$$

(3) The following expression, which is again easily demonstrated by insertion of the expressions above:

$$P_{1'1'}\,P_{2'2'} - P_{1'2'}^2 = P_{11}\,P_{22} - P_{12}^2 \qquad [68]$$

On the other hand, it is important to note that the coherence γ_{12} and the phase angle Φ between 1 and 2:

$$\gamma_{12}^2 = (P_{12}^2 + Q_{12}^2)/P_{11}\,P_{22}; \qquad \tan\Phi = -\,Q_{12}/P_{12} \qquad [69]$$

are not invariant, in other words they depend on the orientation of the axes. However, from the formulas given, it can be demonstrated that for each frequency ω there is a certain orientation θ such that γ_{12} attains its maximum value, and this maximum value is independent of the axes. For details, see e.g. Fofonoff (1969).

Seismology offers familiar examples of coordinate axes rotation. With x and y corresponding to the E and N directions of recording, it is frequently of advantage to rotate the axes such that for example x' points to the source and thus y' is perpendicular to the source direction.

Chapter 6

PRINCIPLES AND GEOPHYSICAL APPLICATION OF FILTERING

Filtering implies separation, e.g. separation of one wave component out of a mixture of waves. Filtering is of dominant significance in geophysics and it enters our problems in two essentially different ways:

(1) Natural filtering, imposed by nature itself. Almost every natural phenomenon, observed or recorded, has had some path of transmission or been subject to other influences. These are all to be classified as filtering effects.

(2) Artificial filtering, imposed by the operator during his data handling. Records generally consist of a mixture of phenomena (different kinds of waves mixed, etc.) and it is desirable to extract just that component which is under study and suppress the others. This is done by filtering of the given data.

There is now a very comprehensive literature on filtering methods and filtering effects. Our purpose here will be only to give so much of fundamental information that the application of filtering in data preparation will be understandable. We shall also aim at explaining the filtering effects in nature, by which we are provided with a means to study various natural properties and phenomena. In the latter respect, we could say that at the same time as filtering in nature poses certain interpretational complications, it also furnishes us with very efficient means to explore nature's properties. Because of the dual way, in which filtering enters our studies, it will form a natural link between the data handling, discussed in previous chapters, and the applications to geophysics, which will occupy following chapters.

6.1 PRINCIPLES OF FILTERING

Our presentation of filters, given here, is only a brief review to the extent that is necessary for spectral analysis. Good presentations in more detail can be found in Laning and Battin (1956), Davenport and Root (1958), Bendat (1958), Papoulis (1962), Panter (1965), Arsac (1966), Robinson (1967a, b), Bendat and Piersol (1971) and Otnes and Enochson (1972). Briefer, but very clear reviews of filtering methods are given by Holloway (1958), M. K. Smith (1958), Alavi and Jenkins (1965), Kertz (1966), Silverman (1967), Wood (1968), Finetti et al. (1971), Zelei (1971), and others. For more details about filtering in geophysical, especially seismic, prospecting, the reader is referred to Grau (1966) as well as to numerous papers, especially in the journals *Geophysics* and *Geophysical Prospecting*. A detailed account of digital filters and methods to design such filters in the frequency domain, including the use of digital computers, has been given by Gold and Rader (1969).

6.1.1 System properties

By system or filter we mean quite generally anything by which an input signal $f(t)$ is modified into an output signal $g(t)$; see Fig.53. Thus, a system or filter has a very wide implication. To fix ideas, we may take a more specific example of a seismic wave traversing the earth's interior. Then, $f(t)$ is the signal at the source, the earthquake, while the system or filter is the earth's interior, and $g(t)$ is the

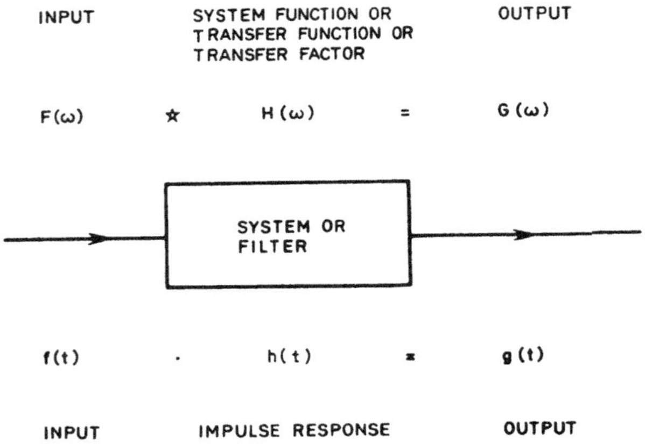

Fig.53. Scheme of a system or filter with input and output.

TABLE XXVII

Review of system properties

System (filter) property	Input	Output	Comment
	if $f(t)$	then $g(t)$	
	if $f_n(t)$	then $g_n(t)$	$n = 1, 2, 3, \ldots$
Linear	if $a_n f_n(t)$	then $a_n g_n(t)$	a_n = constant
Linear	if $\sum_n a_n f_n(t)$	then $\sum_n a_n g_n(t)$	
Time-invariant (stationary)	if $f(t - \tau)$	then $g(t - \tau)$	
Stable	if $\lvert f(t)\rvert < M$	then $\lvert g(t)\rvert < MI$	M and I constants
Causal	if $f(t) = 0$ for $t < t_1$	then $g(t) = 0$ for $t < t_1$	

signal received at a seismograph station. The system properties are known if the relation between input and output is known.

The systems we shall be dealing with have two fundamental properties: they are linear and stationary. The *principle of linearity* can best be described by the relations between input and output of a system, as given in Table XXVII. The principle of linearity is also called the *principle of superposition. Linear operator* is synonymous with linear system. A system is *stationary*, time-invariant or time-independent, when its response is time-independent. In other words, for a certain input $f(t - \tau)$ we get an output $g(t - \tau)$, for any time-shift τ.

Both properties of our systems, linearity and stationarity, depend on the differential equations with govern our systems. These equations are linear and have constant coefficients. The latter condition implies that the properties of the system are time-independent. The conditions of linearity and stationarity are independent of each other.

Considering application to waves in solid media we can distinguish the following three cases from the viewpoints just given:

(1) Perfectly elastic case: wave equations exhibit both linearity and stationarity.

(2) Imperfectly elastic or anelastic case: wave equations also exhibit both linearity and stationarity, only that a damping term, corresponding to anelasticity (viscosity), is added to the equations under 1.

(3) Plastic case: equations for plastic waves and shock waves are non-linear and coefficients become time-dependent. See further Kolsky (1953). In spectral applications to seismology, we are only concerned with items 1 and 2, whereas item 3 does not enter the usual problems encountered in seismology.

Conditions close to an earthquake focus may constitute an exception. Any effects from this may be avoided considering that the system corresponding to the source has no input but only an output. Moreover, it is always possible to replace the focal mechanism by its projection on a focal sphere with such a radius that the projected motions are down to the elastic range. Another system which enters all cases where seismic records are used, is the seismograph. The equation for a seismograph exhibits both linearity and stationarity, and thus the corresponding system is also linear and time-invariant.

However, in several other geophysical applications caution is called for. For example, ocean surface wave phenomena are linear only for small amplitudes, but for finite amplitudes, which are encountered as a rule, non-linearity enters. This is true for ocean waves on the open ocean. Non-linearity occurs also for waves in shallow water, as mentioned in section 3.6.4. Wave-breaking and energy transfer between different wave components are examples of non-linear effects. In the book edited by Vetter and Bretschneider (1963) there are several papers on non-linear properties of ocean surface waves, including extensive discussions. Similarly, meteorological turbulence phenomena are non-linear. As described in section

3.6.4, analysis by means of bispectra would be appropriate for non-linear phenomena, whereas linear-system concepts are inapplicable.

6.1.2 *Convolution formula*

The relation between the input $f(t)$ to a linear, time-invariant filter and its output $g(t)$ is given by the convolution integral, where $h(t)$, the impulse response, is the output for a $\delta(t)$-function as input. The demonstration is best seen by following the steps outlined in Table XXVIII. That is:

$$g(t) = \int_{-\infty}^{\infty} f(\tau)\, h(t-\tau)\, \mathrm{d}\tau$$

or:

$$g(t) = f(t) \star h(t) \qquad [1]$$

or in the frequency domain by virtue of [11] in Chapter 3:

$$G(\omega) = F(\omega) \cdot H(\omega) \qquad [2]$$

Terminology for $h(t)$ and $H(\omega)$ is given in Fig.53. In addition to the impulse response $h(t)$, with $\delta(t)$ as input, also the step response $j(t)$ of a filter is of importance, being the output for a unit Heaviside step $u(t)$ as input.

A summary is given in Table XXIX of basic filter characteristics with their formulas. The *impulse response* $h(t)$ is obtained as the inverse Fourier transform

TABLE XXVIII

Proof of convolution integral theorem for time-invariant, linear systems

Input	Output	Comment
$f(t)$	$g(t)$	given
$\delta(t)$	$h(t)$	definition of $h(t)$
$\delta(t-\tau)$	$h(t-\tau)$	time-invariance
$f(\tau)\,\delta(t-\tau)$	$f(\tau)\,h(t-\tau)$	linearity
$\int_{-\infty}^{\infty} f(\tau)\,\delta(t-\tau)\,\mathrm{d}\tau =$	$\int_{-\infty}^{\infty} f(\tau)\,h(t-\tau)\,\mathrm{d}\tau$	linearity (extended to an infinite number of terms)
$= f(t)$	$= f(t) \star h(t) = g(t)$	

TABLE XXIX

General properties of linear, time-invariant filters[1]

Input		→ Output	
time domain ↔	frequency domain	time domain ↔	frequency domain
$f(t)$	$F(\omega)$	$g(t)$	$G(\omega)$
$\delta(t)$	1	$h(t) = \frac{1}{\pi}\int_0^\infty \lvert H(\omega)\rvert \cdot \cos[\omega t - \Phi(\omega)]\,d\omega$	$H(\omega) = \lvert H(\omega)\rvert\, e^{-i\Phi(\omega)}$
$u(t)$	$U(\omega) = \pi\,\delta(\omega) - \frac{i}{\omega}$	$j(t) = \frac{\lvert H(0)\rvert}{2} + \frac{1}{\pi}\int_0^\infty \frac{\lvert H(\omega)\rvert}{\omega}\sin[\omega t - \Phi(\omega)]\,d\omega$	$J(\omega) = \pi\lvert H(0)\rvert\delta(\omega) + \frac{\lvert H(\omega)\rvert}{\omega}\, e^{-i[\Phi(\omega)+\pi/2]}$

[1] Corresponding formulas hold for the space–wavenumber domain.
$h(t)$ = *impulse response*, $j(t)$ = *step response*
$H(\omega)$ = *system function* or *transfer function* or *transfer factor*, occasionally called the *admittance function* (e.g. by Wunsch, 1972).

of $H(\omega)$. And $u(t)$ is the Heaviside unit-step impulse. Its Fourier transform $U(\omega)$ was derived in example no. 15 in Table V. The corresponding output in the frequency domain $J(\omega)$ is obtained by [2] as the product $U(\omega) \cdot H(\omega)$, and finally by Fourier-transforming this expression we get the corresponding formula $j(t)$ in the time domain. The output exhibits both an amplitude distortion, a phase distortion and a delay as compared to the input. The formulas collected in Table XXIX are of general validity for linear, time-invariant filters, and they will permit the calculation of the output for any given input.

In the time domain, the filtering operation is equivalent to convolution, i.e. taking a weighted mean over the input function $f(t)$ with $h(t)$ as weight function. In the frequency domain, the same operation is effected by multiplication. The relation $G(\omega) = F(\omega) \cdot H(\omega)$ was met with already in section 2.3.6, where it was demonstrated that it corresponds to a differential equation with constant coefficients. Therefore, the relation between input $f(t)$ or $F(\omega)$ and output $g(t)$ or $G(\omega)$ can be expressed in the following alternative forms: (1) in the time domain as a differential equation; (2) in the time domain as a convolution; and (3) in the frequency domain as a multiplication.

In general, the functions entering [2] are complex. We can separate equations of this type into two parts, representing amplitudes and phases, respectively:

$$\left.\begin{aligned}&|G(\omega)| = |F(\omega)| \cdot |H(\omega)|\\&\Phi_G(\omega) = \Phi_F(\omega) + \Phi_H(\omega) + 2n\pi\\&n = 0, \pm 1, \pm 2, \ldots\end{aligned}\right| \qquad [3]$$

Squaring [3.1] we find the following filtering equation for power, relating output and input powers:

$$|G(\omega)|^2 = |F(\omega)|^2 \, |H(\omega)|^2$$

where $|H(\omega)|^2$ is termed the *power transfer function* of the system. Quite naturally, only the absolute value of $H(\omega)$ enters into the power transfer function, and its phase does not matter. Similarly, by multiplication of the following two equations for two signals 1 and 2:

$$\left.\begin{aligned}&G_1^*(\omega) = F_1^*(\omega) \, H_1^*(\omega)\\&G_2(\omega) = F_2(\omega) \, H_2(\omega)\end{aligned}\right| \qquad [4]$$

we get an expression for the cross-power filtering equation:

$$G_{12}(\omega) = F_{12}(\omega) \, H_{12}(\omega) \qquad [5]$$

Let us assume the input function $f(t) = \mathrm{e}^{i\omega t}$. Then we calculate the corresponding output $g(t)$ from [1.1]:

$$g(t) = \int_{-\infty}^{\infty} \mathrm{e}^{i\omega\tau} \, h(t - \tau) \, \mathrm{d}\tau = \int_{-\infty}^{\infty} \mathrm{e}^{i\omega(t-\xi)} \, h(\xi) \, \mathrm{d}\xi$$

(substituting the variable $\xi = t - \tau$ for τ)

$$= \mathrm{e}^{i\omega t} \int_{-\infty}^{\infty} \mathrm{e}^{-i\omega} \, h(\xi) \, \mathrm{d}\xi = \mathrm{e}^{i\omega t} \, H(\omega) = |H(\omega)| \, \mathrm{e}^{i[\omega t - \Phi(\omega)]} \qquad [6]$$

(by [21], Chapter 2)

From this we can immediately write down the following expression for the *frequency response function* of the system:

$$|H(\omega)|\, e^{-i\Phi(\omega)} = \int_{-\infty}^{\infty} h(\xi)\, e^{-i\omega\xi}\, d\xi \qquad [7]$$

or:

$$\left.\begin{aligned} |H(\omega)| \cos \Phi(\omega) &= \int_{-\infty}^{\infty} h(\xi) \cos \omega\xi \, d\xi \\ |H(\omega)| \sin \Phi(\omega) &= \int_{-\infty}^{\infty} h(\xi) \sin \omega\xi \, d\xi \end{aligned}\right| \qquad [8]$$

Thus we find that the output $g(t)$ has a different amplitude and a different phase as compared with the input $f(t)$, but it has the same frequency. Then, this result can be generalized, as we can consider any input as a summation of functions $f(t)$ of the type given, but with different frequencies and different amplitudes, i.e. a Fourier series. The output will be a corresponding superposition of terms of the type just deduced. For examples, see Jenkins (1965).

From [6] we see that the time delay or phase delay is equal to $\Phi(\omega)/\omega$ or more general $[\Phi(\omega) \pm 2n\pi]/\omega$, where n is an integer. If the filter corresponds to a system through which waves propagate, then this delay corresponds to unit distance. Hence, the phase velocity $= \omega/\Phi(\omega)$ or in general $\omega/[\Phi(\omega) \pm 2n\pi]$, where only one value of n will apply, and the group velocity $= d\omega/d\Phi(\omega)$. Therefore, velocity dispersion curves can be interpreted as being spectra.

These principles are applicable to any kind of waves propagating over an array of stations. The relative phase shifts permit determination of the phase-velocity spectrum. Among applications may be mentioned especially those to seismic waves (section 7.2), geomagnetic pulsations (section 10.2), radio waves, as well as atmospheric and oceanic wave propagation.

6.2 CLASSIFICATION OF FILTERING PROCESSES

A classification of filtering can be made according to various principles, as follows:

(1) Filtering property, i.e. according to characteristic properties of the filters. This item concerns system properties.

(2) Filtering mechanism, i.e. according to the manner in which the filtering enters the process. Like (1), this concerns system properties, but in another way.

(3) Filtering purpose, i.e. according to the aim of the filtering process. This concerns the output properties.

(4) Filtering parameter, i.e. according to the data property that is made the basis for the filtering process. This concerns input properties.

6.2.1 Filtering property

According to the special properties of a filter, we differ between several types, of which the most significant are described here.

(1) *Distortionless filter*. This is defined by the following equation:

$$g(t) = H_0 f(t - t_0) \qquad [9]$$

i.e., the input is multiplied by a constant factor H_0 and delayed a time t_0. By the time-shifting theorem (2.3.4) applied to [9] we find:

$$G(\omega) = H_0 \, e^{-i\omega t_0} F(\omega)$$

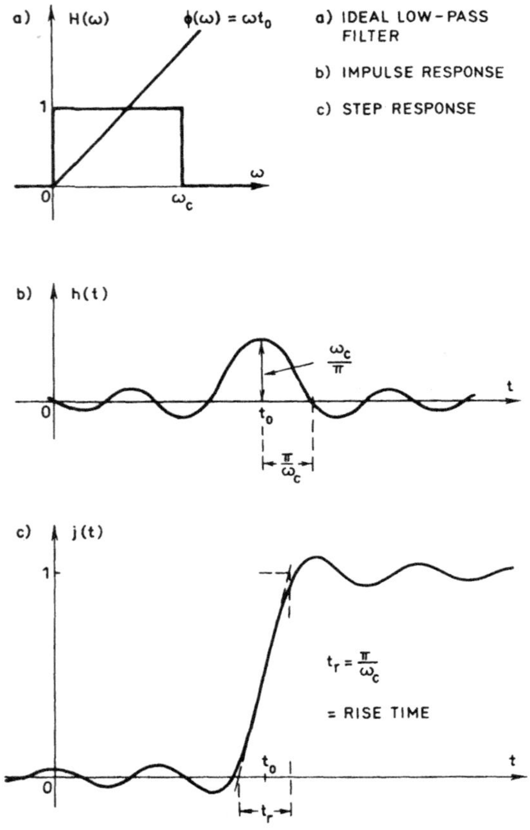

Fig.54. An ideal low-pass filter with its impulse response and step response. Modified and redrafted after Papoulis (1962).

i.e. by comparison with [2]:

$$H(\omega) = H_0\, e^{-i\omega t_0} \qquad [10]$$

or $|H(\omega)| = H_0 = \text{constant}$ and $\Phi(\omega) = \omega t_0$, linear.

It is sometimes said that this filter preserves the form of the input signal $f(t)$. This is not quite true, as slopes are changed by the operation [10]. To preserve the form of the curve $f(t)$ completely, we should not *multiply* by a constant H_0 but *add* a constant.

(2) *Amplitude-distorting filters*: $|H(\omega)|$ not constant, $\Phi(\omega) = \omega t_0$. Depending on the definition of $|H(\omega)|$ we distinguish a number of filters.

(a) Ideal low-pass filter:

$$|H(\omega)| = \begin{cases} H_0 & |\omega| < \omega_c \\ 0 & |\omega| > \omega_c \end{cases} \qquad [11]$$

where ω_c = cut-off frequency (Fig.54). In this case we get:

$$H(\omega) = H_0\, \Pi(\omega/2\omega_c)\, e^{-i\omega t_0} \qquad [12]$$

and corresponding expressions for other quantities can be derived from the general formulas in Table XXIX.

In particular, the case $t_0 = 0$ arises with $h(t) = \text{sinc } t$. From the convolution theorem [11] in Chapter 3 and example 3, Table V, we have:

$$f(t) \star \text{sinc } t \leftrightarrow F(\omega) \cdot \Pi(\omega/2\pi)$$

i.e.: [13]

$$H(\omega) = \Pi(\omega/2\pi)$$

The function sinc t thus provides for an ideal low-pass filter with cut-off frequency $\omega_c = \pi$. See also Table XXX. As an exercise, we can then immediately form the filter function for two spatial dimensions (x,y) with wavenumbers k_x and k_y, respectively. In correspondence to [13], we then find:

$$f(x,y) \star \text{sinc } x \star \text{sinc } y \leftrightarrow F(k_x,k_y) \cdot \Pi\left(\frac{k_x}{2\pi}\right) \cdot \Pi\left(\frac{k_y}{2\pi}\right) \qquad [14]$$

See further section 10.2.1.

(b) Other definitions of $|H(\omega)|$. Quite a number of different filters have been defined for different expressions for $|H(\omega)|$. In general, they do not cause any special mathematical problems.

TABLE XXX

Corresponding operations in time and frequency domains

Case	Time domain		Frequency domain	Remark
1 filtering	convolution: $f_1(t) \star f_2(t)$	$\leftrightarrow$	multiplication: $F_1(\omega) \cdot F_2(\omega)$	effect eliminated by *deconvolution* or *inverse filtering*
	example: truncation for $\lvert\omega\rvert > \omega_c$			
	$f(t) \star \frac{\omega_c}{\pi} \operatorname{sinc} \frac{\omega_c t}{\pi}$	$\leftrightarrow$	$F(\omega) \cdot \Pi\left(\frac{\omega}{2\omega_c}\right)$	
2 windowing	multiplication: $f(t) \cdot w(t)$	$\leftrightarrow$	convolution: $\frac{1}{2\pi} F(\omega) \star W(\omega)$	effect usually modified by smoothing in the frequency domain
	example: truncation for $\lvert t\rvert > t_c$			
	$f(t) \cdot \Pi\left(\frac{t}{2t_c}\right)$	$\leftrightarrow$	$F(\omega) \star \frac{t_c}{\pi} \operatorname{sinc} \frac{\omega t_c}{\pi}$	
3 superposition	addition: $f_1(t) + f_2(t)$	$\leftrightarrow$	addition: $F_1(\omega) + F_2(\omega)$	effect eliminated by *selective filtering*: (a) frequency filtering, (b) velocity filtering, (c) polarization filtering
	example: added noise, false zero-line etc			

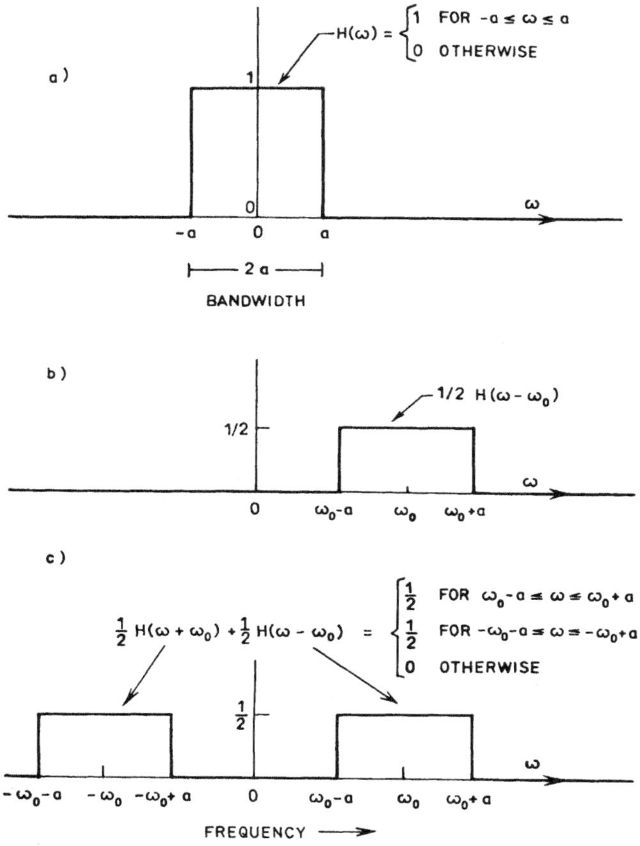

Fig.55. Band-pass filters. a) Ideal low-pass filter. b) Band-pass filter obtained by shifting $H(\omega)$ to ω_0 and halving the amplitude. c) Ideal band-pass filter equal to the sum of two such filters, shifted towards left and right, respectively, and corresponding to [21] in Chapter 6. After E. A. Robinson in Crampin and Båth (1965).

(3) *Phase-distorting filters*: $\Phi(\omega)$ not linear, $|H(\omega)|$ constant $= H_0$. Such a filter uses to be called "all-pass filter", with a slight inconsistency in the terminology. For these filters there is direct proportionality between the energies of the output and input signals:

$$\int_{-\infty}^{\infty} |G(\omega)|^2 \, d\omega = H_0^2 \int_{-\infty}^{\infty} |F(\omega)|^2 \, d\omega \qquad [15]$$

as found directly from [2].

(4) *Band-pass filters*: these are general in the sense that they have both amplitude and phase distortion, but they consist of specially arranged pass-bands.

A usual case is the symmetrical band-pass filter (Fig.55):

$$H(\omega) = H_1(\omega) + H_2(\omega) = |H_1(\omega)|\, e^{-i\Phi_1(\omega)} + |H_2(\omega)|\, e^{-i\Phi_2(\omega)} \qquad [16]$$

which is assumed to be Hermitian. Shifting H_1 towards left or H_2 towards right so that they are symmetrical around the origin, we get a low-pass filter:

$$H_l(\omega) = H_1(\omega_0 + \omega) = H_2(-\omega_0 + \omega) \qquad [17]$$

The impulse function of this low-pass filter is denoted $h_l(t) \leftrightarrow H_l(\omega)$. Applying the frequency-shifting theorem (2.3.5) to [17] we find:

$$\left.\begin{aligned} h_l(t)\, e^{i\omega_0 t} &\leftrightarrow H_l(\omega - \omega_0) = H_1(\omega) \\ h_l(t)\, e^{-i\omega_0 t} &\leftrightarrow H_l(\omega + \omega_0) = H_2(\omega) \end{aligned}\right| \qquad [18]$$

Adding these two equations we get:

$$2h_l(t) \cos \omega_0 t \leftrightarrow H(\omega) \qquad \text{or:} \qquad h(t) = 2h_l(t) \cos \omega_0 t \qquad [19]$$

This important equation states that the impulse response of any system with an equivalent low-pass system can be found from this low-pass system. This theorem reduces the problem for such systems to the one for low-pass systems, discussed above.

If in particular we assume:

$$H_l(\omega) = 1/2\ \Pi(\omega/2a) \qquad [20]$$

we have by example 2 in Table V, Chapter 2, and the formula [19] for $h(t)$ that:

$$h(t) = (\sin at)/\pi t \cdot \cos \omega_0 t \qquad [21]$$

This is the kind of digital band-pass filter that was described by E. A. Robinson in an appendix to Crampin and Båth (1965). This filter is symmetric, i.e. $h(t) = h(-t)$, and thus no phase shift is introduced. t should assume all integer values from $-\infty$ to $+\infty$. In practice, restriction to a finite interval has naturally to be adhered to, and then a weighting function is applied to $h(t)$; in the paper cited a triangular weighting function was used; see further section 4.4.5. Bloch and Hales (1968) used the same filter function weighted with a cosine function, which produces a sharper filter. See also Brooks (1969), who emphasizes the importance of the shape of $H(\omega)$, especially that it should pass amplitudes corresponding to its center frequency unattenuated. See also Landisman et al. (1969) for a discussion of such filters.

On the basis of the fundamental equations [1] and [2] it is clear that multiplication of $F(\omega)$ with any frequency-dependent function is equivalent to filtering and to convolution in the time domain. Let us take two examples:

(1) Time-shifting is equivalent to filtering and convolution. From the time-shifting theorem (2.3.4) and example 11, Table V, Chapter 2, we have:

$$f(t-a) \leftrightarrow F(\omega)\,e^{-ia\omega} \leftrightarrow f(t) \star \delta(t-a) \qquad [22]$$

(2) Derivation is equivalent to filtering and convolution. By the derivation theorem (2.3.6) and example 9, Table V, we have:

$$\left.\begin{aligned} f'(t) &\leftrightarrow i\omega\, F(\omega) = \omega e^{i\pi/2} F(\omega) \leftrightarrow \delta'(t) \star f(t) \\ f^{(n)}(t) &\leftrightarrow (i\omega)^n F(\omega) = \omega^n e^{in\pi/2} F(\omega) \leftrightarrow \delta^{(n)}(t) \star f(t) \end{aligned}\right| \qquad [23]$$

Derivation in the time domain is evidently equivalent to high-pass filtering. Correspondingly, integration of $f(t)$ is equivalent to low-pass filtering in the frequency domain.

6.2.2 Filtering mechanism

Filtering processes enter geophysics in a number of ways. In seismology, the most important ways can be summarized in the following points:

(1) *Natural filters:* the earth's interior consists of filters beyond human control. In this case, the filtering action is used for the extraction of useful information about the physics of the interior, such as of attenuation and of velocity dispersion.

(2) *Instrumental (electric, mechanical) filters:* seismographs are filters under human control. By their setting, certain frequency bands can be suppressed and others emphasized. Filtering apparatuses, many of them nowadays available on the instrumental market, belong to this group.

(3) *Mathematical filters:* special filtering techniques, usually by digital computers, may be used to separate waves on a record from each other, such as similar waves of different origin or waves of the same origin but of different types.

Of these, item 1 represents an unavoidable filtering, which carries useful information about the earth's interior that we want to isolate and to identify. Items 2 and 3 constitute means to accomplish such studies, and often one has to make a choice between these two procedures, i.e. whether one wants to filter waves (electrically and/or mechanically) before they are recorded (item 2) or after (item 3). Item 2 may be preferred when an instrumental installation is set up just for a special purpose, whereas item 3 permits a wider application of a given record to a number of different problems. In permanent recording in seismology, a combination of 2 and 3 is used, i.e., a fully equipped station has a series of instruments with different characteristics, but in addition mathematical filtering has to be frequently applied to the records. Similar situations exist in other branches of geophysics.

Let us consider the natural filtering also from a wider, geophysical viewpoint. This kind of filtering is always bound to exist as soon as we do not measure a phenomenon right at its source, i.e., in any case where there is a medium (a system) separating the recorder from the source. When the system function is frequency-dependent, which it is in practically all cases, then it is especially important to be aware of the effects of the system on the observations. In general, a system acts as a low-pass filter, and this holds true whether it concerns propagating waves or just observations of a phenomenon at some distance. Examples of these two cases can easily be found:

(1) For propagating waves, the lower frequencies generally propagate to larger distances. Typical examples are provided by seismic surface waves and ocean waves.

(2) Phenomena observed at a larger distance have a higher power ratio of long/short wavelengths than if observed at smaller distance. Typical examples are provided by sea-surface observations of bottom magnetism (the system function is then $e^{-2\pi h/L}$, where h is water depth and L wavelength). Similar conditions prevail for gravity observations. Also, a pressure detector on the ocean bottom records proportionally more of low frequencies of the ocean surface waves than a level recorder on the surface.

The efficient filtering of seismic records made possible by using galvanometers as band-rejection filters has been demonstrated by Pomeroy and Sutton (1960); this procedure comes under item 2 above. An even more versatile filtering technique is described by De Bremaecker et al. (1962). However, a general experience is that galvanometer recording is simpler and more dependable in operation than electronic equipment. Combination of strain and inertial seismographs may under certain circumstances lead to efficient instrumental filtering (Shopland and Kirklin, 1969, 1970).

6.2.3 Filtering purpose

Filtering may be made for a number of different purposes, such as for extraction of signals from noise, as a preparation of records for spectral analysis, etc. Filtering methods are applied for such purposes in practically every field where time series enter, such as in electrical engineering, optics, acoustics, geophysics, economy, etc. Geophysical prospection, notably seismic prospection, is a field where filtering plays a great role. We can conveniently group the fields of application of filtering under a number of sub-headings to which we now turn.

(1) *Filtering of noise from a signal.* In principle, this filtering is easy to express

as follows. Given a record $f(t)$ consisting partly of a signal $s(t)$ and partly of noise $n(t)$, we have:

$$\begin{aligned} &\text{in the time domain} && f(t) = s(t) + n(t) \\ &\text{in the frequency domain} && F(\omega) = S(\omega) + N(\omega) \end{aligned} \qquad [24]$$

For efficient elimination of the noise, we need a filter with a transfer function $H(\omega)$ such that:

$$[S(\omega) + N(\omega)] \cdot H(\omega) = S(\omega)$$

or:

$$H(\omega) = S(\omega)/\{S(\omega) + N(\omega)\} \qquad [25]$$

Just as easy as this appears in principle, as difficult it may be in the actual design of appropriate digital filters and much effort has gone into this business. Much of this development is due to intensive efforts in the last decade to increase sensitivity in recording small seismic signals, e.g. from small nuclear explosions, amidst the

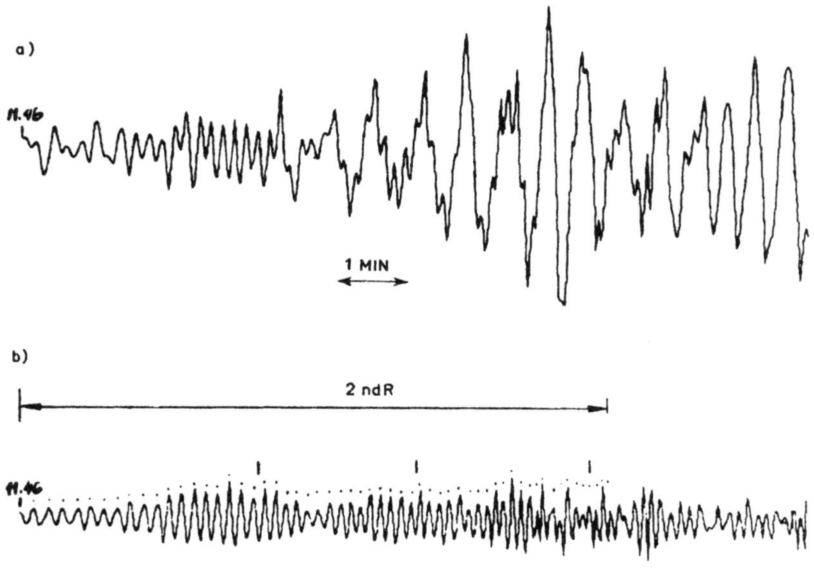

Fig.56. a) The seismogram of an earthquake in Ryukyu Islands, August 17, 1963, recorded at Uppsala by a long-period vertical-component seismograph (digital reproduction, cf. Fig.32). b) The same trace after being passed through a band-pass filter 6–15 sec period, this trace clearly exhibiting higher-mode Rayleigh waves. After Crampin and Båth (1965).

ambient noise. Moreover, in any study of the signal spectrum $S(\omega)$ it is desirable to be able to eliminate noise from the spectrum $S(\omega) + N(\omega)$ which the record yields without filtering. Stationarity (in time) of the analyzed wave group is assumed to hold when digital filters are designed to filter out noise from seismic signals. Stationarity is usually tested by taking extended time-lengths of the signal and observing the due changes in the computed spectra (Kulhánek, 1971). Space-stationarity is generally assumed in the correlation analysis of seismic vibrations, such as microseisms, etc.

Of particular importance are filters whose impulse response functions $h(t)$ are chosen such that the mean-square error between a desired output and the actual output is minimized. Such filters appear frequently in the literature and are generally called *Wiener optimum filters* (for a clear exposition, see Robinson and Treitel, 1967). As examples among many others, we may mention Kulhánek (1967, 1968) who used this technique for detection of seismic signals amidst noise and Clarke (1969) who applied two-dimensional Wiener filters to gravity maps.

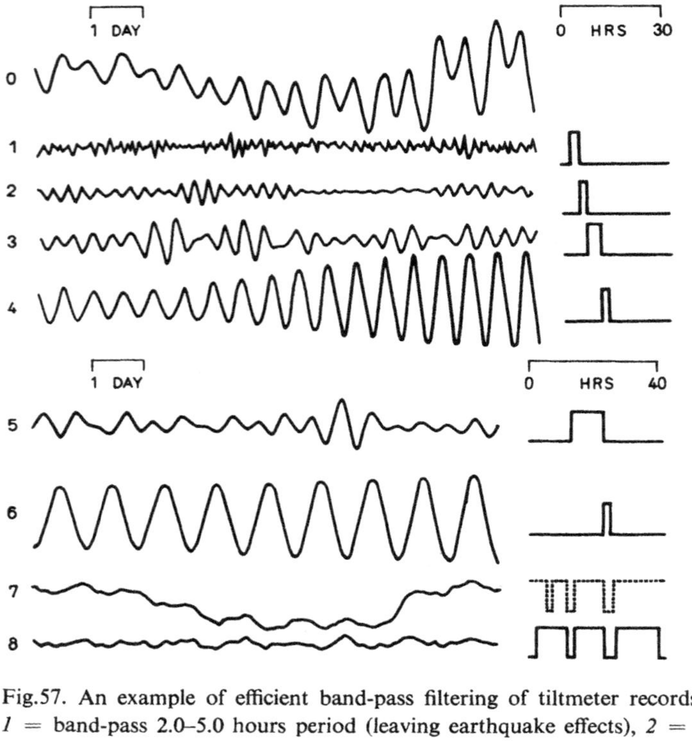

Fig.57. An example of efficient band-pass filtering of tiltmeter records. 0 = original record, *1* = band-pass 2.0–5.0 hours period (leaving earthquake effects), *2* = 5.0–7.0 hours (quarter-diurnal tide), *3* = 7.0–11.0 hours, *4* = 11.0–13.0 hours (semi-diurnal tide plus meteorological effects), *5* = 13.0–23.0 hours, *6* = 23.0–25.0 hours (diurnal tide plus meteorological effects), *7* = subtraction of quarter-diurnal, semi-diurnal and diurnal effects, leaving those of longer period (meteorological effects), *8* = 2.0–11.5, 13.5–23.0, 26.5– 40.0 hours. After T. Tanaka (1966).

Related to these are the *maximum-likelihood filters*, used to suppress noise without signal distortion in a record, using information on the noise structure prior to the signal arrival. See Capon et al. (1967), who both give the theory of the method and apply it to LASA data, and Lacoss (1971), who describes two adaptive methods for power spectrum estimation. For other applications of [25], see for example Tsujiura (1967).

(2) *Filtering of different signals from each other.* This is based on the same principle as in [25]. An example is shown in Fig.56, where different modes of surface waves are filtered, using a band-pass filter from 6–15 sec period. Another example of efficient separation by means of band-pass filtering is offered by Tanaka (1966), who applied the technique to tiltmeter records (Fig.57; cf. also section 10.1.3).

(3) *Filtering of different microseisms from each other.* This is also based on the same principle as in [25]. An example is shown in Fig.58 where by means of a band-pass filter 15–25 sec, long-period microseisms (with periods around

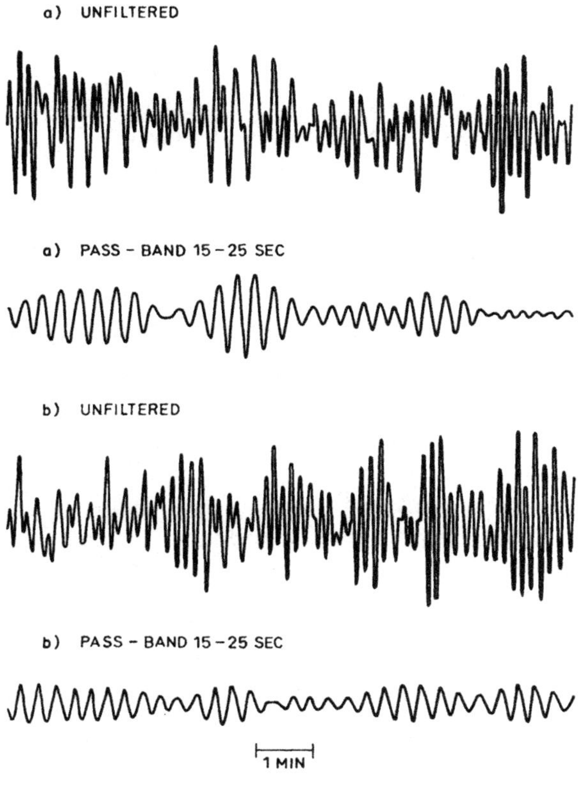

Fig.58. Band-pass filtering (15–25 sec period) of two samples (a and b) of long-period N–S component seismic records from Umeå, Sweden.

15–20 sec) are filtered out from superimposed microseisms of shorter period.

In addition to these cases, where we separate two or more signals or noise of different origin, we may want to "clean" one and the same signal from its deformation encountered during the passage through various systems, as earth media or seismographs. As the effect of any system on a signal $f_1(t)$ involves convolution of $f_1(t)$ with the impulse response of the system (filter), the original signal $f_1(t)$ can be recovered by *deconvolution* or *inverse filtering*. This technique uses special filters designed such that the influence of any given medium is eliminated. The deconvolution can be effected either in the time domain or by multiplication in the frequency domain. The latter procedure is the one generally applied in seismology, when spectral ratios are formed in order to eliminate one or more common factors. See section 6.5. A review is given in Table XXX. For further detailed discussion of this, the reader is referred to Robinson (1967a), Ko and Scott (1967) and Ulrych (1971).

6.2.4 Filtering parameter

In order to be successful, the filtering technique requires that the wanted and the unwanted wave components differ in some respect. This property can then be made the basis for the filtering process. From this point of view we differ in geophysics and especially seismology between four types of filtering.

(1) *Frequency filtering*. This type of filtering is applicable as soon as the wave components to be separated exhibit enough frequency difference. The success of the filtering depends on the degree of frequency separation between the components. If frequency ranges partially overlap, separation will not be complete and the filtering action will affect the spectrum to be constructed. On the other hand, if frequency separation is virtually complete, then this kind of filtering will be very effective. It is the most usual filtering method in geophysics, and is the type which has been described hitherto in this chapter.

(2) *Velocity filtering*. As distinct from frequency filtering, the velocity filtering is based on different wave velocities for the components to be separated, e.g. between body waves and microseisms. This method has found application in array-data processing and will be the subject of section 6.3.1.

(3) *Polarization filtering or mode filtering*. This is based upon differences in particle motion of signal and noise or of different signals. It is applied only in the time domain and will not be further discussed here. See for example Key (1968), Simons (1968), Basham and Ellis (1969) and Montalbetti and Kanasewich (1970) for details with applications.

(4) *Amplitude filtering*. Based on amplitude differences, e.g. of signal and noise, this method is in fact applied in devices used for earthquake alarms or for light-intensification in photographic recording. This filtering naturally also works in the time domain and will not be further discussed here.

6.2.5 Correlation functions and power spectra for signal and noise

Under special conditions it is possible to calculate correlations and power spectra for signal and noise separately, without first performing any filtering operation. In the discussion of correlations and power spectra we have hitherto been concerned with functions $f_1(t)$ and $f_2(t)$ without any further consideration of how these are built up. In general, they consist partly of signals, partly of noise. We shall treat the case of two time series $f(t)$ and $g(t)$, which consist of the same signals $s(t)$ but of different noise components $n(t)$ and $m(t)$:

$$\left.\begin{aligned} f(t) &= s(t) + n(t) \\ g(t) &= s(t) + m(t) \end{aligned}\right| \qquad [26]$$

The signal $s(t)$ is assumed to be uncorrelated to the noise components and also the two noise components are uncorrelated to each other. $f(t)$ and $g(t)$ are the recorded traces, and our problem is now to deduce autocorrelation functions for $s(t)$, $n(t)$ and $m(t)$.

The autocorrelation for $f(t)$ becomes:

$$C_{ff}(\tau) = \int_{-\infty}^{\infty} f(t) f(t + \tau)\, \mathrm{d}t = \int_{-\infty}^{\infty} [s(t) + n(t)]\, [s(t + \tau) + n(t + \tau)]\, \mathrm{d}t$$

$$= C_{ss}(\tau) + C_{sn}(\tau) + C_{ns}(\tau) + C_{nn}(\tau) = C_{ss}(\tau) + C_{nn}(\tau) \qquad [27]$$

because $C_{sn}(\tau) = C_{ns}(\tau) = 0$, as $s(t)$ and $n(t)$ are uncorrelated. Similarly, we find:

$$C_{gg}(\tau) = C_{ss}(\tau) + C_{mm}(\tau) \qquad [28]$$

and a similar development for the cross-correlations leads to the following expression:

$$C_{fg}(\tau) = C_{gf}(\tau) = C_{ss}(\tau) \qquad [29]$$

Then we find from [29]:

$$C_{ss}(\tau) = \frac{1}{2} [C_{fg}(\tau) + C_{gf}(\tau)]$$

and from [27] and [28]: [30]

$$C_{nn}(\tau) = C_{ff}(\tau) - \frac{1}{2} [C_{fg}(\tau) + C_{gf}(\tau)]$$

$$C_{mm}(\tau) = C_{gg}(\tau) - \frac{1}{2} [C_{fg}(\tau) + C_{gf}(\tau)]$$

which solves our problem.

Power spectra of signals and noise, respectively, are then obtained from [30], using the general formula for power spectra. There will be no cross-power between signal and noise or between the two noise components. The power spectrum of any of the given records $f(t)$ or $g(t)$ will by [27] and [28] be equal to the sum of signal and noise power, which is also immediately obvious. The cross-power of $f(t)$ and $g(t)$ will by [29] be equal to the signal power alone.

These conclusions are valid under the assumptions made for the two series [26], and it has to be observed in each application if these assumptions are fulfilled or not. This method of extracting signal and noise statistics from seismic prospection records has been outlined by Dash and Obaidullah (1970). It is suggested as an exercise for the student to perform the corresponding development for the convolution functions. Shaub (1963) made similar developments but defined also a modified cross-correlation function, which would suit certain geophysical demands better. Other cases of composite wave motions lend themselves to similar treatment, under appropriate assumptions. One example is the analysis of noise consisting of several Rayleigh-wave modes (Douze, 1964b).

6.3 EXAMPLES OF FILTERING OPERATIONS

The list of examples given below is far from complete, but encompasses some of the more common filtering procedures. The reader will from these examples be able to examine any filtering method, its effects both in time and frequency domain, especially on a computed spectrum.

6.3.1 Low-pass filtering – smoothing

The term smoothing implies elimination, partial or complete, of short-period components in a record, i.e. equivalent to low-pass filtering, before its spectrum is calculated. Smoothing is effected by forming *moving or floating averages*, for example over three successive values. One example is provided by a seismograph record, which exhibits long-period surface waves with superimposed waves of shorter periods. If one wants to make a spectral study of the long-period waves and for that purpose smoothes out the shorter waves, then this could affect also those Fourier components which we are interested in. An obvious meteorological analogy consists of investigations of the annual temperature variation from records on which the daily temperature variation is superposed. Low-pass filtering may be effective in eliminating aliasing (section 4.3.3).

Averaging or smoothing is a process perfectly equivalent to filtering, which is seen as follows. Let us apply equal-weight averaging over successive intervals, each of length Δt, then we write the average $\bar{f}(t)$ as:

$$\bar{f}(t) = \frac{1}{\Delta t} \int_{t-\Delta t/2}^{t+\Delta t/2} f(\tau)\,d\tau = f(t) \star \frac{1}{\Delta t}\,\Pi\left(\frac{t}{\Delta t}\right) \leftrightarrow F(\omega) \cdot \operatorname{sinc}\frac{\omega \Delta t}{2\pi} \qquad [31]$$

applying example 1, Table V, and the time-scaling theorem (section 2.3.3). The filter or "smoother" is the rectangular function:

$$\frac{1}{\Delta t}\,\Pi(t/\Delta t) \qquad \text{which is } = 1/\Delta t \text{ for } |t| \leq \Delta t/2 \text{ and } = 0 \text{ otherwise.}$$

The sinc-function in [31] is the same as the "sigma factor" of Lanczos (1966, p.65), which he used as a multiplier in the Fourier series expansion to reduce the Gibbs phenomenon (section 2.1.2). See also F. B. Smith (1962), who derives formulas relating to the sampling and averaging of observations for the study of meteorological turbulence. Similar filtering operations, also of the same mathematical form, are encountered in the spatial domain, when averages over a certain distance range are used instead of point values. This is the case for some measurements in meteorology, for example Silverman (1968). Application to the space domain is immediately possible with all following examples, and therefore it will be enough to restrict the formulation here to the time domain.

The smoothing of $f(t)$ described so far is only one example of smoothing a time series, and may serve as an introduction. Different smoothing functions have to be used, depending upon the purpose of the smoothing procedure. Some of the most common methods have been compiled in Table XXXI, to which we give the following comments.

Case 1. Running means over N consecutive readings, all with equal weight, is an extension to N readings of our discussion just made. The frequency response function is maximum $= 1$ for $\omega = 0$, and it is $= 0$ for $\omega = \infty$, i.e., this kind of smoothing corresponds to low-pass filtering. As the smoothing function is real and even, the frequency response is real and even, i.e., there is no phase shift.

The frequency response has been derived under the assumption of a continuous function $f(t)$. In fact, $f(t)$ is given instead as discrete digits, and then integrations have to be replaced by summations. The Fourier transform then becomes for N digits:

$$\begin{aligned} H(\omega) &= \frac{1}{N}\left[1 + 2\sum_{k=1}^{(N-1)/2} \cos k\omega\Delta t\right] \\ &= \frac{1}{N}\left(1 + 2\cos\omega\Delta t + 2\cos 2\omega\Delta t + \ldots + 2\cos\frac{N-1}{2}\,\omega\Delta t\right) \\ &= (\sin N\omega\Delta t/2)/(N \sin \omega\Delta t/2) \qquad [32] \end{aligned}$$

which is easily proved by using a summation formula for the series of cosine terms.

TABLE XXXI

Some usual filtering operations and their properties

Case	Filtering procedure $f(t) \star h(t)$	Amplitude response $\lvert H(\omega)\rvert$	Phase response $\Phi(\omega)$	Characteristics
1	running means over N readings, equal weight: $f(t) \star \frac{1}{N\Delta t} \Pi\left(\frac{t}{N\Delta t}\right)$	$\left\lvert\frac{\sin(N\omega\Delta t/2)}{N\omega\Delta t/2}\right\rvert$	0, π	low-pass filtering, no phase shift, only phase reversal in alternating side-lobes ($-1 = e^{i\pi}$)
2	running means, unequal weights (cosine-tapered): $f(t) \star \frac{\pi}{2N\Delta t} \cos \frac{\pi t}{N\Delta t}$ $\lvert t\rvert \leq \frac{N\Delta t}{2}$	$\left\lvert\frac{\pi^2 \cos(N\omega\Delta t/2)}{\pi^2 - N^2\omega^2(\Delta t)^2}\right\rvert$	0, π	low-pass filtering, no phase shift, only phase reversal in alternating side-lobes
3	running means, unequal weights (Gaussian curve): $f(t) \star e^{-at^2}$ $a > 0$	$\sqrt{\frac{\pi}{a}}\, e^{-\omega^2/4a}$	0	low-pass filtering, no phase shift

TABLE XXXI (continued)

Case	Filtering procedure $f(t) \star h(t)$	Amplitude response $\lvert H(\omega)\rvert$	Phase response $\Phi(\omega)$	Characteristics
4	instrumental inertia smoothing:			
	$f(t) \star u(t)\, a\, e^{-at}$	$a/\sqrt{a^2 + \omega^2}$	$\tan^{-1}\left(-\frac{\omega}{a}\right)$	low-pass filtering, phase shift
	$a > 0$			
5	running differences:			
	$f(t) - a\, f(t - \Delta t)$	$(1 + a^2 - 2a \cos \omega\Delta t)^{1/2}$	$\tan^{-1} \frac{a \sin \omega\Delta t}{1 - a \cos \omega\Delta t}$	high-pass filtering, phase shift
	$a > 0$			
6	running differences, $a = 1$:			
	$f(t) - f(t - \Delta t)$	$2\left\lvert\sin \frac{\omega\Delta t}{2}\right\rvert$	$\pm \frac{\pi}{2} - \frac{\omega\Delta t}{2}$	high-pass filtering, phase shift
7	combination of 1 and 6:			
	$[f(t) - f(t - \Delta t)] \star \frac{1}{N\Delta t}\, \Pi\left(\frac{t}{N\Delta t}\right)$	$2\left\lvert\sin \frac{\omega\Delta t}{2}\right\rvert \left\lvert\frac{\sin (N\omega\Delta t/2)}{N\omega\Delta t/2}\right\rvert$	$\pm \frac{\pi}{2} - \frac{\omega\Delta t}{2}$	band-pass filtering, phase shift

As we see, the formula [32] for the discrete case is not exactly the same as for the continuous case. The latter can be considered as an approximation to the true expression [32], but the approximation is in general sufficiently accurate for practical purposes. Of course, for $\omega\Delta t$ small, we have:

$$N \sin \omega\Delta t/2 \rightarrow N\omega\Delta t/2$$

i.e. as in the continuous case. Smoothing (low-pass filtering) by an amplitude-frequency function of this type, $(\sin x)/x$, has been applied among other things to secular vertical movements (Mizoue, 1967). Smoothing (low-pass filtering) by two-dimensional running averages is applied to gravity observations by Hagiwara (1967). See further section 10.2.1 and Table LIII.

Frequently, the low-pass filtering may be further accentuated by *decimation*, which means that not all values are included in the further spectral calculations, but, for instance, only every second, third, ... value. In addition, this procedure reduces the demands for computer storage.

For application to power spectra, the amplitude response $|H(\omega)|$ should be squared. Moreover, we note quite generally that subtraction of a low-pass filtering function from the original spectrum implies high-pass filtering. Thus, in the present case, the function:

$$1 - [\sin (N\omega\Delta t/2)/(N\omega\Delta t/2)]^2 \qquad [33]$$

corresponds to high-pass filtering of power spectrum. One example is Benioff et al. (1961), who used this high-pass filter to eliminate the earth tides in studies of the free oscillations. A similar operation was applied by Roden (1968) to certain oceanographic observations.

Case 1 could also be immediately inverted: a rectangular amplitude response function will correspond to a sinc-function as smoother in the time domain (this is obvious from examples 2 and 3, Table V, Chapter 2). This case has an obvious significance, because by such a response in the frequency domain we will exclude all contributions above a certain cut-off frequency and include everything undistorted below this frequency. Cf. section 4.3.1. The narrow band-pass filtering, which is applied in some cases, for example to construct spectra (section 1.6.1), belongs to this category. Cf. Archambeau and Flinn (1965). Then integration is extended over a narrow frequency band $\Delta\omega$, which is equivalent to multiplication with a sinc-function in the time domain. This method was applied by Kanamori and Abe (1968) for group-velocity studies of surface waves. However, the sinc-function with its infinite side-bands entails certain inconveniences when applied in the time domain. An objection to smoothing according to Case 1 is that it may give too low weight to the central value. Then an operation as the following (case 2) will be preferable.

Case 2. Running means can be made with unequal, but generally symmetric weights, i.e. symmetric around the central or *principal weight*. The cosine-tapered weights are one example of such smoothing, which among others have found useful application to sea-level records (Groves, 1955) and marine gravity measurements (Boyarskiy and Kogan, 1968), being superior to the equal-weight smoothing. Comparing 1 and 2 we find from $|H(\omega)|$ that 2 leads to more efficient low-pass filtering, i.e. passes more of low frequencies and less of higher frequencies than 1. Case 1 is especially unfavourable at:

$$\omega = n\pi/N\Delta t, \qquad n = 1, 3, 5, \ldots$$

As an exercise, the reader can test that a smoothing function:

$$h(t) = \frac{1}{N\Delta t}\left(1 + \cos\frac{2\pi t}{N\Delta t}\right)$$

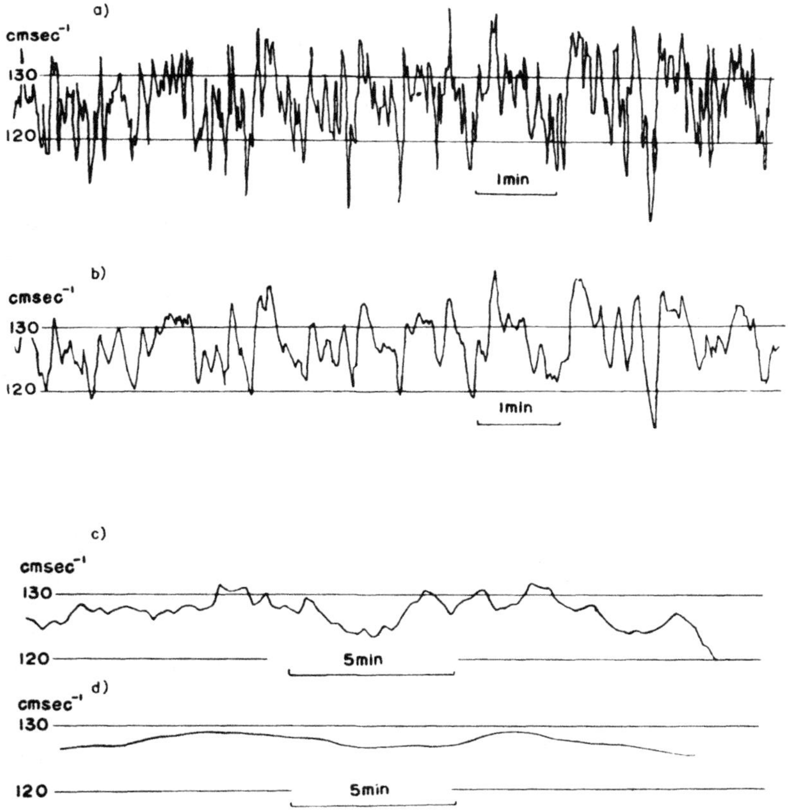

Fig.59. Smoothing of records of velocity in a river for different averaging intervals. a) Instantaneous values. b) Averaging interval 6 sec. c) Averaging interval 1 min. d) Averaging interval 5 min. After Yokosi (1967).

would be even better than 2 from these viewpoints. This was used for example by Yokosi (1967), who demonstrates the result in an instructive figure, applied to river turbulence (Fig.59).

Symmetric filtering can also be achieved by replacing a function $f(t)$ by its value, smoothed symmetrically over its adjacent values, like this:

$$f(t) + n[f(t + \Delta t) + f(t - \Delta t)] + m[f(t + 2\Delta t) + f(t - 2\Delta t)] \quad [34]$$

In this case, the frequency response is:

$$|H(\omega)| = 1 + 2n \cos \omega\Delta t + 2m \cos 2\omega\Delta t \quad [35]$$

By suitably adjusting the values of n and m, it is possible to obtain good filtering (applied, for example, with $n = m = \frac{1}{2}$ by Aki (1960a), for smoothing Love-wave records, and with a larger number of terms by Cisternas (1961) for recovering long-period Rayleigh waves from short-period records). In evaluating such expressions, the summation formulas for:

$$\sum_{n=1}^{N} \cos n\alpha, \qquad \sum_{n=1}^{N} \sin n\alpha$$

are most convenient, which can be found in mathematical tables. A most common procedure is to smooth over three consecutive values, e.g. with weights 0.25, 0.50, 0.25 (elementary binomial smoothing). It is left as an exercise to the student to investigate theoretically the properties of a general binomial smoothing. See also Table XVI for other commonly used weights.

Case 3. Another example of unequal, symmetrical smoothing is offered by the Gaussian-curve smoothing (Table XXXI, with reference to example 25 in Table V). Again, such smoothing corresponds to low-pass filtering without phase change. In practice, smoothing is generally done so that the sum of the weights, applied in each smoothing calculation, equals one. In this way, the same ordinate scale applies as for the original series. This explains the normalizing factors of $h(t)$ (Table XXXI), by which the area under the $h(t)$-curve equals unity.

Comparing the Gaussian curve (case 3) with rectangular smoothing (case 1), we find that case 3 has the advantage of no side-lobes in the frequency domain (no "ringing"), but that, on the other hand, case 3 has generally a slower cut-off than case 1, i.e., case 3 does not represent any ideal filter. As examples of Gaussian smoothing, we may mention Agarwal and Lal (1972b) who applied it to space smoothing of gravity data, Bilham et al. (1972) who used it to smooth strainmeter records and their power spectra, and Mikumo and Nakagawa (1968) who applied it to earth tide records.

Obviously, the smoothing operators dealt with hitherto under 1, 2, and 3, are just special cases of more general operators with similar properties (see for example Chan and Leong, 1972).

Case 4. Instrumental inertia smoothing. This is a one-sided smoothing, as an instrumental setting is the sum of the influences from past effects, whereas future effects will have no influence. Examples are: (1) thermal inertia: a mercury thermometer always lags behind the temperature fluctuations in the ambient atmosphere; (b) mechanical inertia: a heavy wheel will need some time before it gets the rotational speed corresponding to an applied force; and (c) electrical inertia: the electric current in a circuit containing an inductor will not reach a constant value until after the magnetic field in the inductor has been built up. In these and many other cases, an exponential smoothing function corresponds to the time effect. $1/a$ is called the *time constant* of the system, i.e., the time interval after which an instrumental setting has decreased to $1/e$.

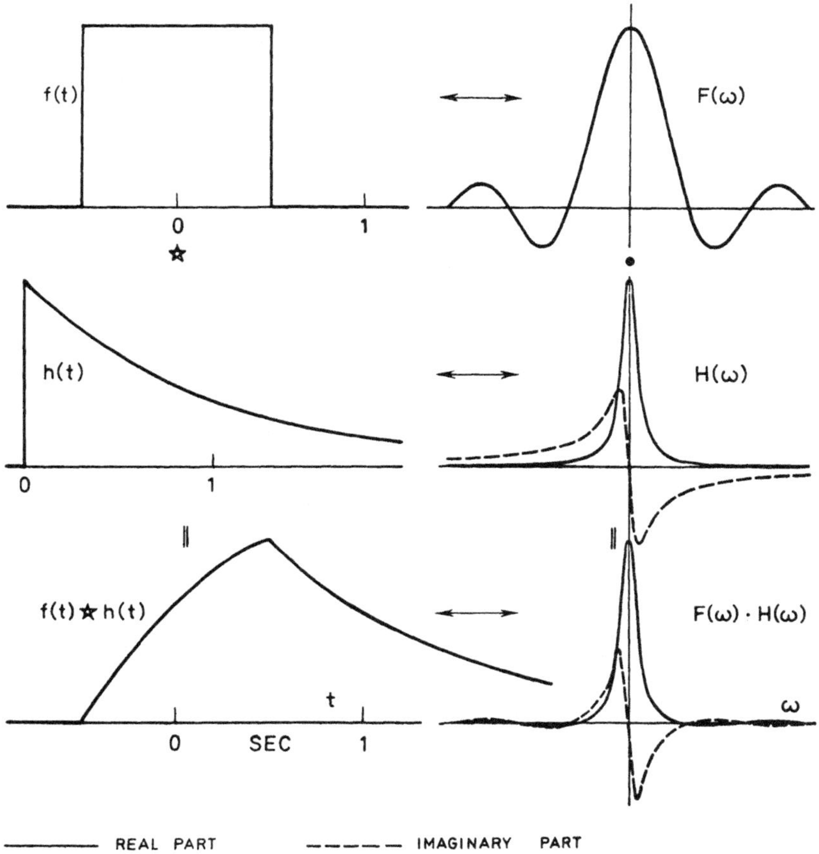

Fig.60. Instrumental inertia smoothing of a rectangular time function. With smaller time-constant ($1/a$), the effects on the resulting record and its spectrum will be less (cf. Fig.21).

As the frequency response (cf. example 23, Table V, Chapter 2) is largest for $\omega = 0$, this smoothing also corresponds to low-pass filtering. But as the smoothing function is not even, there is also a phase shift, in fact a phase delay. As the phase shift is frequency-dependent, *phase distortion* is introduced, which completely modifies the signal shape. Fig.60 shows an example of this smoothing operation when $f(t)$ is a rectangular function.

For a review, see Hall (1950). Dutton (1962) has applied and expanded such considerations to the measurement of radiation from the ground by instruments mounted on airplanes, and Volkov (1971) applied the method to a study of phase characteristics of tidal recording instruments. The amplitude response of this filter makes it a special case of a group of filters, called *Butterworth filters*; cf., for example, Gold and Rader (1969, p.59).

6.3.2 High-pass filtering—prewhitening

Case 5 (Table XXXI). Running differences are another type of smoothing which may be applied to eliminate longer periods in the time series $f(t)$. The frequency response, here given for a continuous curve $f(t)$ by straight-forward Fourier transforming, is seen to be minimum $= 1 - a$ for $\omega = 0$ and to increase as ω increases. This corresponds to high-pass filtering, and is therefore suitable to remove trends from observational series before further analysis (section 4.2.4). Moreover, as the smoothing procedure is not even, there is also a phase shift, equal to $\pi/2$ for $\omega = 0$. We have assumed that $a > 0$, but if instead $a < 0$, this filter obviously becomes a low-pass filter.

Case 6. This is the same as 5 but with $a = 1$, by which the formulas simplify, but the general properties (high-pass filtering and phase shift) remain.

Substituting $t + \Delta t/2$ for t, this operation is changed into a difference:

$$f\left(t + \frac{\Delta t}{2}\right) - f\left(t - \frac{\Delta t}{2}\right)$$

which is symmetric with regard to $f(t)$. Using the time-shifting theorem (section 2.3.4), it is easy to deduce its properties. This can be extended to include symmetric differences of more than two terms (cf. Schaub, 1961).

Case 6 is equivalent to taking time derivatives, which, as we have seen in section 2.3.6, corresponds to high-pass filtering. Cases like 5 or 6 are suggested as effective methods of *prewhitening*, i.e. methods to get a flatter spectrum (cf. Tukey, 1967). Prewhitening is made on data showing rapid spectral variations, in order to avoid side-lobe effects and to get a more reliable spectrum. The true spectrum is then obtained by compensation for the prewhitening. This compensation ("post-colouring") is effected by dividing the amplitude spectrum by the amplitude response, listed in Table XXXI (or dividing the power spectrum by the square of the amplitude response in Table XXXI).

The geophysical literature contains numerous examples of prewhitening, and there is no need to enumerate these here. May it just suffice to mention Bolt and Marussi (1962), who applied it on tiltmeter records, Whitham and Andersen (1965), who prewhitened electric and magnetic records, and Dyer (1970), who applied it to a Markov process (Fig.22), represented by snowfall intensities. The great importance of prewhitening before calculation of geophysical spectra stems to a great extent from the fact that many such spectra exhibit a power or amplitude significantly decreasing with increasing frequency (section 9.3.3). For a power $E(\omega)$ varying with frequency ω as ω^{γ}, $\gamma < 0$, the power will become infinite at zero frequency and very high at low frequencies. As emphasized by Hinich (1967), this will lead to leakage of low-frequency energy through the side-lobes of the filter used. See Fig.61. To avoid such spectral distortion, prewhitening is necessary, especially if the recording apparatus has no high-pass filter to eliminate the effect of $E(\omega) \sim \omega^{\gamma}$. Hinich (1967) gives an alternative, simplified method for calculation of exponential spectra.

At this point it may occur to the reader that the operations of detrending (section 4.2.4) and of prewhitening are closely related, almost identical, both consisting in certain modifications to a given record before its spectrum is calculated. Still, it is preferable to distinguish between these two processes, mainly for the following reasons:

(1) Detrending means elimination of some unwanted drift of the observations, usually of natural or instrumental origin or introduced during the analysis. Trend effects are not restored by any process after a spectrum has been calculated. Pre-

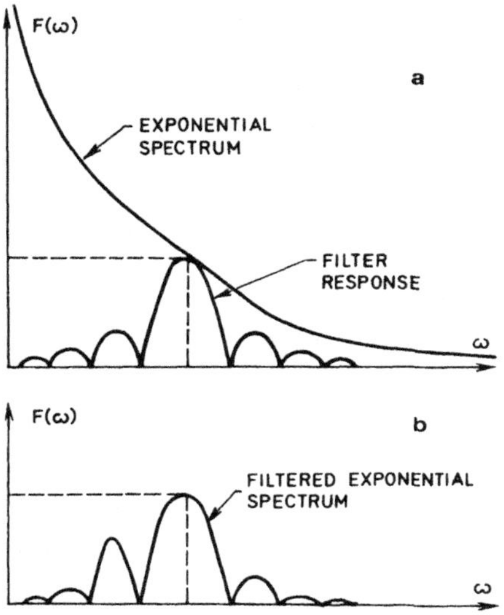

Fig.61. a) Exponential spectrum and filter response. b) Side-lobe distortion due to low-frequency leakage. After Hinich (1967).

whitening, on the other hand, means elimination of some dominating spectral components before the spectrum is calculated, as these would otherwise distort the spectrum due to their side-lobe effects. Afterwards, the prewhitening is compensated for by "post-colouring", which reinstalls the temporarily suppressed frequencies.

(2) Detrending is intended to be complete, for full elimination of unwanted components. Detrending is therefore required even for small trends. Prewhitening, on the other hand, does not need to be complete, as only a partial prewhitening may be enough to suppress low frequencies so as to eliminate their distortion of the spectrum. Prewhitening may therefore be limited to elimination of components of only greater influence.

(3) Detrending is usually executed by means of some formula, which has to be determined from the observations themselves, as the drift is generally not exactly known beforehand. Prewhitening, on the other hand, is usually made with a given formula, based on some knowledge of the analyzed phenomenon. The formula for "post-colouring" executes the inverse operation in the frequency domain.

(4) It follows that detrending and prewhitening may both be involved in the same calculation process, only that they are kept separate.

As no compensation is made for detrending after a spectrum is calculated (as opposite to prewhitening, point 1 above), it is of great significance that the detrending has been made correctly. Otherwise, it may have bad consequences for the calculated spectrum.

6.3.3 Cascade filtering

Often it may be advantageous to apply several successive filterings to the same time series $f(t)$, and then the different operations may be identical (i.e., the same operation is carried out repeatedly) or there may be a combination of two or more different operations. Such successive operations correspond to repeated convolutions in the time domain and to multiplication of the respective frequency response functions in the frequency domain. As convolution is commutative, it is clear that the order in which we perform different kinds of smoothing on a given time series $f(t)$ is immaterial.

As an example of repeated operation of the same smoothing function, we may take the running averages over N consecutive terms. Carrying out this operation M times, using the expression for the discrete time series, we obtain the response function from [32] by M-time multiplication:

$$H(\omega) = [(\sin N\omega\Delta t/2)/(N \sin \omega\Delta t/2)]^M \qquad [36]$$

The result is repeated low-pass filtering.

A suitable combination of low-pass and high-pass filtering, as by combination of cases 1 and 6 (Table XXXI), leads to passage of intermediate frequencies, i.e. band-pass filtering. As soon as one of the component operations is not an even function of t, this will entail phase shifts. Naturally, it would be possible to eliminate an introduced phase shift by further smoothing.

Frequently it is necessary to subject an observed series to a sequence of filtering operations to isolate one particular phenomenon for further study. We may mention tidal studies from ocean current measurements as one example (Godin, 1967). In brief and schematic form, we can demonstrate the operations as follows. The given record consists of three essential parts:

$$f(t) = f_1(t) + f_2(t) + f_3(t) \qquad [37]$$

where $f_1(t)$ is turbulence effects (high frequency), $f_2(t)$ is steady flow (low frequency) and $f_3(t)$ is tides (daily and semidaily). The successive filtering operations are then as follows, written only in a schematic form:

(1) Removal of $f_1(t)$ by low-pass filtering (Godin, 1967, applies a filter of type 1 in Table XXXI, three times over), which results in a new time series:

$$\bar{f}(t) = f(t) - f_1(t) = f_2(t) + f_3(t) \qquad [38]$$

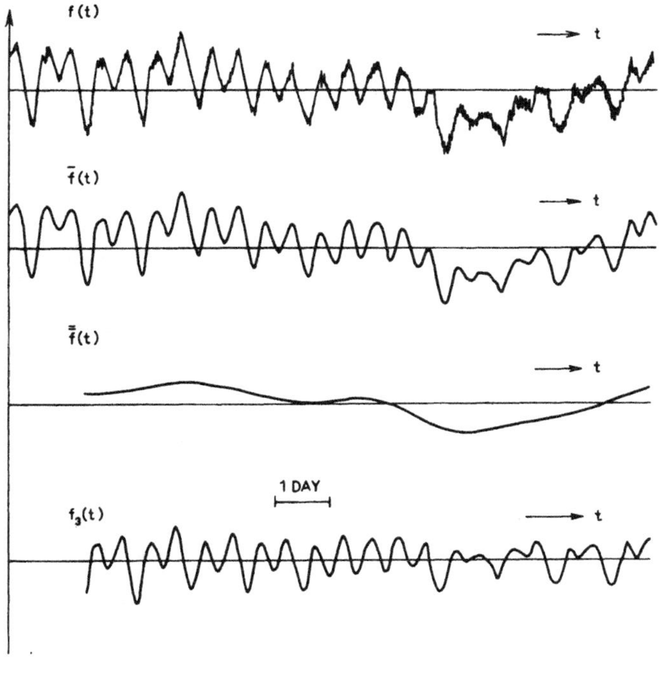

Fig.62. Successive filtering of records of oceanic tides. Notation explained in the text (section 6.3.3). After Godin (1967).

(2) Removal of $f_3(t)$ by another low-pass filtering (again of the same type as 1 in Table XXXI), leaving:

$$\bar{\bar{f}}(t) = f(t) - f_1(t) - f_3(t) = f_2(t) \qquad [39]$$

(3) Finally, subtracting 2 from 1 leaves the tidal components for further analysis:

$$\bar{f}(t) - \bar{\bar{f}}(t) = f_3(t) \qquad [40]$$

The procedures are illustrated in Fig.62.

6.3.4 General discussion

Great care is required in performing filtering procedures. Especially it is necessary to see what the effect a chosen procedure in the time domain will have in the frequency domain, in other words it is necessary to be clear about the spectral shape of the amplitude response or transfer function, Table XXXI. If the amplitude response has a pronounced peak at some frequency, this peak will be introduced in the resulting spectrum as a spurious peak. Such effects have to be taken carefully into account in interpretations of spectra. A striking example from geomagnetism is given by Currie (1966), where this phenomenon is referred to as the *Slutzky effect.*

By tracing the filtering functions in Table XXXI versus frequency, the reader can easily convince himself that most of the filters are not very efficient, i.e., they do not produce any sharp separation of low and high frequencies. In other words, they are often far from being *ideal filters*. Much effort has gone into the construction of more efficient filters, by elaborating on the function $|H(\omega)|$ by multiplication with certain frequency-dependent factors and otherwise (see, for example, Gol'tsman, 1957).

Sometimes, it may be necessary to develop special filtering formulas which suit a particular problem, so as to eliminate unwanted effects without considerably affecting wanted effects. One example is provided from meteorology by balloon observations of wind profiles (Armendariz and Rachele, 1967). Balloon oscillations introduce spurious high-frequency effects in wind power spectra. The best method to eliminate these was found to expand the given observations in a Fourier series and to truncate this at a certain frequency (0.05–0.15 cps), and then subject this series to spectral analysis. Cf. section 4.2.4 for some related, though different, procedures.

Quite clearly, any filtering done on recorded data can be compensated by application of filter response (Table XXXI) to the calculated spectrum. For this purpose, divide calculated amplitudes by amplitude response, and subtract phase response from calculated phases. This is done in "post-colouring" to remedy the effects of prewhitening, but the procedure has naturally a general validity, if wanted.

The spectrum $F(\omega)$ can also be subject to smoothing, usually corresponding to cases 1 or 2 in Table XXXI. The purpose of this smoothing is to remove larger oscillations, which may have been introduced by using a rectangular window in the transformation. See further Chapter 4. This smoothing, being a convolution in the frequency domain, corresponds to multiplication in the time domain.

6.4 TWO-DIMENSIONAL FILTERING

Our discussion so far has been restricted to one-dimensional (time) filtering, but can be immediately applied to one-dimensional space filtering as well as be extended to two or more dimensions. We shall consider the two most important cases of two-dimensional filtering.

6.4.1 Frequency–wavenumber filtering or velocity filtering

This method, which is based on different apparent velocities of wave components to be separated, appears also under the names of *fan filtering* or *moveout filtering* (Savit et al., 1958). The principle of velocity filtering is explained in Fig.63, which provides for plotting observations from a linear seismic array in a frequency–wavenumber graph. The figure has been restricted to positive values of ω and k and shows complete separation between the velocities of the signal and the ambient noise. In such a case it is possible to design a filter with a transfer function $H(\omega,k)$ defined by:

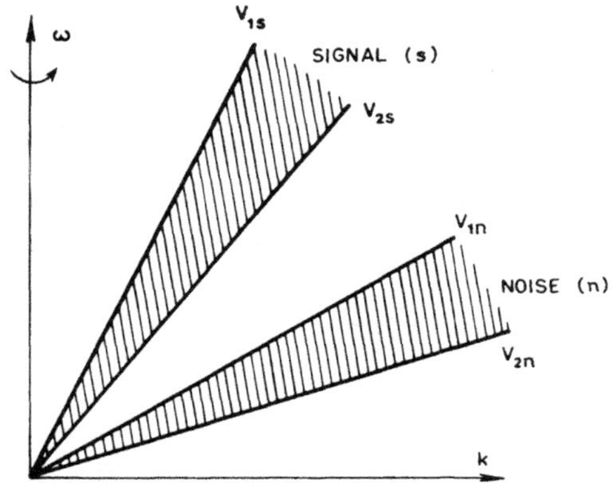

Fig.63. The principle of velocity filtering.

$$H(\omega,k) = \begin{cases} 1 & \text{for } \omega/V_1 < k < \omega/V_2 \\ 0 & \text{otherwise} \end{cases} \qquad [41]$$

This will effectively remove the noise from the signal.

For body waves, the two limits V_1 and V_2 of the moveout or apparent horizontal velocity correspond to certain distance limits Δ_1 and Δ_2, which can be obtained from travel-time data. For $V_1 > V_2$ as in Fig.63 we get $\Delta_1 > \Delta_2$ for P and for PKKP (over the larger arc) but $\Delta_1 < \Delta_2$ for PcP. By letting V_1 coincide with the ω-axis, we have V_1 infinite, which corresponds to vertical incidence. The velocity discrimination makes it possible to separate even closely arriving phases, such as in the group of core phases (PKP, PKS, SKP, etc.). See e.g. Hannon and Kovach (1966). For surface waves, the velocities V in Fig.63 are phase velocities, and in this case the velocity filtering is distance-independent. Thus, velocity filtering implies distance filtering for body waves but not for surface waves.

In other cases, the velocity ranges (hatched in Fig.63) may overlap partially, and then no complete separation is possible even with velocity filtering. Frequency filtering, on the other hand, on a case as shown in Fig.63, would only be of little effect, as the frequency ranges overlap to a very large extent.

Velocity filtering requires determination of wavenumber and thus needs a profile or array of seismometers. Moveout-velocity differences, on which this method is based, may be due to differences in wave type (e.g. signal versus noise) or, for a given wave, due to different dips, the latter case being significant in seismic

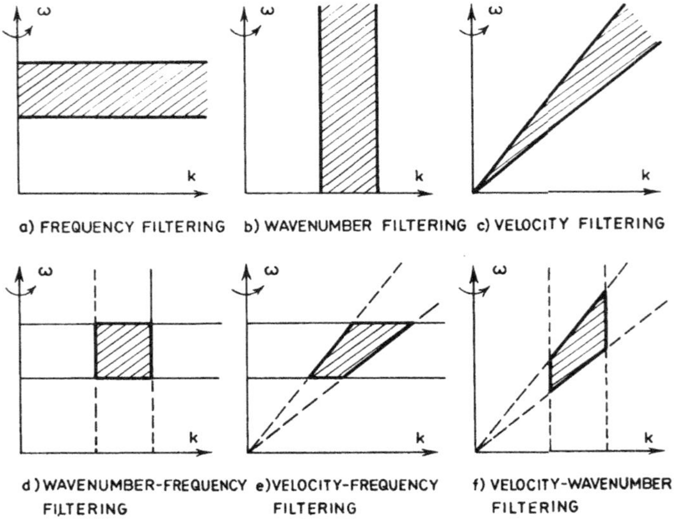

Fig.64. Different modes of filtering in the frequency–wavenumber domain. Each mode can be combined with directional filtering by letting the rotation around the ω-axis include only a limited section of arrivals.

prospecting. For more details about velocity filtering, the reader is referred to Embree et al. (1963), Fail and Grau (1963), Robinson (1967a), Nakhamkin (1969).

In the frequency–wavenumber domain we may easily envisage other filtering transfer functions than the one of Fig.63. A review is given in Fig.64. For example, Wiggins (1966) designed filters with a limited pass-band both in velocity and in frequency (Fig.64e). The filtering schemes shown in Fig.64 are only the most simple ones. None of these would for instance be effective in filtering out a certain mode of dispersed waves (surface waves). Then, special filters are needed which fit the respective dispersion curves, that can be placed into the frequency–wavenumber diagram (Laster and Linville, 1966).

6.4.2 Wavenumber–wavenumber filtering or two-dimensional spatial filtering

Two-dimensional spatial filtering has special significance for various kinds of potential fields (gravity, geomagnetism, cf. Dean, 1958), in seismology (array-station processing), and in structural geology (e.g., to filter out certain linear structures, strikes). We shall here essentially follow Fuller (1967), only with notation chosen so as to conform with the usage in this book.

With input function $f(x,y)$ and impulse response $h(x,y)$ in two spatial directions x,y, the output function $g(x,y)$ is as follows:

$$g(x,y) = f(x,y) \star h(x,y) = \int_{-\infty}^{\infty}\int_{-\infty}^{\infty} f(x-\xi,\, y-\eta)\, h(\xi,\eta)\, \mathrm{d}\xi\, \mathrm{d}\eta$$

$$= \int_{-X}^{X}\int_{-Y}^{Y} f(x-\xi,\, y-\eta)\, h(\xi,\eta)\, \mathrm{d}\xi\, \mathrm{d}\eta \qquad [42]$$

if $h(x,y) = 0$ for $\begin{cases} |x| \geq X \\ |y| \geq Y \end{cases}$

The corresponding equation in the wavenumber domain is the following:

$$G(k_x,k_y) = F(k_x,k_y) \cdot H(k_x,k_y)$$

where:

$$H(k_x,k_y) = \int_{-X}^{X}\int_{-Y}^{Y} h(x,y)\, \mathrm{e}^{-i(k_x x + k_y y)}\, \mathrm{d}x\, \mathrm{d}y$$

$$= 4\int_{0}^{X}\int_{0}^{Y} h(x,y) \cos k_x x \cos k_y y\, \mathrm{d}x\, \mathrm{d}y \qquad [43]$$

provided $h(x,y)$ is even both in x and y. The inverted transform of $H(k_x,k_y)$ gives $h(x,y)$:

$$h(x,y) = \frac{1}{4\pi^2} \int_{-k_{0x}}^{k_{0x}} \int_{-k_{0y}}^{k_{0y}} H(k_x,k_y)\, e^{i(k_x x + k_y y)}\, dk_x\, dk_y \qquad [44]$$

$$\text{if} \quad H(k_x,k_y) = 0 \qquad \text{for} \begin{cases} |k_x| \geq k_{0x} \\ |k_y| \geq k_{0y} \end{cases}$$

k_{0x} and k_{0y} being the cut-off wavenumbers, chosen as 0.5 cycles/data interval, i.e. equal to the Nyquist wavenumber. Note that in two- or multi-dimensional sampling, the sampling interval may be different along the different coordinate axes, with the consequence that the corresponding Nyquist frequencies or wavenumbers will not be the same.

These are the formulas for continuous functions $f(x,y)$, $h(x,y)$ and $H(k_x,k_y)$. The corresponding expressions for discrete data are (m, n, l and j are integers):

$$g(x,y) = \sum_{m=-X/\Delta x}^{X/\Delta x} \sum_{n=-Y/\Delta y}^{Y/\Delta y} f(x - m\Delta x,\, y - n\Delta y)\, h(m\Delta x, n\Delta y)\, \Delta x\, \Delta y$$

$$= \sum_{m=-X}^{X} \sum_{n=-Y}^{Y} f(x - m,\, y - n)\, h(m,n) \qquad [45]$$

putting $\Delta x = \Delta y = 1$;

$$H(k_x,k_y) = 4 \sum_{m=0}^{X} \sum_{n=0}^{Y} h(m,n) \cos k_x m \cos k_y n \qquad [46]$$

and:

$$h(m,n) = \frac{1}{\pi^2} \sum_{l=0}^{0.5/\Delta k_x} \sum_{j=0}^{0.5/\Delta k_y} H(l\Delta k_x, j\Delta k_y) \cos(l\Delta k_x m) \cos(j\Delta k_y n)\, \Delta k_x\, \Delta k_y \qquad [47]$$

The procedure in applications is as follows: (1) first select a function $H(k_x,k_y)$ which satisfies the demands for wavenumber separation; (2) calculate $h(m,n)$ from [47]; and (3) convolve $h(m,n)$ with the given data $f(x,y)$, [45], which gives the desired output $g(x,y)$.

The two formulas for $H(k_x,k_y)$ are not used here, but have to be applied in case we instead start with an assumed form of the function $h(x,y)$.

Obviously, it is possible to perform any wavenumber-dependent operation on the given data by following this scheme. This suggests a unified treatment of a number of problems. Such a discussion, which thus comprises quite a number of different papers, has been published by Fuller (1967). His treatment includes cases where $H(k_x,k_y)$ is chosen so as to get various derivatives of potential fields, also analytic continuation of such fields, as well as filtering (low-pass, high-pass, band-pass) in the wavenumber domain. Similar discussions are presented by Meskó (1965), Carrozzo and Mosetti (1966), Darby and Davies (1967) and Mihail and Nicolae (1972). It should be noted that upward and downward continuations of potential fields (gravity, geomagnetism) are equivalent to low-pass and high-pass spatial filtering, respectively (cf. Hagiwara, 1965). *Optimum filtering* is of particular value in emphasizing effects of certain wanted formations (signals) and suppressing other effects (noise). The optimum filtering is based on minimizing the average square difference between desired and actual output. Scollar (1970) has given a non-mathematical, but extensive and very elucidative account of two-dimensional filtering methods, with special application to geomagnetic fields measured in archaeology.

By suitable combination of the treatments in sections 6.4.1 and 6.4.2, these can be generalized to the three-dimensional case: space–time x,y,t, wavenumber–frequency k_x, k_y, ω. For example by rotation around the ω-axis in Fig.63 and 64, we achieve an extension to three dimensions. With two dimensions, the wavenumber (k) axis either points in the direction of wave arrival, or else represents only one component of the total wavenumber. In three dimensions, the axes are ω, k_x, k_y. Then it is possible to add direction of arrival to the other three filtering parameters (velocity, frequency, wavenumber). This multiplies the number of cases in Fig.64 by two, and for all of these the transfer functions can be formally written down in analogy to [41] above.

The three-dimensional ω,k_x,k_y-diagram has important application to array-station data. The simplest procedure applied to different array sensors is some kind of record summation, with the purpose of increasing the signal/noise ratio (for detection of weak events) with as little signal distortion as possible. In order of increasing sophistication, summation methods may be listed as follows (cf. section 2.5.4):

(1) Straight summation.

(2) Delay and sum method ("beam-forming").

(3) Weighted delay and sum method.

(4) Filter and sum method: (a) maximum likelihood filtering, and (b) Wiener filtering.

Filtering (item 4) may be of value to decrease side-lobe effects. Three-dimensional filtering has been dealt with by Holzmann (1960a), Vinnik (1963), and others. These extensions do not involve any special difficulties or any basically new principles, and the reader should not find any difficulty in following this literature with the background provided here. For other extensions including filtering on a spherical surface, see Hannan (1966).

6.5 FILTERING AND EQUALIZATION IN GEOPHYSICS

6.5.1 Cascade filtering and equalization in general

With the filter properties laid down in Table XXVII, especially linearity and stationarity, we could immediately extend the scheme in Fig.53 to any number of cascaded filters as in Fig.65. The filter properties assumed permit them to be taken in any order, corresponding to the commutative law for convolution (in the time domain) or for multiplication (in the frequency domain). Also, it is possible for any number of the filters to be active simultaneously, with no other change.

The relation in Fig.65 for the frequency domain:

$$S(\omega) \cdot H_1(\omega) \cdot H_2(\omega) \cdot \ldots \cdot H_n(\omega) = X(\omega) \qquad [48]$$

can be immediately split into its two parts, i.e. the amplitude spectrum:

$$|S(\omega)| \cdot |H_1(\omega)| \cdot |H_2(\omega)| \cdot \ldots \cdot |H_n(\omega)| = |X(\omega)| \qquad [49]$$

and the phase spectrum:

$$\Phi_S(\omega) + \Phi_{H_1}(\omega) + \Phi_{H_2}(\omega) + \ldots + \Phi_{H_n}(\omega) + 2m\pi = \Phi_X(\omega) \qquad [50]$$

$$m = 0, \pm 1, \pm 2, \ldots$$

The corresponding equation for power is immediately obtained by squaring [49], whence $|H_1(\omega)|^2 \cdot |H_2(\omega)|^2 \cdot \ldots \cdot |H_n(\omega)|^2$ is the total power transfer function. In the following we will restrict the discussion to amplitude and phase, the discussion for power being fully analogous to the one for amplitude, only with squared quantities.

Usually, only the output $X(\omega)$ is available for direct observation (after spectralizing a record $x(t)$), while the functions constituting $X(\omega)$ according to

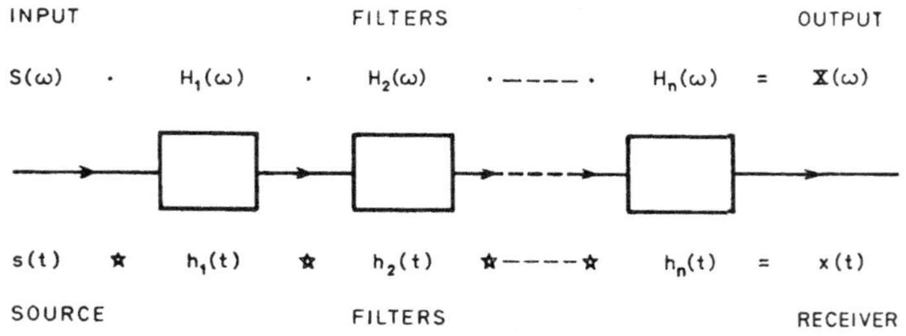

Fig.65. Scheme of a cascade filter.

[48], i.e. both the source $S(\omega)$ and the medium properties $H_1(\omega)$, $H_2(\omega), \ldots, H_n(\omega)$ are unknown. Equation [48] expresses in a nutshell the standing of spectroscopy in geophysics and in general. That is, the dependence of an observed phenomenon on a whole series of factors is both a weakness and a strength of the spectral method. The weakness is that by a single observation $X(\omega)$ it is impossible to make deductions about source or medium properties, because of the ambiguity in such an interpretation. On the other hand, the strength lies in the fact that by clever combination of several factors, it is possible to isolate just one of the factors on the left-hand side of [48] which will then be determined. Applied in such a way, the spectral method is usually more powerful than other methods.

Obviously, due to the multiple influence on $X(\omega)$, [48], the spectral method works with comparisons of measurements taken under similar conditions, preferably where only one factor is varied at a time. This procedure goes under the name of *equalization*, which is simply explained as follows. Suppose we have two observations I and II (denoted by superscripts), then the amplitude ratio and the phase difference are as follows from [49] and [50]:

$$\frac{|X^{\mathrm{I}}|}{|X^{\mathrm{II}}|} = \frac{|S^{\mathrm{I}}| \cdot |H_1^{\mathrm{I}}| \cdot |H_2^{\mathrm{I}}| \cdot \ldots \cdot |H_n^{\mathrm{I}}|}{|S^{\mathrm{II}}| \cdot |H_1^{\mathrm{II}}| \cdot |H_2^{\mathrm{II}}| \cdot \ldots \cdot |H_n^{\mathrm{II}}|}$$

$$\begin{aligned}\Phi_X^{\mathrm{I}} - \Phi_X^{\mathrm{II}} = \Phi_S^{\mathrm{I}} - \Phi_S^{\mathrm{II}} + \Phi_{H_1}^{\mathrm{I}} - \Phi_{H_1}^{\mathrm{II}} + \Phi_{H_2}^{\mathrm{I}} - \Phi_{H_2}^{\mathrm{II}} + \ldots + \\ + \Phi_{H_n}^{\mathrm{I}} - \Phi_{H_n}^{\mathrm{II}} + 2m\pi\end{aligned} \qquad [51]$$

Still, the formulas are in an intractable shape, but if we have chosen the cases I and II such that $S^{\mathrm{I}} = S^{\mathrm{II}}$, $H_1^{\mathrm{I}} = H_1^{\mathrm{II}}$, as well as the rest of the H functions, with just one exception:

$$H_2^{\mathrm{I}} \neq H_2^{\mathrm{II}}$$

then the formulas simplify as follows:

$$\frac{|X^{\mathrm{I}}|}{|X^{\mathrm{II}}|} = \frac{|H_2^{\mathrm{I}}|}{|H_2^{\mathrm{II}}|}$$

$$\Phi_X^{\mathrm{I}} - \Phi_X^{\mathrm{II}} = \Phi_{H_2}^{\mathrm{I}} - \Phi_{H_2}^{\mathrm{II}} + 2m\pi \qquad [52]$$

These equations provide a relative information of H_2^{I} and H_2^{II}. If one of them is known in some other way, also an absolute information about the other is possible.

In most geophysical procedures, it is not possible to reach such a perfect equalization as sketched here, where all factors except one are equal. The general rule is rather that equalization procedures are less perfect, with one factor differing far more than the others, and then the others are set approximately equal, if no other information is available.

The functions $x(t)$ and $s(t)$ above may represent any physical quantity. Starting from the convolution formula:

$$x(t) = s(t) \star h_1(t) \star h_2(t) \star \dots \star h_n(t) \quad [53]$$

where we now let $x(t)$ and $s(t)$ be displacements, then by the theorem for derivation of convolutions (section 3.2.3), we find for the velocities:

$$x'(t) = s'(t) \star h_1(t) \star h_2(t) \star \dots \star h_n(t) \quad [54]$$

and for the accelerations:

$$x''(t) = s''(t) \star h_1(t) \star h_2(t) \star \dots \star h_n(t) \quad [55]$$

Corresponding formulas hold for the frequency domain.

6.5.2 Cascade filtering and equalization in geophysics, especially seismology

The ideas laid down in section 6.5.1 have an immediate application to practically every observation in geophysics. We are almost never observing the source directly, but observations are made at some distance and also by means of some instrument. The propagation of a seismic wave from its source (earthquake, explosion) to the seismic record will be taken as a very illustrative example of filtering effects along the travel path.

The factors affecting a seismic wave and thus its record can be best grouped into three main items:

(1) Source properties, corresponding to $S(\omega)$ in [48]. These include the source time function and the source space function (source mechanism and focal depth), magnitude, coupling, surrounding geology, stress field, reverberation, etc in the source crust.

(2) Path properties, corresponding to $H_1(\omega)\ H_2(\omega) \dots$ in [48]. These include absorption and scattering, reflection, refraction and diffraction, dispersion, interference, geometrical spreading.

(3) Receiver properties, also corresponding to the $H(\omega)$-factors in [48]. These include reverberation and other wave interactions, as signal with noise, resonance effects, in the receiver crust. The seismograph represents also a filter, constituting the last factor before the record is obtained.

To this system, we apply the general equation [48], which becomes for body waves:

$$X(\omega,r) = S(\omega)\ B(\theta)\ C_s(\omega)\ M(\omega,r)\ G(r)\ C_r(\omega,r)\ I(\omega) \quad [56]$$

and for surface waves:

$$X(\omega,r) = S(\omega)\,B(\theta)\,C(\omega,r)\,G(r)\,I(\omega) \qquad [57]$$

where $X(\omega,r)$ = received spectrum at distance r
$S(\omega)$ = source spectrum, corresponding to source time function
$B(\theta)$ = source space function, where θ stands for direction from source
$C_s(\omega)$ = source crust effect on spectrum
$M(\omega,r)$ = mantle effect on spectrum
$G(r)$ = geometrical spreading
$C_r(\omega,r)$ = receiver crust effect on spectrum
$C(\omega,r)$ = crust and upper mantle effect on surface waves
$I(\omega)$ = instrumental response.

As in the general case, we can split [56] and [57] into their amplitude and phase forms. For example, from [56] we get immediately:

$$|X(\omega,r)| = |S(\omega)|\,|B(\theta)|\,|C_s(\omega)|\,|M(\omega,r)|\,G(r)\,|C_r(\omega,r)|\,|I(\omega)|$$

and:

$$\Phi_X = \Phi_S + \Phi_P + \Phi_I + 2m\pi \qquad [58]$$

where Φ_S and Φ_P stand for source and path effects, respectively.

Equations of this general shape constitute the basis for any equalization procedure in geophysics. For the purpose of illustration, let us consider an equalization of surface-wave spectra observed at two distances r_1 and r_2 along one and the same azimuth from an earthquake. The amplitude equalization leads to the following equation from [57]:

$$\frac{|X(\omega,r_2)|}{|X(\omega,r_1)|} = \frac{|C(\omega,r_2)|}{|C(\omega,r_1)|}\,\frac{G(r_2)}{G(r_1)} \qquad [59]$$

assuming same instrumental response at the two places, and the phase equalization yields:

$$\Phi_{X_2} - \Phi_{X_1} = \Phi_{C_2} - \Phi_{C_1} + 2m\pi \qquad [60]$$

This example will be treated in more detail in section 7.2 and 7.4.

Another form of equalization implies the reduction of an obtained spectrum to a certain fixed epicentral distance or to the source. With application to surface waves, [57], we could formally write the reduction of $X(\omega,r)$ to the source as:

$$\frac{X(\omega,r)}{C(\omega,r)\,G(r)\,I(\omega)} = S(\omega)\,B(\theta) \qquad [61]$$

This can be immediately split into its amplitude and phase forms, which would yield source amplitudes and initial phases, respectively. Dividing the obtained spectrum $X(\omega,r)$ by one or several of the filtering factors implies elimination of the filtering effect. It is equivalent to *deconvolution* or *inverse filtering*, carried out in the frequency domain.

The amplitude equalization provides a method to investigate the dynamic properties. i.e. all properties that influence amplitudes, while the phase equalization is a method to investigate kinematic properties, such as phase and group velocities. We will find many occasions to see applications of these equations in the solution of various problems in following chapters. The development was made here with special reference to the propagation of seismic waves, but fully analogous relations and equalization procedures hold in other cases. Therefore, equalization is a method of general importance in interpretation of geophysical spectra, provided conditions for linearity and stationarity are fulfilled. Cf. section 6.1.1.

For high efficiency of the method, it is necessary to have reliable information on factors entering the equations, beyond those which are actually being determined. This is obvious from the circumstance that so many factors enter every problem, and also from the fact that hardly any equalization is perfect, and then it is necessary to investigate validity limits. A complete exploration of all entering factors, both for body and for surface waves, and both for their amplitude and phase effects, encompasses practically the whole of seismology. Here we have given an outline of various influencing factors. For more detailed accounts, reference has to be made to special studies, e.g. Kogeus (1968), Berckhemer and Jacob (1968) and others. Both theoretical spectra and laboratory spectra (model spectra) are of great importance for interpretation of spectra obtained by observations in nature.

Many geophysical phenomena are interrelated to a greater or lesser extent, and often such interrelations extend over both the solid, liquid and gaseous parts of the earth. Investigations are then mostly concerned with relations between cause and effect, or in other words, between input and output. This immediately suggests filtering operations and this is exactly what is taking place. Then, working in the frequency domain is advantageous because of the simple relation between output and input: $G(\omega) = F(\omega)\ H(\omega)$, cf. section 6.1.2. If the properties of G and F have been observed, then the properties of H can be calculated. The geophysical literature offers numerous examples of this or similar approach. As just one example involving both the atmosphere and the solid earth, let us refer to a paper by Munk and Hassan (1961). The general shape or fluctuations in any spectrum or spectral ratio may be partly due to real effects, discussed in this section, partly due to effects of the data handling, discussed in Chapter 4. In any interpretation, it is significant to be aware of this fact.

6.5.3 Attenuation and geometrical spreading

The attenuation during the passage of waves through the earth's interior is partly due to inelastic properties, partly due to other effects, especially scattering at inhomogeneities. The combined effect is expressed in the quality factor Q of the medium. The following relation holds between the attenuation coefficient $\varkappa$ and the quality factor Q:

$$\varkappa = \pi/QTV \qquad [62]$$

The attenuation $e^{-\varkappa r}$ can thus be written as $e^{-\pi r/QTV}$ or, referred to one wavelength, as $e^{-\pi/Q}$ (the latter expression holding true only for non-dispersive waves). As the attenuation properties vary along the path through the earth's interior, the exponent has to be written in integral form, i.e., we get the following expression for $M(\omega,r)$ (Kurita, 1968):

$$M(\omega,r) = \exp\left[-\frac{\omega}{2}\int\frac{\mathrm{d}r}{Q(\omega,r)\,V(r)}\right] \qquad [63]$$

where the integration is extended over the whole path.

The expression [63] is the same whether applied to body waves or to surface waves. For dispersed wave trains (surface waves), the velocity V should be the group velocity (Brune, 1962), which is seen from the fact that energy (proportional to amplitude squared) and thus also amplitude propagate with the group velocity, not the phase velocity (Lamb, 1945, p.383). Methods for attenuation studies by means of spectra will be described in more detail in section 7.4.

The attenuation coefficient $\varkappa$ as used here refers to amplitude. For power the corresponding attenuation coefficient becomes $2\varkappa$, equivalent to the square of [63].

The factor $G = G(r) = G(\Delta)$ takes account of the amplitude variation with distance due to the geometrical spreading of the seismic rays (expansion of the wave front). For body waves, the geometrical spreading or divergence coefficient for amplitude can be calculated from:

$$G = \frac{1}{r_0}\left|\frac{\rho_h\,V_h}{\rho_0\,V_0}\,\frac{\sin i_h}{\sin\Delta\,\cos i_0}\,\frac{\mathrm{d}i_h}{\mathrm{d}\Delta}\right|^{\frac{1}{2}} \qquad [64]$$

where values with subscript h refer to the source (at depth h) and those with subscript 0 refer to the receiving station. Shimshoni and Ben-Menahem (1970) have indicated a simple and sufficiently accurate method to calculate G from a given travel-time table by applying cubic spline interpolation.

For surface waves, we have to consider spreading over a spherical surface, and then the expression for G for amplitude simplifies to the following:

$$G = 1/(r_0 \sin\Delta)^{\frac{1}{2}} \qquad [65]$$

except near the epicenter and anticenter (for these, see for example Brune and King,

1967). Lateral inhomogeneities of the earth may easily obscure the purely geometrical shape of the G-function. For power, the geometrical spreading factors will be the square G^2 of [64] and [65], respectively.

In conclusion, both attenuation and geometrical spreading affect only amplitude and not phase, and attenuation is frequency-dependent while geometrical spreading is frequency-independent.

6.5.4 Instrumental response

Most instruments used for recording geophysical phenomena have their own response for amplitude and phase and thus constitute the last filter before the recording. Seismographs offer a good example of such filters. An accurate knowledge of the instrumental characteristics is necessary for any calculation of ground motion. Trace spectra, without instrumental correction, can be used only for comparative purposes between records of instruments with identical characteristics, or in cases where amplitudes are not read, e.g., when only frequencies of spectral maxima and minima are of interest.

Denoting the ground spectrum by $X_0(\omega,r)$, equal to the product of all factors ahead of $I(\omega)$ in [56] and [57], we have:

$$X(\omega,r) = X_0(\omega,r)\, I(\omega)$$

or, for calculation of the ground spectrum:

$$X_0(\omega,r) = X(\omega,r)/I(\omega) \qquad [66]$$

This can then be immediately split into its amplitude and phase forms. The ground trace in the time domain $x_0(t,r)$ can then be obtained by inverse transformation of $X_0(\omega,r)$. Dividing $X(\omega,r)$ by $I(\omega)$ is evidently inverse filtering or deconvolution in the frequency domain. In the time domain, the same operation reads symbolically as follows (cf. Robinson, 1967b):

$$x(t) \star i^{-1}(t) = x_0(t) \star i(t) \star i^{-1}(t) = x_0(t) \star \delta(t) = x_0(t) \qquad [67]$$

For a critically damped seismograph, $I(\omega)$ has the following expression:

$$\left.\begin{aligned} |I(\omega)| &= \frac{q\omega^3}{(\omega_0^2 + \omega^2)\,(\omega_g^2 + \omega^2)} \\ \Phi_I(\omega) &= \frac{1}{\pi}\left(\tan^{-1}\frac{\omega_0}{\omega} + \tan^{-1}\frac{\omega_g}{\omega}\right) - \frac{1}{4} \end{aligned}\right| \qquad [68]$$

where $\omega_0 = 2\pi/T_0$, $\omega_g = 2\pi/T_g$ (T_0 = seismometer period, T_g = galvanometer

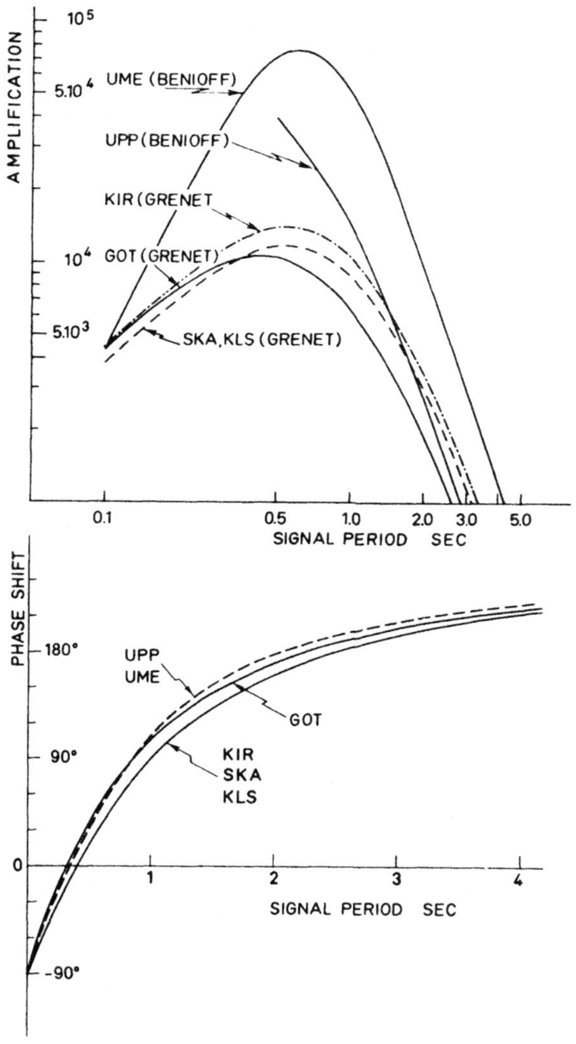

Fig.66. Examples of short-period vertical-component seismograph amplitude and phase response. UME = Umeå, UPP = Uppsala, KIR = Kiruna, GOT = Göteborg, SKA = Skalstugan, KLS = Karlskrona. After Husebye and Jansson (1966).

period) and q is a normalizing constant taking care of the magnification of the individual instrument (examples are shown in Fig.66).

For the determination of $I(\omega)$ for electromagnetic seismographs, three main methods are available:

(1) Using theoretical formulas and measuring necessary constants (seismometer and galvanometer periods and damping and coupling). A drawback is that

hardly any theory is fully exact. Cf. Jenschke and Penzien (1964) for application to accelerographs.

(2) Using shaking-table tests. These do not assume any special properties of the instruments, but are generally only applicable to sinusoidal vibrations of the table. Moreover, a number of technical considerations have to be taken into account in the vibration tests.

(3) Applying special test pulses and frequency-analyzing the recorded pulses. This method is generally applicable and eliminates the troubles with (1) and (2), and can easily provide a daily check on the operation. Test pulses can be of unit impulse type (Matumoto, 1958) or of unit step type.

Nowadays, test pulses are frequently used in instruments. The test pulse $f(t)$ for an applied step displacement $u(t)$ will be:

$$f(t) = u(t) \star i(t)$$

or in the frequency domain: [69]

$$F(\omega) = U(\omega) \cdot I(\omega)$$

The test pulse spectrum instead corresponding to a unit impulse displacement is obtained from the relation $\delta(t) = u'(t)$ and the derivation theorem (2.3.6):

$$\delta(t) \star i(t) = u'(t) \star i(t) \leftrightarrow i\omega\, U(\omega) \cdot I(\omega) = I(\omega) \qquad [70]$$

thus yielding the sought $I(\omega)$, because the transform of $\delta(t)$ is equal to 1. In other words, using the delta-function is equivalent to an input spectrum = 1 for all frequencies. The output reflects the instrumental response $I(\omega)$. Cf. Espinosa et al. (1962). Here we have assumed that $u(t)$ and $\delta(t)$ are *displacements* of the pendulum, but in actual test pulses, these are instead *forces* exerted on the pendulum, i.e. proportional to the *acceleration* of the pendulum. This will lead to corresponding modifications and the results are summarized in Table XXXII. This means that in order to obtain $I(\omega)$, we should multiply the spectrum of the test pulse by

TABLE XXXII

Spectrum $F(\omega)$ of test pulse under various conditions

Input test pulse	Displacement	Force
$u(t)$	$U(\omega)\, I(\omega)$	$\frac{1}{(i\omega)^2}\, U(\omega)\, I(\omega)$
$\delta(t)$	$I(\omega)$	$\frac{1}{i\omega}\, U(\omega)\, I(\omega)$

$(i\omega)^3$ if the unit step force $u(t)$ is used (cf. Shima, 1962) and by $(i\omega)^2$ if the unit impulse force $\delta(t)$ is used. That is, amplitudes are multiplied by ω^3 and ω^2, respectively, and phases advance by $3\pi/2$ and π, respectively.

Of the different systems which a seismic wave necessarily has to pass in order to permit human studies, the attenuation (expressed in Q) implies the greatest uncertainty and the instrument (expressed in I) implies the greatest deformation. The latter point needs careful consideration, not only because this is the only point where the operator influences a sequence of phenomena, which is otherwise wholly controlled by nature, but also because this system may have a most drastic deformation of the input spectrum as its result. This is particularly true when narrow-band seismograph responses are used, and also, but to a lesser degree, in case of broad-band seismographs. As the narrow-band seismographs, generally represented by the common short-period seismographs, are very much used, also in seismic spectroscopy, it is important to realize this distortion of an incoming spectrum.

More fully equipped seismograph stations run a series of instruments with partially overlapping response curves. However, such a series does not yield a series of comparable spectra, usually due to uncertainties in the response curves far from their maxima. According to experiments by Savarenskii et al. (1963), it is advisable to use each instrument only out to periods corresponding to 0.1 of its maximum magnification (cf. Kulhánek and Klíma, 1970, section 4.2.2), and to construct a wider spectrum sectionwise, by combination of spectra from a series of instruments with different response.

Most geophysical spectra cover a wide range in frequency or wavenumber and it is impossible to cover the entire range by one observational series or by one instrument. Then, different parts of the spectrum have to derive from different observational sets, i.e. the spectrum is constructed by "patching" pieces together. It is then important to be aware of the degree of homogeneity in the derived spectrum, both considering that different sets of observations and different instruments are involved. Even for overlapping parts of the spectrum, the possible influence of different instruments and measuring techniques on the spectrum has to be considered.

Concerning the relative order of applying a window function and the instrumental response function, there are two alternatives, symbolically:

Method 1:

$$[x(t) \cdot w(t)] \star i^{-1}(t) \leftrightarrow \frac{1}{2\pi}[X(\omega) \star W(\omega)] \cdot I^{-1}(\omega)$$

Method 2:

$$[x(t) \star i^{-1}(t)] \cdot w(t) \leftrightarrow \frac{1}{2\pi}[X(\omega) \cdot I^{-1}(\omega)] \star W(\omega) \qquad [71]$$

Method 1 is the usual procedure, i.e. first applying a window in the time domain, and then applying the instrumental response function in the frequency domain. However, it can be argued that method 2 would be more correct, i.e. first applying the instrumental response and the window function after that. The end results, i.e. the right-hand sides above, are in general not equal. Some comparisons on calculated examples are given by Kurita (1969a), which show that differences may be appreciable under certain circumstances.

6.6 GENERAL COMMENTS ON GEOPHYSICAL SPECTROSCOPY

6.6.1 Development of spectral analysis

Without going into detail, we may point to some milestones in the development of spectra and period research:

(1) Harmonic analysis or Fourier series expansion has a long history (Chapter 1). For a review of studies of periodicities and of harmonic analysis in geophysics up to the spectral era, the reader is referred to Chapman and Bartels (1940, 1951) and Conrad and Pollak (1950).

(2) The book by Blackman and Tukey (1959) led to an enormously increased application of power spectra in geophysics, calculated by the indirect method (section 4.6.1), as well as a general recognition of the importance of spectral studies in geophysics.

(3) The introduction of the Fast Fourier Transform (section 4.6.3) meant improved methods, especially concerning computer time and storage.

(4) Newer developments (sections 3.6.2 and 4.6.4 in particular) are creeping up at the geophysical horizon, some of which may be of great significance within the next few years ...

Comparing the use of different kinds of spectra in the different geophysical disciplines, it is of interest to note that amplitude spectra (in addition to velocity and acceleration spectra) dominate in seismology, whereas in meteorology, oceanography, gravity and geomagnetism, power spectra have the dominant position. This is in part a result of the influence of item 2 above on the last-mentioned fields, while item 3 (FFT) probably got an earlier application in seismology than in most other fields.

The history of spectroscopy is very well reflected in the geophysical literature. Used methods depend both on the type of problem studied but also considerably on available spectral techniques and computer facilities. As just one example, let us mention a most complete joint presentation of solar-terrestrial relationships, edited by Fairbridge (1961) as a result of a conference in 1961. This volume contains not less than 54 papers covering a wide range of possible variations and their solar relations, extending over astrophysics, geomagnetism, meteorology, climatology, paleoclimatology, glaciology, oceanography, and other fields. It is probably characteristic of the time of that report that not a single paper attempts to make harmonic analysis of data, whereas, on the other hand, only four papers have applied newer spectral methods (Brier, 1961; R.Y. Anderson, 1961; Bryson and Dutton, 1961; Ward and Shapiro, 1961a).

6.6.2 *Advantages of spectral analysis*

The main significance of spectral analysis in geophysics, as well as in other fields, is that it permits an evaluation of scales (both in time and space) and their relative importance. In earlier analyses, mostly all data were lumped together and the notion of scales was lost. Just one example is offered by the traditional over-all correlation coefficient versus the auto- and cross-correlation coefficients. Cross-correlation and coherence are used extensively in investigations of possible parallelism between time or space series. This technique could probably be applied with advantage also to global phenomena, presented in the form of spherical harmonic functions (for example, for the treatment of problems discussed by Horai and Simmons, 1968).

Determination of relations (regressions) between observational series occurs frequently in geophysics. Such relations may be useful for comparison of observations with theory or for purposes of prediction. The coefficients in regression equations are usually determined without regard to spectral concepts. However, use of spectra for such calculations has certain advantages over the traditional methods, especially because regression coefficients are obtained as functions of frequency (or wavenumber). This is of special value when the signal/noise ratio varies with frequency; moreover, any frequency-dependent operation (such as filtering, instrumental effects) on the data can be accounted or corrected for, when results are expressed in a frequency-dependent form. For details of spectral methods of calculation of regression coefficients with applications, the reader is referred to Hamon and Hannan (1963), Kemp and Eger (1967), Jenkins and Watts (1968), and Groves and Hannan (1968).

Any quantity which can be expressed as a function of one or several variables, can be subject to Fourier spectral analysis. However, it is only in case of a few variables that such analysis is of any practical advantage. The dominating case occurs when the independent variable is time, followed next by cases where one or more space variables exist. The most important application of Fourier spectral techniques in seismology is to the analysis of seismic waveforms. As we have seen, conditions are not easy in seismology. The method's power and weakness at the same time lie in the fact that we observe a combination of effects, for instance we do not observe source spectrum isolated. By and large, source and path may carry equal influence on the observed spectrum. It is for such reasons that seismological spectral techniques have reached their highest degree of sophistication just in the analysis of waves and waveforms.

When we turn our attention to spectral methods in some other fields, we shall find that situations are somewhat different, even though the same basic techniques are employed. We can summarize some of these differences in the following points:

(1) In many applications, we do not need any equalization method to isolate various effects. For instance, the spectrum of a temperature–time curve is usually taken as the final result. It may naturally be compared with other similar results, but usually no further elimination of various factors is aimed at. Another example is provided by spectral (or harmonic) analysis of water-level fluctuations, especially to investigate seiche phenomena. The obtained spectrum is usually taken as the final product of an investigation. This is mostly the case also in engineering seismology where vibrations of structures are investigated.

(2) Another simplification prevailing in several applications of spectral analysis is that basic periods (frequencies) may be given by the nature of the problem—which is generally not the case in the analysis of seismic waves. One example is again the temperature–time curve for the atmosphere, where one year and one day provide periods of fundamental significance. Even investigations of earthquake periodicity belong in this category.

6.6.3 Geophysical fields covered

Spectral methods have penetrated the whole of geophysics within the last 10–20 years. In the following chapters we shall study the achievements of spectral methods in seismology, meteorology, oceanography, gravity and geomagnetism. These represent those geophysical fields where spectroscopy has scored its most extensive applications and greatest success. We restrict our review to mechanical spectra in these fields, while electromagnetic spectra, of significance in ionospheric research, will be omitted.

In addition to the fields mentioned, mechanical spectra have found application to a smaller extent in a variety of geophysical branches, which we summarize here in brief, including also a few cases where Fourier series expansion has been applied:

(1) Volcanology: partly by spectral analysis of seismic records of volcanic microearthquakes (section 8.1.6), partly by frequency and wavenumber spectral analysis of gravity and geomagnetism over volcanoes (Chapter 10).

(2) Variation of latitude or pole motion (Rudnick, 1956; Barber, 1966; Proverbio and Quesada, 1972), Fennoscandian uplift (McConnell, 1968), geological structures, folds, etc. (Stabler, 1968; Hudleston, 1973), topography, especially of circular structures (Johnson and Vand, 1967), lunar topography (Jaeger and Schuring, 1966), meteor rates (Keay et al., 1966), ground temperature variation (Gold, 1964).

(3) Ice-island oscillations (Hunkins, 1967), deep-sea sediment rates (Kemp and Eger, 1967).

(4) Hydrology (recent examples being Yokosi, 1967; Thakur and Scheidegger, 1970; Ghosh and Scheidegger, 1971; Reed, 1971; and others), properties of porous media (Scheidegger, 1960; Fara and Scheidegger, 1961).

One factor which may affect a geophysical spectrum refers to the well-known Doppler principle, i.e., when there is relative motion of source and receiver, this will influence the observed frequencies (Mangiarotty and Turner, 1967). If such an effect is present, it would be necessary to correct for it, in order to deduce the true spectrum. A few examples from geophysics related to the Doppler principle may be given:

(1) Meteorology: sound observations with relative motion of source and receiver (this being a common experience); spatial distribution of wind as observed on aircraft (Pinus, 1963).

(2) Oceanography: wave measurements on a moving ship.

(3) Geomagnetism: motion relative to the sun affecting measurements of a magnetic storm.

(4) Seismology: surface waves from a moving source (section 8.1.3); body waves from a finite dislocation (Moskvina, 1971).

In addition to modification of periods, also modification of amplitudes may arise in observations from moving platforms – as for example in ship-borne observations of ocean surface waves or of gravity. Much effort has gone into instrumental improvements and precautions for such effects, which, if uncorrected, will influence any deduced spectrum.

Chapter 7

SPECTRAL STUDIES OF THE EARTH'S STRUCTURE

As outlined in section 6.5.2, seismic waves depend both on source properties and on the structure of the media traversed by the waves. A received wave spectrum bears the imprint of both. By proper equalization techniques, it is possible to isolate one effect from the other.

Structural properties can conveniently be divided into two main groups:

(1) Kinematic properties, such as layering, velocities, etc., which are determined from kinematic measurements, such as frequencies of spectral maxima and minima, spectral phases, etc. For convenience we split this large subject into three categories according to scale: local scale (based on body waves), regional scale (based on surface waves), and global scale (based on the earth's free oscillations).

(2) Dynamic properties, such as quality (Q), which are determined from dynamic measurements, especially of spectral amplitudes.

Nearly all of these problems have been attacked already in an early stage of seismology by time-domain measurements. But it is clear that spectral methods, working with frequency as independent parameter, are much more powerful. The phenomena mentioned depend on frequencies and not on time. For example, a

TABLE XXXIII

Structural problems investigated by spectra

Type of medium	Phenomenon	Waves used	Spectra used	Spectral parameters
Layered (especially crust)	reverberation interference	body waves	amplitude	frequency of minima
	dispersion	surface waves	phase	phase shift: phase velocity
	attenuation	body and surface waves	amplitude	amplitude ratio: attenuation
Continuous (mantle)	dispersion	surface waves	phase	phase shift: phase velocity
	attenuation	body and surface waves	amplitude	amplitude ratio: attenuation
Earth as a whole	free oscillation	surface waves	amplitude	frequency of maxima
	slowness	body waves	phase	phase shift: apparent velocity, source distance and azimuth

given body or a given layered structure favours certain frequencies and suppresses others, by its own free oscillation modes or reverberations. Likewise, attenuation in a body is a phenomenon which is generally assumed to be frequency-dependent. A review is given in Table XXXIII.

In our applications of spectral methods to seismological studies, we find it most logical to study the earth's structure first (this chapter) and to proceed to source studies in the next chapter. The reason is that structural studies can be done without detailed source information, whereas source studies depend heavily on structural information. The latter is a prerequisite for the former.

The ultimate goal of such studies, as listed in Table XXXIII and their time-domain counterparts, is to learn something about the earth's structure. By and large, two methods are then followed:

(1) deduction of structures from the given observations; and (2) comparison of observations with theoretically deduced properties, the latter based on certain assumptions about structures.

In every case, the question of uniqueness of any given interpretation deserves careful consideration. In addition, a knowledge of structural properties between stations and their influences on spectra and records will make it possible to better compare individual station records. This is especially significant when the structure varies among the sensors compared, as for example for some of the existing array stations (cf. Rygg, 1971).

7.1 LOCAL-SCALE PROPERTIES—BODY WAVES

7.1.1 Body-wave velocities from arrays

In order to get meaningful results, it is necessary to make comparative studies of two or more stations, as single-station data do not suffice. This means that velocities are based on time differences (phase shifts) between stations, and amplitude absorption on amplitude ratios between stations. From this viewpoint, Table XXXIII lists studies which can be made with arrays or networks of stations. The simplest (but not the most informative) approach is to use two stations, preferably with instruments with identical characteristics, located along the same direction from the epicenter.

The significant factor about the spectral methods, as listed in Table XXXIII, is that they use frequency as discriminant. In other words, all observed phenomena are studied in relation to the frequency of the waves. Earlier in seismology, most phenomena were studied only in the time domain, i.e. directly from seismograms, then using period, as measured directly, as discriminant. For example, the time-domain phase-velocity method for surface waves was developed in the 1950's by Press (1956b) and then applied in the USA, Japan, Scandinavia and elsewhere.

As another example, referring to the bottom line in Table XXXIII, determinations of body-wave apparent velocities across arrays or networks of stations can be done in at least three different ways:

(1) By time-domain readings of corresponding peaks and troughs. This is still a sufficiently reliable method when traces exhibit almost perfect similarity, examplified by the Uppsala array (Brown, 1973; Kulhánek, 1973b).

(2) By phase spectra, either from individual spectra or from cross-power spectra. This method was studied and applied by Shima et al. (1964) and Iyer (1971).

(3) By cross-correlation and reading of the time for maximum correlation. Narrow band-pass filtering followed by cross-correlation would then be the best procedure. In some cases, this has been found to be the most reliable method, as, for example, for a network on Hawaii (Sokolowski and Miller, 1967). See also Aki et al. (1958) and Aki and Tsujiura (1959) for application to near earthquakes in Japan, McCamy and Meyer (1964) for application to crustal research, Rykunov (1961) for application to microseisms, and Hales and Roberts (1970) for testing on S and SKS. See also Landisman et al. (1969), Dziewonski and Landisman (1970), and White (1969).

For arrays of stations, the two-dimensional cross-correlation function is most natural to use. This method of determining motion vectors (speed and direction) has found promising application also to other similar problems, e.g. determination of cloud motion vectors from satellite pictures by FFT methods (Leese et al., 1971, from whom Fig.67 was stimulated, demonstrating the principle of the method).

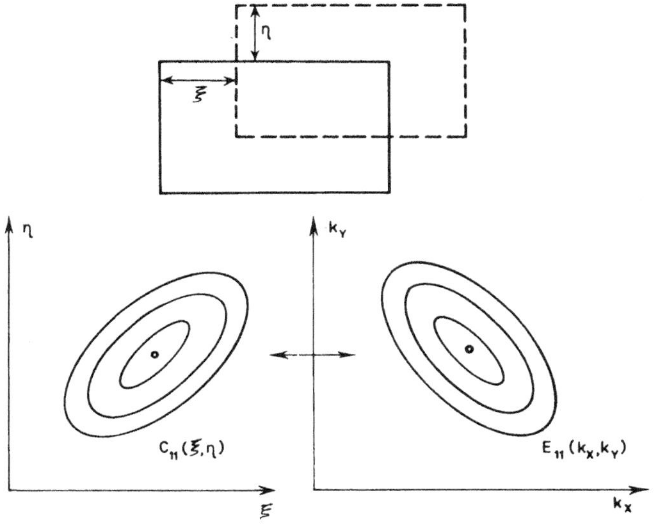

Fig.67. Schematic depiction of two-dimensional autocorrelation and its transform. The upper figure demonstrates the definition of the lags, while the lower are fancy figures not exactly corresponding to the rectangular case. Partly after Leese et al. (1971).

We should observe that Fig.67 demonstrates the space-domain counterpart to the representation of power as isolines in graphs with wavenumbers (k_x and k_y) as coordinates (cf. section 2.5.5), and which is likewise used for determination of motion vectors.

7.1.2 Crustal transfer method—basic principles

The application of spectral analysis of seismic body waves in the study of crustal structure is only of a relatively recent date. The theoretical background is the Thomson-Haskell matrix formulation, which easily permits calculation of responses of any number of plane, horizontal layers to incident plane waves for any angle of incidence. The reader is referred to the following papers to get

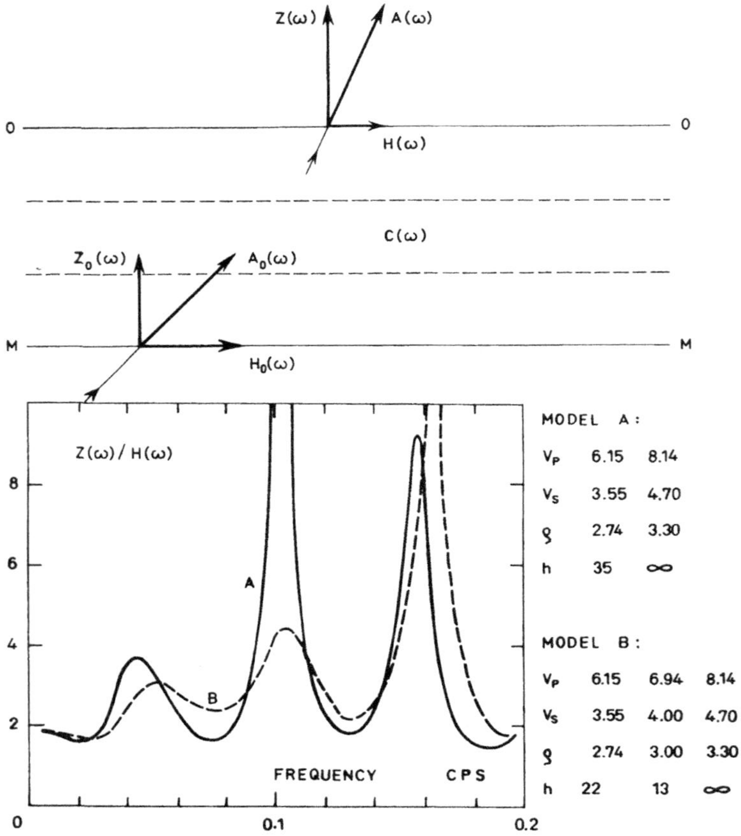

Fig.68. Principle of crustal transfer method (above) and application to two special cases (below). OO = earth's surface, MM = Mohorovičić discontinuity, common units used for model specification (km/sec, g/cm^3, km). Apparent velocity = 20 km/sec.

acquainted with this technique: Thomson (1950), Haskell (1953, 1960, 1962, 1966b), Båth (1968, chapter 9), Harkrider (1964) with extension to a point source in the layered medium; for application, see, for example, Kogeus (1968). An analogous theoretical study was made by Jensen and Ellis (1970), who consider the layered model as a linear system and apply the equivalence between time-domain convolution and frequency-domain multiplication (which reduces to Haskell's formulation), and also allow for attenuation within the layers.

Among various techniques to eliminate all spectral-shaping factors except just the crustal effect, the use of ratios of vertical-to-horizontal components of the same body wave has become of rather great significance (Phinney, 1964). With reference to Fig.68 we can write the crustal filtering equation for the vertical and horizontal amplitude components of a body wave, respectively, as follows:

$$\left.\begin{aligned} Z(\omega) &= A_0(\omega)\, C_Z(\omega) \\ H(\omega) &= A_0(\omega)\, C_H(\omega) \end{aligned}\right| \qquad [1]$$

where for brevity we put $A_Z(\omega) = Z(\omega)$ and $A_H(\omega) = H(\omega)$. Dividing the two equations with each other, we get a ratio, called the *crustal transfer ratio*, which is independent of the spectrum $A_0(\omega)$ of the incident signal and depends exclusively on the layered structure:

$$Z(\omega)/H(\omega) = C_Z(\omega)/C_H(\omega)$$

or, taking the amplitudes:

$$|Z(\omega)|/|H(\omega)| = |C_Z(\omega)|/|C_H(\omega)| = T(\omega) \qquad [2]$$

$T(\omega)$ is the frequency-dependent tangent of the (apparent) angle of emergence.

The application of this method consists of calculating the spectral amplitude ratio $|Z(\omega)|/|H(\omega)|$ from observations and comparing with theoretically calculated $T(\omega)$ for a number of crustal models, trying to find the closest match by some error-minimizing procedure. The observational side is quite straight-forward and proceeds suitably along the following steps:

(1) From the given records, get the spectra, $Z(\omega)$ and $H(\omega)$, e.g. by FFT method, and apply instrumental corrections.

(2) Smooth the spectra (to avoid spurious fluctuations in the ratios).

(3) Divide the amplitude spectra to get $|Z(\omega)|/|H(\omega)|$.

Alternative, but in essence equivalent procedures are possible, such as applying some suitable window in the time domain, instead of smoothing the spectra, or using power spectra instead of amplitude spectra, taking the square root of the ratio of the powers. Bakun (1971), for example, found that a "modified Hanning window", equivalent to a half cosine bell with its maximum at the P onset (section 4.4.4), will provide for better consistency of the spectral ratios than a rectangular window (even a cosine-tapered rectangular window). The reason is excessive leakage through side-lobes in the latter cases.

The theoretical side, i.e. calculation of $T(\omega)$ for assumed crustal models, proceeds by the Thomson-Haskell matrix formulation or related method, for incident plane waves and plane, horizontal layering (Fig.68). Due to resonance effects, $T(\omega)$ will exhibit a series of peaks, with crustal peaks fairly broad and mantle peaks narrow and dense, corresponding to the different separation of boundaries in the two regimes. Theoretical studies of spherical waves in a layered crust offer another approach, which can also serve for comparison with observational amplitude and phase spectra, as applied by Gurbuz (1970) to explosion-generated body waves.

The method was introduced by Phinney (1964), and has since then been modified and also applied to a great extent. One advantage with this method is the more localized crustal information it is able to furnish, as compared to surface-wave methods. This is especially true when teleseisms are used, a requirement also considering that the theory assumes plane wavefronts incident on MM (Fig.68). To facilitate interpretation of P-wave spectra in terms of crustal structure, Fernandez (1967) published a comprehensive set of master curves, displaying the transfer functions for vertical and horizontal P and their ratio, for one- and two-layered crusts. Among modifications, the one due to signal truncation is most important (Leblanc, 1967; McCamy, 1967), which we shall deal with next.

The equation:

$$A(\omega) = A_0(\omega) \cdot C(\omega) \qquad [3]$$

holds only for an infinitely long seismogram. In any seismogram analysis, we have to apply a time window $w(t)$ of finite length which yields the truncated signal $A_T(t)$ of length T from the infinitely long signal $A(t)$:

$$A_T(t) = A(t) \cdot w(t)$$

or in the frequency domain:

$$A_T(\omega) = \frac{1}{2\pi} A(\omega) \star W(\omega) \qquad [4]$$

where $A_T(\omega)$ is called the *truncated spectrum*. Inserting [3] into [4] we obtain:

$$A_T(\omega) = \frac{1}{2\pi} [A_0(\omega) \cdot C(\omega)] \star W(\omega) = \frac{1}{2\pi} A_0(\omega) \cdot [C(\omega) \star W(\omega)]$$

$$= A_0(\omega) \cdot C_T(\omega) \qquad [5]$$

where the transformation from the second to the third stage is permitted if: (a) $c(t)$ approximates a weighted sum of delayed delta-functions; and (b) $A_0(t) = 0$ outside a time interval short compared to T, as demonstrated by Leblanc (1967). $C_T(\omega)$ is called the *truncated crustal transfer function.*

In applications, we have: (a) an assumed crust, for which we calculate $C_T(\omega)$ by the following steps:

$$\left.\begin{aligned} c(t) &= \frac{1}{2\pi}\int_0^{\omega_1} C(\omega)\, e^{i\omega t}\, d\omega \\ C_T(\omega) &= \int_0^T c(t)\, e^{-i\omega t}\, dt \end{aligned}\right| \qquad [6]$$

and we have (b) an observed spectrum $A_T(\omega)$. Combining the two, we calculate $A_0(\omega)$. This is compared with observed or known $A_0(\omega)$, determined, for example, by averaging $A_T(\omega)$ for a number of stations. Then, crustal parameters, i.e. $C_T(\omega)$, are varied until good agreement is found. The method, as applied to short-period P is able to elucidate the fine details of crustal structure. The method has been presented by Leblanc (1967).

The condition above that $A_0(t) = 0$ outside a time interval short compared to T, implies that reverberations from the *source* crust are absent in the incoming signal. This limits the applicability of the method to deep-focus earthquakes, where effects of the crust in the source region will be sufficiently well separated from the direct wave. If not, the method has to be modified as indicated by Glover and Alexander (1969), which would require a detailed knowledge of the source crust for numerical application.

The *length of the time window* to be used in this method (including its original version) is a question of importance. While it is desirable to include as much of reverberated P as possible, due consideration has to be taken of effects of included pP and PcP. Some seismologists have avoided this problem by applying the technique to deep earthquakes in such distance ranges that both pP and PcP were beyond the window used. Such a restriction of the applicability of the method appears undesirable. Considering that pP and/or PcP, when included, for large enough distances approach the receiver crust with incidence angles close to those of P, no significant effect would be expected by their inclusion. In fact, as demonstrated by Bakun (1971), this is true, and it is especially significant that spectral peaks and troughs are not altered in their frequency position by including pP and/or PcP in such cases. However, Hasegawa (1971b) emphasizes that the advantage of more data by including pP and/or PcP is more than offset by difficulties in interpreting observed transfer ratios in terms of crustal structure.

In general, it is found from applications that the method has met with much more success when applied to long-period P than to short-period P. The reason is that increased scattering at inhomogeneities in the latter case, P–S conversion, etc. limit the applicability of the method. On the other hand, it has been shown that short-period P can be used by this method to detect thin, surficial low-velocity layers, not detected by long-period P (Hasegawa, 1971b).

There is also another point to note in these calculations. What a seismograph records is not just the incoming signal, but the ground motion resulting from the incoming wave and the waves reflected against the earth's surface at the point where the seismograph is located. By application of response characteristics, we recover the true ground motion from the recorded motion on the seismogram. But we do not recover the incoming signal alone. In order to do that, it is necessary in addition to correct the obtained ground motion for the contribution from reflected waves. Or else, to include a corresponding correction in the theoretical calculations which are used for comparison with the observations.

Interpretations of observed crustal transfer ratios are made by comparison with theoretical ratios, calculated for assumed crustal models. In the latter, plane, horizontal and homogeneous layering is assumed. Therefore, failure to find acceptable crustal models in such terms may be due to deviations from such simple structures. *Effects of dipping layers* will then have to be considered. Bakun (1971) examplified this for the Berkeley area, and Rogers and Kisslinger (1972) developed and applied a method for dip determination from crustal transfer determinations for P-waves. Dips of crustal layers are a very common phenomenon, in fact they are the rule, and the use of the plane, horizontal layering hypothesis for the theoretical models is an acceptable approximation only within certain limits, say for dips of 10° or less and for long-period body waves. For larger dips and shorter periods, the method will fail unless special precautions are taken. In some applications, for example Kurita (1969b and c, 1970), a number of earthquakes in different azimuths have been used, and the results combined into an azimuthally varying structure, which by necessity includes dips.

Horizontal, plane layers represent the first approximation to real earth structures. Dips, constant over some area, represent a second approximation. To proceed to higher-order approximations, one should include varying dips, i.e. undulatory interfaces. The literature contains several papers dealing with this case, not reviewed here. Anyway, it is clear that interpretational difficulties increase rapidly as we proceed to more complicated structures, also the uniqueness of interpretation is difficult to assess.

Signal-generated noise due to structural inhomogeneities in the vicinity of an observatory is another reason for difficulties in the application of the crustal transfer ratio method. For example, Key (1967) demonstrated such noise at the Eskdalemuir array, generated by P- to Rayleigh-wave conversion near the station. As feared by Key (1967), such a second phase could easily be misread for pP, if only one record is available, but this can be avoided by an extended array or a network of stations.

The method as developed so far refers only to amplitude ratios (Z/H). Additional information can be obtained by including the *phase difference* ($Z - H$) in the analysis and comparing with results from theoretical models. This combined study has been developed and applied by Kurita (1969b and c, 1970) to structural studies in Japan.

The method, so far developed for P-waves, suggests also its *extension to other waves*, whose amplitude spectral ratios would vary with structure in different ways from P, thus furnishing valuable constraints on the structure deduced from P only. Such an extension was made by Kurita and Mikumo (1971) and Kurita (1973) to S-waves. Their method consisted in evaluating both amplitude ratios and phase differences for (1) Z/H for SV, and (2) SH/SV (horizontal component of SV). They presented a detailed study of the influence of various structural properties on the quantities measured, and the technique was applied to regions of Japan (Kurita, 1971). Even though undisturbed S-wave records may be harder to find than undisturbed P-wave records, especially the phase difference $Z - H$ for SV proved very valuable in discriminating between various structural models, which would fit P-wave data equally well. The second method, SH/SV, was applied to both Japan and North America by Mikumo and Kurita (1971). By including SH, this method provides further restriction on possible structures.

Quite a natural extension of Phinney's (1964) method is to apply the amplitude spectral ratio $|Z(\omega)|/|H(\omega)|$ also to surface Rayleigh waves. This ratio is an expression for the ellipticity of the Rayleigh-wave particle motion, which is dependent on crustal layering and is particularly sensitive to sedimentary layers with large contrast. This has been demonstrated both theoretically and observationally many years ago, especially with relation to microseisms at different localities. Together with phase velocity, the ellipticity is able to serve as an additional constraint in crustal structure determinations (Boore and Toksöz, 1969).

7.1.3 Crustal transfer method—applications

Since around 1964 quite a number of papers have been published where the crustal transfer method is applied. Most of them are listed in Table XXXIV. We shall restrict this section to some comments, essentially on the methodical side.

Applying Phinney's method (1964) to long-period P-waves, Fernandez and Careaga (1968) determined total crustal thickness and average crustal P-velocities for central USA and La Paz, by matching observed amplitude spectral ratios $|Z(\omega)|/|H(\omega)|$ plotted versus ω against master curves of Fernandez (1967). The results in terms of structural parameters are in good agreement with independent results by other methods. Similar studies were made by Denham (1968) for New Guinea and the Solomon Islands, and by Glover and Alexander (1969) for in-

TABLE XXXIV

Examples of the application of the spectral amplitude ratio Z/H to crustal research

Reference	Wave used	Locality studied	Remark
Phinney (1964)	long-period P	Albuquerque, Bermuda	
Utsu (1966)	short- and long-period P	Canada	
McCamy (1967)	short-period P	East coast, USA (explosions)	
Denham (1968)	long-period P	New Guinea, Solomon Islands	
Ellis and Basham (1968)	short-period P	Canada	
Fernandez and Careaga (1968)	long-period P	Central USA, La Paz	
Boore and Toksöz (1969)	Rayleigh waves	LASA, Montana, USA	
Glover and Alexander (1969)	long-period P	LASA, Montana, USA	
Hasegawa (1969, 1970)	short-period P	Yellowknife, Canada	poorer agreement with theory for earthquakes than for nuclear explosions
Kurita (1969b, 1970)	long-period P	Japan	also phase difference used, detailed methodical discussions
Kurita (1969c)	long-period P	Kanto Plain, Japan	phase difference ($Z - H$) used in addition to amplitude ratio (Z/H)
Bonjer and Fuchs (1970)	long-period P	SW Germany	
Bonjer et al. (1970)	long-period P	East African rift	
Bakun (1971)	long-period P (also pP and PcP)	Berkeley area	
Hasegawa (1971b)	short- and long-period P (also pP and PcP)	Yellowknife, Canada	
Kosarev (1971)	long-period P	Moscow	discussion on the method
Kurita (1971)	long-period SV and P	Japan	also phase difference used
Rogers and Kisslinger (1972)	long-period P	South America	method for dip determination

vestigation of lateral crustal heterogeneity beneath the LASA array in Montana, USA. It is quite a general experience that the method is much more successful when applied to long-period than to short-period waves (see, for example, Utsu, 1966).

In an application of the spectral ratio Z/H, based on Haskell's matrix formulation and Phinney's (1964) method, to short-period P-waves recorded in Canada, Ellis and Basham (1968) in general found relatively poor agreement between theoretical and observational curves, although the structure is considered well known. The discrepancies are ascribed to scattering and P–S conversion, factors which thus limit the applicability of this method for crustal exploration.

Using short-period P-wave records (0.4–2 cps) at numerous stations from the Peruvian earthquake of April 13, 1963 (depth = 125 km), Leblanc and Howell (1967) emphasize an oscillatory character of the amplitude spectra as the most striking feature, varying in an apparently irregular fashion from station to station. Such oscillations tend to make short-period spectra almost useless, masking attenuation effects and radiation patterns, and they have to be removed before any

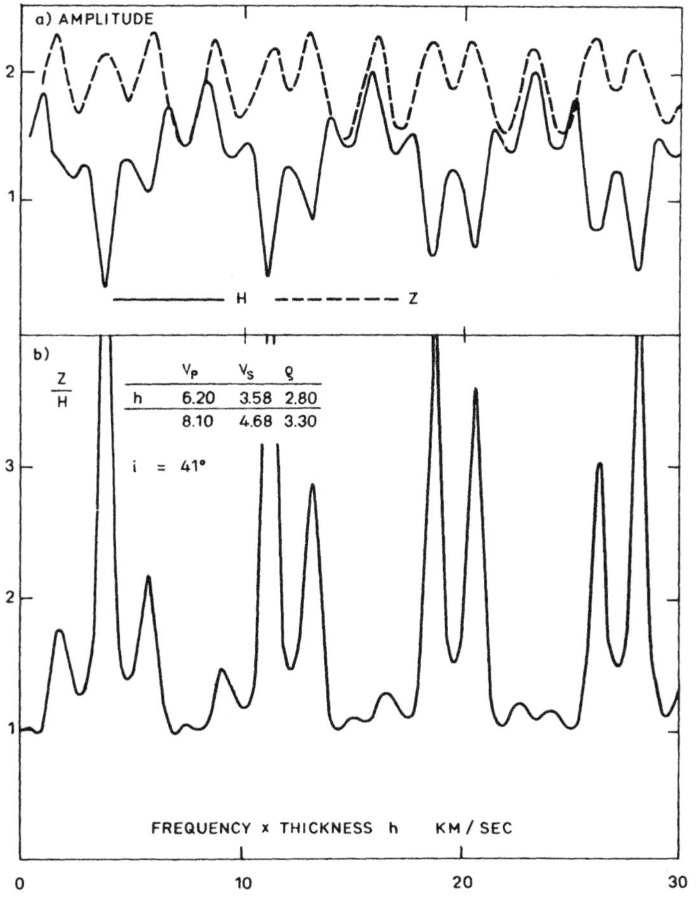

Fig.69. Theoretical P-wave vertical and horizontal amplitudes (a) and the spectral ratio Z/H (b) plotted versus frequency × layer thickness. After Utsu (1966).

reliable study can be made with the spectra. For such removal of oscillations, the authors suggest partly averaging of spectra of several stations, partly applying crustal inverse filtering, even though uncertain in itself. The oscillatory character is ascribed to reverberations in the receiver crust. Even though the authors do not report any information on crustal structures beneath the stations used, the lesson is that stations located on as simple a crust as possible will give spectra permitting the most reliable interpretation. Sedimentary layers, which may vary strongly even over short distances, should be avoided, and stations located on granite or similar rock selected. In comparing stations, it should also be considered that the crust beneath the stations is so closely similar as possible. Long-period P-waves are less likely to appear so erratic as short-period P. This is in fact supported by Belotelov and Kondorskaya (1964), who found spectra of P-waves in the period range 3–15 sec to vary only little from station to station.

Results closely agreeing with those of Leblanc and Howell (1967) were earlier reported by Ichikawa and Basham (1965) for Canadian stations. Their short-period P-wave amplitude spectra (about 0.5–4 cps) exhibit apparently irregular peaks in addition to a general decrease with increasing frequency, proportional to ω^{γ}, where $\gamma = -2.2$ to -3.5 varying from station to station. Local structures beneath the stations are again referred to as the cause of the spectral variabilities. In any such comparison, it is important to keep time windows constant, both in length and shape, as emphasized in section 4.4.8. That structure beneath the stations is the major factor in characterizing spectra was further confirmed by Utsu (1966), using both short- and long-period P-waves. Comparison of observed spectral ratios Z/H (vertical to horizontal amplitudes) with theoretical curves (Fig.69) was attempted to determine crustal thickness. As evidenced by Fig.69, such comparison requires high resolution in the observed spectra, but even then, good agreement is often hard to reach, probably because theoretical models were too simple (especially horizontal layering).

In applications of the spectral ratio method of Phinney (1964) to short-period P of teleseismic events, Hasegawa (1969, 1970) finds that this is capable of resolving only thin crustal layers, whereas a combination of short- and long-period P is needed to resolve both thin and thick crustal layers. The record time window used for analysis should, on the one hand, be long enough to include practically all of the local crustal reverberations in the P-coda and to guarantee a good spectral resolution, on the other hand, short enough to avoid much signal-generated noise. A record length of 20–25 sec for short-period P is considered adequate.

In an effort to apply the crustal transfer ratio to explosion seismology, McCamy (1967) found this ratio to depend heavily on layer thicknesses and P-velocities, less on S-velocities and hardly at all on density. Quantitative inversion from observed ratios Z/H to crustal structures is considered impossible, but the method can nevertheless be useful in discriminating between various trial models.

Using the Thomson-Haskell matrix formulation to plane P-waves incident on thin, alluvial layers, Hannon (1964) synthesizes the surface motion in the time-domain from the formula:

$$\begin{matrix}\text{horizontal component: } h(t) \\ \text{vertical component } \quad z(t)\end{matrix}\Bigg\} = \frac{1}{2\pi}\int_{-\infty}^{\infty}\begin{Bmatrix}\bar{H} \\ \\ \bar{Z}\end{Bmatrix} A_0\, e^{i\omega t}\, d\omega \qquad [7]$$

where $\bar{H} = H/A_0$ and $\bar{Z} = Z/A_0$ denote the ratios of surface motion to incident total amplitude at the base, for given frequencies and given crustal models. $\bar{H}$ and $\bar{Z}$ are calculated by the matrix formulation, and various assumptions are made for A_0 as rectangular frequency functions.

7.1.4 Crustal reverberation

As soon as a semi-infinite medium is covered by one surface layer, spectra of incident waves will be affected. Reflections, refractions and subsequent constructive and destructive interference between different wave trains, in short, reverberations, will make their imprint on the spectra obtained on the surface. At the same time as recordings and spectra become complicated, a means is provided to explore the layering. The crust of the earth consists in general of several layers and with increasing number of layers, we can easily envisage that complication will increase rapidly.

Simpler cases, however, lend themselves to interpretation quite easily, and they may serve the purpose to understand the principles of reverberation. In order to acquire familiarity with the problem, it is suggested to the reader to work out for himself various cases, starting with the simplest and proceeding to more complicated cases: (1) one layer, normal incidence; (2) one layer, oblique incidence; (3) two layers, normal incidence; (4) two layers, oblique incidence.

We shall begin with the case of one layer embedded between two homogeneous media at some depth below the surface. Amplitude spectra of waves reflected from such a layer (Fig.70) will exhibit a sequence of minima, corresponding to destructive interference between waves reflected from the upper and lower sides of the layer. Considering that there will be a phase change of π upon reflection against a medium of lower acoustic impedance, but no phase change upon reflection against a medium of higher acoustic impedance, we arrive at the scheme summarized in Table XXXV. Derivations are simple and are suggested as an exercise to the reader. The case of normal incidence is obtained therefrom putting $i_2 = 0$. It is clear from Table XXXV that spectral minima will be close to each other for L/h small (L = wavelength) and/or i_2 small (near normal incidence), and conversely when these quantities are large. In fact, this is nothing but echo analysis (section 8.4.2), applied to a special case. It can also be looked upon as a case of resonant vibrations of a layer or sequence of layers, especially for $i_2 = 0$, as the dominant frequencies are determined by the layer properties (thicknesses and wave velocities).

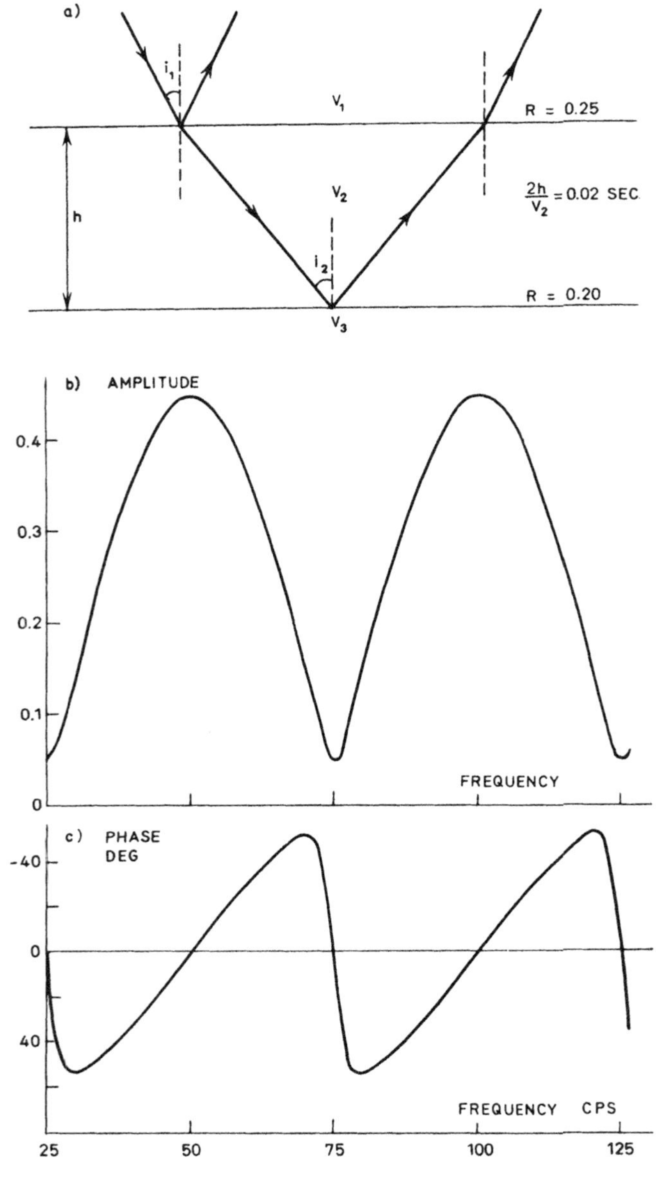

Fig.70. Reflection from an embedded homogeneous layer. a) Model used. b) Amplitude and c) Phase spectra for a given model: normal incidence of unit amplitude with reflection coefficients (R) and double transit time in the layer as given in a) and refractions ignored. Modified after Khudzinskii (1961, 1966).

TABLE XXXV

Minima in amplitude spectra of waves reflected from a layer (Khudzinski, 1961)

Spectral parameter	Layer velocity intermediate: $V_1 < V_2 < V_3$ or $V_1 > V_2 > V_3$	Layer velocity larger or smaller: $V_1 < V_2 > V_3$ or $V_1 > V_2 < V_3$
Frequencies of minima[1]	$\frac{(n - 1/2)\, V_2}{2\, h \cos i_2}$	$\frac{n\, V_2}{2\, h \cos i_2}$
Frequency range between adjacent minima	$\frac{V_2}{2\, h \cos i_2}$	$\frac{V_2}{2\, h \cos i_2}$
Frequency of first minimum	$1/2\, \frac{V_2}{2\, h \cos i_2}$	$\frac{V_2}{2\, h \cos i_2}$

[1] Cyclic frequencies are used. Notation is explained in Fig.70. $n = 1, 2, \ldots$

Khudzinski (1961), who proposed and applied this method for investigation of layer properties (h, V_2) from spectra of reflected waves, investigated various factors which could disturb the regular sequence of minima. Noise is considered the most serious disturbance. Later, Khudzinskii (1966) extended his study also to phase spectra of reflected waves from single, homogeneous layers at normal incidence. The spectra in Fig.70 illustrate the spectral properties and their formulas can easily be deduced, starting from interference between waves reflected at top and bottom of the layer. Multiple reflections lead only to minor modification. The phase spectra can by themselves be used for determination of certain layer parameters, and to facilitate such interpretation, Khudzinskii (1966) gives a set of master curves for selected models.

Berson (1959) demonstrates theoretically that the spectrum of P-waves reflected from thin layers depends on: (1) the velocity contrast (layers with lower velocities producing poorly correlated reflections); (2) the angle of incidence (greater change for higher velocity in layer); and (3) the absorption in the overlying medium. Berzon (1965) demonstrates that for P-waves reflected from a thin, homogeneous layer between halfspaces, a combined study of amplitude and phase spectra yields a unique determination of layer properties (velocity, thickness), whereas use of amplitude spectra alone will not yield a unique solution.

Complications, both in calculations and in obtained spectra, increase rapidly with increasing number of layers. Contrary to the one-layer case, amplitude maxima are in general unequal and at unequal frequency separation for two or more layers, also the number of extrema per frequency range increases. Already with two layers, modification is in general so strong, that a simplifying assumption of one layer is not permitted (Mikhaĭlova et al., 1966).

Especially for the purpose of exploration seismology, spectral properties of waves reflected from thin, homogeneous layers have been investigated in numerous papers, theoretically, following the Thomson-Haskell matrix method, as well as observationally. There are numerous Russian papers on this topic, e.g. theoretical by Berson (1959, 1965), Starodubrovskaya (1964), Mikhaĭlova et al. (1966), Ratnikova and Levshin (1967), Lossovskiy (1968), Berzon and Ratnikova (1971), and others, and observational by Khudzinsky (1962) and others. Likewise, there are numerous both theoretical and experimental contributions to these studies especially by American and other research workers, as evidenced by the numerous articles in this field in the journal *Geophysics*, also with particular reference to its importance in interpretation of seismic prospection results. Among the early contributions from *Geophysics*, I like to mention, as examples, those of Shugart (1944), and Clewell and Simon (1950). Among other, later theoretical investigations, also with numerous references, I like to mention Hirasawa and Berry (1971). Reverberation conditions are similar or even more exaggerated in a surface water layer (Backus, 1959).

Reverberation is a property common to any layered medium and the theoretical foundations are identical whether applied to thick, crustal layers or to thin, sedimentary layers. Assuming a one-layered horizontal crust with given velocity ratio crust/mantle, Ibrahim (1969) deduced the crustal thickness from amplitude minima in the SH-wave spectrum, applying the Thomson-Haskell matrix formulation. Also phase spectra were used, which have zero phase coinciding with the amplitude minima. The good agreement between crustal thicknesses determined in this way and by other methods demonstrates the applicability of the method also to large-scale crustal exploration, not only to thin layers encountered in prospection work.

Detailed theoretical studies are indispensable for correct interpretation of observed spectra. The most striking feature from all these investigations is the strongly oscillatory character of any calculated or observed spectra. Especially with thin layers, the spectral methods become indispensable, because arrivals are too close to be reliably separated in the time domain. In all cases, the layering effect (reverberation) is equivalent to a filtering effect. Based on certain models, it is possible to work out methods for inverse filtering or deconvolution to eliminate the reverberation effects on the recorded signal. Backus (1959) demonstrates this for the case of an upper water layer and Lindsey (1960) for seismic ghost reflections, but the same technique is applicable to any layering, leading to reverberations.

Concerning layered media and their spectral effects, our attention has been concentrated on crustal and near-surface layering, as being of inevitable significance in any seismological observation. But within the earth, layering is probably of significance also in connection with other boundaries, notably the Mohorovičić discontinuity and the core–mantle boundary. The nature of these boundaries, whether sharp in relation to the wavelengths used or transitional, with several layers, will also make its imprint upon seismic waves. Spectra will then again offer a useful tool for such investigations. The nature of the Moho boundary has been compared with model experiments (section 7.1.8) and for the core–mantle boundary we shall present more details in sections 7.4.2 and 7.4.3.

7.1.5 Near-surface geological structure

Considering the fact that many, perhaps most, seismograph stations are located on complicated geological structures, frequently sedimentary layers, it will be necessary to consider their effect on observed spectra. Even thin sedimentary layers may be expected to be influential because of large contrast to underlying granitic rock. Both amplitudes and periods will be affected by reverberation and resonance. But spectral methods with regard to superficial structures have also other, more practical aspects. These concern earth structures investigated in seismic prospecting work, and there is an enormous amount of literature on filtering, inverse filtering (deconvolution) and spectral techniques to achieve the maximum output in such work. Also, earthquake engineering is concerned with vibrations in superficial layers, based on the old experience of greater damage on soil than on bedrock, to which also spectral methods have been extensively applied. Only an outline can be given here, and the interested reader must be referred to professional journals, as *Geophysics*, *Geophysical Prospecting*, *Izv. Akad. Nauk SSR*, *Bulletin of the Earthquake Research Institute*, *Tokyo*, and others.

Resonance effects. As this problem concerns the behaviour of seismic body waves in a layered medium, it can be treated by the same techniques as outlined in the preceding section. In connection with layers of sediment and soil, the emphasis has been on vertically incident waves, which excite free (standing) vibrations in the overburden. The problem has been approached both by theory, by field observations and by model experiments.

Tsai (1970), in a theoretical investigation of layered media, defined the *spectral amplification* AMP(ω) as the ratio of the surface motion to twice the motion of the bedrock. In case of no superficial layers, AMP (ω) $= 1$, otherwise it is > 1. His theoretical analysis treats any number of horizontal, linearly elastic layers (viscoelastic or nonviscous) overlying a halfspace (nonviscous), with sinusoidal plane waves vertically incident from below as a steady-state excitation. The expected sequence of spectral maxima and minima is very clear in Fig.71 due to reverberation, also the strong dependence on α_N, i.e. the acoustic impedance ratio at the bedrock:

$$\alpha_N = \rho_{N-1} V_{N-1}/\rho_N V_N \qquad [8]$$

(N = bedrock, $N - 1$ next overlying layer). Obviously, the smaller the contrast becomes, the more is the amplification spectrum smoothed out. In case the in-

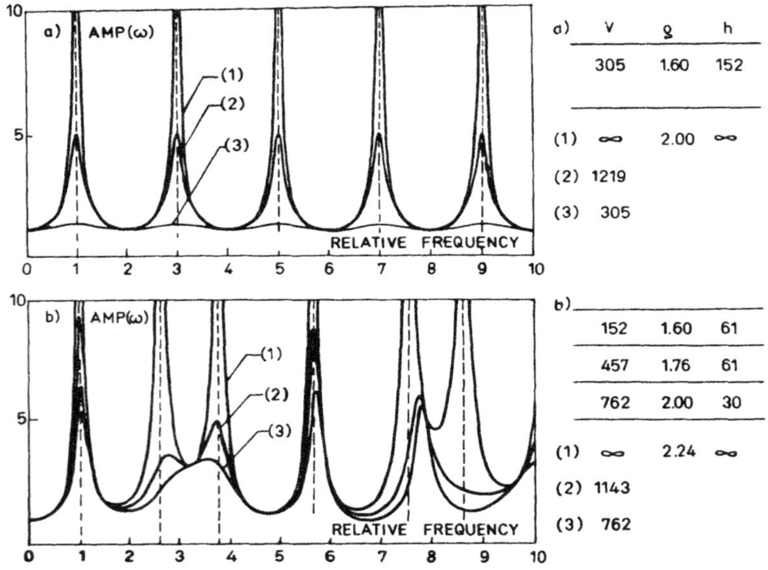

Fig.71. Amplification spectra for different values of α_N, i.e. the acoustic impedance ratio at the bedrock. V is velocity in m/sec, ρ is density in g/cm^3 and h layer thickness in m. After Tsai (1970).

cident wave has large energy in frequency bands corresponding to AMP(ω)-maxima, large surface motion will naturally result. Maximum amplitudes observed at any frequency can either depend on the input spectrum or on response peaks in the layered system. Moreover, it follows from the theory that the maximum values of AMP(ω) are bounded as follows:

$$1/\alpha_N \leq \mathrm{AMP}(\omega) \leq 1/\alpha_1 \, \alpha_2 \, \ldots \, \alpha_N \qquad [9]$$

which can also be easily verified from Fig.71.

Inserting expressions for the acoustic impedances into [9] we find:

$$\rho_N \, V_N/\rho_{N-1} \, V_{N-1} \leq \mathrm{AMP}(\omega) \leq \rho_N \, V_N/\rho_1 \, V_1 \qquad [10]$$

Applying this equation, we get immediately the following:

(1) One layer: $\mathrm{AMP}(\omega) = \rho_2 \, V_2/\rho_1 \, V_1$ [11]

(2) Two layers: $\rho_3 \, V_3/\rho_2 \, V_2 \leq \mathrm{AMP}(\omega) \leq \rho_3 \, V_3/\rho_1 \, V_1$ [12]

(3) Three layers: $\rho_4 \, V_4/\rho_3 \, V_3 \leq \mathrm{AMP}(\omega) \leq \rho_4 \, V_4/\rho_1 \, V_1$ [13]

Quite generally, for three or more layers, the limits of AMP(ω) do not depend on intermediate layers. Computational techniques for layered models were presented by Tsai and Housner (1970).

Tsai's (1970) analysis is able to explain many features of local geology found by several other authors, especially in Japan, and incorporates their findings. Numerous such papers can be found especially in the *Bulletin of the Earthquake Research Institute, Tokyo*, particularly under the names Sezawa, Kanai and their coworkers, ever since the 1930's. Reviews of Japanese papers in this field are given by Kanai (1957) and Shima (1962). The numerous spectra for ground motion amplitude in downtown Tokyo published by Shima (1969) clearly demonstrate the sequence of maxima and minima typical of subsoil layers. For application to sites in California, see for example Tanaka et al. (1966). From acceleration spectra at numerous locations in the Dushanbe area, Skorik (1969) found in the mean an increasing acceleration from rock to till and loess for wave periods in the range 0.1–1.1 sec. Resonance maxima at 0.2 sec period are found in relative spectra.

The method allows generalizations as follows:

(1) Any wave, P, SV, and SH, can be treated for any incidence. The Thomson-Haskell matrix formulation is naturally applicable also to thin sedimentary and soil layers, considered here.

(2) Various types of anelasticity can be introduced.

(3) Application to varying geological structures, i.e. not only plane, horizontal layers, is possible by the finite-element technique, in which the region of interest is partitioned into a discrete set of elements.

(4) Variation in the base-rock conditions, here assumed as a homogeneous half-space, may be introduced. As demonstrated by Lysmer et al. (1971), such variations have no important effect on observed accelerations as long as the base rock is homogeneous, whereas layering may lead to significant changes in the surface accelerations.

Reverberation and resonance effects depend on the layering and the type of material. Considering the fact that at least larger buildings and other constructions extend to some depth under the surface, we understand that surface records alone may not give the correct information. Preferably, they should be supplemented with recordings underground. Comparison of records and spectra obtained on the surface and underground of near earthquakes and blasts has proved of great importance, especially for engineering purposes. The generally found increase of amplitudes at the surface is ascribed to multiple reflections. Numerous investigations of this kind have been produced in Japan, for example by Kanai (1961, 1962a), Kanai et al. (1959, 1963, 1965, 1966a and b), Shima (1962), Yoshikawa et al. (1967), Yoshizawa et al. (1968). See Fig.72. Such amplitude and phase studies are able to give information on ground vibrational characteristics, necessary for engineering purposes, as distinct from travel-time studies in prospection work, which are only able to determine the structure but give no immediate information on its vibrations. Also for engineering purposes, investigation of cracks by means

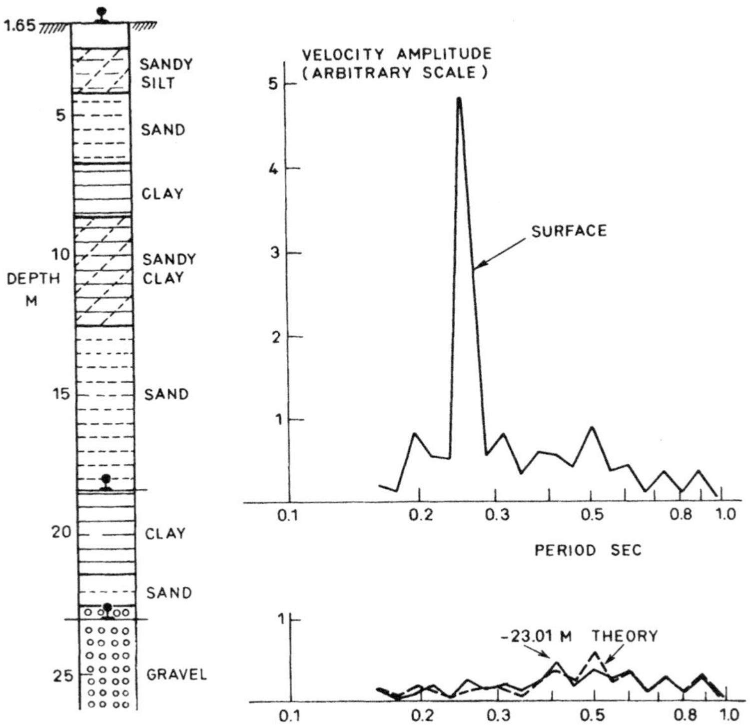

Fig.72. Spectra of microtremors observed simultaneously at the surface and underground at 23 m depth. After Kanai et al. (1965).

of spectral analysis of microearthquake records is of importance, cracks in firm ground yielding shorter predominant periods than those in soft ground (Wada et al., 1972).

It has been maintained, especially in Japan, that spectra of microearthquakes exhibit dominant periods, which are characteristic of the ground (soil) conditions. Hence, observations of microearthquakes could furnish valuable information concerning expected earthquake damage and be a guide in formulating building codes. Even if there is general agreement on these points, there is some disagreement about the exact nature of the waves in microearthquakes (microtremors) in the presence of soil layers: steady-state body waves or surface waves. In the first case, observed dominant periods should coincide with the maxima as shown in Fig.72, in the second case they should coincide with minima in the group dispersion curves. However, some tests of this kind carried out in Japan (Allam and Shima, 1967) proved inconclusive, as the amplitude maxima and dispersion minima coincided. Theoretical investigations if this is a general property or not would be welcome.

Differential effects. The theories developed above refer to horizontal layering. However, in comparing two sites with different layering, we may also be concerned with sloping layers, at least when the sites compared are close to each other.

In an attempt to estimate quantitatively the effect of local geology on ground motion, Borcherdt (1970) applied a spectral-ratio method based on the concept of linear systems:

$$A(\omega) = S(\omega) \cdot M(\omega) \cdot C(\omega) \cdot I(\omega) \qquad [14]$$

where $S(\omega)$, $M(\omega)$, $C(\omega)$ and $I(\omega)$ represent the spectral contribution of the source, the wavepath, local geology and the recording instrument, respectively. At the two observation sites, with amplitude $A_s(\omega)$ on sediment and amplitude $A_b(\omega)$ on bedrock, the source function $S(\omega)$ is the same (as the same signal is compared), and the wavepath function $M(\omega)$ is also the same (with good approximation for closely located sites), whereas the local-geology function $C(\omega)$ is different. Then we get, assuming identical instruments or that corrections for instrumental functions $I(\omega)$ have been made, considering only the amplitude spectra:

$$|A_s(\omega)|/|A_b(\omega)| = |C_s(\omega)|/|C_b(\omega)| \qquad [15]$$

This ratio, as a function of frequency, is called *spectral amplification* or *ground amplification factor* and serves as a useful ground characteristic. Obviously, this definition is analogous to the one of Tsai (1970), discussed above. It is important to note, that [15] can only be applied to sites which are close to each other, and whose difference is almost entirely in the local geology. For greater mutual distance, the total wavepaths will diverge too much to permit assumption of equal $M(\omega)$.

From unanimous results of seismic recordings of earthquakes, nuclear explosions and microseisms in the San Francisco Bay region, Borcherdt (1970), using the spectral amplification method, found that amplitudes and particle velocities, especially of the horizontal component, were many times larger on soil than on bedrock. Results were essentially independent of source characteristics but showed great dependence on local geology. Therefore, spectral amplification can serve very well for predicting intensities in earthquakes. Similar results are reported by Espinosa (1969).

Murphy et al. (1971) defined an *experimental amplification factor*, equivalent to the one used by Borcherdt (1970), but using pseudo-relative velocity spectra (see section 7.1.7) instead of Fourier amplitude spectra. The advantage is that amplification factors become smoother and easier to interpret with no loss of information. The need to take account of more complicated structures and not only consider flat-layer approximations was demonstrated by Boore et al. (1971). Spectral effects of topography with larger amplitudes on mountains were reported by Davis and West (1973).

The conclusions to be drawn from this section are of the greatest significance to any spectral analysis of seismic waves. The surficial layering has such a great influence on waveforms and spectra, that it may easily make further studies impossible. The difficulties are not only the great influences but also that they vary considerably from place to place due to variation in the layering. To take this into account and possibly to eliminate its effect would require such a detailed knowledge of the structure as is only seldom available. We may cite one example from array-data processing methods. These require similar waveforms, for example, in such a simple operation as addition of seismogram traces from different seismometers. If the geology varies strongly, such summation would be ineffective due to dissimilarity of the component traces. One example is provided by a station network in central California for which tests proved that this simple summation technique does not work. The main reason is the varying sedimentary layering from station to station. On the other hand, within Fennoscandia, where nearly all stations are located on good bedrock (granite, gneiss, or similar), the summation procedure

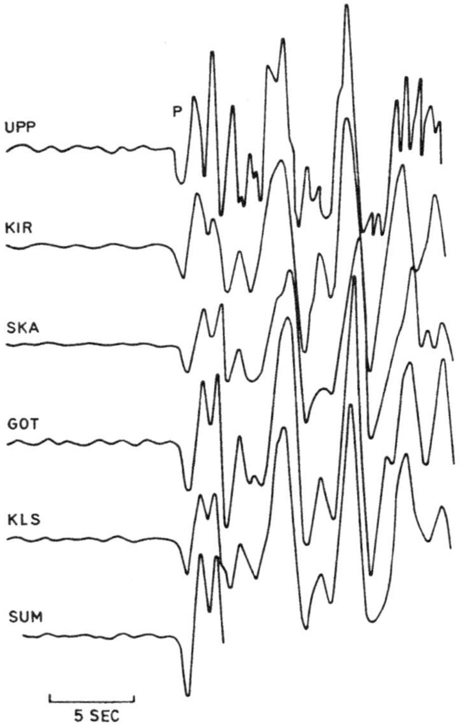

Fig.73. P signals from a Panama earthquake (July 26, 1962) recorded by short-period vertical-component seismographs in Sweden, demonstrating high signal similarity. UPP = Uppsala, KIR = Kiruna, SKA = Skalstugan, GOT = Göteborg, KLS = Karlskrona. After Husebye and Jansson (1966).

works with good success (Husebye and Jansson, 1966; Jansson and Husebye, 1968; Whitcomb, 1969), even though the mutual station distances are much larger than in the California network. The waveform similarity is demonstrated by a striking example in Fig.73. Another example is provided by signal variation across the LASA array (Mack, 1969) and for the NORSAR array as compared to the high signal similarity across the Uppsala array (Kulhánek, 1973b), the latter with an aperture of 100 km and located on the homogeneous Baltic shield. Stuttgart and Strasbourg with similar mutual distance (107 km) exhibit much greater variation even for long-period P-waves (Pho, 1971). The lesson to be learnt from these and many other similar examples is that when seismic wave spectra are to be used for investigation of source or path properties, stations located on good bedrock (preferably granite or similar) should be chosen, whereas stations located on sedimentary layers should be avoided.

7.1.6 Seismic prospection and crustal research

The purpose of geophysical prospection of all kinds is to reveal structures near the surface which have economic importance, such as structures which may contain oil, certain minerals, ore bodies, etc. Filtering techniques and spectral methods have gained an extensive and fruitful application in all kinds of geophysical prospection. This is witnessed by the large number of papers devoted to such problems, especially filtering techniques, in the professional journals, such as *Geophysics* and *Geophysical Exploration*. Also several books, especially those of recent date, contain much information on spectral calculation and spectral properties of seismic waves encountered in field investigations of the crust and upper mantle. As examples of such literature, Gurwitsch (1970) and Kosminskaya (1971) may be mentioned. There is naturally no possibility to review this vast field in this book. For application to gravity and geomagnetism, we refer to Chapter 10.

In seismic prospection, where the problem is to detect superficial (thin) layers, great requirements on accuracy in measurements have to be combined with sophisticated data handling. The problem consists to a large extent in identifying closely spaced significant, primary signals among their reverberations or multiples and noise. The efficient use which can be made for this purpose of autocorrelation and convolution has been clearly demonstrated by Anstey (1966) and Anstey and Newman (1966). Retro-correlation, as used by these authors, is synonymous with convolution. In their methods, multiple reflections are turned into useful information. For instance, a primary reflection is identified as such if followed by a sequence of multiples. The sectional autocorrelogram permits determination of depth of water or shallow layers and of the reflection coefficient at the bottom, with considerable precision. The convolution clearly distinguishes primaries from multiples, due to the fact that only the latter show up on the sectional retro-correlogram. Probably, the same technique could be applied not only to seismic prospection records, but also to ordinary station records of earthquakes, which are also complicated by crustal reverberation.

Not only seismic prospecting but also deep seismic sounding of the earth's crust is often complicated by closely packed arrivals and by reverberation and noise. A method devised by Merkel and Alexander (1969) is based on cross-correlation between a given signal and a record to be analyzed, assuming the source function for both to be identical. Those time lags are sought which maximize the cross-correlation. The success of the method depends on the frequency spectra (amplitude, phase) being different for different arrivals as well as for noise. The method provides for more accurate evaluation of seismic field records than visual techniques.

For maximum output of deep seismic sounding efforts, source factors, yield, emplacement, etc are of significance as well as the signal-to-noise ratio. Based on USSR measurements, Kosminskaya and Zverev (1968) find that spectra of explosion-generated waves are independent of yield but that explosions on land have higher frequencies than explosions at sea, about 10 cps versus 5 cps for the spectral peak.

Field studies of deep crustal reflections in the Canadian shield are reported by Clowes et al. (1968). Power spectra reveal that most of the reflected energy is concentrated to the 5–15 cps band, which would require sharp boundaries of less than 0.5 km in transition thickness. It is emphasized that by its higher resolution, the reflection technique is able to reveal more structural details than even a comprehensive refraction survey. For Alpine explosions, O'Brien (1967a) reports spectral peaks at 4.0 and 2.3 cps, probably produced by multiple reflections of P and S in a waveguide (the ratio $4.0/2.3 = 1.74$, i.e. the velocity ratio of P to S; cf. Table XXXV).

An equation for the seismic wave transmission, including source, path and receiver properties, can be written down for explosions as well as we did for earthquakes in section 6.5. Similarly, equalization methods are applicable to deduce various relations, such as between amplitude and yield, between amplitude and distance, etc. There is no need to repeat these matters here, as procedures are analogous to those used for teleseismic events.

Vibroseis methods in prospection. Besides impulsive wave sources (explosions, impacts) also vibratory systems, emitting a continuous wavetrain for some time, have found important application in seismic prospection. As this method affords a good example of correlation techniques, we shall explain its principles.

We let the emitted signal be:

$$f_1(t) = s(t) \qquad [16]$$

and the received (recorded) signal be:

$$f_2(t) = g(t) \star s(t) + n(t) \qquad [17]$$

where $g(t)$ is the impulse response of the ground and $n(t)$ is noise. Cross-correlating $f_1(t)$ with $f_2(t)$ we have by section 3.1.2:

$$\begin{aligned} C_{12}(t) &= f_2(t) \star f_1(-t) = [g(t) \star s(t) + n(t)] \star s(-t) \\ &= g(t) \star C_{11}(t) + n(t) \star s(-t) = g(t) + n(t) \star s(-t) \simeq g(t) \end{aligned} \quad [18]$$

This holds true if $C_{11}(t)$ approaches a Dirac delta function $\delta(t)$, i.e., that it has a high central value $C_{11}(0)$, corresponding to a broad power spectrum (remembering that the spectrum of $C_{11}(0)$ represents the total power, cf. section 3.3.3). Signal/noise ratio is made high partly by high power as mentioned, partly by choosing the signal $s(t)$ such that effective filtering of noise (second term above) is achieved. Under such conditions the cross-correlation between the known input signal and the recorded signal provides the impulse response function $g(t)$. For more detail, see for example Grau (1966). The Fourier transform of $g(t)$ yields the transfer factor $G(\omega)$ of the system. $G(\omega)$ provides information both on the amplitude response (absorption) and phase response (phase velocities) of the system, as usual.

It is instructive to compare this case with the impulsive wave source. For the latter the emitted signal is, at least in an idealized case:

$$f_1(t) = s(t) = \delta(t) \quad [19]$$

and the received signal is:

$$f_2(t) = g(t) \star s(t) + n(t) = g(t) \star \delta(t) + n(t) = g(t) + n(t) \simeq g(t) \quad [20]$$

provided that sufficiently high energy is concentrated to $s(t)$ so as to make $n(t)$ negligible. Then, $f_2(t)$ would yield the same information as $C_{12}(t)$ in the other case. For similar achievements it is necessary to impart very large energy in the impulsive source, whereas with a vibratory system it is only necessary that the total energy be large, whereas the energy at any moment may be much lower. In the vibroseis method, velocities are calculated only from phase differences between stations along some profile (there is no "shot moment"). In the impulsive method, the same attack can be used, or alternatively, velocities can be calculated from knowledge of the shot moment (the latter method being most common).

In a related, though different method, use is made of a succession of regularly spaced impulses, electrically generated. The high frequencies employed are particularly useful for investigation of smaller details of earth structures. A spectral presentation of this method is given by Nash and Barnes (1968), which is a direct application of example 15, Table V.

Besides the vibroseis method, a number of other methods have been developed for seismic prospection work on land. A review of such techniques, describing source and receiver properties, pulse shapes and spectral properties is given by

Wardell (1970). For instance, underground explosions represent broad-band spectra (example 9, Table V) and surface impacts (e.g. weight drop) give low-frequency spectra.

Atmospheric explosions (above ground) have interest both for possible application in seismic exploration and in connection with studies of nuclear explosions. In such cases, the explosion exerts a pressure over a certain, circular area on the surface. The pressure-distance function $p(x)$ along a diameter could at least approximately be represented by a box-car function (or better by some of the tapered window functions of section 4.4), and the wavenumber transform $P(k_x)$ would then be a sinc-function (example 1, Table V). The larger the horizontal extent of $p(x)$ is, the more will low wavenumbers dominate in $P(k_x)$. From teleseismic records of nuclear explosions, it is also very clear that low frequencies dominate for atmospheric explosions as compared to underground ones.

7.1.7 Earthquake engineering

Earthquake engineering or engineering seismology is a field of science directly concerned with vibrations and damage to various constructions, buildings, dams, bridges, etc, caused by earthquakes. This field, which is of the greatest practical significance, is intimately concerned both with ground vibrations and with structural vibrations and their interplay. It is immediately obvious that spectral analysis plays a great role in such studies. A useful review was given by Penzien (1965).

The engineering side, especially structural vibration and damage, is outside the realm of geophysics and will therefore not be dealt with in this book. May it suffice to say that the interested reader can find numerous papers on structural vibrations in the *Bulletin of the Seismological Society of America*, in the *Bulletin of the Earthquake Research Institute, Tokyo*, as well as in numerous more special engineering journals. On the other hand, we will give some information on measurement of ground vibrations, which ties in closely with preceding sections on surface geology.

A compact and informative description of the ground motion, especially for engineering purposes, is provided by the acceleration spectrum. As a typical example, Housner et al. (1953) published numerous strong-motion spectra for earthquakes in the United States. The spectra depend strongly on the damping n of the motion, expressed in a factor of the type $e^{-n\omega t}$. This means that $n\omega = a_i$ in the formula for attenuated sinusoidal motion (section 3.6.1). The damping n is due to the structure, and is expressed as the ratio of actual damping to critical damping. However, there is evidence that acceleration may not provide the best means to judge damage, or as stated by Richter (1958, p.88): "In relation to damage it is probable that the velocity is more significant than either acceleration or displacement taken alone."

Without entering into details, it may be stated that a structure acts as a system (a filter), in the simplest case as a linear system with one degree of freedom. The motion at any point of the structure (even in the basement of a building) has a spectrum which is the product of the ground motion spectrum (input) and the system function (transfer factor); cf. section 6.1. The latter would be the response or response spectrum of the structure, in conformity with the definition of response of a seismograph. However, in engineering seismology, response or response spectrum is frequently identified with the actual output. This naturally entails a dependence of the response also on the input function. Response spectra appear both as displacement, particle velocity, and acceleration spectra.

In relation to our discussion in section 2.2, there is also another point to note. An acceleration spectrum, defined in fact as an *acceleration density spectrum*, i.e. acceleration per unit frequency interval, is a velocity spectrum. Or equivalently expressed, for an oscillation $y = A \sin \omega t$, $y' = A\omega \cos \omega t$, $y'' = -A\omega^2 \sin \omega t$, we have amplitude $= A$, velocity $= A\omega$, acceleration $= A\omega^2$. Hence, acceleration per unit frequency interval:

$$\sim A\omega^2/\omega = A\omega = \text{velocity}.$$

Similarly, velocity per unit frequency interval:

$$\sim A\omega/\omega = A = \text{displacement amplitude}.$$

With an acceleration $a(t)$ as input and a building system function $e^{-n\omega t} \sin \omega t$, ω being the natural undamped frequency of the system, then the output will obviously be:

$$a(t) \star e^{-n\omega t} \sin \omega t = \int_{-\infty}^{\infty} a(\tau)\, e^{-n\omega(t-\tau)} \sin \omega(t-\tau)\, d\tau$$

$$= \int_{0}^{t} a(\tau)\, e^{-n\omega(t-\tau)} \sin \omega(t-\tau)\, d\tau \qquad [21]$$

$$\text{if} \quad a(\tau) \neq 0 \quad \text{for } 0 \leq \tau \leq t; \qquad = 0 \quad \text{otherwise}$$

The output has the dimension of velocity and is a function of n (damping, expressed as the ratio between actual damping and critical damping), ω (frequency) and t (time). The outputs maximized with regard to t constitute so-called *velocity response spectra*, of which an example is shown in Fig.74. Other examples are given by Cloud and Hudson (1961), Hudson et al. (1961), Housner and Trifunac (1967), and many, many others.

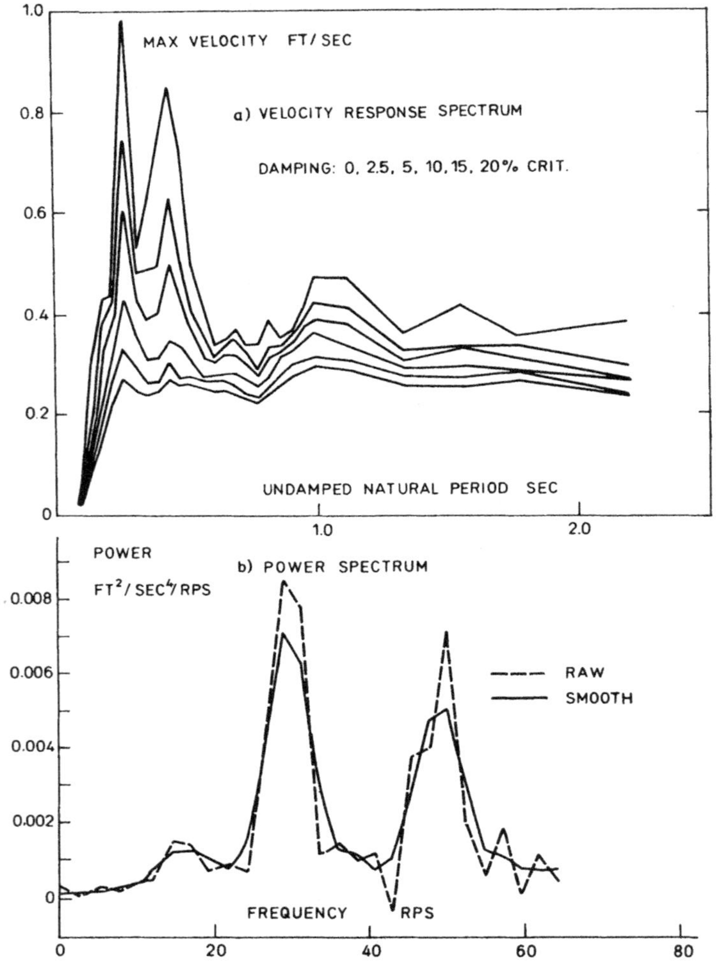

Fig.74. Spectra from accelerogram at Golden Gate Park in San Francisco earthquake of March 22, 1957. a) Velocity response spectrum after Hudson and Housner (1958). b) Power spectrum after Liu and Jhaveri (1969). RPS = radians per second.

The procedure is then the following: (1) given a recorded accelerogram $a(t)$; (2) get the velocity response spectrum from [21], calling this V; (3) get the displacement response spectrum from V/ω and the acceleration response spectrum from ωV.

These simple relations between the three spectra make it possible to present all three together in so-called tripartite plots (section 5.4.3). However, these relations are only approximate, but sufficiently accurate for most purposes of earthquake engineering (Hudson, 1962; Zeevaert, 1964; Bustamante, 1964). Sometimes, the term *pseudo-spectra* is used to denote approximate spectra, e.g. pseudo-velocity spectra for V as calculated above. For structures with more than one degree of freedom, approximate responses can be obtained by superposition of response spectra (Clough, 1962).

In handling accelerograms $a(t)$ for analysis, the usual care has to be taken, such as follows:

(1) Correction for false zero-line (section 4.2.4; see also Wiggins, 1962; Schiff and Bogdanoff, 1967; Trifunac, 1971b; Irikura et al., 1971).

(2) Consideration of record length to be included. Bustamante (1964) found that for damping as low as 2% of critical, it is sufficient to use only the strongest part of $a(t)$, whereas for zero damping the whole accelerogram is needed.

(3) Instrumental response correction. This is especially important for frequencies higher than the natural frequency of the instrument (seismograph or accelerograph), as demonstrated with examples by Jenschke and Penzien (1964).

(4) Digitizing errors (section 4.2; see also Schiff and Bogdanoff, 1967; Nigam and Jennings, 1969; Irikura et al., 1971; Trifunac et al., 1973).

From a detailed analysis of various error sources in the evaluation of strong-motion accelerograms, Schiff and Bogdanoff (1967) conclude that these are inadequate in many respects for determination of response spectra. Nigam and Jennings (1969), in presenting a new method for calculation of response spectra, emphasize that digitization errors are the main limitation on accuracy of calculated spectra. These may easily lead to errors of 15–20% in the spectra. A fast and accurate digital method for calculation of response spectra was given by Beaudet and Wolfson (1970). Even though velocity response spectra have found their widest application to strong-motion earthquakes, they are naturally applicable to any strong-motion data. A recent and important example is the ground shaking in the vicinity of underground nuclear explosions. For instance, Lynch (1969) determined the dependence of velocity response spectra on explosion yield and distance.

Another application of strong-motion spectra is to define *intensity* in a more satisfactory way than is possible by the usual intensity scales. The usual procedure is then to define intensity as the area under the curve, usually the velocity response spectrum, between some specified frequencies. Intensities thus defined have been studied in relation to other variables, such as local geology, depth below surface, etc., and the method has placed such studies on a more reliable basis than was possible by traditional intensity ratings.

As examples of such studies, using the refined intensity measure, we may mention Hudson and Housner (1958), who found clear dependence of intensity on local geology for the San Francisco earthquake of March 22, 1957, and Chopra (1966) who found the vertical intensity to be about 20–30% of the horizontal intensity. Similar intensity determinations in Italy are reported by Console and Peronaci (1971). The intensity as defined here is an integrated value over a certain frequency range. Alternatively, it is possible to take this range relatively short, which may be especially useful for relative intensity estimates, displayed as function of frequency. This method has been applied by Zeevaert (1964), who found a clear intensity decrease with depth and that a soft clay layer amplifies ground motion by resonance effects, and by Cloud and Perez (1967), who found an intensity ratio alluvium/rock less than 3. An engineering intensity scale, also based upon velocity response spectra, but more elaborated and clearly defined, has been worked out by Blume (1970).

Recently, Riznichenko (1970) proposed the use of certain spectral parameters for mapping of seismic risk, instead of intensity according to some intensity scale. The proposal still needs to be worked out in detail and especially tested on actual cases in order to assess its reliability.

The undamped velocity response spectra differ from Fourier spectra of accelerograms only in the way that maximum values are used for the response spectra, otherwise there is close agreement. The two kinds of spectra show only minor differences, as demonstrated by Hudson (1962) and Cloud and Carder (1969). Damping has the effect of reducing and smoothing the spectra. See also Jenschke and Penzien (1964). Fourier spectra of accelerograms, instead of response spectra, have been used among others by Duke et al. (1970) and Lastrico et al. (1972). Thomson (1959) derived expressions for the power spectral density of the acceleration $a(t)$ and emphasized that these expressions are sufficient to describe the spectral characteristics of the ground motion (apart from damping effects) with no need to introduce response spectra. Liu (1969) prefers autocorrelation and power spectra of accelerograms before response spectra, especially considering the earthquakes as random processes. It is of interest to compare power spectra and velocity response spectra, calculated from identically the same accelerograms (Fig.74). Power density spectra are preferred also by Power (1969).

A certain drawback with all spectra, mentioned so far, is that they give only an over-all picture and do not display the time history, which may be important, considering the non-stationary nature of the motion. To remedy this shortcoming, Liu (1970) devised a method for calculation of running spectra or evolutionary power spectra from accelerograms; cf. section 3.6.5. It is thus demonstrated that power spectral density can vary in time (over the time span of the accelerogram) both in amplitude and peak locations. Likewise, Trifunac (1971a and c) has developed methods for calculation of time-dependent amplitude and acceleration spectra of strong motion, using a display equivalent to that used for spectrographs (sections 1.6 and 7.2.2). He introduced the *response envelope spectrum* (RES) $A(t,\omega)$ of ground displacement or $\omega A(t,\omega)$ of ground velocity or $\omega^2 A(t,\omega)$ of ground acceleration, obtained as the envelope of the records of a series of single-degree-of-freedom systems with different responses, a procedure corresponding to narrow band-pass filtering. This technique permitted Trifunac (1971a) to conclude that surface waves are the most dangerous ones from the engineering point of view—a warning against overestimating the engineering significance of vertically incident waves, discussed in earlier sections of this chapter. On the other hand, from a good agreement between calculated and observed ground motions on various kinds of soil and on bedrock, Schnabel et al. (1972) conclude that in the period range of 0 to 2 sec the engineering characteristics are determined primarily by waves propagating upward from underlying rock formations. Too optimistic hopes for success in such enterprises are, however, demolished in a significant paper by Hudson (1972), demonstrating large variations in response spectra from point to point. To explain these, reference to local geological structure is insufficient, and in addition surface topography and subsurface configuration seem to be of significance, with the consequence that the response spectra also depend on the direction of wave propagation. This clearly reflects the problems encountered in making a reliable map of seismic risk.

In summary, we can state that the following types of spectra of strong-motion accelerograms have been used:

(1) Acceleration density spectra, Fourier spectra.
(2) Velocity response spectra, pseudo-spectra.
(3) Power density spectra.
(4) Evolutionary power density spectra.
(5) Response envelope spectra.

7.1.8 Model experiments

In addition to theoretical and observational studies, also ultrasonic model experiments have found useful application to investigation of spectral modification of waves, especially of waves reflected and refracted at internal boundaries. Structures investigated in this way include: (1) intermediate velocity layer with sharp boundaries; (2) linear-transition layer; (3) saw-tooth boundaries.

In all cases, for wavelengths comparable to layer thicknesses or irregularities, the internal structure is found to act as a low-pass filter, with a marked difference in spectral content of the refracted (head) waves compared to the direct wave. In case 3, the filtering effect is ascribed to scattering at irregularities in the boundary.

The model experiments confirm a critical relation between wavelengths used and thicknesses of embedded layers. For large wavelengths, the effects of embedded layers are small, but they become appreciable when the wavelength has decreased to about the same value as layer thickness. Among field applications may be mentioned the search for a possible thin transition layer at the Mohorovičić boundary. Model experiments will help in selecting the right field procedures for discovery of any such transition layer.

Among papers using spectral techniques (partly including both amplitude and phase spectra) to study model experiment measurements and some related field studies, we mention those of Nakamura (1964), Nakamura and Howell

(1964), Siskind and Howell (1967), Howell and Baybrook (1967), Nakamura (1968), Behrens et al. (1969).

Another example of spectral methods in model experiments is offered by the resonant sphere technique, by which a sphere of the test sample is made to oscillate (analogously to the earth's free oscillations). From the observed spectra, the elastic constants of the sample can be calculated (Soga and Anderson, 1967). Other examples of the application of spectral methods to seismic model experiments can be found in the literature. In fact, equally varied and extensive applications can be envisaged as in the large-scale seismology which one tries to reproduce in the laboratory.

7.2 REGIONAL-SCALE PROPERTIES—SURFACE WAVES

Extensive application of surface-wave dispersion has been made over more than twenty years for studies of the structure of the crust and the upper mantle. Most of these studies, especially the earlier ones, were made without the use of any spectral analysis. However, more recently various spectral techniques have got increased application also in dispersion investigations. For easier review, we shall distinguish between methods for phase-velocity and those for group-velocity determinations, although there is some overlapping between the two approaches.

7.2.1 Phase-velocity dispersion

Phase-shift method. The use of Fourier analysis in surface-wave dispersion studies was introduced by Valle (1949) and Satô (1955, 1956a and b, 1958, 1960). In this method use is made of records of the same wave, at two stations, or alternatively, of successive passages of mantle waves at one station.

Equation [58.2] in Chapter 6:

$$\Phi_X = \Phi_S + \Phi_P + \Phi_I + 2n\pi \qquad [22]$$

will now be written in more explicit form with relation to surface waves. The source phase is written as:

$$\Phi_S = \varphi_S + \pi/4 \qquad [23]$$

where φ_S is the *initial phase* and $\pi/4$ is a phase advance at the source (Aki, 1962; Båth, 1968, p.47). Path effects on phase are:

$$\Phi_P = \underset{\substack{\uparrow \\ \text{phase advance}}}{\omega t} \; - \; \underset{\substack{\uparrow \\ \text{phase delay}}}{\frac{\omega}{c} r} \; + \; \underset{\substack{\uparrow \\ \text{phase advance}}}{\frac{m\pi}{2}} \qquad [24]$$

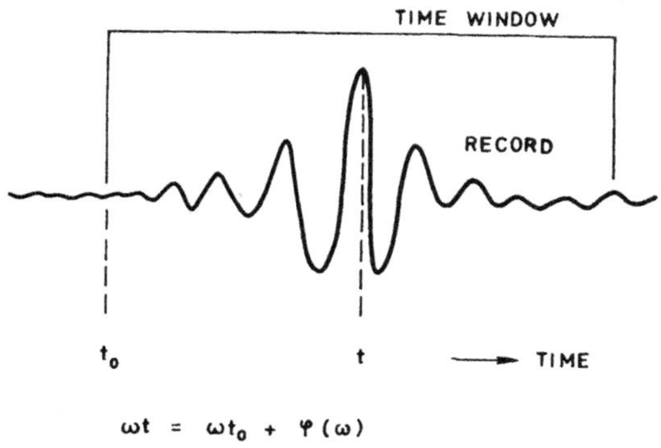

Fig.75. Arrival time of a certain phase (t) and beginning of Fourier time window (t_0).

The last term refers to phase advances upon epicentral and anticentral ("polar" and "antipodal") passages, $m = 0, 1, 2, \ldots$ (Brune et al., 1961a). The phase shift of $\pi/2$ is exactly true only for a spherically symmetric earth, but can be taken as approximately true for the real earth, at least for long-period surface waves (Aki, 1966a). The instrumental response enters as a phase delay, and for consistency we put $\Phi_I = -\varphi_I$. Making these substitutions in [22], we have:

$$\Phi_X = \varphi_S + \frac{\pi}{4} + \omega t - \frac{\omega}{c} r + \frac{m\pi}{2} - \varphi_I + 2n\pi \qquad [25]$$

where all phases are counted from the same zero ($t = 0$) and t refers to the time of a certain phase, e.g. a peak. In this form, solved for initial phase φ_S or for phase velocity c, the equation has been frequently used for time-domain measurements (Brune, 1960; Brune et al., 1960, 1961b; Aki, 1962).

For frequency-domain measurements, however, we make the following substitution (Fig.75):

$$\omega t = \omega t_0 + \varphi(\omega) \qquad [26]$$

where t_0 refers to the beginning of the time window used in Fourier analysis and $\varphi(\omega)$ is the phase counted from t_0. Then we get:

$$\Phi_X = \varphi_S + \frac{\pi}{4} + \omega t_0 + \varphi(\omega) - \frac{\omega}{c} r + \frac{m\pi}{2} - \varphi_I + 2n\pi \qquad [27]$$

Considering now two stations 1 and 2 located along the same azimuth from an earthquake and identifying the phases at these two stations, we have:

$$\Phi_{X_1} = \varphi_S + \frac{\pi}{4} + \omega t_{01} + \varphi_1(\omega) - \frac{\omega}{c} r_1 + \frac{m_1\pi}{2} - \varphi_I + 2n_1\pi$$

$$= \Phi_{X_2} = \varphi_S + \frac{\pi}{4} + \omega t_{02} + \varphi_2(\omega) - \frac{\omega}{c} r_2 + \frac{m_2\pi}{2} - \varphi_I + 2n_2\pi \qquad [28]$$

Assuming the same initial phase and the same instrumental response for the two stations, moreover putting $m = m_2 - m_1$ and $n = n_2 - n_1$ we obtain after solving for the phase velocity:

$$c(\omega) = \frac{\omega(r_2 - r_1)}{\omega(t_{02} - t_{01}) + [\varphi_2(\omega) - \varphi_1(\omega)] + m\pi/2 + 2n\pi} \qquad [29]$$

Note that the added term $2n\pi$ reflects the phase ambiguity in the frequency domain, equivalent to the peak identification ambiguity between two stations in the time domain (Nafe and Brune, 1960).

The method is applicable to any kind of wave: for surface waves the phase shift yields the true phase velocity, for body waves it gives the apparent velocity along the array, i.e. usually along the earth's surface. The procedure is identical in both cases. Because of the multi-valued term $2n\pi$, [29] does not give a unique determination or $c(\omega)$. Other independent velocity information is necessary in order to choose the right value of n. From $\varphi_1(\omega)$ and $\varphi_2(\omega)$ obtained from spectra (corrected for instrumental response if unequal instruments are combined), we plot $c(\omega)$ versus frequency ω or period for trial values of n. Usually, these curves are so well separated that only one value of n yields a reasonable result. The superiority of the Fourier technique over time-domain readings is particularly obvious in application to irregular and noisy records or records of low amplitudes (i.e. records of low signal/noise ratio) or of long periods, when exact readings of peaks and troughs in the time domain may be uncertain.

In application of [29] to mantle surface waves, use is frequently made of only one station and successive trains of mantle waves. It is then of importance to use waves which have left the focus in the same direction (e.g., combine only odd numbers G1, G3, G5, ..., R1, R3, R5, ... or only even numbers), as otherwise additional phase difference depending on initial phases may be introduced.

In this method we have taken the Fourier transform of $f_1(t)$ and $f_2(t)$ separately, and obtained the phase difference between the two transforms. An alternative but equivalent procedure is the following:

(1) Get the cross-correlation of $f_1(t)$ and $f_2(t)$.

(2) Get the cross-power $E_{12}(\omega)$ of $f_1(t)$ and $f_2(t)$ by Fourier-transforming their cross-correlation.

(3) The phase of $E_{12}(\omega)$ gives the wanted phase difference between $f_1(t)$ and $f_2(t)$. This is seen by section 3.3.4:

$$E_{12}(\omega) = F_1^*(\omega)\,F_2(\omega) = |F_1(\omega)|\,\mathrm{e}^{-i\Phi_1(\omega)}\,|F_2(\omega)|\,\mathrm{e}^{i\Phi_2(\omega)}$$

$$= |F_1(\omega)|\,|F_2(\omega)|\,\mathrm{e}^{i[\Phi_2(\omega) - \Phi_1(\omega)]} \qquad [30]$$

Among numerous applications of Satô's phase-shift method, we list the following:

(1) Regular surface waves: Satô (1956b) for a South Atlantic earthquake in

August, 1953; Satô (1958) for earthquakes in New Guinea in February, 1938, and in Kamchatka in November, 1952; Aki and Kaminuma (1963) for Love waves from the Aleutian earthquake in March, 1957; Noponen (1966) for surface waves from Greek earthquakes recorded in Finland. Particularly for crustal structure studies, the method was used in a simplified manner by Berckhemer et al. (1961), also by McEvilly (1964) for structural studies of the USA using fundamental-mode Rayleigh and Love waves, and Wickens and Pec (1968) applied the method to segmental phase-velocity determination of Love waves in North America combined with structural interpretations. Bozhko and Starovoit (1968) found that phase velocities of Rayleigh waves across the Russian platform determined by the spectral method agree within 1% with those found by the peak-and-trough method. Tarr (1969) used the method by FFT-calculation and got group-velocity dispersion from differentiated phase velocities, for the purpose of a ray-path network technique.

(2) Mantle surface waves have been extensively studied by this method: Båth and López Arroyo (1962) for mantle Love waves from a Peru earthquake in January, 1960; Toksöz and Ben-Menahem (1963) for five earthquakes recorded at two stations; Toksöz and Anderson (1966) for the Alaska 1964 earthquake; López Arroyo (1968) for mantle Rayleigh waves from three earthquakes recorded at Toledo and Malaga; Kanamori (1970c) for mantle Rayleigh and Love waves from Kurile Islands, 1963, and Aleutian Islands, 1964, earthquakes recorded at numerous stations; and Starovoit (1971) also for mantle Love and Rayleigh waves. Aki (1966a) applied the method to G-waves over complete great circles, but expresses also some doubt on the significance of such results in view of the spectral variability. He demonstrates velocity differences for periods as high as 200 sec, depending upon path properties.

(3) Explosion-generated waves were studied by this method by Satô (1955, 1956a), as recorded on an ice sheet, and by Kisslinger (1960), for surface waves from small, near explosions. Shima et al. (1964) applied the method to determination of apparent velocities of explosion-generated P-waves, finding no indication of dispersion in the 5–10 cps frequency band, and Satô (1960) used it for method testing by model experiments with flexural waves.

Improved accuracy in phase determinations is achieved by applying filtering before the Fourier analysis, by which disturbances are mostly eliminated, such as due to noise, to surface waves travelling over longer paths than the great circle arc, and to surface waves of other modes. Toksöz and Ben-Menahem (1963) used low-pass filtering before Fourier-analyzing Love and Rayleigh mantle wave records and Noponen (1969) combined band-pass filtering with the peak-and-trough method. Even more efficient are time-variable filters, as applied by Landisman et al. (1969) and Starovoit (1971).

While the phase-shift method is based on phase comparisons, the methods to be described next are based on amplitudes at two stations.

Methods of summation and multiplication of records. These methods start from the obvious fact that the sum or product of two related records will give maximum amplitudes when the two records are in phase. If the two records derive from two stations, some distance apart, then by time-shifting one record in relation to the other and forming sums and/or products stepwise, the point when these show maxima will give information on the corresponding time-shift and thus on the phase velocity. The two methods developed by Bloch and Hales (1968) are based on these principles and can be summarized as follows, where numbering indicates the successive steps in the procedures, starting from digitized records.

Method of summation:

(1) Time-variant filtering, shaped with guidance from independently determined group-velocity dispersion curves, to get smoother phase-velocity dispersion curve.

(2) Time-shifting in steps, corresponding to steps in phase velocity.

(3) Sums or differences between records formed:

$$A_1 \cos \omega t \pm A_2 \cos (\omega t + \varphi) = \text{maximum} \qquad [31]$$

for $\varphi = n \cdot 2\pi$ and $\varphi = (2n + 1)\pi$ for the plus and minus sign, respectively ($n = 0, \pm 1, \pm 2, \ldots$).

(4) Fourier analyzing sums and/or differences.

(5) Plotting amplitudes as isolines in graph with phase velocity and period as axes.

(6) The "ridge", corresponding to maximum amplitudes, defines the phase-velocity dispersion curve. This follows by the stationary phase method (Båth, 1968, pp.43–49).

Method of cross-multiplication:

(1) Records are windowed, for example with $\cos^2$-window.

(2) Narrow band-pass digital filtering, centered at the various periods.

(3) Time-shifting (= 2 in the other method).

(4) Cross-multiplication of records:

$$A_1 \cos \omega t \cdot A_2 \cos (\omega t + \varphi) = (A_1 A_2/2) [\cos (2\omega t + \varphi) + \cos \varphi] \qquad [32]$$

(5) Determining the d.c. level, which is maximum when records are in phase.

(6) Plotting d.c. level in graphs with same coordinates as in the other method.

(7) Same as 6 in the other method.

The latter method does not involve any Fourier analysis, and according to Bloch and Hales (1968), it also provides the better results. See Fig.76a. Both methods are used as two-station methods, which means that events have to be selected such that azimuths are nearly the same to both stations.

The next extension of phase-velocity methods will be to arrays or networks of stations, beginning with the triangular array.

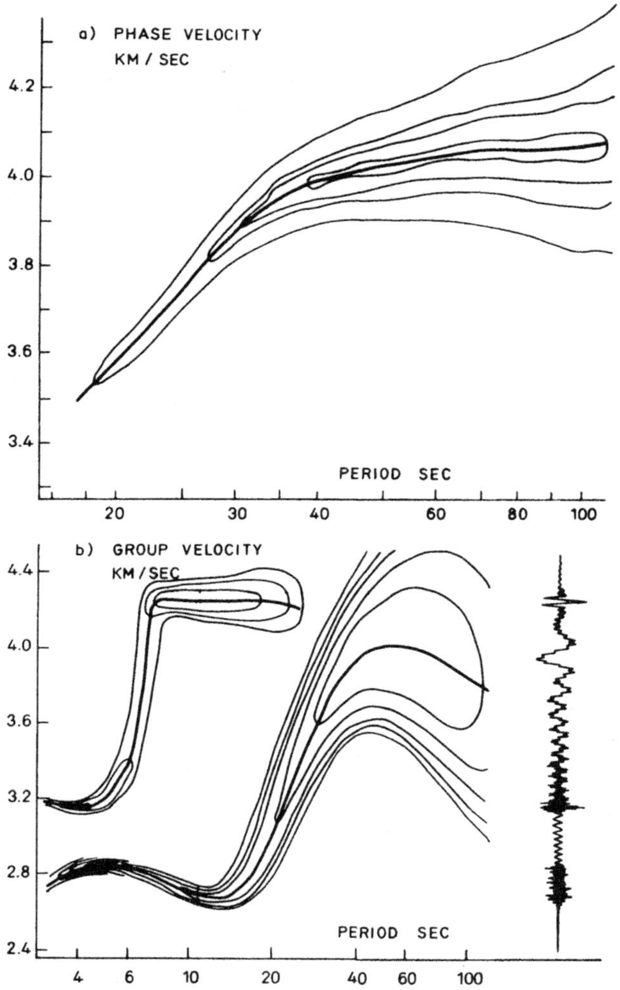

Fig.76. a) Phase-velocity contour diagram resulting from cross-multiplication process, applied to fundamental-mode Rayleigh waves. After Bloch and Hales (1968). b) Multiple-filter group-velocity analysis of vertical-component seismogram consisting of fundamental and first higher Rayleigh modes. After Dziewonski et al. (1969).

Triangular-array method. This was introduced in the middle of the 1950's for phase-velocity determinations (Press, 1956b). The method was applied at several places, e.g. USA, Japan and Scandinavia. In these works, direct readings were generally used and no recourse was taken to spectral analysis. Later, spectral techniques were introduced in evaluating triangular-array measurements by Pilant and Knopoff (1964); Knopoff et al. (1966); Knopoff and Pilant (1966); and others. Knopoff et al. (1966) use phase velocity of Rayleigh waves for the determination of the structure of crust and upper mantle in the Alps. The procedure in data handling and computation, including the use of time-variant filters, is described in detail, and has proved to yield significant results even in an area with such a complicated and varying structure as the Alps. It is emphasized that great care has to be exercised on solutions using only two stations in line with the epicenter, because of the influence of lateral heterogeneities. See also Seidl et al. (1966) and Knopoff et al. (1967). Payo (1970) made a similar study of Rayleigh-wave phase velocities on the Iberian peninsula. It was found that the direction of arrival of the waves depends on frequency, which is quite natural to expect when lateral inhomogeneities of different depth extent play a dominant role. More detailed information is furnished by array stations with more than three sensors, to which we now proceed.

Frequency-wavenumber method. With the advent of array stations it has been practically possible to apply two-dimensional Fourier transforms to observed data. The time–space observations are thus transformed into the frequency–wavenumber domain:

$$F(\omega,k) = \int_{-\infty}^{\infty} \int_{-\infty}^{\infty} f(t,x)\, \mathrm{e}^{-i(\omega t + kx)}\, \mathrm{d}t\, \mathrm{d}x \qquad [33]$$

(cf. section 2.5.1). For graphical display, it is then customary to show amplitudes $|F(\omega,k)|$ or corresponding powers $|F(\omega,k)|^2$ as isolines with frequency and wavenumber as coordinates (see Linville and Laster, 1967). This is the same kind of graph as we have encountered earlier, in discussing velocity filtering (section 6.4.1). In addition to serve filtering purposes, such presentations are useful for dispersion studies. Both phase and group velocity are closely related to ω and k:

$$\text{Phase velocity: } c = \omega/k \qquad \text{Group velocity: } U = \mathrm{d}\omega/\mathrm{d}k \qquad [34]$$

A curve along the "ridges", i.e. along maximum values of amplitude or power, in the ωk-diagram gives the phase-velocity dispersion, and the tangents to this curve give the group-velocity dispersion. The amplitude or power varies in a striking way along the dispersion curve, which depends on unequal source excitation of different frequencies and on unequal attenuation. This is therefore a kind of *dynamical dispersion curve*, in a simpler fashion introduced already by Båth and Crampin (1965).

The frequency–wavenumber method permits in principle a better resolution into different wave modes (fundamental and higher modes) than time-domain readings do, but a complicating factor arises from the side-lobes of the space- and time-windows used in transforming the given data. For an investigation of such problems, see Linville and Laster (1966).

In a practical application of this method to near-surface dispersion of explosion-generated Rayleigh waves, Linville and Laster (1967) find the best computational method to be first to prewhiten the records, then to form two-dimensional correlation functions and to transform these to power spectra. For interpretation of dispersion curves in terms of structure in a rough topography, it is significant to use only in-line stations, and even then, difficulties may arise because of structure (and elevation) varying between stations.

A modified form of frequency-wavenumber spectra uses wavenumbers (k_x,k_y) as axes and plots of power or amplitude as isolines, one graph for each frequency. This method is well applicable to array-station records, and besides dispersion information, such graphs yield information on arrival direction of surface waves. This method was used by Capon (1970) on LASA-data, with the result that most of multi-path propagation of Rayleigh waves is due to lateral refraction at continental margins. The same method, as applied by Capon and Evernden (1971), has made it possible to discriminate between superimposed surface-wave trains of different origin.

7.2.2 Group-velocity dispersion

As distinct from most methods used for phase-velocity determination, group velocities can be determined from records at only one station. This is quite natural as group velocity is simply obtained by dividing epicentral distance by the respective travel time. Numerous such studies have been made of crust and upper-mantle structure by direct seismogram readings, without involving any spectral analysis.

As group velocity can always be obtained by differentiating phase velocity, all methods described above for phase-velocity determination can also be considered as applicable to group-velocity determination, just by including also the differentiation procedure. For instance, from [29] we find:

$$U(\omega) = \frac{\mathrm{d}\omega}{\mathrm{d}k} = \frac{c}{1 + \dfrac{T}{c}\dfrac{\mathrm{d}c}{\mathrm{d}T}} = \frac{r_2 - r_1}{t_{02} - t_{01} + \dfrac{\mathrm{d}\varphi_2(\omega)}{\mathrm{d}\omega} - \dfrac{\mathrm{d}\varphi_1(\omega)}{\mathrm{d}\omega}} \qquad [35]$$

Formulas of this type have been applied among others by Toksöz and Ben-Menahem (1963), Brune (1965), and Dewart and Toksöz (1965).

Among methods, aiming at a direct determination of group velocities and employing spectral techniques, we shall differ between the following.

Spectrograph methods. Ewing et al. (1959) used an electrical analogue spectrograph which displays the spectral variation of the analyzed seismogram with time. Their method gives a means of directly recording group velocities as a function of frequency. This technique, being non-analytical, was mentioned already in section 1.6 together with some other related methods.

Moving-window method. In the moving time-window method, intervals of suitable length along the record are chosen and each interval is Fourier-analyzed. The results are plotted as power or amplitude isolines in a graph with frequency as one coordinate and group velocity (average for each time interval) or group arrival time as the other coordinate. It is thus a representation of a time-varying spectrum, as for non-stationary processes (section 3.6.5), and the method is in principle similar to the spectrograph technique, just mentioned. The isolines display the dispersive properties of the waves. In mathematical form, the method consists in calculating the so-called *moving Fourier amplitude* $F(\omega,\tau)$ from:

$$F(\omega,\tau) = \int_{-\infty}^{\infty} f(t)\, w(t-\tau)\, \mathrm{e}^{-i\omega t}\, \mathrm{d}t \qquad [36]$$

where τ is the time lag of the moving window $w(t - \tau)$. For any given, constant frequency ω or period T, the calculation of $F(\omega,\tau)$ is done for a series of τ-values, so that the variation of F with τ can be easily displayed. From this we pick that value of $\tau = \tau_{max}$ for which F attains its maximum value. Then τ_{max} is the group arrival time corresponding to the chosen period T. From this, the corresponding group velocity is calculated. This method was applied by Iyer (1964) to surface-wave records of a Chilean earthquake written at Kiruna in Sweden, and by Jacob and Hamada (1972) to a structural study of the Aleutian area from Rayleigh-wave group velocities.

The length of the moving window must be carefully chosen. A too short window will lead to a poor period resolution, while a too long window will lead to poor resolution of the time lag, in other words it will lead to difficulties in fixing the arrival time of the maximum amplitudes. Jacob and Hamada (1972) found that in their case a window about 3 to 4 times the period would be the best compromise. See further Inston et al. (1971) for a suggested improved method to decide the filter bandwidth.

Essentially the same method was employed by Landisman et al. (1969), using a rectangular, cosine- or (cosine)^2-window (cf. section 4.4) as the moving window, and displaying the results in an analogous way. Windows of variable length (about 4 to 6 times the period of interest) are used, in order to maintain the same frequency resolution at all periods.

Multiple-filter method. While the method just described uses a moving-window technique in the time domain, a similar procedure can be applied to the frequency domain, equivalent to filtering. This is the multiple-filter method (Dziewonski et al., 1969) and the steps in the calculation procedure are as follows:

(1) Fourier transform the given record.

(2) Apply instrumental corrections.

(3) Select a set of center frequencies.

(4) Select group arrival times.

(5) Apply narrow-band windows in the frequency domain, centered at the frequencies selected under 3.

(6) Get the quadrature spectrum from the windowed spectrum (by multiplying 5 by $i = e^{i(\pi/2)}$), which is needed in order to recover phase information.

(7) Get the inverse transforms of the windowed spectrum 5 and its quadrature 6.

(8) Get the amplitudes and phases from 7, corresponding to the chosen group arrival time 4.

An example of the result of the analysis is shown in Fig.76b. The method has several advantages over the simple peak-and-trough method in the time domain, such as permitting separation of several, partly simultaneous wave modes, and of nearly simultaneous events, as well as extending dispersion curves over a much longer period interval. In the frequency-time grid (Fig.76) amplitudes would be zero except along the ridges (the dispersion curves), provided filters of infinitely high resolution (i.e. of infinite length) were used for both frequency and time. In practice, filters have limited length, and thus limited resolution, resulting in a more "blurred" amplitude–frequency–time graph. Further application can be found in Bloch et al. (1969); Blum and Gaulon (1971); and Mitchell (1973). For a discussion of errors, see Der et al. (1970).

Related techniques were applied by Kanamori and Abe (1968); Abe (1972d); and Levshin et al. (1972). The similarities are obvious from the following steps involved in the method by Levshin et al. (1972), termed FTAN method (frequency–time analysis method):

(1) Fourier transform the given record:

$$F(\omega) = \int_0^T f(t)\, e^{-i\omega t}\, dt \qquad [37]$$

$f(t) \neq 0 \qquad$ for $0 \leq t \leq T$ only.

(2) Multiply by a narrow-band filtering function $H(\omega)$:

$$F(\omega)\, H(\omega) = F(\omega)\, e^{-\alpha[(\omega-\omega_i)/\omega_i]^2} \qquad [38]$$

where ω_i are a series of center-frequencies and the numerical value of α critically decides the property of $H(\omega)$ as a filter.

(3) Get the inverse transform of 2, i.e.:

$$G(t,\omega_i) = \frac{1}{\pi} \int_0^\infty F(\omega)\, e^{-\alpha[(\omega-\omega_i)/\omega_i]^2}\, e^{i\omega t}\, d\omega \qquad [39]$$

Both amplitude and phase of $G(t,\omega_i)$ are determined.

$|G(t,\omega_i)|$ is plotted in graphs with t and ω_i as coordinates. For a single record, the "ridge" again defines the group velocity. For two stations recording the same wave train (or alternatively, for one station with successive recordings of the same wave train), the amplitude ratio $|G_2(t_2,\omega_i)|/|G_1(t_1,\omega_i)|$ defines the attenuation and the phase difference arg $G_1(t_1,\omega_i)$ − arg $G_2(t_2,\omega_i)$ gives the phase velocity.

Band-pass digital filtering in the time domain has also been applied in group-velocity dispersion studies of higher-mode surface waves, combined with visual readings by Crampin and Båth (1965) and combined with Fourier analysis by Brooks (1969).

Several useful reviews have been published on these newer techniques for analysis of surface waves (e.g. Dziewonski et al., 1968; Landisman et al., 1969; Dziewonski and Hales, 1972) to which the reader is referred for more detail. Further method development is given by Dziewonski et al. (1972). The methods lend themselves to efficient analysis of any type of dispersed wave trains, e.g. explosion-generated atmospheric pressure waves recorded on microbarographs (Pfeffer and Zarichny, 1963). Recently, similar analysis and the same mode of presentation of the results have proved useful in disentangling the complicated wave motion close to an earthquake source, mainly for engineering purposes (Trifunac, 1971a); see also section 7.1.7.

7.3 GLOBAL-SCALE PROPERTIES—FREE OSCILLATIONS

The free oscillations of the earth constitute a relatively new field in seismology.

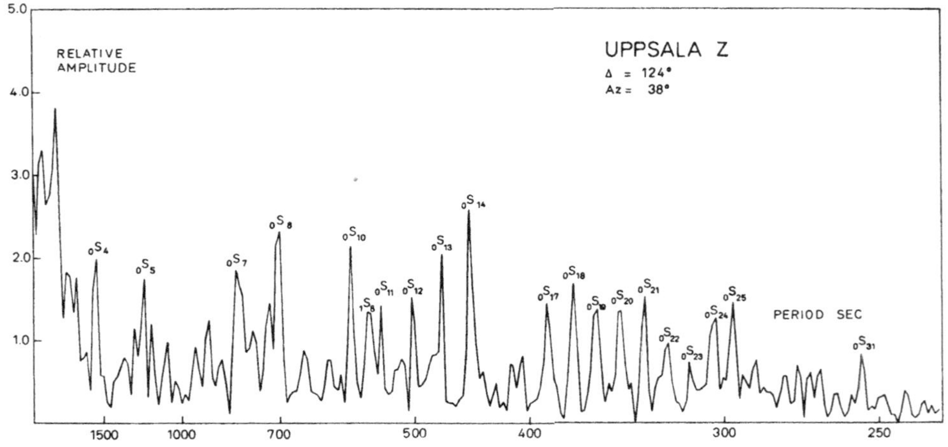

Fig.77. Spectrum showing spheroidal oscillations from the Chilean earthquake on May 22, 1960, recorded at Uppsala, Sweden. After Alsop (1964a).

Early searches for free oscillations, using records of gravimeters and extensometers and power spectrum techniques, were made by Benioff et al. (1959), although not quite successful. This field of study got started around 1959–1960, partly because fundamental theories were then published (Alterman et al., 1959), partly because the Chilean earthquake in May, 1960, provided the first more complete observational material on such oscillations. The papers by Benioff et al. (1961) and Ness et al. (1961) are among the early ones in this field and are particularly valuable for their discussion of methods for the Fourier analysis. Since then, quite a number of studies have been made on the *earth's spectroscopy*. Reviews were published by Bolt (1964), Press (1964) and more recently by Derr (1969a). A typical spectrum is shown in Fig.77 and a summary is given in Table XXXVI.

7.3.1 Methodical problems

Spectral techniques have found important application in sorting out the various components of the oscillations in any observational study, mostly by constructing power spectra of seismic and gravimetric records. In fact, spectral analysis has proved to be an indispensable method to extract useful source and structural information from records of the free oscillations. In most other branches of seismology, there is a corresponding time-domain technique, working directly with the obtained records, whereas for free-oscillation studies, there is hardly any such correspondence, at any rate none of any simple nature. Therefore, spectral analysis naturally dominates this field. The reason for this dominance is that the oscillations consist of numerous closely spaced peaks in a spectrum, which can be separated only in this way.

For the separation and identification of the great number of closely packed peaks in spectra of free oscillations, recourse has to be taken to a combination of methods, such as follows:

(1) Statistical significance of spectral estimates; use of longer time windows to increase spectral resolution.

(2) Combination of spectra from different instruments, different components and different stations; including studies of ground displacement and ground strain (Nowroozi, 1972b).

(3) Application of pertinent seismological theory, for checking frequencies of observed peaks as well as phase velocities.

S. W. Smith (1966) outlines some methods for reliable identification. In published spectra, often around 40 maxima are read and interpreted as significant, even though spectral confidence limits are often omitted. With indication of significance levels, Nowroozi (1972b) identifies no less than nearly 90 spheroidal peaks and nearly 70 toroidal peaks, both fundamental, and in addition a number of overtones. Without recourse to items 2 and 3 just mentioned, it would probably have been impossible to arrive at such an amount of significant results; in other

TABLE XXXVI

Examples of observational spectra of the earth's free oscillations

Reference	Earthquake	Instrument	Frequency spectrum	Oscillation	Remark
Alsop et al. (1961a)	Chile 1960	LP seismographs strain seismographs	amplitude	spheroidal toroidal	
Alsop et al. (1961b)	Chile 1960	strain seismograph	amplitude	spheroidal toroidal	Q-determination
Benioff et al. (1961)	Chile 1960	LP seismographs strain seismographs	power	spheroidal toroidal	includes Q-study
Bogert (1961b)	Chile 1960	LP seismographs	power	spheroidal	
Buchheim and Smith (1961)	Chile 1960	horizontal earth-tide instruments	power	spheroidal toroidal	
Ness et al. (1961)	Chile 1960	gravimeter	power	spheroidal	includes Q-study
Bolt and Marussi (1962)	Chile 1960	tiltmeter	power	spheroidal toroidal	
Connes et al. (1962)	Chile 1960	tiltmeters	power	spheroidal toroidal	includes Q-study
Takeuchi et al. (1962)	Chile 1960	gravimeters	power	spheroidal	
Satô et al. (1963b)	Chile 1960	gravimeters	amplitude	spheroidal	includes Q-study
Winch et al. (1963)	Chile 1960	magnetometers	power	toroidal	
Alsop (1964a)	Chile 1960	Z-component seismographs	amplitude	spheroidal	
Alsop (1964b)	Kurile Islands 1963	Z-component seismographs	amplitude	spheroidal	
Alsop and Brune (1965)	Peru-Bolivia 1963	H-component seismographs	amplitude	spheroidal	
Nowroozi (1965)	Alaska 1964	LP seismographs	amplitude (harmonic analysis)	spheroidal toroidal	
Slichter (1965)	Chile 1960 Alaska 1964	gravimeters	power	spheroidal	
Wirth and Byl (1965)	Chile 1960	gravimeter	power	spheroidal (toroidal)	
Wirth and Skalský (1965)	Alaska 1964	tiltmeters	power	spheroidal toroidal	
Wirth et al. (1965)	Alaska 1964	tiltmeters, gravimeter	power	spheroidal toroidal	
Nowroozi (1966)	Aleutians 1965	strain seismograph	amplitude power	spheroidal toroidal	
S. W. Smith (1966)	Chile 1960 Alaska 1964	strain and pendulum seismographs	power	spheroidal toroidal	
Zadro and Marussi (1967)	Chile 1960	tiltmeter	power	spheroidal toroidal	

TABLE XXXVI (continued)

Reference	Earthquake	Instrument	Frequency spectrum	Oscillation	Remark
Nowroozi (1968)	Alaska 1964 Aleutians 1965	LP seismographs strain seismograph	strain	spheroidal toroidal	Q-determination
Nowroozi and Alsop (1968)	Chile 1960	horizontal seismographs	amplitude	toroidal	
Zadro and Caputo (1968)	Chile 1960 Alaska 1964	horizontal pendulums gravimeter	power power	spheroidal and toroidal spheroidal	includes Q-study
Abe et al. (1970)	Kurile Islands 1963	LP seismographs	power	spheroidal toroidal	includes Q-study
Dziewonski and Landisman (1970)	Chile 1960 Kurile Islands 1963	LP seismographs	power	spheroidal toroidal	
Dratler et al. (1971)	Colombia 1970	gravimeter accelerometer	power	spheroidal (overtones)	includes Q-study
Gaulon (1971, 1972a, b)	Chile 1960 Kuriles 1963 Alaska 1964 Aleutians 1965 Peru 1966 Japan 1968 Talaud Isl. 1969 N. Atlantic 1969 Sumatra 1969 Peru 1970 Colombia 1970 New Guinea 1971	H-component seismographs	power	spheroidal toroidal	includes Q-study
Sterling and Smets (1971)	Japan 1968	deep-well tide-recording installation	power	spheroidal (toroidal)	
Block and Dratler (1972)	Sumatra 1969 Colombia 1970 Mindanao 1970 Chile 1971 Shetland Islands 1971	quartz torsional accelerometers	power	spheroidal toroidal	
Nowroozi (1972b)	Colombia 1970	LP seismographs strainmeters	amplitude	spheroidal toroidal	
Prothero and Goodkind (1972)	Kamchatka 1969	gravimeter	amplitude	spheroidal	
Wiggins and Miller (1972)	Alaska 1964	gravimeter	amplitude	spheroidal	new method for noise suppression presented

words, special combination of spectra and theory is necessary for correct interpretation of individual spectra. Even then, spectral methods may fail to resolve higher-order, i.e. shorter-period, oscillations, especially for orders above 30. Then, recourse is taken instead to surface-wave phase-velocity techniques (Derr, 1969a). Order numbers of free oscillations refer to the number of nodal lines on the earth's surface, while their overtones or higher modes refer only to the number of nodes along the earth's radius.

For empirical determination of periods of free oscillations, there are at least two methods available:

(1) From power spectra of recorded oscillations. This is the customary method.

(2) From phase spectra of surface waves.

The latter method is based on the fact that the free (standing) oscillations arise because of constructive interference between travelling waves. Successive wave trains of the same kind, propagating in the same direction past a station, should exhibit phase differences equal to 2π or integer multiples of 2π for constructive interference. The phase of the cross-power between these two trains gives directly their phase difference. The periods at which this phase difference is equal to $2n\pi$ give the periods of free oscillations. Ignoring lateral inhomogeneities, the relation between phase velocity c and free period T can be easily formulated as follows. For wavenumber k the phase difference for one passage around the earth (radius $= r_0$) is, taking the phase shifts of $\pi/2$ at epicenter and anticenter into account:

$$k \cdot 2\pi r_0 - \pi = 2n\pi$$

Replacing k by $\omega/c = 2\pi/Tc$ we get the well-known Jeans' (1923) formula:

$$c = 2\pi r_0/[T(n + 1/2)] \qquad (n = 1, 2, \ldots) \qquad [40]$$

Gaulon (1972a) demonstrates good agreement between periods determined by the two methods outlined, as applied to oscillations of order number higher than 50. The phase method is particularly useful for higher orders, partly because of closely packed free periods, partly because of low power.

As a certain complication in interpretations of observed spectral peaks, also non-linear effects (especially between different spheroidal modes or between spheroidal and toroidal modes) produce observed peaks, with frequencies equal to sums and differences between theoretical peaks. For the analysis of such peaks, bispectral analysis has been used (Zadro and Caputo, 1968; Zadro, 1971; cf. also section 3.6.4).

While spherical harmonic analysis, including depth z as variable, is the natural approach in theoretical studies of free oscillations of the earth (section

2.5.1 and 2.5.4), the practical analysis usually proceeds by spectralizing individual station records and identifying observed peaks. As a preparation for spectral analysis, it is customary first to eliminate the earth tides from the records. A method for this is described by Satô et al. (1963b). Cf. section 4.2.4. The general procedure in studies of free oscillations can be put in the following scheme:

Record → Spectrum ↗ Periods → Phase-velocity dispersion → Comparison with earth models
Record → Spectrum ↘ Amplitudes → → Comparison with source models

This is the standard empirical approach. As a check on the analysis, it is suitable to start from an assumed theoretical model, and see if the results of the analysis agree with the observations, i.e. the procedure is as follows (Satô et al., 1963a):

Theoretical model → Synthetic seismogram → Spectrum ↗ Periods ↘ Check with observations
Theoretical model → Synthetic seismogram → Spectrum ↘ Amplitudes ↗ Check with observations

7.3.2 Structural studies

Comparing theoretically calculated and observed periods of free oscillations, Pekeris et al. (1961a) found that the agreement is in most cases better than one per cent. They also found a distinct preference for the Gutenberg earth model (with a low-velocity layer in the upper mantle) as compared to other models proposed at that time. They also reported an observed frequency splitting, i.e. that a spectral line is replaced by two lines located symmetrically on either side of the calculated frequency. This is explained as due to the earth's rotation, such that waves travelling in the same direction as the earth's rotation will have lower frequencies than those travelling in the opposite direction. The frequency splitting corresponds in a certain way to the Doppler shift and it has been studied theoretically by Backus and Gilbert (1961), Pekeris et al. (1961b), and Gilbert and Backus (1965). See also MacDonald and Ness (1961). The apparent width of a spectral line depends both on (partially overlapping) frequency-split components and on the quality Q of the medium. As a consequence, the width does not permit an immediate determination of Q. Gilbert and MacDonald (1960) refer the observed line splitting to the ellipticity of the earth, the rotation of the earth (of minor importance) and to crustal inhomogeneity, while imperfect elasticity causes a general broadening of the lines. A similar but vanishingly small effect exists also for ordinary seismic surface waves.

In an extensive theoretical and experimental study of toroidal oscillations,

MacDonald and Ness (1961) emphasize that the resonant frequencies depend on shear-wave velocity distribution in the mantle and on core–mantle boundary properties. Observations yield good confirmation of Gutenberg's upper-mantle low-velocity zone. Theoretical studies by Satô (1964) and Usami et al. (1967) also open up the possibility of more accurate core studies by observations of free oscillations.

In his study of the Chilean earthquake of 1960, Alsop (1964a) found good agreement with the Gutenberg-Bullen A model for periods less than 1100 sec (also found by Bolt and Marussi, 1962, and Nowroozi, 1965) and with the Gutenberg-Bullen B model for higher periods. His spectra from eight different stations use time windows ranging from 15.0 to 24.2 hours. The ensuing effects on spectral resolution are quite clear from his spectra. He emphasizes that spheroidal and toroidal modes can be distinguished simply by their different periods for lower orders than 10, but not for higher orders, for which it is recommended to use vertical-component seismographs or gravimeters for positive identification of spheroidal modes.

Slichter (1965) found an excellent agreement between power spectra deduced from gravimeter records of the spheroidal free vibrations excited by the Chilean 1960 and the Alaskan 1964 earthquakes. The data have led to modified density distribution in the earth, with higher gradients in the upper mantle and lower gradients in the lower mantle than assumed before. From power spectra of both spheroidal and toroidal oscillations, deduced from strain and pendulum instruments, also S. W. Smith (1966) found good agreement between these two events.

In addition to comparing observed free oscillations with existing earth models, efforts have also been made, notably by Derr (1969b), to deduce new earth models from observed free oscillations. Abe et al. (1970) and Nowroozi (1972b) found best agreement with the earth model of Haddon and Bullen (1969), which may perhaps not be so surprising considering that this earth model, as distinct from earlier ones, has been deduced with the help of free oscillation data.

Alsop and Brune (1965) report observations of fundamental and higher-mode spheroidal oscillations (period range 100–700 sec), displayed as an amplitude–frequency spectrum, which are remarkable partly because they were generated by a deep earthquake (depth 540 km in Peru–Bolivia border region), partly because the observations permitted a determination of faulting mechanism, confirmed by P-wave observations.

Even though studies of the earth's free oscillations have mostly been focussed on checking and modifying models for the earth's internal structure, including its attenuation properties, the studies may thus be used for exploring source mechanisms of earthquakes (Ben-Menahem et al., 1972). Future application of free oscillation observations to all such studies will probably be much encouraged by the assistance rendered by extensive tables and graphs recently published by Ben-Menahem et al. (1971).

7.4 SEISMIC WAVE ATTENUATION

Seismic wave attenuation, expressed by the quality factor Q, is another field where spectral analysis is the natural approach. This is clear as we want to see attenuation in relation to wave frequency.

Starting from [58.1] in Chapter 6, we can write the amplitude A of a recorded seismic wave as:

$$|A(\omega,r)| = |S(\omega)|\,|B(\theta)|\,|C_s(\omega)|\,|M(\omega,r)|\,G(r)\,|C_r(\omega,r)|\,|I(\omega)| \qquad [41]$$

Taking the amplitude ratio between two general cases, labelled 1 and 2, we have, also introducing the expression [63] from Chapter 6 for the attenuation effect:

$$\frac{|A_1(\omega,r)|}{|A_2(\omega,r)|} = \frac{|S_1(\omega)|\,|B_1(\theta)|\,|C_{s1}(\omega)|\,G_1(r)\,|C_{r1}(\omega,r)|\,|I_1(\omega)|\exp\left[-\dfrac{\omega}{2}\displaystyle\int\frac{\mathrm{d}r}{Q_1(\omega,r)V_1(r)}\right]}{|S_2(\omega)|\,|B_2(\theta)|\,|C_{s2}(\omega)|\,G_2(r)\,|C_{r2}(\omega,r)|\,|I_2(\omega)|\exp\left[-\dfrac{\omega}{2}\displaystyle\int\frac{\mathrm{d}r}{Q_2(\omega,r)V_2(r)}\right]} \qquad [42]$$

Here we let the mantle effect M consist in attenuation; if in addition reflection or refraction enters the propagation, the corresponding coefficient must be included on the right-hand side. Equation [42] represents the basis for equalization, but in a very unsatisfactory form. For successful equalization, one seeks to find cases such that on the right-hand side all ratios except one will cancel.

The determination of attenuation from spectral amplitude ratios offers numerous examples of spectral equalization techniques. As a rule, every equalization method is strictly speaking imperfect, and no fully exact method exists for determination of attenuation. They are all more or less loaded with simplifying assumptions, necessary to be able to reach some numerical results.

Because of the exponential functions on the right-hand side of [42], it is more convenient in application to use the logarithms of the expression [42]. It is to be observed that A in our formulas refers to amplitude. If instead powers $E(\omega)$ are used, as is not uncommon, then we have to remember that $A^2 \sim E$ and $2 \ln A \sim \ln E$. Likewise, all formulas in this section are immediately applicable to power spectra, just by squaring the formulas given here.

According to the way in which the equalization is made, we differ between three methods for attenuation measurements (Table XXXVII):

(1) the frequency-ratio method; (2) the station-ratio method; and (3) the wave-ratio method.

As elsewhere in this book, we shall restrict our treatment essentially to methods employed, which here means leaving out the many numerical results

TABLE XXXVII

Review of spectral amplitude ratios for determination of attenuation

Method	Exact notation for ratio	Simplified notation for ratio[1]	Number of frequencies compared	Number of stations compared	Number of waves compared
General	$\frac{A_1(\omega_1,r_1)}{A_2(\omega_2,r_2)}$	$\frac{A_1(\omega,r)}{A_2(\omega,r)}$	two or more	two or more	two or more
Frequency-ratio	$\frac{A(\omega_1,r)}{A(\omega_2,r)}$	$\frac{A(\omega_1)}{A(\omega_2)}$	two or more	one	one
Station-ratio	$\frac{A(\omega,r_1)}{A(\omega,r_2)}$	$\frac{A(\omega)_1}{A(\omega)_2}$	one (at a time)	two or more	one
Wave-ratio	$\frac{A_1(\omega,r)}{A_2(\omega,r)}$	$\frac{A_1(\omega)}{A_2(\omega)}$	one (at a time)	one	two or more

[1] The principle has been always to keep ω as a parameter, and for the rest either drop symbols or simplify.

for Q-factors which have appeared in the literature in recent time. Attenuation methods are naturally applicable to both body waves and surface waves. We shall deal with body waves first, for the three methods, where applications are more complicated, referring to application to surface waves near the ends of the respective sections. Very useful reviews have been published on attenuation theory, methods, and results up to 1967 by Anderson (1967) and Sato (1967) and more recently by Payo (1969), Jackson and Anderson (1970) and S. W. Smith (1972).

7.4.1 Frequency-ratio method

Equations [41] and [42] are considerably simplified by lumping all factors S, B, C_s, G, C_r, I together to one factor, denoted $|A_0(\omega)|$. Thus the observed spectrum $|A(\omega)|$ becomes:

$$|A(\omega)| = |A_0(\omega)| \exp\left[-\frac{\omega}{2}\int\frac{\mathrm{d}r}{Q(r)\,V(r)}\right] = |A_0(\omega)| \exp\left[-\frac{\omega}{2}\frac{t}{\bar{Q}}\right] \quad [43]$$

where we have introduced travel time t and an average Q, denoted $\bar{Q}$. Taking the ratio of $|A(\omega)|$ for two frequencies ω_1 and ω_2 and then the natural logarithms, we get:

$$\ln\frac{|A(\omega_1)|}{|A(\omega_2)|} = \ln\frac{|A_0(\omega_1)|}{|A_0(\omega_2)|} - \frac{\omega_1 - \omega_2}{2}\frac{t}{\bar{Q}} \quad [44]$$

Plotting the left-hand side of [44] versus travel time t for a series of stations (alternatively, a number of events recorded at one station), we get an approximately straight line, corresponding to [44], whose slope gives $\bar{Q}$. Slight deviations from a straight line may, besides measuring errors, be due to the assumption that the first term on the right-hand side is a constant, independent of distance, which is not exactly true.

Among applications of this method we mention the following: Ōtsuka (1963) for spectra of ScS-waves from deep earthquakes recorded in Japan; Tsujiura (1966, 1969) for P-waves from near and distant earthquakes and from nuclear explosions recorded in Japan; Passechnik (1968) for Pn and P from two underground nuclear explosions; and Choudhury (1972) for P-waves from distant earthquakes and one nuclear explosion. Okada et al. (1970) applied this method to explosion-generated head waves (P) in a detailed investigation of the structure at Matsushiro in Japan, which revealed high attenuation and low velocity coinciding with the highest seismic activity. With the assumption $|A_0(\omega_1)| = |A_0(\omega_2)|$, Okano and Hirano (1971) applied the same method to P and S near Kyoto, Japan, using maximum amplitudes recorded by seismographs of different frequency response.

In the method given in formula [44] we compare the amplitude ratio for two fixed frequencies at a number of stations along a profile (a certain combination of the frequency-ratio and the station-ratio method, discussed in the next section). As an alternative method we could consider one station and the amplitude ratio $|A(\omega)|/|A(\omega_1)|$ between a running and a fixed frequency. Then, [44] can be written:

$$\ln \frac{|A(\omega)|}{|A(\omega_1)|} = \ln \frac{|A_0(\omega)|}{|A_0(\omega_1)|} - \frac{\omega - \omega_1}{2} \frac{t}{\bar{Q}} \qquad [45]$$

where ω_1, t, $|A(\omega_1)|$ and $|A_0(\omega_1)|$ are constant. A difficulty is the dependence on ω of A_0 and $\bar{Q}$. Asada and Takano (1963) who applied this technique to short-period P-waves for calculation of $\bar{Q}$, made various assumptions about A_0, especially $A_0 \sim 1/\omega$, and assumed $\bar{Q}$ independent of ω. Related technique was followed by Terashima (1968) in estimating Q from short-period P-waves generated by microearthquakes. Further studies of mantle attenuation of short-period P- and S-waves by this method were published by Takano (1971a), who claimed a dependence of Q both on period and wave type.

The slope and general shape of the amplitude spectra can in general give some indication of the possible frequency-dependence of A_0 under the assumption of constant $\bar{Q}$. A complicating factor is naturally local geological conditions, especially with sedimentary layers, which may emphasize certain frequencies and suppress others, and thus modify the spectral slope.

Absorption coefficients in dependence on frequency were determined by similar methods for P and S in the range 1–30 cps for depths down to 150 km in the south Kurile Islands by Fedotov and Boldyrev (1969) from over 700 records at one near station. In the treatment, amplitude spectra were averaged with regard to energy, depth and distance. Spectra (computed by a multi-filter analyzer) were correlated with various S-P time lags. High absorption, especially for S, was found in the depth range 70–100 km.

The frequency-ratio method is essentially concerned with spectral slopes, but the method could easily be enlarged to include spectral shapes on the whole. Then, comparisons would be made between spectral shapes of a given wave at a given station. It has proved fruitful to make comparisons between observed spectra and calculated ones, the latter based on certain assumed structural properties, including attenuation. Successful experiments of this kind are reported by Clowes and Kanasewich (1970), using power spectra of waves reflected from the Conrad and Mohorovičić discontinuities at near-vertical incidence.

The frequency-ratio method can also be easily applied to studies of attenuation of surface waves. It is then to be observed that in any attenuation study of dispersed waves, the velocity used should be the group velocity, not the phase velocity (section 6.5.3).

7.4.2 Station-ratio method

This method is based on spectral amplitude ratios at two distances, usually at two stations, but it could also be applied to one and the same station for two epicenters at different distances. A more general term for the method would then be *distance-ratio method.*

$|A_1(\omega)|$ and $|A_2(\omega)|$ which appear in [42] represent amplitude spectra of the same signal observed at two stations, 1 and 2 respectively. For a profile of stations extending in one direction from the source, we can have at least approximately: $|S_1(\omega)| = |S_2(\omega)|$, $|C_1(\omega)| = |C_2(\omega)|$. In special cases, however, these may not be permitted assumptions. For identical instruments or after instrumental correction has been made, the general formula then simplifies to the following:

$$\ln \frac{|A(\omega)_1|}{|A(\omega)_2|} = \ln \frac{G_1}{G_2} \frac{|C_1(\omega)|}{|C_2(\omega)|} - \frac{\omega}{2} \left[\int_1 \frac{dr}{QV} - \int_2 \frac{dr}{QV} \right]$$

or:

$$\ln \frac{|A(\omega)_1|\,|C_2(\omega)|}{|A(\omega)_2|\,|C_1(\omega)|} = \ln \frac{G_1}{G_2} - \frac{\omega}{2} (t_1^* - t_2^*) \qquad [46]$$

where $t^* = \int dr/QV$.

The ratio on the left-hand side may be called a *reduced spectral ratio* and the difference $t_1^* - t_2^*$ is called *differential attenuation.* A formula essentially of this type was applied by Teng (1968) to records at a number of stations from two deep-focus earthquakes, taking ratios between various stations and one base station. Plotting the left-hand side of [46] versus ω yields an approximately straight line, whose slope gives the differential attenuation. Teng (1968) gives a detailed discussion of methods to determine the Q-depth distribution from this information. The uncertainty of the crustal response functions $|C(\omega)|$ is a major limitation to this method.

If, as a further simplification, we consider an average Q between two stations, the general formula simplifies as follows:

$$\ln \frac{|A(\omega)_1|}{|A(\omega)_2|} = \ln \frac{G_1}{G_2} \frac{|C_1(\omega)|}{|C_2(\omega)|} + \frac{\omega(t_2 - t_1)}{2Q} = \ln \frac{G_1}{G_2} \frac{|C_1(\omega)|}{|C_2(\omega)|} + \frac{\omega(r_2 - r_1)}{2QV} \qquad [47]$$

Essentially this form of the equation was applied by Frasier and Filson (1972) to two stations recording one and the same event.

The formula could be rewritten as follows, with amplitudes corrected for geometrical spreading and crustal effects on the left-hand side:

$$\ln \frac{|A(\omega)|\, G_1\, |C_1(\omega)|}{|A(\omega)_1|\, G\, |C(\omega)|} = \frac{\omega(r_1 - r)}{2QV} \qquad [48]$$

The form [47] is suitable for application to only two stations, when the first term on the right-hand side is a constant. Then $\ln |A(\omega)_1|/|A(\omega)_2|$ could be plotted versus ω, giving a line whose slope yields a value of Q. Clearly, this Q will be an average between the two stations, also for the depth range covered by the rays to the two stations, as well as for the frequency interval used. In the form [48], the method is applicable to any number of stations along a straight-line profile from the epicenter. The amplitude ratio (corrected) is then plotted versus distance r with one line for each ω, whose slope again gives Q. Reverse profiles will, if available, provide useful checks on results.

Crust and mantle. Sumner (1967) applied the station-ratio method to P-waves along the Andes, essentially in the form [47]. He discussed the importance of relative station effects, which may be considerable in an area with strongly varying geology. From observations and theory, Kurita (1966) emphasized that not only amplitudes decrease but also periods of spectral peaks increase with distance. Applying this finding to long-period P-waves by the station-ratio method, he was able to deduce mantle Q-values. Therefore this constitutes another approach to the attenuation problem, not using amplitude measurements. Applying a formula equivalent to [46], Kurita (1968) demonstrated that Q-values for short-period P-waves increase both with frequency and with depth in the mantle. By similar techniques, Mikumo and Kurita (1968) studied the mantle distribution of Q for long-period P-waves.

Lateral heterogeneity in the crust and mantle exhibits itself in a corresponding lateral variation of attenuation. Our method can be easily extended to account for lateral variation of Q. We define for brevity:

$$t^* = \int \mathrm{d}r/QV \quad \text{and} \quad \delta t^* = \int \mathrm{d}r/(\delta Q \cdot V) \tag{49}$$

where δQ denotes a lateral variation of Q. Equation [46] then becomes the following:

$$\ln \frac{|A(\omega)_1|}{|A(\omega)_2|} + \frac{\omega}{2}(t_1^* - t_2^*) = \ln \frac{G_1}{G_2} \frac{|C_1(\omega)|}{|C_2(\omega)|} + \frac{\omega}{2}(\delta t_2^* - \delta t_1^*) \tag{50}$$

In a graph of the left-hand side versus ω, the slope will give a measure of $\delta t_2^* - \delta t_1^*$, i.e. the relative variation of attenuation. This technique was applied by Solomon and Toksöz (1970) to P- and S-waves in the USA.

Most investigations of this kind have for obvious reasons been concentrated to island arc areas, where large lateral variations are natural to expect. On the whole, it has been found that the dipping zone (the underthrusting oceanic slab) has a high Q, while adjacent parts of the upper mantle (at least down to 200–300 km depth) have a lower Q. Among others, this has been evidenced by Asada and Takano (1963) and Utsu and Okada (1968) for Japan by means of the spectral-ratio method. For Mt. Tsukuba in Japan, Takano (1970) finds significantly higher Q for waves from northeast than from southwest, depending partly on different source spectra, partly on lateral inhomogeneities.

An inherent difficulty in such investigations is that ground amplitudes depend not only on path properties (their Q-value) but also on local ground conditions as well as source radiation. The complicated distribution of Q in the crust, especially in seismic areas, necessitates the use of sophisticated methods. An example is given by Suzuki (1972) who investigated the lateral variation of Q in the Matsushiro area for P-waves, using the distance variation of the spectral amplitude ratio $(A)_{20\,\mathrm{cps}}/(A)_{10\,\mathrm{cps}}$. This is again a certain combination of the frequency-ratio method and the station-ratio method, but unlike the methods in section 7.4.1, the present approach is multi-directional and not uni-directional. Magma chambers of partially molten material exhibit themselves by disappearance of S-waves and attenuation especially of high-frequency P-waves, the latter effect clearly displayed in amplitude spectra (see, for example, Matumoto, 1971). Probably closely related are findings of low Q in the upper mantle near the Tonga arc as demonstrated by P-wave spectra by Barazangi and Isacks (1971).

Further evidence for lateral variations of attenuation has been presented by Pomeroy (1963), Ichikawa and Basham (1965), Utsu (1966), Tsujiura (1969), Bufe and Willis (1969), using spectral-amplitude ratios. All these phenomena have been amply summarized by Utsu (1971), with special reference to the Japanese region. In connection with a study of magnitude relations $m - M$ for earthquakes vs. underground explosions (Chapter 8), Ward and Toksöz (1971) apply the amplitude spectral-ratio method to short-period P-waves, taking the finiteness of the earthquake sources into account. Variations in wave attenuation in the upper mantle under the source and receiver sites are found to be a major reason for variations in m.

For the study of attenuation in the upper mantle, and especially in the crust, much use has been made of *explosion-generated waves*. Among many such studies where spectral methods have been applied, a straight-forward procedure is to derive frequency spectra for various distances along a profile. This was done among others by Willis (1960), Wright et al. (1962), Frantti (1965), Howell (1966b), O'Brien (1967b), Schenk (1971), Sengupta and Ganguli (1971), Duclaux (1971), Bodoky et al. (1972). It is a general procedure to assume an amplitude-distance dependence of the following form:

$$A = (A_0/\Delta^n)\,\mathrm{e}^{-\varkappa(\omega)\Delta} \qquad [51]$$

where A = amplitude, A_0 = a constant, Δ = distance, n = exponent defining the geometrical spreading, $\varkappa(\omega)$ = attenuation coefficient. There is not only an ambiguity in the direct application of [51], as either n or $\varkappa$ has to be assumed in order to determine the other (usually n is assumed, being more reliable, and $\varkappa$ is calculated), but there is also to be noted that [51] is a simplification of the complicated wave-propagation phenomena in the crust. A discussion was given by Yoshiyama (1959).

One method to avoid the apparent dilemma with simultaneous ignorance of attenuation and exact form for geometrical spreading for crustal waves was devised by Long and Berg (1969). Starting from [47], this can be written as follows, after differentiation with respect to ω and assuming $C_1(\omega)$ = constant $\cdot$ $C_2(\omega)$ and Q independent of ω:

$$\frac{\partial}{\partial\omega}\ln|A(\omega)_1| - \frac{\partial}{\partial\omega}\ln|A(\omega)_2| = \frac{t_2 - t_1}{2Q} \qquad [52]$$

From this equation, Q may be calculated by the least-squares method from given travel times t_1 and t_2 and the spectral slopes at the two stations 1 and 2. This therefore represents a certain combination of the frequency-ratio method and the station-ratio method. In essence, it is clear that to eliminate uncertainty in geometrical spreading it was necessary to make some other sacrifices, in this case assuming $|C_1(\omega)| \sim |C_2(\omega)|$ and Q independent of ω. No doubt, in every investigation of such a kind it is essential to try to weigh the different methods and their relative uncertainties against each other. The velocity layering in the crust and upper mantle causes focussing and defocussing of seismic wave energy, producing more complicated $A-\Delta$ curves, which can only be interpreted by combination with travel-time data (see, for example, Archambeau et al., 1966).

Willis and DeNoyer (1966) applied the partial transform $f(x,t) \leftrightarrow F(x,\omega)$, displaying P and S particle velocities, i.e. $F(x,\omega)$, as isolines in a rectangular coordinate system with distance x as abscissa and frequency ω as ordinate. With such a three-component presentation, any vertical line gives the velocity spectrum at any assigned distance, and any horizontal line gives the attenuation with distance for any given frequency. Residual iso-particle–velocity graphs, displaying differences between observations and theory, were subject to minimizing procedures to determine the most likely coefficients for attenuation and geometrical spreading.

Among other methods for Q-determination, we note one proposed by Dorman (1968), which may be attractive from the analytical point of view, but whose general usefulness in applications remains to be investigated.

The station-ratio method can also be well applied to *surface waves*. Examples are given by Tryggvason (1965) who used a form equivalent to [48] with $C_0 = C$ for Rayleigh waves, and by Tsai and Aki (1969), who used the same form for Love and Rayleigh waves, modified such that the left-hand side is the ratio of observed and theoretical spectra (for given source model). Mitchell (1973) also used the method on Rayleigh waves. Attenuation studies by spectral means have also been extended to smaller scale including soil layers, which may have engineering significance. Thus, attenuation of shear and Love waves in soil layers has been determined experimentally (Kudo and Shima, 1970, Kudo et al., 1970).

Core-mantle boundary. The station-ratio method is also well applicable to body waves diffracted around the core boundary. Using a profile of stations in nearly the same azimuth from the source, we eliminate the effects of the source and the mantle propagation down to the core. Moreover, selecting stations with similar crustal and mantle properties, we also eliminate effects of the upgoing branch through the mantle and crust. We are left only with effects along the core boundary, i.e. partly geometrical spreading on a sphere, partly attenuation along this boundary. The formula [47] is directly applicable to this case:

$$\ln \frac{|A(\omega,\Delta_1)|}{|A(\omega,\Delta_2)|} = \frac{1}{2} \ln \frac{\sin \Delta_2'}{\sin \Delta_1'} + \frac{\omega(\Delta_2' - \Delta_1')}{2QV} \qquad [53]$$

where Δ_1' and Δ_2' denote distances of respective paths along the core boundary only, $\Delta_2' - \Delta_1' = \Delta_2 - \Delta_1$. Plotting the left-hand side, observed, in [53] versus distance for given frequencies, we determine Q from the slope.

The method was developed by Alexander and Phinney (1966) and Phinney and Alexander (1966) and applied to long-period P-waves. The results indicate considerable lateral variation of core-boundary properties. More discussion and more data are given by Phinney and Cathles (1969); great scatter seems to prevent firm conclusions. Part of the trouble may arise because of too simple assumption for the geometrical spreading. The assumption of a spreading $\sim 1/(\sin \Delta)^{\frac{1}{2}}$ is valid only with no lateral variations, but such are suggested from the attenuation measurements. The situation is similar to the one for surface-wave attenuation determinations, for which this factor was pointed out by Yoshiyama (1960). Combination with travel-time data would provide for some remedy of such effects, for surface waves also more direct investigation of multi-path propagation (direction of arrivals etc.). The problem of attenuation along the core–mantle boundary, its nature and especially the diffraction effects, were studied in detail by means of model experiments by Shimamura (1969). It is demonstrated that especially uncertainty of the diffraction properties may easily invalidate determinations of Q near the core–mantle boundary. In fact, Sacks (1966) considers that diffraction is the most likely cause for the amplitude decrease, i.e. more significant than pure attenuation effects or the velocity structure near the core–mantle boundary. Further relevant data and problems are raised by Bolt et al. (1970). At the core–mantle boundary we may point at least to three factors which may influence measured amplitudes, i.e. diffraction, attenuation due to non-elasticity, and resonance effects in a possibly layered medium. Careful analyses are needed to separate such factors from each other.

Combination of methods 1 and 2. In efforts to eliminate as many factors as efficiently as possible and to isolate just one, involving Q, combinations of two of the methods in Table XXXVII have also been used. Besides some such procedures, already mentioned, we give here also a combination of 1 and 2, with special application to Pn-waves. We use the following expressions:

Frequency-ratio at distance Δ:

$$\frac{|A(\omega_1,\Delta)|}{|A(\omega_2,\Delta)|} = \frac{|S(\omega_1)|}{|S(\omega_2)|}\exp\left[\frac{\omega_2}{2}\int_{\Delta}\frac{\mathrm{d}r}{Q_2V} - \frac{\omega_1}{2}\int_{\Delta}\frac{\mathrm{d}r}{Q_1V}\right] \qquad [54]$$

Frequency-ratio at distance δ, assumed small:

$$\frac{|A(\omega_1,\delta)|}{|A(\omega_2,\delta)|} = \frac{|S(\omega_1)|}{|S(\omega_2)|}\underbrace{\exp\left[\frac{\omega_2}{2}\int_{\delta}\frac{\mathrm{d}r}{Q_2V} - \frac{\omega_1}{2}\int_{\delta}\frac{\mathrm{d}r}{Q_1V}\right]}_{\simeq 1} \qquad [55]$$

Division of these two equations yields, with notation η introduced just for brevity:

$$\eta = \exp\left[\frac{\omega_2}{2}\int_{\Delta}\frac{\mathrm{d}r}{Q_2V} - \frac{\omega_1}{2}\int_{\Delta}\frac{\mathrm{d}r}{Q_1V}\right]$$

$$\ln\eta = \frac{\omega_2-\omega_1}{2}\int_{\Delta}\frac{\mathrm{d}r}{QV} = \frac{\omega_2-\omega_1}{2}\sum_{i=1}^{N}\frac{t_i}{Q_i} \qquad [56]$$

where η is observed and the frequency-dependence of Q has been ignored in the last part. ln η is plotted versus Δ and trial Q-values are fitted to the observations. This is the technique applied to explosion-generated Pn-waves (head waves) by Archambeau et al. (1969).

7.4.3 Wave-ratio method

In this method use is made of spectral amplitude ratios between different waves from one event recorded at one station (Table XXXVII). The advantage is that with proper choice of waves for comparison, source and path effects are to a great extent eliminated. For attenuation measurements, mostly core-reflected waves have been used in this method.

The principle of determining Q-values from multiply reflected core phases was introduced by Press (1956a), who used time-domain measurements of the ratio 2 ScS/ScS for near-normal incidence (Table XXXVIII). If the condition of near-normal incidence is dropped, then we have to revert to a more complete form of the filtering equation, like this:

$$\ln \frac{|A_1(\omega)|}{|A_2(\omega)|} = \ln \frac{G_1}{G_2} + \ln \frac{S_1}{S_2} + \ln \frac{C_1}{C_2} + \ln \frac{R_{s1}}{R_{s2}} + \ln \frac{R_{c1}}{R_{c2}} - \omega \int_M \frac{\mathrm{d}r}{QV} \quad [57]$$

including source mechanism (S), crustal transmission (C), reflection coefficients at the surface (R_s) and at the core (R_c). Moreover, the mantle (M) absorption will depend on the slope of the ray. Sato and Espinosa (1967) applied this method for the distance range 7–80° to multiply reflected ScS and sScS waves from a deep

TABLE XXXVIII

Review of attenuation measurements by the wave-ratio method

Reference	Amplitude ratios used
Press (1956a)	2ScS/ScS
Anderson and Kovach (1964)	$(n+1)$ScS/nScS, $(n+1)$sScS/nsScS, nsScS/nScS
Kovach and Anderson (1964)	
Kovach (1967)	
Suzuki and Sato (1970)	SKS/ScS
Kanamori (1967b)	ScS/ScP, PcS/PcP
Sato and Espinosa (1967)	$(n+1)$ScS/nScS, $(n+1)$sScS/nsScS
Choudhury and Dorel (1973)	ScS/ScP
Balakina et al. (1966)	ScS/S
Kanamori (1967a)	PcP/P
Niazi (1971)	pP/P
Sacks (1972)	PKIKP/PKP

South American earthquake, although with several simplifying assumptions, especially regarding C, R_s, and R_c, as pointed out by Solomon and Toksöz (1970). From their measurements, Sato and Espinosa (1967) determined mantle Q-values for shear waves and viscosity and rigidity for the outer part of the core.

The formula [57] above can be appropriately modified to take account of any comparison of core-reflected waves. For instance, for multiply reflected ScS for near-normal incidence (reflection coefficients = 1 both at the free surface and at the core boundary), we have:

$$\ln \frac{|A_1(\omega)|}{|A_2(\omega)|} = \ln \frac{G_1}{G_2} - \omega \int_M \frac{dr}{QV} = \ln \frac{G_1}{G_2} - 2\varkappa h \qquad [58]$$

where the integral with subscript M refers to the mantle + crust, of total thickness $= h$, and where $|A_1(\omega)|$ should be identified with amplitude of $(n + 1)$ScS and $|A_2(\omega)|$ with amplitude of nScS. Using this method, also on multiply reflected sScS waves, Anderson and Kovach (1964) and Kovach and Anderson (1964) were able to determine average mantle Q for shear waves from records of two deep-focus earthquakes. Including also the ratio nsScS/nScS, for which the difference lies entirely in the region above the focus, it was possible to separate this region from the one below, if the effect of focal mechanism was ignored. Using spectral amplitude ratios for SKS/ScS with waves selected such that they differ only by the part K through the outer core, Suzuki and Sato (1970) were able to estimate the viscosity of the outer core.

In order to compare average mantle Q-values for S- and P-waves, use can be made of the amplitude spectral ratios ScS/ScP and PcS/PcP, which differ essentially in their second legs (S vs. P). Identifying these waves with subscripts 1 and 2, respectively, we can simplify our general formula as follows, assuming the same source function and ignoring the receiver crustal effect:

$$\ln \frac{|A_1(\omega)|}{|A_2(\omega)|} = \ln \frac{G_1}{G_2} \frac{|R_{c1}(\omega)|}{|R_{c2}(\omega)|} - \frac{\omega}{2} \left[\int \frac{dr}{Q_1 V_1} - \int \frac{dr}{Q_2 V_2} \right] \qquad [59]$$

(R_c = core reflection coefficient)

Putting:

$$Q_2 = \xi Q_1 \qquad \text{and} \qquad V_2 = 1.82\, V_1$$

we get:

$$\ln \frac{|A_1(\omega)|}{|A_2(\omega)|} = \ln \frac{G_1}{G_2} \frac{|R_{c1}(\omega)|}{|R_{c2}(\omega)|} - \frac{\omega}{2\bar{Q}_1} \left(1 - \frac{1}{1.82\xi}\right) \int \frac{dr}{V_1}$$

$$= \ln \frac{G_1}{G_2} \frac{|R_{c1}(\omega)|}{|R_{c2}(\omega)|} - \frac{\omega}{2} \frac{t_1}{\bar{Q}} \qquad [60]$$

where t_1 = S-wave travel time in the second leg, and $\bar{Q}$ is a mantle-averaged quality factor, depending on the difference between S- and P-wave attenuation. The left-hand side plotted versus ω will be an approximately straight line, whose slope permits calculation of $\bar{Q}$. This method was applied by Kanamori (1967b).

In an attempt to determine core-boundary properties, Balakina et al. (1966) compared amplitude and phase spectra of S with ScS for angles of incidence at the core = 70–85° in order to assure that differences between S and ScS were solely due to the core reflection. Ratio of amplitude spectra yielded reflection coefficients and differences between phase spectra yielded phase shifts upon reflection, in both cases as functions of frequency. However, comparison with theoretical models failed to give a unique solution, probably due to shortcomings of the models used. In addition, there is an inevitable shortcoming of the method of comparing S and ScS under these conditions. Even though it may be admitted that they leave the source as practically identical waves (the angles of incidence at the source being nearly equal), there is above the core an interval, easily several hundred kilometers in thickness, which is traversed by ScS but not by S. Frequency-dependent attenuation in this layer would influence the comparison of amplitude spectra. Due to great uncertainty about quality factors (Q) in this range, it is very hard to assess this effect quantitatively. The interference of P and PcP under similar conditions was investigated by Johnson (1969).

In the formula [57], we assumed comparison of two core-reflected waves. When instead we compare two waves of which only one is core-reflected, the formula has to be modified. Say that 1 and 2 correspond to the pair P and PcP or to the pair S and ScS. If epicentral distances are not too short, we could assume $S_1 = S_2$ and $C_1 = C_2$. Considering also that there is only one reflection (at the core), the formula [57] becomes:

$$\ln \frac{|A_1(\omega)|}{|A_2(\omega)|} = \ln \frac{G_1}{G_2} - \ln |R_{c2}(\omega)| - \frac{\omega}{2} \int_1 \frac{dr}{QV} + \frac{\omega}{2} \int_2 \frac{dr}{QV} \qquad [61]$$

A formula of this type was applied by Kanamori (1967a) for short-period P–PcP comparison, with various assumptions for R_{c2} depending upon the nature of the core–mantle boundary.

In a comparative study of amplitude spectra of PcP and P from earthquakes and explosions, Buchbinder (1968) finds no reason to infer any detailed structure of the core–mantle boundary, whereas Ibrahim (1971a and b), also basing his study on PcP, without using spectral methods, finds convincing evidence of a layered structure at the core–mantle boundary and a finite rigidity in the upper part of the core. From the slope of the spectral ratio of PKIKP/PKP, Sacks (1972) finds some evidence for a low-Q layer just outside the core, even though this layer is only 2–8% of the earth's radius and the determination relies upon knowledge of Q in the rest of the earth.

In general, all methods employing core-reflected waves for the study of mantle attenuation suffer from the ambiguity that the spectral ratios used depend not only on this attenuation but also on the core–mantle boundary reflection. A combination of several different ratios between core-reflected waves will probably be of assistance. Recent theoretical and observational studies of PcP and other core-reflected waves and their spectral behaviour are expected to place such studies on a more firm basis (Singh et al., 1972; Kogan, 1972; and others).

By appropriate reformulation of the general spectral amplitude formula, it may be applied to any spectral ratio. Applied to the spectral ratio pP/P, it will give information on the Q-value in the zone above the focus, and the following formula can then be easily derived:

$$\ln \frac{|A_1(\omega)|}{|A_2(\omega)|} - \ln \frac{G_1}{G_2} - \ln |R_s(\omega)| = \ln \frac{S_1}{S_2} - \frac{\omega}{\bar{Q}} \int_h \frac{dr}{V} \qquad [62]$$

where 1 refers to pP and 2 to P, and the integration is extended over h, i.e. from the focus to the surface reflection and down to the focal depth, along the ray. R_s is the reflection coefficient at the free surface. This method was developed and applied by Niazi (1971). In the application the left-hand side is plotted versus ω and the slope of the line gives $\bar{Q}$.

Still another application of the wave-ratio method is offered by taking spectral amplitude ratios between waves multiply reflected from crustal layers as well as from thin sedimentary layers. A complication is again that in general both attenuation and reflection coefficients are unknown, but this can be circumvented by including more than two waves, i.e. having two or more spectral ratios. Such methods are described by Berzon (1961) with special application to thin sedimentary layers.

Reflections from crustal layerings have attracted much interest, both observational and theoretical. Considering the ratio PP/P, where PP at least in part has been reflected at the base of the crust, this can be written as follows, with application of general filtering equations:

$$\frac{|A_{PP}(\omega)|}{|A_P(\omega)|} = \frac{|S_{PP}(\omega)|}{|S_P(\omega)|} |R(\omega,i)| \frac{|C_{PP}(\omega,i)|}{|C_P(\omega,i)|} \exp\left[- \int_{PP} \varkappa \, dr + \int_{P} \varkappa \, dr \right] \qquad [63]$$

Here $|R(\omega,i)|$ = reflection coefficient, depending upon layering at reflection point, $|C(\omega,i)|$ = respective crustal transfer functions, i = angle of incidence. In [63] the geometrical spreading has been assumed the same for P and PP, which is a permissible but unnecessary simplification. Obviously, the amplitude ratio PP/P cannot be used for information on any factor in [63], unless all the others are known. This was emphasized by Wu and Hannon (1966), who calculated $|R(\omega,i)|$

for structures including liquid layers using the Thomson-Haskell matrix method. Free-surface reflections lead to the largest reflections, as expected, which is of significance for correct reading of pP, even though their theory does not apply strictly to pP. For the latter, curved (approximately spherical) wavefronts will have to be assumed.

Like the other two methods for attenuation determinations, also the wave-ratio method is applicable to *surface waves* by proper choice of waves to be compared. The best method is to compare mantle waves which have left the focus in the same direction, e.g. G1—G3—G5— etc., G2—G4—G6— etc., and similarly for R mantle waves. Then in fact we compare the same wave leaving the focus in the same direction, but with path differences equal to one or several circumferences of the earth. The method is in fact a wave-ratio method, which is equivalent to a station-ratio method. The formula is the same as [48], except that the geometrical spreading is the same for the compared wave trains:

$$\ln \frac{|A(\omega,\Delta_1)|}{|A(\omega,\Delta_2)|} = \frac{\omega}{2} \frac{\Delta_2 - \Delta_1}{QU} \qquad [64]$$

where U = group velocity, or equivalently:

$$\ln \frac{|A(\omega,t_1)|}{|A(\omega,t_2)|} = \frac{\omega}{2} \frac{t_2 - t_1}{Q}$$

The results yield average mantle values of Q for frequencies covered by the mantle waves. This method has been developed by Satô (1958) and applied by him, by Båth and López Arroyo (1962), Ben-Menahem (1965), Aki (1966a), López Arroyo (1968), Kanamori (1970c), and others.

In a related method, also using records of only one station, the time rate of decay of free oscillations of the earth will yield values of Q. However, in this case certain conditions have to be applied to the length of the time intervals analyzed and their time separation, imposed by the rotation of the standing wave pattern with respect to the earth (Alsop et al., 1961b). For applications, see for example Nowroozi (1968), Abe et al. (1970), Gaulon (1971, 1972a and b), Dratler et al. (1971). Considerably higher Q for some overtone modes than for the fundamental modes has been demonstrated by Dratler et al. (1971) and Gaulon (1972b).

Chapter 8

SPECTRAL STUDIES OF SEISMIC SOURCE PROPERTIES

Seismic sources are characterized by a number of source parameters:

(1) Dynamic parameters, for whose determination use is made of amplitudes of seismic waves:

(a) Source mechanism (source time and space function). The traditional method to determine mechanism is based on direction of first swing (compression or dilatation) of P- and other waves. With spectral methods, use is made of whole wave groups and one does not depend on readings of first motion. Moreover, the spectral methods have played a decisive role in the study of finite moving sources, probably the most significant advancement in source studies over the last 10 to 15 years.

(b) Magnitude. Traditionally this is based on point measurements of maximum amplitude and period in a record. Spectral methods use whole wave groups instead and are therefore expected to yield a higher accuracy.

(2) Kinematic parameters, for whose determination use is made of kinematic properties, such as arrival times, velocities, distances:

(a) Epicentral coordinates. No spectral studies are envisaged.

(b) Focal depth. The traditional method is to use depth phases, pP–P or similar. The spectral methods work also in cases when readings of single phases like pP may be difficult or impossible, again for the reason that they are based on whole wave groups instead of single measurements.

(c) Origin time. The only way this enters into our subject is via studies of earthquake periodicity. As distinguished from the rest of this chapter, such studies do not work with spectral analysis of seismic waveforms but use times of earthquake occurrence.

The list above will be applied to different types of waves, i.e. both body waves and surface waves, and to different types of sources, i.e. both earthquakes and explosions. Here as well as in other chapters, we shall essentially restrict the discussion to methods, whereas results will be only briefly dealt with. For more discussion of results, the reader is referred to the literature quoted.

On the whole, it is evident from our list above, that spectral methods are powerful compared to traditional time-domain methods in analyzing seismograms above all for the fact that they are based on whole wave groups. They are not only less dependent on isolated, sometimes unreliable, point readings, but also they extract a much larger amount of information from the records. Moreover, in some cases the spectral methods have furnished information, which the traditional methods are unable to, especially concerning finite moving earthquake sources. The best procedure is to combine traditional and spectral methods in any study, for mutual checking.

8.1 EARTHQUAKE MECHANISM

8.1.1 Brief review of mechanism studies

Basically two different methods are available for investigation of earthquake mechanism: (1) field investigations of permanent near-source effects; and (2) seismogram analysis of radiated waves.

Comparing these two methods for focal mechanism studies, we can state that the first method is quite limited in application. It can only be used for earthquakes on land, excepting the very rare cases where investigation of sea-bottom topography is possible. Moreover, it requires a detailed knowledge of the landscape before the rupture has occurred, just to provide a possibility to estimate the effect of an earthquake in relation to existing structural features. Finally, method 1 is applicable only to earthquakes large enough to produce clearly visible surface effects, i.e., the method is restricted to large, shallow earthquakes. Even so, care is required in interpreting a focal mechanism from surface expressions, as even normal-depth earthquakes easily have their foci around 10–30 km below the surface. This inference about differences between surface displacements and true displacements in earthquake foci is substantiated by Aki (1968). In conclusion, method 1 has certainly great value in geological and structural engineering connections, whereas method 2 is the only one to provide an accurate knowledge of focal mechanism. And it is only in connection with method 2 that spectral analysis of seismic waves will enter as a useful tool.

Method 2 has already been used for many years, mostly based on first-motion studies of P- and other waves and polarization of S, with no need for spectral methods. Great developments in our studies of focal mechanism have taken place in the last 10 to 15 years, based partly on theoretical investigations of radiation patterns from different models, partly on application of spectral techniques, especially amplitude and phase equalization. As distinct from earlier, first-motion studies, the spectral methods imply use of the total signals, which has no doubt permitted more detailed and reliable mechanism studies than hitherto.

In the following, it will also be important to consider separately the space and time functions of the source. The space function describes the geometrical force system acting at the source, such as a single couple, double couple etc. (Table XXXIX), and constitutes what is usually covered by the term "focal mechanism". But, in addition we shall have to consider the source time function, i.e., a function

TABLE XXXIX

Explanation of some terms as used in connection with earthquake mechanism[1]

Type of source	Direction of faulting	Direction of slip motion
Single couple (a dipole): one couple of forces, generally with a moment	*unilateral:* fault propagating in only one direction, usually from one end of the fault	*unidirectional:* one component of slip only, e.g. only strike slip or only dip slip
Double couple (double dipole): two couples of forces, perpendicular to each other, generally each with moments but with zero net moment	*bilateral:*[2] fault propagating in two opposite directions	*bidirectional:*[2] two components of slip motion, i.e. both strike slip and dip slip

[1] All combinations, in total eight, of the items listed in the table are possible
[2] There is some confusion in the literature in the use of these terms. For example, I have seen bidirectional sometimes used in the sense of bilateral, as defined here.

which gives the time dependence of the source action, as for example a step function, a build-up step, a pulse, etc. Recent extensive reviews of source mechanism studies by means of seismic records, both from theoretical and observational points of view, have been given by Ben-Menahem and Singh (1972) and Khattri (1973). In principle, comparative studies of mechanisms derived in different ways (different methods applied to all kinds of waves, both body and surface waves) are of great significance, both to place mechanism studies on a reliable basis and to elucidate wave generation and wave propagation.

8.1.2 Equalization methods—surface and body waves

Equalization methods, as outlined in section 6.5, based on reduction of amplitudes and phases to the source or to a fixed distance from the source, play a great role in source mechanism studies. Such equalization was earlier mostly done in the time domain, but with the more general application of spectral methods, the procedure has been shifted over to the frequency domain. The theoretical discussions underlying amplitude and phase equalization have been presented in section 6.5. In the present context we are concerned with application to studies of source mechanism and source time functions, whereas corresponding structural studies were dealt with in Chapter 7. There is no need to repeat the theoretical basis here, and it may suffice to give examples from the literature with some comments. Table XL summarizes pertinent source studies in the frequency domain.

Phase equalization was applied by Aki (1960b, c, d) already at an early stage in the newer development of source studies. Phase equalization means adjusting the observed phases at one station to those at another station, correcting for distance and instrumental phase shifts, or else reducing phases at all stations to a common near-source distance. The phase characteristics of Fourier compo-

TABLE XL

Examples of observational source mechanism studies, based on various kinds of spectral equalization

Method	Body waves	Surface waves
Directivity function	Ben-Menahem et al. (1965) Hirasawa and Stauder (1965) Schick (1968) Khattri (1972)	Press et al. (1961) Ben-Menahem and Toksöz (1962, 1963a, b) Filson and McEvilly (1967) Schick (1968) Niazi (1969) Udías (1969) Udías and López Arroyo (1970) Udías (1971) Ben-Menahem and Rosenman (1972) Ben-Menahem et al. (1972) Canitez (1972) Furumoto (1972) López Arroyo and Udías (1972) Rodriguez-Portugal and Udías (1972)
Differential phase	Khattri (1972)	Press et al. (1961) Ben-Menahem and Toksöz (1963a, b)
Absolute amplitude equalized to source or fixed distance	Pollack (1963b) Hirasawa and Stauder (1964) Ben-Menahem et al. (1965) Hirasawa (1965) Teng and Ben-Menahem (1965) Berckhemer and Jacob (1968) Davies and Smith (1968) Mino et al. (1968) Khattri (1969b) Chandra (1970b, c) Randall and Knopoff (1970) Fukao (1972) Hanks and Wyss (1972) Maasha and Molnar (1972) Thatcher (1972) Wyss and Hanks (1972)	Brune (1960) Chander and Brune (1965) Wu and Ben-Menahem (1965) Aki (1966a, b) Wu (1968) Tsai and Aki (1969, 1970a) Udías (1969) Kanamori (1970a, b, 1971a, b) Udías and López Arroyo (1970) Ben-Menahem and Aboodi (1971) Mendiguren (1971) Abe (1972a, b, c) Ben-Menahem and Rosenman (1972) Ben-Menahem et al. (1972) Lambert et al. (1972) Maasha and Molnar (1972) Rodriguez-Portugal and Udías (1972) Sudo (1972) Massé et al. (1973) Mitchell (1973)

TABLE XL (continued)

Method	Body waves	Surface waves
Absolute phase equalized to source (initial phase) or to fixed distance	Ben-Menahem et al. (1965) Teng and Ben-Menahem (1965) Berckhemer and Jacob (1968)	Brune (1960) Aki (1962, 1966a) Ben-Menahem and Toksöz (1962, 1963a, b) Chander and Brune (1965) Odaka and Usami (1970) Udías and López Arroyo (1970) Ben-Menahem (1971) Ben-Menahem and Rosenman (1972) Ben-Menahem et al. (1972)
Relative amplitude[1] (equalized to source)	Hirasawa and Stauder (1964)	Aki (1960b, c, d; 1964a) Ben-Menahem (1971) Marshall and Burton (1971) Lambert et al. (1972)
Relative phase[1] (equalized to source)		Aki (1960b, c, d; 1964a) Press et al. (1961)

[1] *Relative* implies comparison of (a) different waves at the same station; or of (b) different components of the same wave at the same station.

nents can in general be determined with higher accuracy than the amplitude characteristics. Uncertainties, especially of phase velocities, limit the accuracy obtainable in equalization of phases (Aki, 1962), whereas uncertainties in attenuation limit the accuracy in amplitude equalization methods.

In several studies, the spectra, equalized to some fixed distance, are inverted to yield the corresponding records, and these are displayed in dependence of azimuth from the source. This is equivalent to a deconvolution of the given records. The deconvolved records are used for comparison with synthetic waveforms, calculated for assumed source models.

Besides equalization of absolute amplitudes and phases, we have also to consider equalization of relative quantities, i.e. of amplitude ratios and/or phase differences, as listed in Table XL. One notable example is Aki (1960b, c, d; 1964a), who used amplitude ratios and phase differences between Love and Rayleigh waves, reduced to the source, and compared the distributions with those derived theoretically for independently deduced faulting mechanisms. As the observed phase (Φ_X) is the sum of the initial phase (Φ_S), the phase shift during propagation (Φ_P) and that due to the seismograph (Φ_I), we have from sections 6.5 and 7.2:

$$\Phi_X = \Phi_S + \Phi_P + \Phi_I + 2n\pi \qquad [1]$$

and the initial phase difference between Rayleigh (R) and Love waves (L) becomes (assuming n identical in the two cases):

$$\Phi_{SR} - \Phi_{SL} = \Phi_{XR} - \Phi_{XL} - (\Phi_{PR} - \Phi_{PL}) = \Phi_{R} - \Phi_{L} + \omega\Delta\left(\frac{1}{c_{R}} - \frac{1}{c_{L}}\right) \quad [2]$$

For the determination of the phase difference $\Phi_R - \Phi_L$ at the station, Aki (1964a) discusses three methods: (1) from Fourier spectra in a straight-forward manner; (2) by the stationary phase method (cf. Båth, 1968, pp.43–49, especially p.47); and (3) by band-pass filtering and correlation analysis.

The advantage of Aki's (1964a) method is that in the amplitude ratios and phase differences, thus formed, the source time function and the source finiteness are eliminated, being the same for R and L, whereas the source spatial function (force system) is kept, as having different effects on R and L. Compared to the methods of Ben-Menahem (1961), this well illustrates different procedures in equalization. In any equalization method one tries to eliminate all factors except just one, which is subject to the study.

A similar strategy as Aki (1964a) used, was applied by Hirasawa and Stauder (1964), i.e. comparison of observed spectral radiation patterns with theoretical distribution for an earthquake with known fault mechanism, determined independently. The differences between these two papers are that Hirasawa and Stauder (1964) used only body waves (P, S, ScS) and only amplitude spectra, and finally they took the ratio between two components E/N of the same wave (S and ScS) as a method of equalization. The S/P spectral ratio yielded a measure of the source volume.

The alternative equalization method of reducing amplitudes and phases back to the source for comparison with theoretical radiation patterns has been developed by Ben-Menahem et al. (1965) to long-period P- and S-waves from deep earthquakes. Correction for crust (by Haskell's matrix method), instrumentation and mantle effects (absorption, geometrical spreading) are dealt with in detail. The crustal correction affects especially shorter periods, and these cannot be used unless the crustal structure under the station is accurately known. In order to correctly retrieve the initial motion it is necessary to include not only the direct P- and S-waves in the record but also converted waves (P to S, S to P) as well as all reverberations in the receiver crust. This calls for time windows of about 45 sec length, and, in order to exclude other phases (pP, etc.), the method was restricted to deep events.

Teng and Ben-Menahem (1965) applied this method to P (and S) in the period range of 20–100 sec recorded at numerous stations well distributed in azimuth around a deep earthquake (h = 350 km) in the Banda Sea on March 21, 1964. Comparison with theoretical models indicated a double-couple source as the best model. A source displacement-time function in the form of a build-up step was found to suit the equalized spectra best, also confirmed by Love and Rayleigh waves, and formally agreeing with Kasahara (1958) and Brune (1970):

$$s(t) = c(1 - e^{-at}); \qquad t \geq 0 \qquad [3]$$

c and a are positive constants, $a > 0.1 \text{ sec}^{-1}$. The source space function was found to be independent of frequency, at least up to 0.1 cps (down to 10 sec period).

The same computational procedure was applied by Berckhemer and Jacob (1968) in a study of fault parameters of 17 deep-focus earthquakes, and by Khattri (1969b) in a study of a deep earthquake in Brazil (November 3, 1965, h = 593 km).

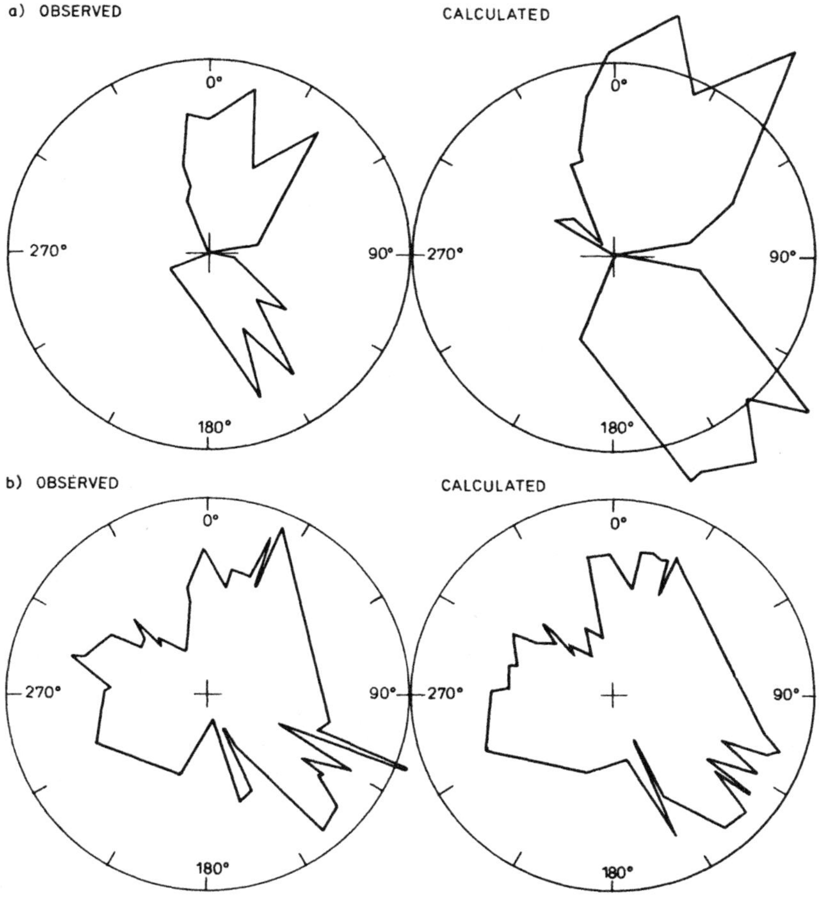

Fig.78. Observed versus theoretical radiation patterns for P-waves. a) Banda Sea earthquake of March 21, 1964. After Teng and Ben-Menahem (1965). b) Sinkiang earthquake of November 13, 1965. After Chandra (1970c).

Chander and Brune (1965) applied amplitude and phase equalization to the source for mantle Rayleigh waves from an intermediate-depth Hindu Kush earthquake. Limitations in the experimental accuracy are due to insufficient knowledge of regional phase velocities and of Rayleigh-wave absorption and to inaccuracies in instrumental characteristics. Love waves and P- and S-waves are also used in a combined study.

Assuming double-couple point sources and applying the amplitude-equalization method, Chandra (1970b) demonstrated for a number of deep earthquakes good agreement among mechanisms derived from P-wave spectra, S-wave polarization and P-wave first motion. Theoretical models were compared with observations and varied until best agreement was obtained. In case of P-wave spectra, theoretical P-wave amplitude distributions were compared with observed distributions of different components (different periods) of the amplitude spectra, after these had been equalized to the source. See also Chandra (1970c) and Fig.78. By similar methods, Randall and Knopoff (1970) found evidence for dominating double-couple source mechanism for deep shocks, a result of interest against the background of discussions on deep foci as possibly different from shallow ones. From body-wave amplitude spectra of South American deep shocks, Linde and Sacks (1972) confirmed the existence of dominant periods in the spectra, which are larger for S than for P and which increase with earthquake magnitude. Cf. section 8.3.3. The source volume dimension, assumed spherical, is calculated from dominant periods. By application of Brune's (1970) theory to the shape of observed equalized amplitude spectra of SH, Thatcher (1972) was able to infer differences in source properties (seismic moment, source dimension, stress drop) between two localities.

Amplitude and phase equalization of free oscillations of the earth may likewise give source information (Abe et al., 1970; Ben-Menahem, 1971), even though in the present state of the art, such efforts seem to be at least as much informative on the free oscillations as on source mechanism.

In general, equalization methods are based upon elimination of several parameters, in the ideal case leaving only one parameter for study and interpretation. Among various such methods, we like to mention also an amplitude-equalization method, called the "common-path method", in which we keep epicenter, path and station constant. From the general equations [56] and [57] in Chapter 6, it is then clear that the following quantity:

$$S(\omega) \cdot C_s(\omega,h)/F(\omega)$$

(h = focal depth) will be constant for a constant frequency. Therefore, from observed amplitude spectra $F(\omega)$ and known values of the constant, e.g. from other events fulfilling the requirements for unchanged epicenter, path and station, we can calculate the product $S(\omega) \cdot C_s(\omega,h)$, the so-called source-layering function, and display the same as spectra. The method promises interesting comparisons to be made between earthquakes, atmospheric and underground explosions (Marshall and Burton, 1971).

The progress on the observational side has been paralleled by corresponding *theoretical developments.* This has in fact been necessary, to provide further theoretical base with which observational results could be compared, in order to deduce source parameters. Among quite a number of such theoretical studies, I like to mention especially Ben-Menahem and Harkrider (1964). They deduced theoretical radiation patterns and spectra for Love and Rayleigh waves in the period range 50–350 sec for various source models (single couple, double couple, arbitrary dip and slip angle), embedded at various depths (to over 600 km) in a realistic flat-earth model (Gutenberg continental earth model). Ben-Menahem (1964) further extended these results to ultra-long surface waves in a spherical earth.

On the basis of the excitation theory of normal modes in laterally homogeneous layered media, Tsai and Aki (1970b) demonstrated the efficient use which can be made of Rayleigh- and Love-wave amplitude spectra (in the period range 10–50 sec) for source studies, including focal depth, even when source finiteness is not of importance. Cf. further section 8.4.4.

A relation between the ruptured fault area and the shape of seismic wave spectra was formulated by Berckhemer (1962). Wu and Ben-Menahem (1965) compared the azimuthal distribution of amplitude mantle surface-wave spectra (reduced to a common distance) of the Iran earthquake of September 1, 1962, $M = 7.0$, with theoretical results by Ben-Menahem and Harkrider (1964). The best fit yielded information on source nature (double couple) and source geometry (strike, dip, slip angle, depth), which agrees closely with results from first P-wave motions. The spectral approach is more efficient than the corresponding time-domain method.

Comparisons between theoretical and observational radiation patterns for deducing source parameters (considerably facilitated by Ben-Menahem et al., 1970) may of course lead to a question of uniqueness, i.e. if an observed pattern can be explained by more than one model. Usually, when a larger number of stations or waves are used, the number of restrictions are also so large, that solutions can be considered as unique. Global application of spectral methods to source studies is envisaged by Ben-Menahem et al. (1968), which is a welcome remedy to the circumstance that modern, spectral methods have so far been applied only to few earthquakes, mostly for method testing.

8.1.3 Finite source methods—surface waves

Spectral studies have been applied to earthquake mechanism studies especially after the introduction of the concept of extended faults with a finite rupture velocity. After some early, more primitive notions of finite sources, usually of spherical shape, a more realistic picture, the *moving-source model*, was introduced in the early 1960's. According to this model, the fault first breaks at one point, usually near one of its ends, and from this point the rupture propagates with a finite velocity, often to the other end of the fault or as far as there is any strain energy

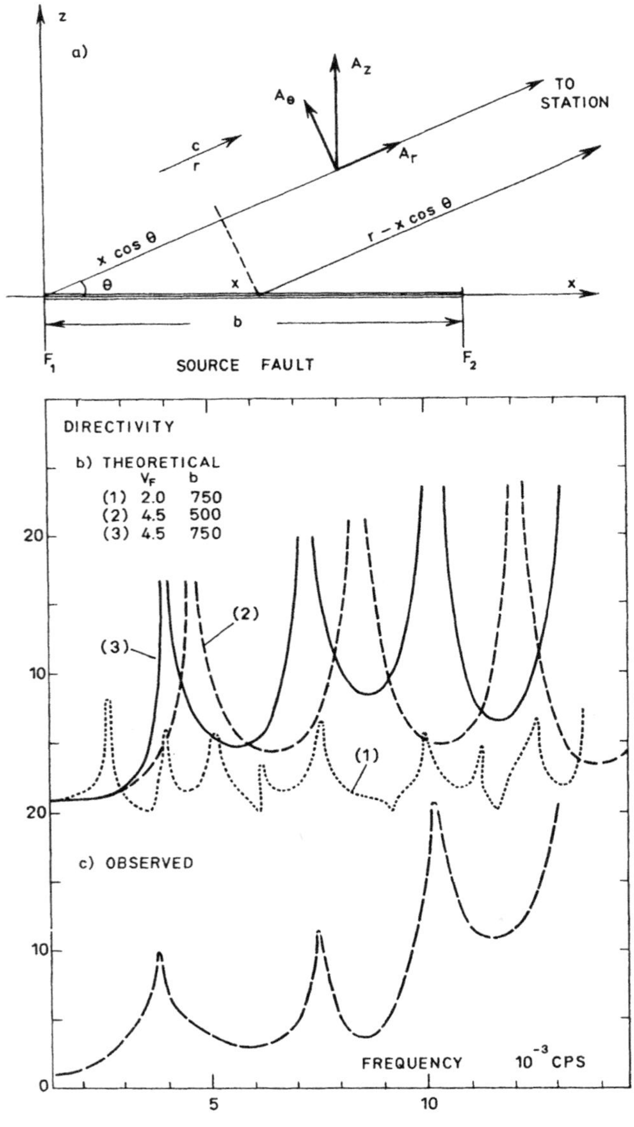

Fig.79. Finite moving source. a) Simple sketch of fault F_1F_2 and radiated waves. b) Theoretical and c) Observed directivity functions for Pasadena R2/R1 from Chilean earthquake in May, 1960. Fig. b) and c) after Press et al. (1961).

to release. The velocity with which the fault propagates is usually referred to as *rupture velocity* or *fault propagation velocity*, and the length over which the motion extends is termed *fault length*. Both surface-wave spectra and body-wave spectra have been applied to studies of this mechanism.

In parallel with theoretical works by Knopoff and Gilbert (1959), Savage (1965) and others, Ben-Menahem (1961) developed a theory for finite moving sources, which became very significant especially for its application to observations. This theory emphasizes the importance of fault dimension and rupture time for the wave radiation field, not the least its azimuthal variation, especially when these are of the same order of magnitude as wavelengths and wave periods. Visualizing the source as a point, or rather a line of points, moving across the fault plane with finite velocity, the far-field displacement spectra of Rayleigh and Love waves are derived for various source models (strike slip and dip slip, both vertical and of arbitrary dip). Ben-Menahem's (1961) paper is of fundamental importance, especially in initiating more intense observational studies of finite source mechanisms by spectral methods.

Fundamentals of finite source theory. At the receiving end, the displacement spectra for surface waves from a strike-slip fault are expressed in cylindrical coordinates for two-dimensional spreading as follows (see Fig.79a and Ben-Menahem and Toksöz, 1963a):

$$A(\omega) = \frac{1}{\sqrt{r}}\,|B(\theta)|\,\mathrm{e}^{i\Psi(\theta)}\,\frac{\sin X}{X}\,\mathrm{e}^{-iX}\,|S(\omega)|\,\mathrm{e}^{i\Phi_S(\omega)}\,\cdot$$

$$\cdot\, M(\omega,r)\exp i\left[\omega\left(t-\frac{r}{c}\right)+\frac{(2m+1)\pi}{4}\right] \qquad [4]$$

with component $A_\theta(\omega)$ for Love waves and components $A_r(\omega)$ and $A_z(\omega)$ for Rayleigh waves.

As in the general equations in section 6.5.2, we have factors that depend on source and on path properties. In [4] the *source function* is represented by the product of three factors:

(1) Spatial factor, $|B(\theta)|\,\mathrm{e}^{i\Psi(\theta)}$, which depends on the force system of the source, whether single force, a single couple or a double couple, etc.

(2) Propagation factor, $(\sin X/X)\,\mathrm{e}^{-iX}$, due to the finite fault propagation, to be explained in more detail shortly.

(3) Temporal factor, $|S(\omega)|\,\mathrm{e}^{i\Phi_s(\omega)}$, which is the Fourier transform of the source time function, whether a step function, an impulse, etc. The total phase angle depending on source time factors, i.e. $\Phi_S(\omega) - X$, is the *initial phase*. It plays a significant role in source mechanism studies. In case of a point source, $X = 0$, and the initial phase reduces to $\Phi_S(\omega)$.

The *effect of the path* is expressed by two factors:

(1) $1/\sqrt{r}$ which gives the geometrical spreading over a plane (cylindrical spreading). This should be replaced by $1/(r_0 \sin \Delta)^{1/2}$ (r_0 = earth's radius, Δ = angular distance) for spreading over a sphere (section 6.5.3).

(2) The factors

$$M(\omega,r) \exp i\left[\omega\left(t - \frac{r}{c}\right) + \frac{(2m+1)\pi}{4}\right]$$

which represent the amplitude response of the medium traversed (incl. source region, mantle, and receiver crust). Cf. section 7.2.1.

The factor $(\sin X/X)\, e^{-iX}$ represents the effect of the finite propagation along the fault at the source. This can be understood in a simple way as follows (see Fig.79a). For a point x on the fault, the emitted wave arrives at the receiver with a time delay of:

$$(x/V_F) + (r - x \cos \theta_0)/c$$

Therefore, the transfer function for the whole, unilateral fault is the following:

$$\frac{1}{b}\int_0^b \exp i\omega\left[t - \left(\frac{x}{V_F} + \frac{r - x\cos\theta_0}{c}\right)\right] dx = \frac{\sin X}{X}\, e^{-iX} \exp i\omega\left(t - \frac{r}{c}\right) \quad [5]$$

Analogous reasoning is applicable also to more complicated faulting, such as listed in Table XXXIX, only by appropriate extensions. For a vertical strike-slip fault, X is given by:

$$X = \frac{\omega b}{2c}\left(\frac{c}{V_F} - \cos\theta_0\right) = \frac{\pi v b}{c}\left(\frac{c}{V_F} - \cos\theta_0\right) = \frac{\pi b}{L}\left(\frac{c}{V_F} - \cos\theta_0\right) \quad [6]$$

For a point source, $b = 0$ and $b/V_F = 0$, we get $X = 0$ and $(\sin X/X)\, e^{-iX} = 1$, as expected. This makes [4] generally applicable, also for point sources as a special case. The same special case ($X = 0$) can happen for a propagating fault for $c = V_F \cos \theta_0$, and is approximated for $L \gg b$. Thus, for wavelengths much larger than the fault length, the source behaves effectively as a point source (Moskvina, 1971).

The propagation factor has its zeros at $X = n\pi$ for $n = 1, 2, \ldots$ which is the condition for destructive interference by superposition of waves radiated at time lags inversely proportional to the rupture velocity. At those frequencies for which $(\sin X)/X$ has its zeros, the phase X changes abruptly by π, due to the corresponding change of sign of $(\sin X)/X$. It should also be noted that spectral minima follow at regular, constant intervals of X, but not of ω, as $c = c(\omega)$ for dispersed surface waves, and then X is not a simple linear function of ω. The basic idea is not new to physics; for instance, a complete analogy exists between this phenomenon and diffraction in optics (see, for example, Crawford, 1965, p.485). The frequencies at which the spectral minima occur permit calculation of source extension (b) and rupture velocity (V_F).

Even though these minima would be expected in the spectra from a single station, the method leads to more reliable results, if all factors other than those depending on X are eliminated. This is accomplished in any of the following ways:

(1) Using records of the same wave from the same earthquake at two stations.

(2) Using records of different waves from the same earthquake at the same station.

(3) Using records of the same wave from different earthquakes at the same station.

In either case, there are two complementary methods, both depending on X, i.e.: (1) amplitude-ratio method (directivity-function method) and (2) phase-difference method (differential-phase method).

In the following paragraphs, we shall develop these approaches for a vertical strike-slip fault, which is the simplest case and fully sufficient to demonstrate principles involved. For other fault models, similar functions depending upon source finiteness appear, also with ensuing spectral minima, which depend on fault extension (b), rupture velocity (V_F), fault orientation (θ_0) in relation to station, and on fault dip.

Application of finite source theory. The amplitude spectral ratio at two stations symmetrically located with respect to the epicenter leads to the *directivity function* D (Ben-Menahem, 1961). For a strike-slip fault, D is given by the following expression for Rayleigh and Love waves:

$$D = \left|\frac{A(\theta_0)}{A(\pi + \theta_0)}\right| = \left|\frac{Y \sin X}{X \sin Y}\right| \qquad [7]$$

where $X = \dfrac{\pi b}{L}\left(\dfrac{c}{V_F} - \cos\theta_0\right)$ and $Y = \dfrac{\pi b}{L}\left(\dfrac{c}{V_F} + \cos\theta_0\right)$

This expression for D is valid for any component in [4] under the assumption that in choosing diametrically opposite stations all factors cancel except the propagation factor. As expressed in [7] for D, this depends only on amplitudes and not on phases. Conventionally, the numerator of D is taken to correspond to waves leaving the source in the direction of rupture.

From the definition of D in [7] we see that D will run through a sequence of minima and maxima as the frequency varies:

$$\left.\begin{aligned} &\text{Minima for } \frac{\pi b}{L_i}\left(\frac{c_i}{V_\mathrm{F}} - \cos\theta_0\right) = n\pi \\ &\text{Maxima for } \frac{\pi b}{L_a}\left(\frac{c_a}{V_\mathrm{F}} + \cos\theta_0\right) = n\pi \end{aligned}\right\} n = 1, 2, \ldots \qquad [8]$$

Subscripts i and a denote minima and maxima, respectively (Fig.79b). To obtain fault parameters by use of the directivity function, observed curves are compared with theoretical spectral curves, calculated for trial V_F and b values, and the closest agreement is sought.

The directivity function D, as defined by Ben-Menahem (1961), has application only to strike-slip faults (vertical or slanting). For a vertical dip-slip fault, it is shown that $D = 1$, and for a slanting dip-slip fault, D is not defined. Later, three different versions of D formulation were given (Ben-Menahem and Toksöz, 1962), corresponding to variable fault-strength (exponentially decaying), bilateral faulting and variable rupture velocity.

Alternatively, it is possible to combine records of even and odd orders of mantle waves, recorded at one station (Press et al., 1961), by appropriate correction for path differences. On the basis of the spectral variability demonstrated by Aki (1966a), he expresses serious doubt on the reliability of the directivity function determined at single stations, and he suggests averaging of spectral densities over several stations in certain azimuth ranges from epicenter to get more stable results. Another more recent extension consists of a generalized directivity function, defined for any two stations, thus not only for diametrically opposite stations (Udias, 1971). This generalization provides for much greater amount of records which can be combined in a minimizing procedure, when amplitudes at compared stations have been reduced to the same epicentral distance.

If, on the other hand, we compare phases at stations diametrically opposite in relation to the epicenter, we find by the same argument that among the four phase factors in [4], only e^{-iX} is expected to lead to a phase difference. There will be a phase difference between these two stations because of the propagation factor and this phase difference is easily seen to be $(2\pi b/L)\cos\theta_0$. This is the *differential phase function* (Ben-Menahem, 1961). Its slope thus depends on b.

Summarizing, we have for two diametrically opposite stations:

(1) Directivity function: amplitude spectral ratio. Depends on source parameters b, V_F, θ_0 and wave parameters L, c.

(2) Differential phase function: phase difference. Depends on source parameters b, θ_0 and wave parameter L.

For given values of phase velocities and fault orientation (θ_0), the differential phase yields the fault length (b), and then the directivity gives the rupture velocity (V_F).

Some results of the method are listed in Table XLI. Values obtained for rupture

TABLE XLI

Examples of source parameter determination from directivity, differential phase functions and finiteness factors (arranged in order of decreasing fault length b)

Earthquake	Fault length b (km)	Rupture velocity V_F (km/sec)	Fault direction	Reference
From surface waves:				
Chile, May 22, 1960, $M = 8.3$[1]	1000	4.5		Press et al. (1961)
	1200	3.5		Wada et al. (1963)
Alaska, Mar 28, 1964, $M = 8.5$	650	3.0	N 45° E	Ben-Menahem et al. (1972)
	800	3.0	N 30° E	Furumoto (1972)
Kamchatka, Nov 4, 1952, $M = 8.4$	700	3.0	N 146° W	Ben-Menahem and Toksöz (1963a)
Mongolia, Dec 4, 1957, $M = 8.3$	560	3.5	N 100° E	Ben-Menahem and Toksöz (1962)
Aleutian Islands, Feb 4, 1965, $M = 8.1$	420	3.0	N 70° W	Ben-Menahem and Rosenman (1972)
Alaska, July 10, 1958, $M = 7.9$	350	3.5	N 40° W	Ben-Menahem and Toksöz (1963b)
Kurile Islands, Oct 13, 1963, $M = 8.1$	250	3.5	N 40° E	Ben-Menahem and Rosenman (1972)
Dasht-e Bayāz, Iran, Aug 31, 1968, $M = 7.4$	80	0.5	N 77° W	Niazi (1969)
	30	2.2		Filson and McEvilly (1967)
Parkfield, USA, June 28, 1966, $M = 6.0$	(40)	(1.3)	(N 33° W)	Wu (1968)
	37	2.2	N 43° W	Tsai and Aki (1969), Aki (1969)
From body waves:				
Banda Sea, Mar 21, 1964, $h = 350$ km, $m = 6.6$	35	5		Davies and Smith (1968)
Brazil, Nov 9, 1963, $h = 600$ km, $m = 7.1$	27	3.8		Davies and Smith (1968)[2]

[1] Magnitudes are from Duda (1965) and Uppsala and Kiruna records.

[2] Also studied by Khattri (1972) who interpreted the source as a bilateral, bidirectional fault, resulting in a fault length about 3 times the one given by Davies and Smith (1968).

velocities have attracted considerable interest in the literature. We notice that they are in the range of shear-wave velocities or lower. Also, we see from the expressions for X and D, that it is the ratio c/V_F that is of decisive importance, rather than the values of V_F alone (Aki, 1968). Well measurable fault extensions exist also for smaller earthquakes, as demonstrated by the same method by Schick (1968) for magnitudes below 4.

The values listed in Table XLI indicate an increase of rupture velocity V_F with increasing fault length b, suggesting that rupture time b/V_F may also be a parameter to consider. In fact, the logarithm of rupture time shows a roughly linear dependence on magnitude M, an approximate solution being:

$$\log (b/V_F) = 0.5\,M - 1.9 \qquad [9]$$

with a standard deviation of ± 0.2 for $\log(b/V_F)$. This formula shows good agreement, even numerically, with the dependence of dominant period T on magnitude M (section 8.3.3). This is not surprising since the larger the rupture time b/V_F, the larger dominant periods are to be expected. Formula [9] has been checked with some other determinations of b and V_F found in the literature, and fairly good agreement has been obtained at least with more reliable determinations (Ben-Menahem and Harkrider, 1964; Aki, 1966a and b; Kanamori, 1970b), including determinations by Schick (1968) for small-magnitude earthquakes. This extends the range of validity of [9] to approximately $2.5 \leq M \leq 8.5$.

Moreover, the theory implies an azimuthal variation in dominant periods, in analogy to the Doppler effect. This finds its expression through the azimuth-dependent factor of X, with the following consequence for unilateral faulting: (1) $\theta_0 = 0$: minimum period; (2) $\theta_0 = \pi$: maximum period.

This is also observed and useful for determination of fault orientation.

Geological survey, aftershock distribution pattern, initial P-wave motion, T-phase duration, and accelerograph records (Aki, 1968) are among the various auxiliary methods used to provide independent supporting evidence for the validity of the method. The method was also put to test by model experiments (Press et al. 1961). The situation is simulated by ultrasonic models using a circular plate and the equivalence of moving-source condition. Agreement between experiment and theory was found to be reasonably good. Further experimental studies are reported by Savage (1967).

Still *another method to effect equalization* employs the ratio of spectral surface-wave amplitudes between the main earthquake and its smallest, well recorded aftershock. The underlying idea is that the smallest aftershock is considered as a point source, in distinction to the main shock and the bigger aftershocks, which are assumed to have finite sources. Then, records at a given station from a given source would differ only by the source finiteness, which could then be determined. The method, suggested by Filson and McEvilly (1967) and applied by them to Love waves in the Parkfield, California, 1966 sequence, is a special case of the general theory, given above. For the point source we have $X = 0$ and $(\sin X/X)\, e^{-iX} = 1$; for the finite source shocks we have:

$$X = \frac{\omega b}{2c}\left(\frac{c}{V_F} + 1\right) \qquad [10]$$

as $\theta_0 = \pi$ in the special case of Filson and McEvilly (1967). Other factors being the same, except that amplitudes are proportional to fault length, the spectral amplitude ratio $G(\omega)/F(\omega)$ between any shock and the smallest one can be written down immediately. Results of the study are given in Table XLI.

The same technique was applied by Niazi (1969), except that in his case $\theta_0 = \pi/2$. With this value, his formulas also follow immediately from the general theory. However, it may be instructive also to follow the development especially adapted to such a case. Thus, we have:

Aftershock: point source $f(t) \leftrightarrow F(\omega)$, where

$$F(\omega) = \int_{t_1}^{t_2} f(t)\, e^{-i\omega t}\, dt \qquad [11]$$

Main shock: finite source $g(t) \leftrightarrow G(\omega)$, where

$$\left.\begin{aligned} g(t) &= \int_0^b f\left(t - \frac{x}{V_F}\right) dx \\ G(\omega) &= \int_0^b F(\omega)\, e^{-i\omega x/V_F}\, dx \end{aligned}\right| \qquad [12]$$

using the time-shifting theorem (section 2.3.4). By partial integration of the expression [12] for $G(\omega)$, we get:

$$\frac{G(\omega)}{F(\omega)} = \frac{iV_F}{\omega}\left(e^{-i\omega b/V_F} - 1\right) = \frac{b \sin(\omega b/2V_F)}{\omega b/2V_F}\, e^{-i\omega b/2V_F} \qquad [13]$$

from which we find:

$$|G(\omega)/F(\omega)| = \min \qquad \text{for } \omega b/2V_F = n\pi, \qquad n = 1, 2, 3, \ldots$$

Reading the frequencies ω for minimum ratios $|G(\omega)/F(\omega)|$, we get b/V_F. Knowing one of these, e.g. V_F, from other information, we can calculate the other.

8.1.4 Finite source methods—body waves

Following the original formulation for surface waves (Ben-Menahem, 1961), the finite source method was later extended in an analogous way to body waves (Ben-Menahem, 1962; Ben-Menahem et al., 1965; Hirasawa and Stauder, 1965). The effect of the source finiteness on body-wave spectra is given quite generally by factors of the form:

$$(\sin X/X)\,e^{-iX}\cdot(\sin Y/Y)\,e^{-iY} \qquad [14]$$

For *unilateral, bidirectional* faulting, X and Y are defined as follows (Khattri, 1969a):

$$\left.\begin{array}{l} X_{\mathrm{P,S}} \\ \\ Y_{\mathrm{P,S}} \end{array}\right\} = \frac{\omega b_{x,y}}{2V_{\mathrm{P,S}}}\left(\frac{V_{\mathrm{P,S}}}{V_{\mathrm{F},x,y}} - \cos\theta_{0,x,y}\right) \qquad [15]$$

where P, S refer to P- and S-waves, respectively, and where b, V_{F}, θ_0 are referred to two perpendicular axes (x, y), corresponding to the two components of the bidirectional motion. The finiteness factors in [14] have zero values for X, $Y = n\pi$, $n = 1, 2, 3, \ldots$ with simultaneous phase changes of X, Y by π. The periods of the minima are, for P-waves:

$$T_n = \frac{b}{n}\left(\frac{1}{V_{\mathrm{F}}} - \frac{1}{V_{\mathrm{P}}}\cos\theta_0\right) \qquad [16]$$

and for bidirectional faulting we have to differ between two sequences of minima,

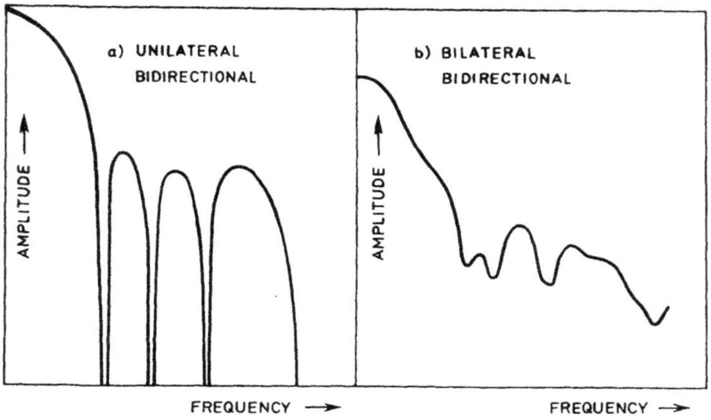

Fig.80. Amplitude spectra for the P-wave for a given bidirectional fault model. a) Unilateral. b) Bilateral. After Khattri (1969a).

corresponding to the two directions: one sequence along the x-axis with specified values of b, V_F, θ_0 and another along the y-axis with in general different values of b, V_F, θ_0. See Fig.80. Readings of minima permit determination of b and V_F. Khattri (1969a) applied this method to P- and S-waves from a deep earthquake in Brazil (November 3, 1965, $h = 593$ km).

For $n = 1$, the expression for T_n gives the total time delay. Considering [16] as a linear relation between T_n and $\cos\theta_0$, we find b and V_F from plots of T_n versus $\cos\theta_0$ for numerous stations. Obviously, b/V_F is the intercept and b/V_P is the slope of the straight line. This corresponds to the method used in the time domain by Bollinger (1968, 1970).

In addition to amplitude spectra, also phase spectra are able to furnish valuable source information, by virtue of the fact that minima ("holes") in amplitude spectra correspond to jumps of π radians in the phase spectra. The directivity function and the differential phase function are important tools for determination of fault length and rupture velocity, just as for surface waves. The power of the method derives from the fact that these functions are independent of a number of parameters, as they are ratios and differences, respectively. An inconvenience of the directivity function is the requirement of having records from stations diametrically opposite to the epicenter. As a remedy of this circumstance, Schick (1970) defined a related directivity function by forming the spectral ratio S/P:

$$D = \left|\frac{A_S}{A_P}\right| = \frac{Y\sin X}{X\sin Y} \qquad [17]$$

where

$$\left.\begin{array}{c} X \\ \\ Y \end{array}\right\} = \frac{\omega b}{2V_{S,P}}\left(\frac{V_{S,P}}{V_F} - \cos\theta_0\right)$$

θ_0, V_P and V_S are obtained from fault-plane solutions and travel-time tables. Comparison with empirically obtained D as function of ω and model graphs is used in efforts to determine b and V_F. Ambiguities may easily enter in such comparisons. To the formula [17] for D it has to be remarked that equal attenuation is assumed for P and S. For unequal attenuation a correction factor has to be entered in the expression for D.

For *bilateral, bidirectional* fault models, more complicated expressions are obtained. Unlike the unilateral, bidirectional fault models, the spectra do not have zero values. However, minima points (of non-zero values) do occur which reflect the source parameters. See Fig.80b. Cf. also Moskvina (1971), Khattri (1972). In general, interpretation in terms of source parameters becomes much more difficult, and Khattri (1972) found P to be more useful than S, because of larger variations in the directivity function for P.

By means of amplitude and phase spectra of long-period P-waves for a number of earthquakes and stations, Kishimoto (1964) made a comparative study of spectral shapes with regard to possible effects of epicentral distance, azimuth, station location and depth of foci. Distance effects were generally negligible, whereas some azimuth effects were found which could be explained by the finite source factor.

Starting from the moving-source concept, invoking the dependence of

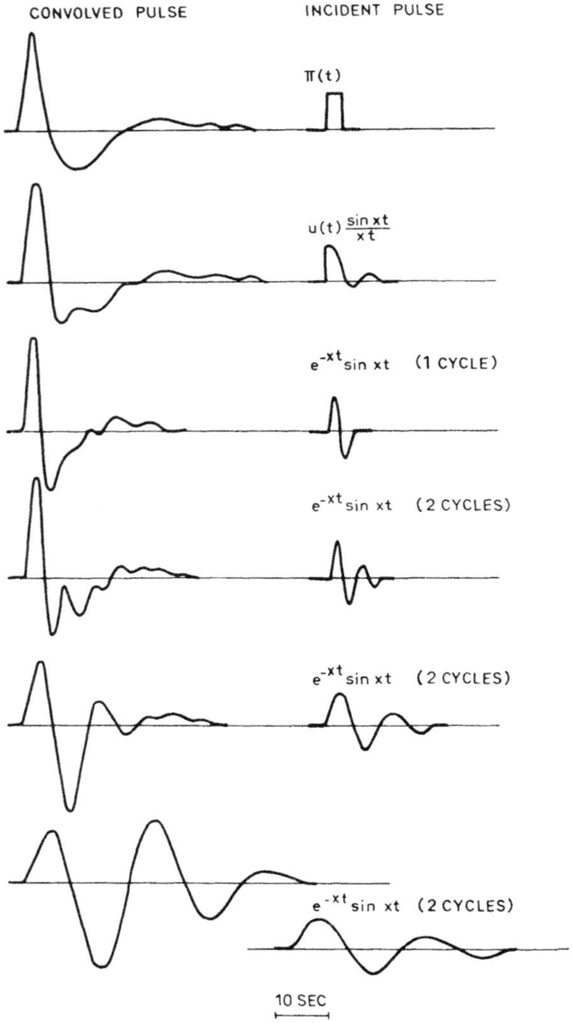

Fig.81. Theoretical P-wave records (left) for various input signals (right), calculated with a system function of Ben-Menahem et al. (1965). Note how an increase of the rupture time (b/V_F) leads to longer dominant periods in the records. After Bollinger (1968).

dominant periods (time duration of recorded P) on rupture time, Bollinger (1968, 1970) was able to get information about source properties (fault length and rupture velocity) by comparing theoretical and observed long-period P-waves. Theoretical P-waveforms were deduced from assumed source time functions by convolving these with the system function (transmission and receiver). See Fig.81. His observations support the general increase of b/V_F with M, expressed in [9], only with some modification of the numerical constants. This research was done entirely in the time domain and constitutes a revival of waveform analysis (section 1.3), but contrary to early descriptive classification of waveforms, his study primarily attempts explanations of waveforms in terms of source parameters.

The oscillatory character of the finiteness factor suggests the application of cepstral techniques (section 3.6.3). The displacement spectrum is proportional to:

$$\frac{\sin X}{X} e^{-iX} = \frac{\sin \omega T/2}{\omega T/2} e^{-i(\omega T/2)} \qquad [18]$$

where T is called the *source duration*, and is given by the expressions for X. The corresponding velocity spectrum is obtained by the derivation theorem ($f'(t) \leftrightarrow i\,\omega\,F(\omega)$, section 2.3.6) as proportional to:

$$\frac{1}{T}(1 - e^{-i\omega T}) \qquad [19]$$

Taking the real part of the cepstrum of the oscillatory portion of [19], we get:

$$\mathrm{Re}\,\bar{f}(\bar{t}) = \mathrm{Re}\int_0^{\Omega} e^{-i\omega t}\, e^{-i\omega \bar{t}}\, d\omega = \frac{\sin \Omega(T + \bar{t})}{T + \bar{t}} \qquad [20]$$

$$(0 \leq \omega \leq \Omega)$$

Obviously, a maximum in the cepstrum will occur at $-\bar{t} = +T$, which determines T. A method following essentially these principles was developed and applied by Davies and Smith (1968) to long-period P-wave spectra equalized to the source; see Table XLI for some results.

The finite source model with accompanying spectral interpretations no doubt means an important step forward in the study of source properties. The finiteness factor of the source leads to a sequence of minima in the spectra both of surface waves and of body waves, from which values of fault length b and fault propagation velocity V_F can be deduced. The idea is based on a continuously moving point source. However, the notions of constant rupture velocity and of constant energy emission during rupture are no doubt simplifications. In connection with an experimental test of the directivity function, Savage and Hasegawa (1965) point out the significance of accelerated motion, which may be of importance also in nature. An accelerated motion would naturally disturb the regular appearance of minima. In an experimental verification of the moving-source theory for S-waves, Savage (1967) gives a modified explanation in terms of interference between initial and stopping phases. Moreover, a sequence of spectral minima could also be obtained for two or more discrete events, occurring in the same or slightly different points with small time shifts relative to each other. The spectral minima do then depend on these time and space shifts. Considering the common occurrence of multiple P-waves, especially in records of larger earthquakes, testifying to discontinuous, rather than continuous source processes, there is no doubt that this alternative solution deserves careful consideration. Cf. Pilant and Knopoff (1964) and Wu (1968) and see further section 8.4.3 on interference analysis. Nagamune (1971) generalized the theory to a series of sources, discrete both in time and space and all of different strength and applied this theory to observations of superimposed surface-wave trains from large earthquakes. He was able to demonstrate multiple sources, especially for the Chilean earthquake of 1960, yielding $b = 856$ km and $V_F = 4.9$ km/sec. However, multiple-path effects, which no doubt also exist, may enter as significant disturbances. Naturally, the different viewpoints are closely related, and in fact the finite moving point source model could be considered as an averaging idealization of a discontinuous point source or of an accelerating point source.

8.1.5 Corner-frequency method

The spectral studies for finite sources so far described in this chapter have been essentially concerned with readings of frequencies of minima for directivity functions (both their modulus and their phase). However, also other spectral parameters may provide useful source information. Starting from the general expressions given above, we can construct an idealized spectrum, i.e., one in which all factors except the source propagation factor are eliminated. Thus, we make an amplitude–frequency spectrum in a log-log plot, in which the amplitude A is proportional to $|\sin X/X|$ and X is proportional to frequency ω (see Fig.82). We see that this idealized spectrum exhibits a flat portion for low frequencies and a certain slope for high frequencies. This shape agrees with observed spectra, after due corrections have been made as mentioned. It is natural that the level of the flat portion depends on released energy, magnitude, seismic moment or stress drop. Also, the "corner frequency" ω_0, i.e., where the curve breaks, provides a method to calculate fault extension. In Fig.82c $\omega = \omega_0$ at $X = 1$, i.e., on the basis of the finite moving source model we get:

$$b = \frac{2V_{P,S}}{\omega_0\,(V_{P,S}/V_F - \cos\theta_0)} \qquad [21]$$

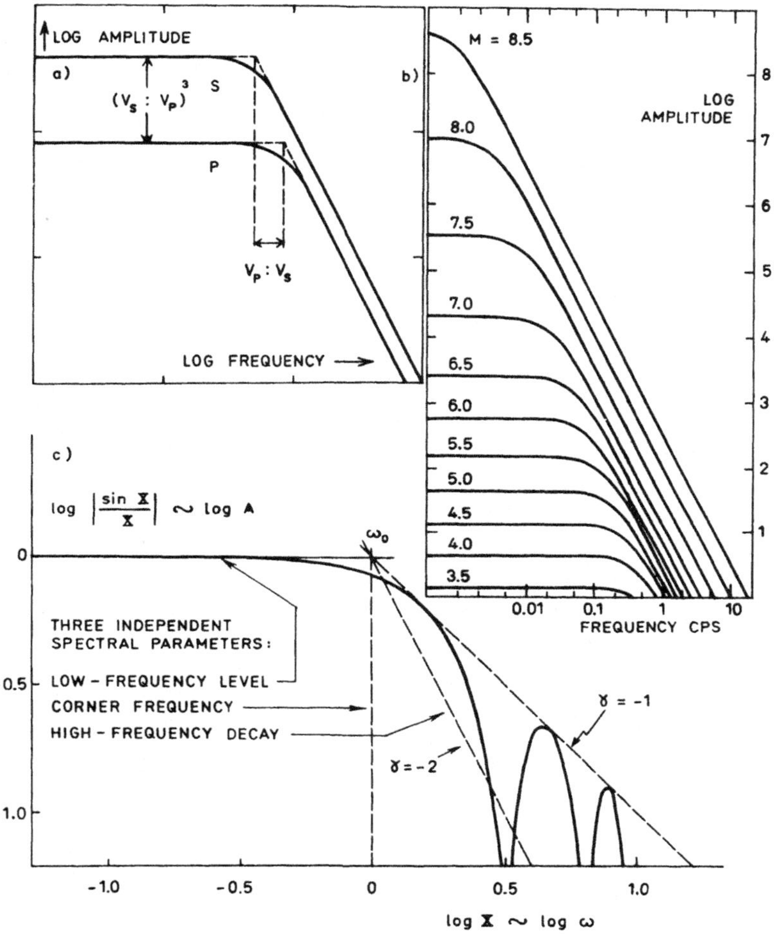

Fig.82. Idealized body-wave spectra. a) After Hanks and Wyss (1972). b) ω-square model after Aki (1967). c) Model calculated from moving-source theory.

After observed spectra have been corrected for instrumental response, radiation pattern, attenuation, geometrical spreading and effects of the crust and free surface, it is thus possible to calculate the seismic moment and the source dimension from readings of the spectral level of the flat portion and the corner frequency, respectively. The seismic moment is proportional to the amplitude spectral level of the low-frequency part and the fault area is inversely proportional to the square of the corner frequency. A full development of the method, with other starting points than here, is given by Brune (1970, 1971).

Hanks and Wyss (1972) and Wyss and Hanks (1972) applied it to four earthquakes and compared their results with field data. There may be some

TABLE XLII

Source parameters of the San Fernando earthquake, February 9, 1971, according to two different sources

Parameter	Wyss and Hanks (1972)	Canitez and Toksöz (1972)
Moment, dyne cm	$7 \cdot 10^{25}$	$16.4 \cdot 10^{25}$
Fault area, km^2	570	196
Dislocation, cm	45	280
Stress drop, bar	14	70

uncertainty in fixing the low-frequency level as well as the corner frequency from observed spectra, but still good agreement is reached. However, comparison between apparently independent determinations for the San Fernando earthquake (Table XLII) reveals the amount of agreement (or the lack of it) between results of different data and different methods. One cannot get away from the impression that further efforts to possibly increase accuracy would not be a waste. Maasha and Molnar (1972) applied Brune's (1970, 1971) method both to body waves (P and S) and to surface waves from some African earthquakes; Douglas and Ryall (1972) applied the method to S-waves from microearthquakes which incidentally permitted calculation of source properties for shocks down to a magnitude $M = 1.0$; and Trifunac (1972) applied the method to the Imperial Valley earthquake of 1940, finding good agreement with field data. Further applications were made by Molnar and Wyss (1972) and Wyss and Molnar (1972) to earthquakes in the Tonga–Kermadec area.

As demonstrated by Hanks and Thatcher (1972), the three independent spectral parameters indicated in Fig.82 are sufficient to calculate the source properties already mentioned. A comparatively extensive study by Berckhemer and Jacob (1968) is also related to this technique in the way that they determine the fault area (and connected quantities) from the level of the low-frequency part of the P-wave spectrum, with application to records of deep-focus earthquakes.

From certain theoretical considerations and experimental results, Hasegawa (1972) advocates the existence of a spectral peak in the low-frequency range of teleseismic P, rather than a flat spectrum, as given above. Moreover, he provides evidence that the frequency of the peak is an order of magnitude less for earthquakes than for underground explosions. Also, he shows that the peak frequency for earthquakes depends primarily on the focal depth, but for explosions mainly on the source function.

8.1.6 Special applications

It is quite natural that the propagating-fault mechanism has been applied mainly to larger earthquakes. It is only in exceptional cases with good observational coverage that it has been possible to apply this model to earthquakes of smaller magnitude. One example is provided by the Parkfield earthquake of June 28, 1966, which had a magnitude of $M = 6.0$ according to Uppsala and Kiruna records (Table XLI). However, spectral analysis techniques based on other models have found various applications also to the study of waves from small earthquakes. Recent installation of high-sensitive long-period seismographs has contributed efficiently to this development.

Microearthquakes, i.e. earthquakes with $M < 3$, have recently received considerably increased attention for their supposed ability to furnish much more detailed information on the stress and strain conditions in the earth and to reproduce secular statistics within relatively short time spans.

Due to their low energy level, special high-magnification seismographs are generally used arranged in dense networks for their observation. Array techniques have often been used with advantage in the evaluations. Aki et al. (1958) and Aki and Tsujiura (1959) used a tripartite station for near earthquake location. They used correlation coefficients corresponding to phase changes as functions of frequency and time delay. From knowledge of station separations, apparent velocities are computed. Special, low-noise recording sites, frequently underground, have often to be used, because of the low magnitudes of the events. In spectral studies of microearthquakes from underground recordings down to 800 m depth in Japan, Takano and Hagiwara (1966, 1968) found signal/noise ratios several times as large as on the surface.

From particle-velocity spectra of microearthquakes recorded in Pakistan, DeNoyer (1964) found that their maxima fall in the general range of 10–50 cps. Such high frequencies, observed also by others, preclude the study of microearthquakes except by special equipment with high-frequency response. The spectra could be grouped into two categories: those with one sharp peak corresponding to a relatively long time duration of source activity, and those with a more extended spectrum corresponding to a short time duration (in agreement with the rules in section 2.4.4).

From sonagram analyses (section 1.6.2) in the range of 0–25 cps of Matsushiro microearthquakes, Srivastava et al. (1971) found a persistent spectral maximum at 5 cps both at the epicenter and at 20 km distance for magnitudes ≥ 1.0. In an extensive spectral study of microearthquakes recorded at Mt. Tsukuba, Japan, Terashima (1968), using both sonagram and Fourier transform methods, found a predominance of higher frequencies for shorter epicentral distances and for smaller magnitudes (cf. section 8.3.3). A detailed study of microearthquakes in Japan is reported by Allam (1970) with special regard to the influence of local structure on the recordings.

Aki (1969) developed a method for determination of seismic moment for small, local earthquakes from the spectrum of the coda, assumed to consist of surface waves. The method is based on a statistical treatment of the coda waves as originating from scattering at uniformly distributed discrete scatterers, such that the coda power spectrum is independent of distance and wave path and depends only on the time lapse since the origin time. This implies a random process, non-stationary in time but stationary in space. The method is applied to Parkfield (1966) aftershocks. A comparison with the Matsushiro swarm (1965–1969) was given by Takano (1971b). The source functions of the Matsushiro swarm earthquakes were investigated by Irikura et al. (1971) by comparing velocity spectra, deduced from accelerograms, with theoretical spectra, based on certain source models.

The controversial question whether the earth tides can release seismic energy has been debated for many years. From a detailed study of microearthquakes in California, Ryall et al. (1968) found evidence of maximum number of earthquakes at gravity maxima, using both cross-correlation and power-spectrum techniques. The probable role of the earth tides has recently got positive confirmation from some observations of moonquakes. A maximum in their number has been found at maximum moon tides. For the moon, however, the question remains whether the tides act as moonquake generators or simply as moonquake triggers.

Earthquake swarms are in many respects related to microearthquakes, as swarms are to their greatest extent built up by microearthquakes. Earthquake swarms are generally encountered in volcanic regions, either of present or of past volcanic activity. The shocks constituting an earthquake swarm are generally recognized as volcanic, i.e., their genesis is more directly connected with volcanic phenomena than the case is for ordinary tectonic earthquakes. Spectral analysis of records of volcanic tremors together with other investigations can yield valuable information on the lava motion, both its approximate locality, motion pattern and intensity. It has been found that more long-period seismic waves (periods of 2 sec and more) have deeper origin under the volcano, whereas short-period waves (periods less than 1 sec) are connected with eruptions and similar superficial phenomena. Spectra which vary with time, indicate corresponding lava motions and other volcanic processes going on. Examples of such investigations are given by Kubotera (1963) for the Aso volcano in Japan; Shimozuru et al. (1966, 1969) for the Merapi volcano in Indonesia and the Kilauea volcano on Hawaii; and Y. Tanaka (1969) for the Mihara-yama volcano in Japan. The latter author distinguished four kinds of volcanic tremors, differing in predominant frequencies, amplitudes and wave types, as displayed by autocorrelations and power spectra. These four kinds of tremors correspond to various stages in the development of the volcanic activity.

Moving-window Fourier spectra (section 3.6.5 and 7.2.2) have proved useful in analyzing closely following shocks, i.e. multiple events (Trifunac and Brune, 1970).

Fore- and aftershocks. Several investigations have been made on fore- and aftershock spectra to test a hypothesis on the possible dependence of P-wave spectra on the stress conditions in the aftershock volume. According to this hypothesis, this volume is characterized by high stress before the main shock and much lower stress after it. And the high-stress condition would probably lead to higher frequencies in the radiated P-waves than the low-stress condition.

Thus, Gostev and Fedotov (1964) studied the spectra of P and S at near stations of fore- and aftershocks of the Kurile Islands earthquake of November 6, 1958 ($M = 8.0$). They find a difference, greater for S than for P, during the first 20 days after the main shock in the sense of a contraction of the aftershock spectrum and a slight shift towards lower frequencies. The effects are, however, small.

Similarly, Korkman (1968) investigated teleseismic P-wave spectra of fore- and aftershocks of the Aleutian Islands earthquake of February 4, 1965 ($M = 8.1$). No difference could be found, either because any such effect were too small to be discovered by such means or that it were totally absent. The frequencies of P-waves observed at teleseismic distances may be too low to exhibit such differences, which can be inferred from Suyehiro's (1968) work. On the other hand, clear regional patterns of the P-wave spectra were displayed, such that different areas exhibit each their characteristic features.

By local P-wave studies of the Matsushiro swarm earthquakes, Suyehiro (1968) found a clear preponderance of high frequencies (over 20 cps) before and of low frequencies after the swarm earthquakes. This effect was explained as due to fracturing of the crust by the shocks, which in turn causes increased scattering, especially of high-frequency waves.

8.2 EXPLOSION-EARTHQUAKE DISCRIMINATION

8.2.1 Spectral properties of explosion parameters

In addition to earthquakes also explosions have attracted very great attention as generators of seismic waves. Part of this is based on the need to be able to discriminate between earthquakes and nuclear explosions by means of teleseismic records. But seismology itself has also profited considerably by a profound study of explosion records, as explosions may provide seismology with accurately known source properties and with a mechanism that offers interesting comparisons with earthquake mechanisms.

Source parameters of explosions can be summarized as follows: (1) space parameters (source volume, depth of source, height of source above the earth's surface, etc.); (2) time parameters (source time function, etc.); (3) yield (magnitude, energy, etc.); and (4) physical and chemical parameters of the surrounding medium (atmosphere, water, underground, geological structure, coupling, etc.).

In any observed spectrum from an explosive source, all these factors in addi-

TABLE XLIII

Examples[1] of spectral evaluation of explosions[2]: $F \sim W^{\alpha} r^{-\beta} f(h,K,m)$

Reference	Type of spectra	F	α	β	$f(h, K, m)$	Type of explosion
Willis and Wilson (1962)	particle velocity: P and surface waves	particle velocity	(1.0)	2.6	decoupling	underground
Willis (1963a)	particle velocity: surface waves	particle velocity	0.70–1.03		ripple firing	underground
Willis (1963b)	particle velocity: P, S, surface waves	particle velocity			increases until a depth around 20 m is reached, then constant; frequencies essentially depth-independent	underwater
Berg and Papageorge (1964)	amplitude: P-waves	amplitude	0.64–1	1.0–1.9	increases with cavity radius	underground
Burkhardt (1964) Burkhardt et al. (1968)	amplitude: P, S	amplitude	0.68			underwater
O'Brien (1965)	amplitude: body and surface waves	amplitude	2/3 (body waves) 1/2 (surface waves)		shot-depth effects investigated	underwater
Gurvich (1967)	particle velocity: P-waves	particle velocity				underground
Hays (1969)	particle velocity: P, S, surface waves	particle velocity	0.59–1.00[3]	1.50–3.02[3]		underground
Power (1969)	power density, velocity, peak velocity: P, S, surface waves	power	1.0	3.55	increases with h	underground
Mueller and Murphy (1971)	pseudo relative velocity: all waves	amplitude	0.45–0.90[4]		increase of frequency and acceleration with h	underground (deep, contained)

[1] In addition, there are numerous useful papers on these problems not specially based on spectra, one notable example being Murphy and Lahoud (1969).

[2] F = quantity measured (amplitude, power, etc.), W = yield, r = distance, h = shot depth, K = coupling, m = miscellaneous source factors.

[3] Depending upon wave type.

[4] 0.45 for high frequencies, and 0.90 for low frequencies.

tion to the path effects are more or less interwoven, and skilled equalization techniques or similar methods are needed to isolate each single source factor. In the following we shall summarize some of the spectral results of explosion records. A summary is also given in Table XLIII.

Early observational and experimental investigations of explosion-generated P-wave spectra were published by Kasahara (1957), who found that dominant periods were only slightly increasing with distance but clearly increasing with yield. A source time function of a unit step and a spherical source were assumed for the theoretical source model. Kasahara (1960) uses the relation between S-wave spectra and azimuth to differentiate fault-origin from spherical-origin models. Further detailed spectral investigations of P, SV and SH generated by small explosions in soft media in Japan were made by Hattori (1972). The spectra exhibit dependence both on source properties, especially yield, and on path properties.

The dependence of amplitude spectra on explosive source conditions (yield, depth, emplacement) is important to know, partly in order to attain optimum conditions in seismic prospecting (among numerous articles on this subject, see for example Molotova, 1964; Gurvich, 1966a and b; 1967), partly to be able to take due account of source-dependent spectral maxima and minima and possibly to eliminate them in structural interpretations. Cf. also Weston (1960) and others.

Space parameters. The explosion as such is spherically symmetric, and this is also what would be observed for an explosion in a homogeneous, isotropic and unstressed medium. Actual observations, however, reveal that in the real earth, explosive source functions are a superposition of the explosive symmetric function and an asymmetric function, the latter depending on tectonic stress released by the explosion and asymmetric structures. This has been found by amplitude-equalization and initial-phase determination applied to spectra of Love and Rayleigh waves from underground nuclear explosions (Brune and Pomeroy, 1963; Aki, 1964b; Toksöz et al., 1965; Aki and Tsai, 1972). Asymmetry in the Rayleigh-wave radiation and the existence of Love waves in records from underground explosions testify to the combined source space function. Love waves have in some cases also been observed from atmospheric nuclear explosions.

The influence of the *depth* of the explosion on spectra will be reserved for a later section in this chapter (8.4), where a general discussion of depth effects is given. For atmospheric explosions, their *height* above the earth's surface may influence spectra. This is due to the fact that the transmission of energy from the explosion to the earth is executed by means of an atmospheric pressure wave. The higher the explosion occurs, the more will the pressure wave action on the ground be extended both in space and time. As a consequence, the higher the explosion occurs, the more will long-period components dominate in seismic wave spectra. This circumstance has been utilized for estimation of heights of atmospheric explosions (Båth, 1962).

Time parameters. The spectral effect of blast duration time was investigated experimentally by Frantti (1963b), who found variations in energy spectra, especially for duration times close to the dominant seismic periods.

Ripple firing or delay-time firing is a common method in quarry blasting. This will lead to destructive interference for some frequencies, depending upon the time and space properties of the firing technique. By suitable set-ups it is thus possible to largely delete any frequency. The method was investigated by Willis (1963a). On the theoretical side, these conditions could be dealt with by the interference analysis, outlined in section 8.4.3. Another advantage of this method, besides deleting undesired frequencies, is that generally more work in terms of broken rock is achieved for the same amount of explosive.

From a series of vertical-component particle-velocity spectra, Willis (1963c) demonstrates higher frequencies for explosions than for earthquakes, higher for explosions than for their following collapse (underground, nuclear; cf. S. W. Smith, 1963), and higher frequencies for lower-magnitude earthquakes (the latter statement agrees qualitatively with the relation [50]). Particle-velocity spectra by Willis (1964) demonstrate the existence of high-frequency seismic waves (10–20 cps) out to distances of several hundred kilometers from explosions and earthquakes in northeastern USA recorded on broad-band instruments. The reason may be partly source effects, partly and especially path effects.

Frequency spectra have also proved useful in identifying the closely packed seismic arrivals from near sources. For instance, Rg is generally recognized by its lower frequency as compared to P- and S-waves. A more detailed discussion of source time functions will be given in section 8.2.2.

Surrounding medium—coupling. The seismic efficiency, i.e., the fraction of the explosive energy that is converted into seismic wave energy, depends heavily on the physical and chemical properties of the medium.

Decoupling of *underground explosions* has been investigated to a great extent both theoretically and experimentally. It is demonstrated that decoupling will affect spectra of radiated waves, in other words, decoupling is a function of frequency (Adams and Allen, 1961b; Willis and Wilson, 1962). Decoupling is generally defined as the ratio of coupled to uncoupled amplitude spectra, and as such it is given as function of frequency. Reduction to the source is effected by amplitude spectral equalization. As the Fourier amplitude spectra of explosions often exhibit a rather erratic character, this is also the case with decoupling factors. More stable estimates can be obtained by using narrow-band filtering or power density. Both from theoretical and observational studies, Herbst et al. (1961) and Werth and Herbst (1963) studied the dependence of decoupling on various factors, such as:

(1) Medium: tuff–salt giving decoupling factors of 300, salt–salt about 100.

(2) Non-elastic effects, by overdriven spherical cavities: decoupling reduces to 10–30 (but is still appreciable to justify cheaper decoupling by smaller cavities).

(3) Frequency: the general range investigated was 10–30 cps and no very

clear frequency-dependence was observed, except that decoupling seems to increase with frequency in overdriven spheres.

For small explosions in salt cavities, Adams and Carder (1960) demonstrated decoupling factors in the general range of 20–100, using both amplitude and power spectra in the frequency range of 0.5–150 cps. Springer et al. (1968) demonstrate a clear dependence of decoupling factors on wave frequency (essentially P-waves) using observations from a shot-generated cavity in salt, i.e., a cavity generated by an earlier shot (Salmon in 1964). For a yield range of about 0.4–5.3 kiloton, they find a decoupling of 70 $\pm$ 20 at 1–2 cps and $\leq$ 40 for frequencies $>$ 15 cps, as deduced from reduced displacement-potential spectra, using the spectral-ratio method. In the usual P-wave range (1–2 cps), such a decoupling would thus lead to a decrease of the P-wave magnitude by 1.8. On the basis of theory and observed spectra for underground nuclear explosions, Mueller (1969) finds the seismic efficiency to vary between 1.8 and 6.1% for contained explosions, but to be considerably lower for cratering or decoupled events. The medium properties may effect spectra also in other ways. For example, DeNoyer et al. (1962) found azimuthal variations of P- and S-wave amplitude spectra of an explosive source, interpreted as a diffraction effect in the geological structure around the source.

Amplitude spectra of surface waves from *atmospheric explosions* bear the imprint not only of the source but naturally also of the path, as shown by Pomeroy (1963). The amplitude maximum near 10 sec period in Fig.83 is no doubt due to higher-mode Rayleigh waves, which propagate well over the continental path

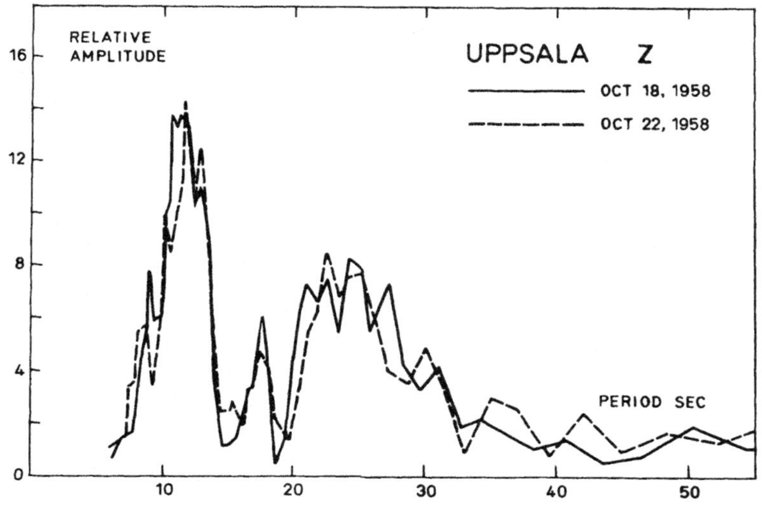

Fig.83. Trace amplitude spectra of Rayleigh waves from Novaya Zemlya atmospheric nuclear explosions recorded at Uppsala, Sweden. After Pomeroy (1963).

TABLE XLIV

Spectral characteristics of signals from underwater explosions (see, for example, Weinstein, 1968; Plutchok and Broome, 1969)

Fundamental cyclic frequency	Harmonics involved	Type of signal
$\frac{V}{4h}$	odd: $\frac{V}{4h}(2n - 1)$	seismic
$\frac{V}{4h_s}$	odd: $\frac{V}{4h_s}(2n - 1)$	seismic
$\frac{1}{T}$	all: $\frac{n}{T}$	seismic, hydroacoustic

V = sound velocity in water, h = water depth, h_s = shot depth, T = first bubble pulse period, $n = 1, 2, 3, \ldots$ Cf. also Table XXXV.

from Novaya Zemlya to Uppsala, while the maximum between 20 and 30 sec is due to fundamental-mode Rayleigh waves.

Signals from *underwater explosions* have also been subjected to spectral investigations. A brief summary is given in Table XLIV. The first two lines in this table refer to reverberations and can easily be understood, considering constructive interference (only for normal incidence), taking into account a phase reversal at the surface reflection and no phase change at the bottom reflection. The shape of the spectrum depends on several factors of the source, and these are important to consider in any interpretation of spectra in terms of structural properties, for example in deep seismic sounding at sea (Zverev, 1962; Bancroft, 1966). Constructive interference between the direct wave through water and reflected waves leads to considerably enhanced amplitudes for certain frequencies, depending on shot depth and water depth (Backus, 1959). Similar principles and the same precautions are naturally valid for explosions on land, especially in the presence of superficial layers with large velocity contrast to underlying medium. Moreover, the results hold also for more continuous wave motion, like microseisms, as evidenced by Bradner et al. (1970).

From spectral studies of underwater explosion records, Burkhardt et al. (1968) demonstrate an increase of the low-frequency portion of the spectrum: (1) with increasing yield W; (2) with increasing distance r; and (3) with increasing shot depth h, while Willis (1963b) found frequencies to be essentially depth-independent.

8.2.2 *Explosion source time functions*

Applying the equalization method to amplitude spectra of Rayleigh waves, Toksöz et al. (1964) determined source parameters for *underground nuclear explosions*, leading to a source time function of the following form:

$$p(t) = P_0\, t\, e^{-at}; \qquad t \geq 0 \tag{22}$$

Here $p(t)$ = pressure time function at the boundary of the elastic zone, P_0 = a scale factor, dependent on yield, a = a positive quantity, dependent on medium and yield, decreasing with increasing yield, thus broadening the pulse, leading to a higher proportion of more long-period waves. The pressure reaches its maximum at time $1/a$. The pressure amplitude spectrum obtained by Fourier transforming [22] is:

$$|P(\omega)| = P_0/(a^2 + \omega^2) \tag{23}$$

It is to be noted that [22] differs from the form used by Mueller (1969), who assumed the following equation for the pressure at the elastic limit:

$$p(t) = P_0\, e^{-at}; \qquad t \geq 0 \tag{24}$$

This is a step function, preferred also by Healy et al. (1971) on the basis of spectral ratios of P-waves and by Aki and Tsai (1972) on the basis of Love-wave spectra from explosions, as distinct from the impulse function in [22]. Still another form for the pressure time function at the elastic limit for a contained underground explosion is used by Mueller and Murphy (1971):

$$p(t) = P_0\, e^{-at} + P_1; \qquad t \geq 0 \tag{25}$$

where P_1 corresponds to steady-state pressure.

It is of interest to compare these expressions with the corresponding function for an *atmospheric explosion*, obtained independently of seismic observations:

$$p(t) = P_0\,(1 - at)\, e^{-at}; \qquad t \geq 0 \tag{26}$$

where P_0 = maximum overpressure and $1/a$ is the duration of the positive phase of overpressure. Obviously, $p(0) = P_0$ and $p(t) = 0$ for $t = 1/a$ and $t = \infty$. The corresponding pressure amplitude spectrum is by Fourier transforming [26]:

$$|P(\omega)| = P_0\omega/(a^2 + \omega^2) \tag{27}$$

Using the equalization of amplitude spectra for two explosions of yields W_1 and W_2 we obtain, identical propagation paths assumed:

$$\frac{|A_1(\omega)|}{|A_2(\omega)|} = \frac{|P_1(\omega)|}{|P_2(\omega)|} = \frac{P_{01}}{a_1^2} \frac{a_2^2}{P_{02}} \frac{1 + \omega^2/a_2^2}{1 + \omega^2/a_1^2} = \frac{W_1}{W_2} \frac{1 + \omega^2/a_2^2}{1 + \omega^2/a_1^2} \qquad [28]$$

noting that P_0/a^2 scales as W. Thus, the amplitude spectral ratio is equal to the ratio of the yields multiplied by a frequency-dependent factor. The correction factor is significant in comparing considerably different yields. Toksöz and Ben-Menahem (1964) applied this method to relative yield estimation of some atmospheric explosions by comparing amplitude spectra of vertical-component Rayleigh waves. From the source time functions given, corresponding source phase spectra can also be easily deduced. This is left as an exercise to the reader. Such formulas can then be used in phase spectra equalization studies, analogously to the amplitude spectral equalization method.

As an alternative to the amplitude spectral equalization method for estimating yields of atmospheric nuclear explosions, another simpler, but equally accurate method can be devised. Next we shall describe this procedure. From the expression [27] we find that $|P(\omega)|$ attains its maximum value for a frequency $\omega = a$, other parameters being unchanged:

$$|P(\omega)|_{\max} = |P(a)| = P_0/2a \text{ scaled as } W^{2/3} \qquad [29]$$

The corresponding amplitude $|A(\omega)|$ also scales as $W^{2/3}$, as W does not enter in any more way than through $|P(\omega)|$. Moreover, this same $|A(\omega)|$, corresponding to the maximum $|P(\omega)|$, is the maximum in the amplitude spectrum, providing that path effects do not vary strongly with frequency within the range considered so as to effect a frequency shift of the amplitude maximum. This is very unlikely to happen, especially considering relatively near stations (as stations in Sweden in relation to Novaya Zemlya tests), whose long-period records are characterized by sharply defined maxima corresponding to the Airy phase. In passing, we note that the frequency of the maximum, $\omega = a$, scales as $W^{-1/3}$, i.e., the larger the yield, the more dominant are the low frequencies, just as found above for underground explosions.

Ratios between amplitude spectral maxima could be replaced by ratios between recorded maximum amplitudes, in the time domain, at least under the favourable conditions of relatively near stations, just mentioned. It results that:

$$A_{1\,\max}/A_{2\,\max} = (W_1/W_2)^{2/3} \qquad [30]$$

where $A_{\max}$ refers to recorded maximum amplitudes. Incidentally, it is of interest to note that this result shows fair agreement with Murphy (1972), who found teleseismic Rayleigh-wave amplitudes A to be proportional to $W^{1/2}$ to $W^{2/3}$, depending upon height of explosions and period of Rayleigh waves. His theory was, however, developed for explosion heights far in excess of those considered here.

The maximum amplitude in the time domain is significant also for its use in magnitude calculations. This permits us to combine the yield W with the total seismic wave energy E. Leaving out common terms we can write this relation as follows:

$$\log E \sim 1.44\,M \sim 1.44 \log A \sim 1.44 \cdot \frac{2}{3} \log W = 1 \cdot \log W \qquad [31]$$

This means that $E_1/E_2 = W_1/W_2$, or $E = K\,W$, where K is a yield-independent seismic coupling factor. We can also say that the seismic wave energy scales as W. Analogous developments could be made also for underground explosions, starting from the expression [22] for the source function and applying appropriate scaling laws.

For determination of the source function, an equalization or calibration method was devised by Trembly and Berg (1968) for application to short-period P-waves from underground explosions. With input $F(\omega)$, output $G(\omega)$, and transfer function $H(\omega)$ for lumped source-receiver system, we have:

Calibration shot:

$$\left.\begin{aligned} |G_1(\omega)| &= |F_1(\omega)|\,|H(\omega)| \\ \Phi_{1G} &= \Phi_{1F} + \Phi_H + 2n\pi \end{aligned}\right| \qquad [32]$$

Other shot, assuming same $H(\omega)$:

$$\left.\begin{aligned} |G(\omega)| &= |F(\omega)|\,|H(\omega)| \\ \Phi_G &= \Phi_F + \Phi_H + 2n\pi \end{aligned}\right| \qquad [33]$$

From these we have:

$$\left.\begin{aligned} |F(\omega)| &= |G(\omega)|\,|F_1(\omega)|/|G_1(\omega)| \\ \Phi_F &= \Phi_G + \Phi_{1F} - \Phi_{1G} \qquad \text{(identical } n \text{ assumed)} \end{aligned}\right| \qquad [34]$$

Here, $F_1(\omega)$, $G_1(\omega)$, Φ_{1F}, Φ_{1G} are known, given by the calibration shot, $G(\omega)$ and Φ_G are measured from the records, and finally $F(\omega)$ and Φ_F are calculated for the new shot. By taking the inverse Fourier transform, we get $f(t)$, the source time function. On the one hand, this method seems reliable by not requiring any knowledge about absorption (lumped into $H(\omega)$), but, on the other hand, its applicability is limited by the request that $H(\omega)$ should be the same as in the calibration.

Comparison of spectra generated by an underground explosion and the following cavity collapse, as observed at one station, gives a unique opportunity to check assumed source time functions. Using spectra of body and surface waves, S. W. Smith (1963) tested observations against theory, suggesting a spherical pressure pulse for the explosion and a downward pulse for the collapse, with appropriate time constants.

Another effort to deduce source parameters of underground explosions from spectra of short-period P-waves was done by Filson and Frasier (1972). They tried to fit theoretical source models (involving three source parameters) to observed spectra in a minimizing procedure. The uncertainties of the wave attenuation put a serious limit to the procedure, and it is emphasized that it is difficult to reach reliable results by having both attenuation and source model as unknowns. For instance, it is found by comparison of several shots recorded in various directions from Semipalatinsk, that path effects dominate over source effects.

8.2.3 Discrimination factors

We have seen how different source properties make their imprints on wave spectra. It is therefore natural to utilize these findings to determine source properties from observed spectra. Source identification has always been a problem of great concern to seismologists. In the last 10–15 years it has got considerably increased significance in connection with the demands to discriminate explosions from earthquakes.

Considering these two kinds of events, we can list the following major differences, upon which discriminants can be based: (1) source time function, (2) source space function (including its location in relation to the earth's surface, etc.).

These two source functions are intimately connected as far as discrimination is concerned and on the whole they act in the same way. This may be considered as a fortunate circumstance with regard to discrimination purposes. Both factors act towards a preponderance of higher frequencies over lower frequencies for explosions as compared with earthquakes. Comparisons can be made in a variety of ways, e.g. by comparing the same wave in explosion and earthquake records, or by comparing wave ratios for the two kinds of events. Quantities compared, in other words discriminants, can be just frequency or amplitude in combination with frequency. Differences in the source space function invite also studies of source mechanism, which affords an additional powerful discriminant.

The different durations of explosion and earthquake phenomena provide for several useful diagnostic tools. The reciprocal spreading phenomenon, as expressed in section 2.3.3, yields one discriminating method:

$$f(at) \leftrightarrow \frac{1}{|a|} F\left(\frac{\omega}{a}\right) \qquad [35]$$

For $a > 1$, i.e. a prolonged action in the time domain, the corresponding spectrum will exhibit both a lower level (by the factor $1/|a|$) and lower frequencies (ω/a instead of ω). Moreover, a prolonged action in the time domain leads to a more sharply peaked spectrum. The reciprocal spreading phenomenon was experimentally verified by Frantti (1963c), comparing energy density spectra for entire seismograms of underground nuclear explosions with their collapses, and of earthquakes with their aftershocks.

Qualitatively the same results are obtained by considering a source time function in the shape of a ramp function, representing displacement:

$$f(t) = (1 - e^{-at})\, u(t) \qquad [36]$$

Its amplitude spectrum is obtained from examples 16 and 23 in Table V:

$$|F(\omega)| = \omega^{-1}(1 + \omega^2/a^2)^{-1/2} \qquad [37]$$

Taking these expressions as general formulas for source time functions (displacements), only with different values of the time constant $1/a$ for different types of sources, we find the following. For explosions, with $1/a$ small, the spectrum contains relatively more of high frequencies, whereas for earthquakes, with larger values of $1/a$, there is relatively more of low-frequency waves. According to Marshall (1970), for periods $T > 20$ sec and $a > 0.1$ sec, the source time effect becomes relatively insignificant, and then the source depth effects dominate: greater depth leading to relatively more of low frequencies (surface waves). Cf. Savino et al. (1971).

Another application of the same principle is furnished by comparing underground and atmospheric explosions. The latter have a longer duration of the seismic source effect (the impact of the atmospheric pressure wave on the ground), and as a consequence atmospheric explosions yield more of long-period body waves and well developed surface waves. The same principle can also be applied to relative estimates of heights of atmospheric explosions, as the seismic source duration increases with explosion height (Båth, 1962).

8.2.4 Spectral discrimination methods

We shall discuss discrimination methods according to the type of wave used.

Body waves. Discrimination methods restricted to body-wave spectra can be grouped as follows: (1) comparison of P-wave spectra of explosions and earthquakes; (2) comparison of P-wave spectral ratios for explosions and earthquakes; and (3) comparison of spectral moments.

The following discussion follows this general order.

Comparison of spectra of the same wave, e.g. P-waves, for underground explosions and earthquakes of the same body-wave magnitude (*m*) has consistently given higher dominant or peak frequencies for explosions than for earthquakes, thus apparently furnishing a reliable method for discrimination (Molnar, 1971; Wyss et al., 1971). It is generally maintained that considerably larger source dimensions of earthquakes than of equal-magnitude underground explosions may provide an explanation. Wyss et al. (1971) compare displacement spectra (period range 0.5–33 sec) of P-waves from underground explosions and earthquakes in the Aleutians (and one Novaya Zemlya underground explosion) recorded at Pasadena. Besides comparing peak frequencies, they also compared spectral slopes with the following results:

(1) for shorter periods (less than 1.5 sec) both kinds of spectra decrease as about ω^{-2}; and (2) for longer periods (over 3 sec) the explosion spectra decrease, as distinct from a general increase of the earthquake spectra. This is also considered as a reliable discriminant.

Another method, also based on comparison of P-wave spectra, utilizes differences in source mechanism. On the basis of the finite moving source method (section 8.1), Davies and Smith (1968) compared P-wave spectra of an underground explosion and an earthquake of same magnitude and practically identical paths. They found that the explosion duration is only 1/2–1/10 of the equivalent earthquake source duration.

Under item 2 above, i.e. comparison between earthquakes and explosions of P-wave spectral ratios between different frequency bands, we note that this method was used by Basham et al. (1970) and others. The method implies use of integrated spectral amplitudes over certain frequency bands, i.e. $A_{\omega_1-\omega_2}/A_{\omega_3-\omega_4}$ and comparison of this quantity for earthquakes and explosions. With clever choice of these limits, a good separation is obtained.

From spectral analyses of Pg in the distance range of about 2–4° from NTS, Bakun and Johnson (1971) demonstrate good discrimination for events of $M \geq 3.2$ based on the following spectral amplitude ratio:

$$\left(\frac{A_{0.6-1.25\ \mathrm{cps}}}{A_{1.35-2.0\ \mathrm{cps}}}\right)_{\mathrm{explosion}} < \left(\frac{A_{0.6-1.25\ \mathrm{cps}}}{A_{1.35-2.0\ \mathrm{cps}}}\right)_{\mathrm{earthquake}} \qquad [38]$$

Aftershocks of explosions were found to give ratios similar to those of explosions, while Basham et al. (1970) found a clear difference for slightly different frequency bands. It is maintained that the discrimination may possibly be an effect of differences in focal depths of the events compared.

Using vertical-component particle-velocity spectra, Willis et al. (1963) demonstrate significantly higher ratios of S(max)/P(max) for earthquakes than for underground nuclear explosions (essentially NTS tests), suggesting a valid discrimination method up to distances of at most 1000 km. As distinct from time-domain measurements, spectra offer the advantage of making such calculations for different frequency bands.

Alternatively, Weichert (1971) found good separation using averaged spectral moments:

$$\sum A_i\, \omega_i^n / \sum A_i \qquad [39]$$

where A_i is the amplitude of frequency ω_i. It was found that $n = 3$ yields somewhat better discrimination than $n = 1$, whereas n higher than 3 did not lead to any improvement. Explosion moments are consistently higher than earthquake moments, corresponding to the fact that an explosion short-period P-wave spectrum is very roughly equal to an earthquake P-wave spectrum but shifted towards higher frequencies, with some difference in spectral slope in addition. Other moment functions with improved discrimination properties have been investigated by Manchee (1972) and others. In such discrimination efforts, comparison of explosions and earthquakes in the same region and recorded at the same station (or array) will generally lead to best results.

Surface waves. Surface waves have been found to provide reliable discriminants between earthquakes and underground nuclear explosions (of equal body-wave magnitude), especially for spectral amplitude ratios of Rayleigh waves within different frequency bands. Thus, the following results may be quoted (subscripts indicate period ranges in sec):

Molnar et al. (1969):

$$\left(\frac{A_{19-22}}{A_{40-60}}\right)_{\text{explosion}} > \left(\frac{A_{19-22}}{A_{40-60}}\right)_{\text{earthquake}}$$

Derr (1970):

$$\left(\frac{A_{15-22}}{A_{22-60}}\right)_{\text{explosion}} > \left(\frac{A_{15-22}}{A_{22-60}}\right)_{\text{earthquake}} \qquad [40]$$

Similar findings are reported by Savino et al. (1971) and the same technique was developed, both theoretically and experimentally, by Von Seggern and Lambert (1970), using ratios of spectral Rayleigh-wave energy averaged over a number of stations. They demonstrate that this technique works better for shallow large-magnitude events than for those of low magnitude. Quite generally, the method is valid only under certain conditions, which need to be further investigated. In line with this finding, also Evernden et al. (1971) conclude that a useful identification criterion simply based upon the ratio of 20- to 50-sec Rayleigh waves cannot be established, but they also suggest that proper use of broad-band long-period data may increase discrimination capability.

The explanations for the differences in spectral ratios obtained are somewhat divergent. There is agreement on the point that these ratio differences depend on source properties, as the ratios are calculated for given source and station locations, which eliminates path and receiver properties. But the ideas diverge concerning exactly which source property is of decisive significance. We could envisage the limited action of an explosion, both in space and time, as distinct from earthquakes as a significant factor. But also differences in focal depths of explosions and earthquakes could lead to differences in the development of surface waves of the kind observed.

Keilis-Borok (1961) demonstrated theoretically that the most efficient excitation of surface waves occurs at wavelengths which are approximately four times the dimension of the source. The different dimension of underground explosions and earthquakes, which generally exists, would thus lead to a significant difference in the surface-wave spectra generated. This is also what has been found repeatedly in examination of seismograms.

However, proceeding to small events (those being of greatest concern for a test-ban treaty), then differences in dimension of earthquakes and explosions (taken out to the elastic zone) may be quite small, as emphasized by Press et al. (1963). The spectral differences which still exist, especially larger amplitudes for earthquake surface waves, are then rather ascribed to the faulting mechanism in earthquakes versus explosions. This is most likely true also for larger events, in addition to possible effects of source dimension.

Molnar et al. (1969) and Savino et al. (1971) consider the different source time functions for earthquakes and explosions to be the main reason for observed differences. However, comparing with spectra of Tsai and Aki (1970b), Fig.87, we find that such differences would also be expected because of differences in focal depth of explosions and earthquakes. From theoretical and observational studies of Rayleigh-wave amplitude spectra in the period range 10–50 sec, Tsai and Aki (1971) conclude that differences between earthquakes and underground explosions are due to differences in focal depth (cf. section 8.4.4) and mechanism rather than in source time functions or source size. They suggest long-period (over 10 sec period) Rayleigh-wave spectra as reliable discriminants, except for very shallow earthquakes. See also Tsai (1972). Similar observations are given by McEvilly and Peppin (1972), who leave the explanation open (suggesting that explosions may have source characteristics of small dimension and short time functions, in addition to effects of source mechanism and focal depth). It appears likely that observed differences in excitation of surface waves may be a result of several factors, but their relative significance may vary from case to case.

A similar technique, i.e. comparison of long- and short-period excitation of surface waves, has also been used to explore "aftershocks" of underground explosions. Observations indicate that such aftershocks produce surface waves with definitely shorter periods than earthquakes in the same area, indicating that these aftershocks are due to spherical radial fracturing around the explosion rather than to fracturing along a linear zone of failure (as in ordinary earthquakes). This result is mostly based on experience from Nevada explosions, which are in a prestressed medium (Archambeau, 1972), while experience from the Aleutians (also prestressed) suggests a combination of an isotropic explosion component and a tectonic component (Toksöz and Kehrer, 1972). Comparison with possible aftershocks of underground explosions in practically non-seismic areas, as e.g. Novaya Zemlya, would be interesting. Some aftershocks have been observed at the Swedish seismograph stations, but only for the largest of the Soviet explosions.

Besides ratios between the same wave in different frequency bands, also the ratio of Love to Rayleigh (L/R) normalized spectral amplitudes in the same frequency band has proved valuable as a discriminant. Lambert et al. (1972) found that $L/R \simeq 1$ for explosions and independent of period, but about 2–3 or more for earthquakes and increasing with period.

Body and surface waves. Relative excitation of surface and body waves for explosions as compared to earthquakes has proved to be a reliable method of discrimination. Both empirical and theoretical investigations have shown that surface waves are underdeveloped in relation to body waves for explosive sources. The relative excitation of body and surface waves in earthquakes and explosions has found its most common expression in the difference $m - M$ between body-wave and surface-wave magnitudes. This is not a spectral method but is considered as quite a reliable discrimination method (see, for example, Liebermann and Pomeroy, 1969, Marshall, 1970). Savino et al. (1971) demonstrated that the discriminant $m - M$ is more reliable for Rayleigh waves of 40 sec period than for those of 20 sec period. However, the $m - M$ relation is not unique, or valid under all circumstances. Quite considerable variations may exist, both in M (partly enhanced by tectonic strain released by explosions) and in m (essentially by regional variation in attenuation both at source and receiver), as demonstrated by Ward and Toksöz (1971). See also section 8.3.

Comparing correlation functions and power spectra of a chemical explosion and an earthquake in the same area of Yugoslavia, Dragasevic (1970) finds simpler and more concentrated curve shapes for the explosion than for the earthquake both for body and surface waves.

8.3 MAGNITUDE

8.3.1 The magnitude formula

In general terms we can write the formula for magnitude computation as follows (see Båth, 1966b, for a review):

$$\left.\begin{matrix} m \\ \\ M \end{matrix}\right\} = q(\Delta,h) + \log\frac{A}{T} \qquad [41]$$

The calibrating function q depends on epicentral distance and focal depth, and it also depends on the type of wave (P, PP, S, R) and the component (Z, H) used. We can look upon [41] as a special case of the more general formula [58.1] in Chapter 6. Every factor which enters that equation is also present in [41]. As magnitude is meant to be a source property, we have to correct for path properties. This is accomplished by the term q in [41]. In nearly all magnitude determinations in practice, use is made only of the maximum ratio of ground amplitude A to its period T within each wave group. Thus, magnitude is calculated from simple point measurements in a seismic record. It is almost surprising that this simple method yields results of sufficient accuracy. The method in fact reaches the limit of accuracy set by the earth itself. And to force this barrier of accuracy is difficult, even with very sophisticated methods.

We understand that to decide upon the maximum A/T we would need broad-band spectra. Gutenberg (1957) made some attempt in this direction, although with simple means. When a narrow-band instrument is used, e.g. a short-period seismograph, the maximum A/T found from its records may be considerably lower than the true maximum A/T for any given wave, leading to too low magnitudes, unless special corrections are applied. Further investigations of the influence of seismograph response characteristics on magnitude calculation were published by Matumoto (1959, 1960).

Solving [41] for A, we can consider the resulting formula as analogous to the displacement spectrum for an earthquake of magnitude M (or m), focal depth h and distance Δ. Such formulas, equivalent to magnitude formulas, have been used particularly in Japan to formulate strong-motion spectra (papers on this subject can be found, especially under the name Kanai and his coworkers, in the *Bulletin of the Earthquake Research Institute, Tokyo*).

8.3.2 Magnitude calculation by spectral methods

It would appear as quite natural to base magnitude calculations on a wave spectrum instead on just point measurements in a seismogram. A spectrum encompasses the whole wave motion, whereas a point measurement may be less representative. Efforts in this direction, with the aim to improve accuracy in magnitude determinations, have not been lacking, even though there has hardly been any increase in accuracy. Howell et al. (1970) discuss two new methods aiming at more precise magnitude determination, which can be summarized as follows, with minor modifications and comments.

(1) *Integrated magnitude.* This is still a time-domain computed magnitude. Instead of using A/T corresponding to its maximum in a given wave group, greater accuracy of m (or M) could be anticipated if instead we integrate over a certain portion $t_2 - t_1$ of the record and replace the maximum A/T by its root-mean-square value, i.e.:

$$m = q_1 + \frac{1}{2}\log \frac{\int\limits_{t_1}^{t_2} \left[\frac{f(t)}{T(t)}\right]^2 \mathrm{d}t}{t_2 - t_1} = q_1 + \frac{1}{2}\log \frac{1}{N}\sum_{i=1}^{N}\left[\frac{f(t_i)}{T(t_i)}\right]^2 \qquad [42]$$

where $f(t)$ is the displacement at time t and the latter formula is in digital form. To conserve the value of m, q has to be changed into q_1.

(2) *Frequency-band magnitude.* This is based on the integrated energy over a prescribed frequency range and can be simply derived as follows. Consider a sinusoidal displacement:

$$f(t) = A \sin(2\pi t/T)$$

and the corresponding particle velocity:

$$v(t) = f'(t) = (2\pi/T) f(t + T/4) \qquad [43]$$

Then we can modify [42] as follows:

$$\frac{1}{t_2 - t_1}\int\limits_{t_1}^{t_2}\left[\frac{f(t)}{T(t)}\right]^2 \mathrm{d}t = \frac{1}{4\pi^2(t_2 - t_1)}\int\limits_{t_1}^{t_2} v^2(t)\,\mathrm{d}t$$

$$= \frac{1}{4\pi^2(t_2 - t_1)}\int\limits_{-\infty}^{\infty} v^2(t)\,\mathrm{d}t$$

(provided $v(t) = 0$ outside the range $t_2 - t_1$)

$$= \frac{1}{(2\pi)^3(t_2 - t_1)}\int\limits_{-\infty}^{\infty} V^2(\omega)\,\mathrm{d}\omega$$

(by Parseval's theorem, section 3.3.2)

$$\simeq \frac{1}{(2\pi)^3(t_2 - t_1)}\int\limits_{\omega_1}^{\omega_2} V^2(\omega)\,\mathrm{d}\omega \qquad [44]$$

(provided $V(\omega)$ is approximately $= 0$ outside the range $\omega_2 - \omega_1$, this naturally being an approximation as a time-limited $v(t)$ cannot have a frequency-limited $V(\omega)$, cf. section 2.4.4). The expression for m then becomes:

$$m = q_1 + \frac{1}{2}\log \frac{\int_{\omega_1}^{\omega_2} V^2(\omega)\,\mathrm{d}\omega}{(2\pi)^3(t_2 - t_1)}$$

$$= q_2 + \frac{1}{2}\log \frac{\int_{\omega_1}^{\omega_2} V^2(\omega)\,\mathrm{d}\omega}{t_2 - t_1}$$

(where q_2 differs from q_1 only by a constant: $q_2 = q_1 - 1.2$)

$$= q_2 + \frac{1}{2}\log \frac{\sum_{i=1}^{N} V^2(\omega_i)}{t_2 - t_1} \qquad [45]$$

(in digital form)

Tentative conclusions from a small material, studied by Howell et al. (1970), suggest somewhat less scatter for these modified methods of magnitude calculation in comparison with standard techniques. Remaining scatter is ascribed to various influences of source, path, and receiver as well as noise. However, refinements in the theory should be investigated. The assumption of $f(t)$ above is no doubt too simple, and could only work if one period were "dominant". More realistic expressions for $f(t)$ should be tried, with ensuing modification in the mathematical development.

An alternative, possibly better way, would be not to assume any special form of the given record $f(t)$, but instead to transform this into ground motion $F(\omega)$ in the frequency domain:

$$f(t) \leftrightarrow F(\omega)$$

$$f'(t) \leftrightarrow i\omega F(\omega) \qquad [46]$$

(by the derivation theorem, section 2.3.6).

Then, the total seismic wave energy E can be written:

$$E \sim \int_{-\infty}^{\infty} [f'(t)]^2\,\mathrm{d}t = \frac{1}{2\pi}\int_{-\infty}^{\infty} \omega^2\,|F(\omega)|^2\,\mathrm{d}\omega \qquad [47]$$

by Parseval's theorem. And the corresponding magnitude would be:

$$m = q_3 + \log \int_{\omega_1}^{\omega_2} \omega^2 \, |F(\omega)|^2 \, d\omega = q_3 + \log \sum_{i=1}^{N} \omega_i^2 \, |F(\omega_i)|^2 \qquad [48]$$

the latter in digital form. Any testing of this formula and its accuracy is beyond our present scope. Chandra (1970a) applied energy formulas equivalent to [47] to P and S with good success, although due correction for absorption poses a problem.

Still another suggestion for spectral evaluation of magnitude is due to Munuera (1969). This is essentially based upon amplitude spectra of body waves, from which A/T is obtained as an average over a number of frequencies. The method is suggested to be less ambiguous than the direct determination of A/T from records, but applied to S-waves from near earthquakes no essential difference was found from the simpler methods. If the averaging is kept to constant frequencies, then the method assumes that the spectral shape would be the same for all earthquakes, only with different scale factors. This will however hardly be met by real conditions, except in special cases.

A similar but apparently more successful technique was applied by Willis (1965) to P-wave records from underground NTS explosions at teleseismic distances (about 22–27°). Just as we are using peak amplitude and corresponding period in the time-domain measurements, he also used the spectral amplitude peak and its corresponding period in calculating magnitude. The spectral magnitudes proved to be much more consistent, showing much less scatter among different stations, than the time-domain magnitudes (see Table XLV). In other words, the spectral peak is more stable than any maximum in the record. However, the method was tested on only two NTS explosions and a relatively small group of stations. It remains to be seen whether this method would yield equally stable

TABLE XLV

P-wave magnitude determinations for two NTS underground explosions according to Willis (1965)

Station	Mississippi explosion (Oct. 5, 1962)		Bilby explosion (Sep. 13, 1963)	
	from seismogram	from spectrum	from seismogram	from spectrum
I	5.2	5.0	5.9	6.0
II	4.6	5.0	5.5	5.8
III	5.1	5.0	5.5	5.8
IV	5.0	5.0		
V	4.5	5.0	5.2	5.8
Mean	4.9	5.0	5.5	5.85
Range	0.7	0.0	0.7	0.2

results for a greater variety of events (including earthquakes), paths and stations. This is necessary before the method can claim any general applicability, superior to time-domain measurements.

In conclusion, it can be stated that not only do the modified procedures require considerably more work than the standard one, but also their possible superiority in accuracy has to be investigated in more detail. In any case, they are justified in routine work only if fully automated. For special investigations, on the other hand, the refined methods may have a given place.

8.3.3 Magnitude as a source parameter

The main significance of magnitude is to provide a classification of seismic sources on the basis of the total energy released in seismic waves. It is natural that magnitude has been related to several other parameters, which in different ways express source dynamics. In several of these studies, spectra of waves have played a significant role. We shall deal with relations of magnitude to the following parameters: (1) dominant wave-period; (2) source mechanism; and (3) seismic wave energy.

Dominant wave-period. Many seismologists have shown that the period of the spectral peak for body and surface waves increases with the magnitude of the event. For example, Terashima (1968) found from spectrograph recordings (section 1.6.2) of microearthquakes ($M < 3$) the following relation for the initial motion:

$$\log T = 0.47\, M - 1.79 \qquad [49]$$

and Kasahara (1957) found for P-wave spectra of large earthquakes ($M \geq 6.3$):

$$\log T = 0.51\, M - 2.59 \qquad [50]$$

Similar results are reported by Linde and Sacks (1972) for South American deep shocks, their graph demonstrating a slope of 0.41 after recalculation to M. Several other, similar relations have been proposed in the literature. The effect has to be explained as a source effect, larger earthquakes generating relatively more of low-frequency motion. As is obvious from given relations, this rule holds over a wide range of magnitudes, perhaps over the whole scale.

From Fourier amplitude spectra of differentiated velocity (acceleration) of 40 Nevada microearthquakes, Douglas et al. (1970) found peak spectral frequencies to decrease with increasing magnitude (approximately agreeing numerically with [49] and [50]) and, in a less pronounced way, to decrease also with increasing epicentral distance (due to differential attenuation), whereas no azimuthal effect was observed.

Not only magnitude, but also ground amplitude, has a direct relation to dominant period. Such a relation follows immediately from the T–M relation [50] and the magnitude formula [41]:

$$\left.\begin{aligned} \log T &= \alpha M - \beta \\ M &= q(\Delta,h) + \log (A/T) \end{aligned}\right| \quad [51]$$

Eliminating M, we find:

$$A \sim T^{(1+\alpha)/\alpha} \quad [52]$$

With the values of α given above, 0.47, 0.51, 0.41, we get $(1 + \alpha)/\alpha \simeq 3$, in fairly close agreement with values summarized by Kanai and Yoshizawa (1958). Vice versa, starting from an empirical A–T relation, it is possible to deduce a T–M relation by combination with the magnitude formula. Even though results regarding period increase with magnitude or with ground amplitude have not always been displayed by spectral means, the results have clear repercussions on any calculated spectra, as demonstrated by Kondorskaya et al. (1967).

Among numerous other papers, dealing with these problems, we might mention the following: Gutenberg and Richter (1956), Aki (1956), Kanai (1958, 1962b), Matumoto (1960), Hatherton (1960), Frantti et al. (1962), Berckhemer (1962), Tsujiura (1966, 1967, 1969).

However, these simple relations are modified when the station is located on one or several superficial layers, soil or sedimentary, which in themselves emphasize certain periods by resonance (section 7.1.5). Then, T tends to assume a constant value, characteristic for each locality, for M exceeding some limit (around M = 3–6 according to Kanai, 1962b) and the equations above then hold only for lower magnitudes. It does not seem completely excluded that such effects could have contributed to the relatively small variation of period found by Suyehiro (1962) for P-waves from Japanese records of deep earthquakes. Evidently, we are up against both source effects, path effects and receiver effects on the dominant periods. The magnitude gives an approximate representation of the source time function.

Source mechanism. In a series of theoretical investigations, Haskell (1964, 1966a) treats an earthquake as a random sequence of dislocations, and he relates the total wave energy radiated from propagating faults to the spectral energy density. In his model, the moving point source is represented by a moving dislocation source. He constructs two theoretical source models, for which the amplitude spectral density $|S(\omega)|$ decreases proportionally to ω^{-2} and ω^{-3} for large values of ω (Fig.82b, cf. also section 8.1.5). These models are commonly referred to as the ω-square and the ω-cube models, respectively. It is of interest to note that already earlier, Berckhemer (1962) arrived at these two models for large ω: ω^{-2} for a dipole and ω^{-3} for a single force.

Aki (1967) made a comparison of these theoretical results and observational data. In gathering the latter, he assumes that one source parameter (surface-wave magnitude M) is sufficient to characterize a source size on the basis of the similarity principle. The latter implies that the source parameters of all earthquakes, large and small, are identical with the exception of those defining the energy and magnitude. Aki (1967) isolated the source factor $|S(\omega)|$ by taking ratios of amplitude spectra of earthquakes of different M but the same propagation paths. At least in part he used Love waves, even though M is strictly defined for horizontal-component Rayleigh waves. Comparing his observed ratios with those calculated theoretically, he found better agreement for the ω-square model than for the ω-cube model. This result, however, rests on the validity of the similarity principle, which may be questioned. The paper has nevertheless significance as an attempt to relate magnitude to seismic spectrum, on the basis of some dislocation models of earthquake sources. The ω-square model has been demonstrated also for teleseismic records of underground nuclear explosions (Von Seggern and Blandford, 1972). More recently, Aki (1972) reports that the ω-square model holds for the far-field seismic spectrum for magnitudes $M > 6$ and periods $T > 10$ sec. For other cases, modified models are suggested.

Seismic wave energy. The total energy radiated as seismic waves can be estimated from energy–magnitude relations, which have been developed by various authors (see Båth, 1966b). In deriving such relations, magnitude is generally calculated in the usual way from [41], whereas the corresponding energy has to be evaluated by independent means. There are two ways to do this:

(1) In the time domain, i.e. on the seismogram, by an integration procedure over the time interval covered by the wave motion. An integration has also to be made over space to cover the energy radiated in all directions from the source.

(2) In the frequency domain, i.e. from a power spectrum of the wave considered. Integrating the power spectrum, i.e. taking the area under the curve (cf. section 5.3.2), will yield a quantity proportional to the total energy of the signal considered. Naturally, also an amplitude spectrum can be used, provided amplitudes are squared (Karapetian, 1964).

Among observational investigations, in which wave energy has been determined by means of spectral integration, we mention Chandra (1970a) who applied the method to spectra of P-, SH- and SV-waves, and Linde and Sacks (1972), who similarly determined body-wave energy by integrating spectra over appropriate frequency bands, allowing for propagation and source effects.

In theoretical investigations, Haskell (1964, 1966a) derived expressions for total radiated energy and energy spectral density of P and S from finite sources with a source time function in the shape of a ramp function. Discrepancies between observed and theoretical total energy values were reconciled by assuming a rough ramp function with discontinuous stick-slip action, rather than continuous motion over the fault plane. This new aspect led to a more statistical treatment of the source action in terms of spatial–temporal autocorrelation coefficients.

In a theoretical study, Harkrider and Anderson (1966) derive spectral energy densities for fundamental and higher-mode Love and Rayleigh waves for different source models, different focal depths and two different earth structures (oceanic and continental). The theoretical power spectra provide useful information for comparison with observed spectra, corrected for propagation effects.

8.4 FOCAL DEPTH

Focal depth is another source parameter of great significance for observed seismic wave spectra, encompassing numerous seismological problems. Moreover, an accurate knowledge of focal depth is of significance for discrimination between earthquakes and explosions (section 8.2). Especially shallower depths, less than 100 km, often present the greatest problem.

The spectral effects of varying focal depth can be summarized in the following points: (1) variation of source function with depth; (2) variation of surrounding medium properties; (3) variation of focal position in relation to the free surface and the crustal layering; and (4) variation of generation of surface waves.

As seen from this subdivision, we are not only concerned with the source itself but also with surrounding structures. Therefore, this section will partially overlap the treatment in Chapter 7. However, we find it most instructive to deal with depth effects under one heading, even more so as a clear distinction between the various depth effects mentioned may not always be possible in observational cases. We shall deal with the depth effects in the order given above.

8.4.1 Source function and surrounding medium

There is almost unanimous agreement among quite a number of experimental results, demonstrating an increasing frequency of body waves with increasing focal depth, both for earthquakes (Gutenberg, 1958; Asada and Takano, 1963; Solov'ev and Pustovitenko, 1964; Tsujiura, 1966, 1967, 1969; Israelson, 1971) and for explosions (Molotova, 1964). Spectral slopes, or equivalently spectral ratios between two fixed frequency components, in dependence on focal depth offer a suitable method for investigation.

As explanations for this behaviour, we could refer partly to the source function, varying with depth, partly to the source medium. Deep events may depend on sudden volume changes as a consequence of phase changes, whereas shallower events depend on the usual fault mechanism. Then, deep earthquakes are able to generate the same energy within much smaller volumes than shallow ones. And smaller dimension of the source region will favour higher frequencies in radiated waves. From this viewpoint, there is a certain similarity between deep earthquakes and explosive sources. The character of the medium surrounding the source, especially its attenuating properties, will be superimposed on the source spectrum. It may naturally be difficult or almost impossible to ascertain if a certain spectral variation with depth should be explained by the source function itself or by the surrounding medium, which may modify the initial source spectrum. In general, both factors may be expected to be active.

If the depth range investigated spectrally is large enough, more complicated depth variations of spectral parameters show up than just a simple increase of high frequencies. This is probably due to the asthenosphere layer and its properties. This layer, extending from just below the crust to about 200 km depth, is characterized by low quality (low Q) and as a consequence by high attenuation, especially of higher frequencies (Mooney, 1970). This layer finds its expression also as a low-velocity layer for seismic body waves. It is to be expected that seismic sources within this layer will lose much more of high frequencies in body waves than sources above or below. This was ascertained by Mohammadioun (1965a and b, 1966), Kondorskaya et al. (1967), and Tsujiura (1966, 1967, 1969).

Al-Sadi (1973) investigated a number of earthquakes at varying depths in the Japanese area. He used several spectral parameters, such as energy-decay parameter, spectral-energy contraction, coordinates of the spectral centroid and the spectral-energy spread, some of them explained in Chapter 5. The two latter were found particularly useful. All parameters gave clear and consistent results, demonstrating much less content of high frequencies for earthquakes in the low-velocity layer (Fig.84). The most likely conclusion from this study seems to be that this result is

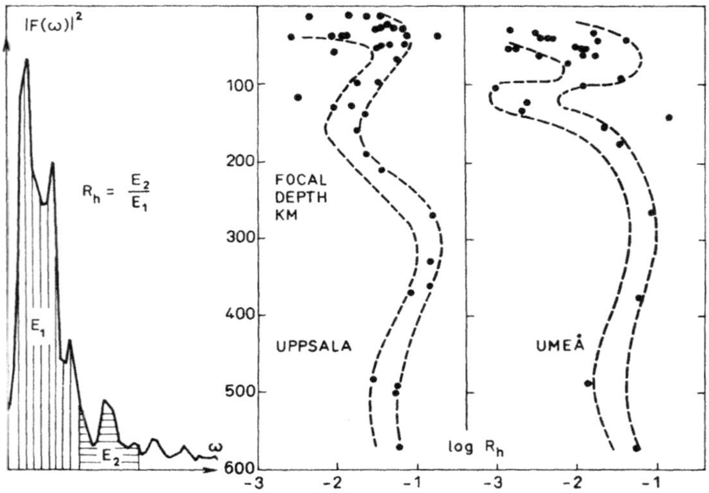

Fig.84. P-wave spectral ratios R_h for earthquakes in the Japanese area recorded at Uppsala and Umeå, demonstrating less content of high frequencies for the upper-mantle low-velocity layer. After Al-Sadi (1973).

an effect of source mechanism rather than high absorption in the low-velocity layer, especially for the following reasons:

(1) Earthquakes above or below the low-velocity zone do not show this effect, even though the P-waves have to pass this zone twice and once, respectively, on their way to the stations.

(2) A rupture in the low-velocity layer, with its more viscous material, is apt to give lower frequencies than a rupture in the more brittle material above or below it. This agrees with results from laboratory tests.

Findings similar to those of Al-Sadi (1973) were reported for the Tonga arc by Barazangi and Isacks (1971). Related studies for the Japanese arc, based on near records of S-waves, were reported by Tsujiura (1972), who inferred a change of source spectrum at around 400 km depth.

8.4.2 Focal position in relation to the free surface

While focal depth influences spectra through source mechanism and medium properties, varying with depth, in our discussion in 8.4.1, our present concern is of a kinematic nature only, i.e. the mere location of the focus in relation to the free surface.

The most reliable method for depth determination rests on an accurate arrival-time difference pP − P. The problem then essentially consists in timing the echo pulse pP amidst the tail of P. Let us approach this problem in a more general way. We assume two echoes and complex reflection coefficients:

$$R(\omega) = |R(\omega)|\, e^{-i\varphi(\omega)} \qquad [53]$$

i.e. at the reflection, both amplitude and phase are altered. The reflection is equivalent to a filtering action on the signal, and as such it is expressed as convolution in the time domain and as multiplication in the frequency domain. In addition, the reflected signal is delayed a time interval τ. Then, we have the following transforms:

$$\left.\begin{array}{ll} \text{Reflection coefficient:} & r(t) \leftrightarrow R(\omega) \\ \text{Time-delayed signal:} & f(t-\tau) \leftrightarrow F(\omega)\, e^{-i\omega\tau} \\ \text{Reflected signal:} & g(t) = r(t) \star f(t-\tau) \leftrightarrow R(\omega)\, F(\omega)\, e^{-i\omega\tau} \end{array}\right| \qquad [54]$$

With two echoes (1 and 2) the received signal is the sum of the direct and the two reflected signals:

$$F(\omega)\,[1 + |R_1(\omega)|\, e^{-i(\omega\tau_1+\varphi_1)} + |R_2(\omega)|\, e^{-i(\omega\tau_2+\varphi_2)}] \qquad [55]$$

Forming the logarithmic power expression from this, and expanding the logarithm of the parenthesis to quadratic accuracy in the reflection coefficients, we get:

$$\ln |F(\omega)|^2 + 2|R_1(\omega)| \cos(\omega\tau_1 + \varphi_1) + 2|R_2(\omega)| \cos(\omega\tau_2 + \varphi_2)$$
$$- |R_1(\omega)|^2 \cos 2(\omega\tau_1 + \varphi_1) - |R_2(\omega)|^2 \cos 2(\omega\tau_2 + \varphi_2)$$
$$- 2|R_1(\omega)|\,|R_2(\omega)| \cos[(\omega\tau_1 + \varphi_1) + (\omega\tau_2 + \varphi_2)] \qquad [56]$$

This can be considered a relatively general expression for a signal with two echoes. Complex reflection coefficients are to be expected as a result of reverberation in a layered crust (cf. Haskell, 1960, 1962). From [56] we can immediately form a series of special cases, in all eight, depending upon whether R is ω-dependent or not, whether there is one or two echoes, and whether we keep terms to quadratic or linear accuracy in R. Let us consider a few of these special cases:

(1) R independent of ω, two echoes. Then [56] simplifies to the following equation:

$$\ln |F(\omega)|^2 + 2R_1 \cos \omega\tau_1 + 2R_2 \cos \omega\tau_2$$
$$- R_1^2 \cos 2\omega\tau_1 - R_2^2 \cos 2\omega\tau_2 - 2R_1R_2 \cos \omega(\tau_1 + \tau_2) \qquad [57]$$

where R_1 and R_2 are positive for $\varphi_1 = \varphi_2 = 0$, and negative for $\varphi_1 = \varphi_2 = \pi$. Obviously, the spectrum will show ripples with "quefrencies" $\tau_1,\tau_2,2\tau_1,2\tau_2,\tau_1+\tau_2$, inviting cepstrum analysis (section 3.6.3). Note that the shorter the delays τ, the longer are the ripple periods, $T = 2\pi/\tau$.

(2) R independent of ω, two echoes, only linear accuracy in R. Then [57] becomes:

$$\ln |F(\omega)|^2 + 2R_1 \cos \omega\tau_1 + 2R_2 \cos \omega\tau_2 \qquad [58]$$

which shows ripples at "quefrencies" τ_1 and τ_2.

(3) R independent of ω, one echo, quadratic accuracy in R. Then, $R_1 = R$, $R_2 = 0$, $\tau_1 = \tau$, $\tau_2 = 0$ and [57] becomes:

$$\ln |F(\omega)|^2 + 2R \cos \omega\tau - R^2 \cos 2\omega\tau \qquad [59]$$

with ripples at "quefrencies" τ and 2τ.

(4) R independent of ω, one echo, linear accuracy in R. Then [59] becomes:

$$\ln |F(\omega)|^2 + 2R \cos \omega\tau \qquad [60]$$

with ripples at "quefrency" τ.

The superposition by the unknown term $\ln|F(\omega)|^2$ distorts the simple picture and may even cause some frequency shifts of the extrema.

It should be pointed out that by the series expansion of the logarithmic function, spurious quefrencies are introduced, not present in the original data. This is easily seen if we consider the expansion:

$$\ln(1 + R \cos \omega\tau) = R \cos \omega\tau - (R^2/2) \cos^2 \omega\tau \qquad [61]$$

where the left-hand side has only the genuine quefrency τ, while the right-hand side, due to the second-degree term, has quefrencies both of τ and 2τ. As in computer calculations, logarithms are calculated using the series expansion, spurious quefrencies might appear.

Applying formula [60] to the pP-reflection, i.e. with $R < 0$, we see that the spectrum [60] will exhibit minima for $\cos \omega\tau = +1$, i.e. for $\omega\tau = 2n\pi$, with $n = 0, 1, 2, \ldots$ Considering the reflection from a spherically symmetric, near-surface event, like an underground explosion, we have (h = focal depth, i = angle of incidence at surface, V = medium P-wave velocity):

$$\tau = (2h \cos i)/V \qquad [62]$$

Thus, the spectrum of P + pP should exhibit a series of minima at the following frequencies:

$$\omega = 2n\pi/\tau = n\pi V/(h \cos i) \qquad [63]$$

Readings of the frequencies for a sequence of minima will thus permit a determination of V/h and of h, if V can be assumed known. The minima correspond to destructive interference of P and pP. This method has been applied by Kulhánek (1971, 1973a) to records of underground explosions both in Nevada and at Semipalatinsk. Typical spectra are shown in Fig.85. In such studies of spectral minima, it is important both that the window length be the same in all spectra compared, and that this length is large enough to include the whole signal for analysis. Otherwise, spurious high-frequency minima will be introduced because of truncation effects, as shown by Gratsinsky (1962). See also Kurita (1969a).

Similar results are reported by Bufe and Willis (1969), Cohen (1970), King et al. (1972), Manchee and Hasegawa (1973), and also Buchbinder's (1968) plot of periods for minima versus magnitude for underground explosions testifies to the same effect. Minima due to interference between direct and surface-reflected waves are not restricted to the pair P + pP, they are equally valid for PcP + pPcP. Interference between P and PcP for waves grazing the core boundary can also be dealt with just as the P + pP interference (Johnson, 1969).

The positive findings can be stated with higher reliability when other reverber-

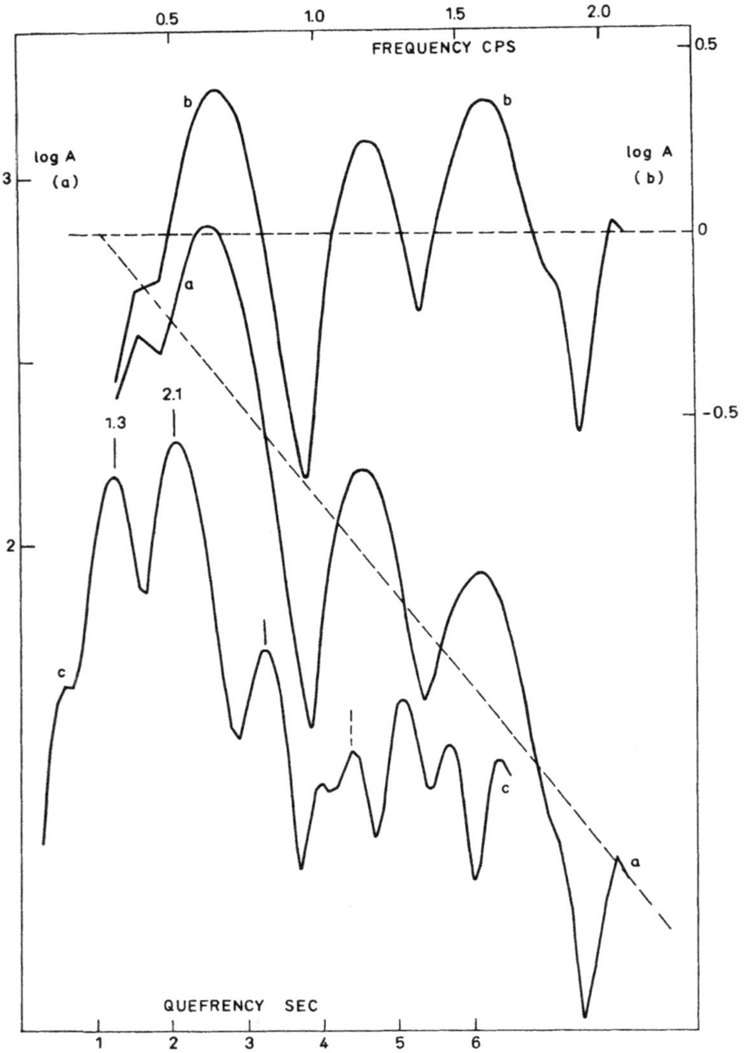

Fig.85. Original P-wave spectrum (*aa*) of an NTS explosion (Bilby) recorded at Uppsala with its linearly detrended spectrum (*bb*) and its cepstrum (*cc*), the latter with arbitrary ordinate scale. After O. Kulhánek (personal communication, 1972).

ations, such as in the receiver crust, can be excluded as reason for observed minima. This was the case for instance with Kulhánek's (1971, 1973a) observations. On the other hand, if the layering in the receiver crust is such as also to produce minima in the same frequency range as the focal depth does, then interpretation will be much more difficult or impossible. One method to eliminate disturbing minima is to average spectra for several stations, located such that structures differ sufficiently but still are so close that the pP-P interference effects are essentially the same (Cohen, 1970). Records from long-period seismographs appear to be less suitable to reveal these minima (Molnar, 1971), while Kishimoto (1964) studied their influence on spectra of long-period P-waves. More improvement in the application of the theory to observations could probably be achieved by taking R as frequency-dependent, following Haskell (1960, 1962).

Frequently, exact time readings of pP are difficult in seismic records, due to reverberation and noise. This is especially true for shallower events, depth less than 100 km, where pP arrives in the coda of P. The cross-correlation technique may here afford a solution, namely by cross-correlating the initial P-waveform with the whole record under consideration (cf. Anstey, 1964). This procedure is analogous to optimum filtering, in which also a certain knowledge of the sought properties is presupposed. High positive or high negative correlation would then be obtained when P coincides with pP, depending upon if pP exhibits no phase reversal or a phase reversal of π in relation to P. This method may prove useful for depths exceeding a few kilometers, about 3 km, to prevent pP to arrive during the first cycle of P. On the other hand, Backus (1966) finds that autocorrelation, applied to whole records, is of little use in discovering pP, probably due to complicated source conditions, and that deconvolution leads to more reliable results. Howell et al. (1967) present a method for finding pP based on inverse filtering of the given record.

Following a theoretical development by Guha (1970), Guha and Stauder (1970) investigated the possibility to determine earthquake focal depth from reduced amplitude spectra. If $A_0(\omega)$ = the reduced P-wave amplitude spectrum, i.e. corrected for the effects of instrument, receiver crust, geometrical spreading and absorption, i.e. reduced back to the source, $S(\omega)$ = the amplitude spectrum of the source and $H(\omega)$ = the transfer function of the source medium, then $H(\omega) = A_0(\omega)/S(\omega)$. The reduced spectra $A_0(\omega)$ present a series of minima, due to destructive interference both from $H(\omega)$ and $S(\omega)$: (1) pP-P interference depends on focal depth: minima appear at longer periods for larger h; (2) pP-P interference depends also on source type, source orientation and azimuth to station (different take-off directions for pP and P), and on the source crust (its reflection coefficient); (3) minima depend also on the finiteness of the source.

Obviously, to be able to determine h from observed minima in $A_0(\omega)$, it is necessary to eliminate other effects influencing the minima, which in general is a difficult task.

8.4.3 Interference or echo analysis

The analysis of echoes displayed in 8.4.2 started from the model pP – P. However, any other system with echo-resembling properties could be analyzed in exactly the same way. One important example is offered by reflections encountered in seismic prospecting, so-called ghost reflections. Assuming just one reflection for simplicity, our equation [55] can be written as follows:

$$\underbrace{F(\omega)}_{\text{input}} \underbrace{[1 + |R(\omega)|\, \mathrm{e}^{-i(\omega\tau+\varphi)}]}_{\text{filter function}} = \underbrace{G(\omega)}_{\text{output}} \qquad [64]$$

The wanted function $F(\omega)$ is obtained by inverse filtering from $G(\omega)$:

$$F(\omega) = G(\omega)/[1 + |R(\omega)|\, \mathrm{e}^{-i(\omega\tau+\varphi)}] \qquad [65]$$

The inverse filtering function is of the same shape as the equation for a feedback system (see textbooks in electronics), and it can be effectuated by such a system (Lindsey, 1960). Overlapping phases are especially to be expected in crustal

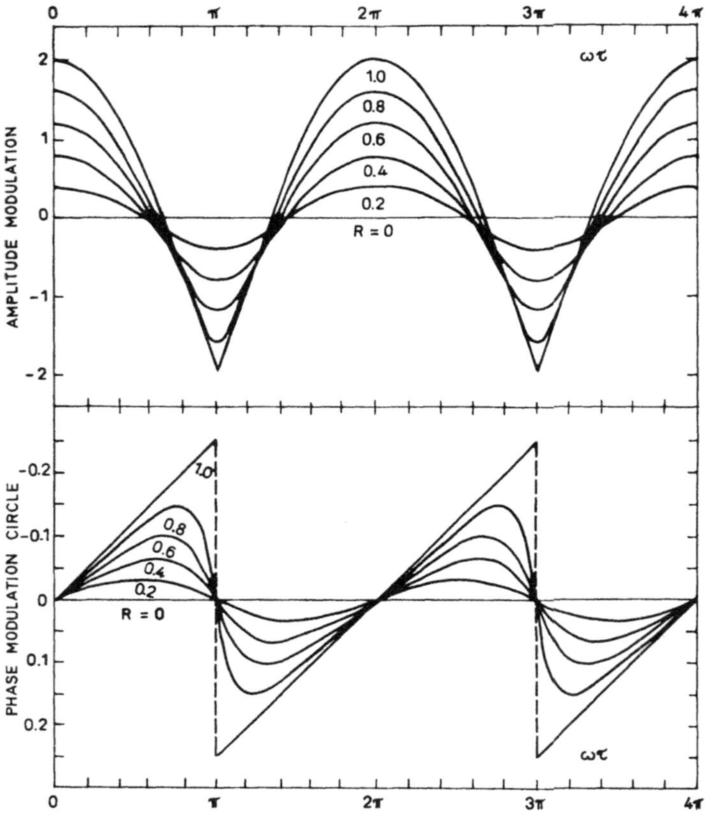

Fig.86. Amplitude and phase spectra of the function $1 + R\,\mathrm{e}^{-i\omega\tau}$. Circles = radians/$2\pi$. After Pilant and Knopoff (1964) and Wu (1968). There is an exact correspondence to Fig.70 for $R = 0.8$, noting that the ordinate π in Fig.86 corresponds to 25 cps in Fig.70.

research, because of the large number of closely arriving phases due to the crustal layering (Kasahara, 1957). Generalization to any number of interfering signals, with application to crustal research, is presented by Nakamura and Howell (1964) and Fernandez (1967).

Another example from seismology is offered by multi-path propagation of surface waves, which is a regular phenomenon in any seismic record. The different echoes then consist of different trains of surface waves, reflected and refracted at (vertical) discontinuities in the crust, each with certain (complex) reflection and refraction coefficients and certain time delays. See Pilant and Knopoff (1964), Boore and Toksöz (1969) and Capon (1970).

Similar effects are to be expected in case of multiple seismic events, with two or more ruptures, following each other with small time intervals, and originating at the same or nearby points (Pilant and Knopoff, 1964; Knopoff and Pilant, 1966; Wu, 1968). The rapid phase shifts (Fig.86) at certain frequencies may cause uncertainties in phase-velocity dispersion studies. Several smoothing methods are discussed by Pilant and Knopoff (1964).

Another striking example of ripples is shown by Kovach et al. (1963), where two interfering signals were caused by two explosions a small time interval apart. In fact, ripple-firing, with a series of shots, is the most frequently used procedure in mining operations. The resulting wave spectra can be deduced in the same way as in the echo analysis above (Pollack, 1963a).

Clearly the "echo analysis" presented above has quite an extensive application,

TABLE XLVI

Interference phenomena: review and some references

Phenomenon	Body waves	Surface waves
1. Multiple event:		
(a) continuous (finite moving source)	section 8.1.4	section 8.1.3
(b) discontinuous	Kovach et al. (1963) Frantti (1963b) Willis (1963a) Pollack (1963a)	Pilant and Knopoff (1964) Filson and McEvilly (1967) Wu (1968) Niazi (1969)
2. Multi-path propagation	pP–P (Bogert et al., 1963; Bufe and Willis, 1969; Cohen, 1970; Guha, 1970; Guha and Stauder, 1970; Kulhánek, 1971, 1973a) crustal reverberation (numerous references; see section 7.1)	Pilant and Knopoff (1964)

and it may more correctly be termed "interference analysis". It is applicable both to body waves and surface waves, both to multiple events and multi-path propagation. For instance, crustal reverberation (section 7.1.4) is included into this interference analysis. A summary review is given in Table XLVI for earthquake and explosion sources. For microseisms, conditions are more complicated, with both multiple sources and multiple paths acting simultaneously, with ensuing interference (beats) in the records.

It is clear from the presentation in this chapter and Chapter 7 that spectral minima play a great role in interpretation of seismic spectra. The main efforts are concentrated on reading the sequence of frequencies at which minima are observed. We have found several reasons for spectral minima which we can summarize in the following points:

(1) Real minima due to destructive interference, including (a) finite source propagation factors (Chapter 8); (b) reverberation in source and receiver crust and in any layered medium (Chapter 7); (c) interference between different waves, such as pP-P interference or of surface waves travelling different paths (Chapter 8).

(2) Fortuitous minima, separating spectral peaks of different origin. These minima do not lend themselves to any useful interpretation in terms of source or structure like those listed under 1. Examples are given in Chapter 9.

It is no doubt clear that in any given spectrum there may be minima caused by more than one of the reasons listed above. It is then necessary to select observations in such a way that the effect under study will be emphasized while others will be suppressed or absent.

8.4.4 Surface-wave generation in dependence on focal depth

The amplitude–depth variation of surface waves has been theoretically investigated as a possible means for source-depth determination. Keilis-Borok and Yanovskaya (1962) found theoretically that fundamental-mode Rayleigh-wave spectra have enhanced amplitudes at high frequencies (periods $T(\text{sec}) < h(\text{km})/8$, h = crustal thickness) for foci located at depths between $h/4$ and $h/2$, and suggest this as a possible method for more accurate depth estimation. Ben-Menahem and Harkrider (1964) demonstrated theoretically that radiation patterns of Rayleigh waves may depend strongly on the depth of an earthquake source. Corresponding calculations, including some evidence from observations, have been published by Levshin et al. (1965), who demonstrate strong attenuation of short periods for increasing depth in the crust, especially in the presence of sediments.

In an attempt to base nuclear-explosion discrimination on source-depth determination, Sherwood and Spencer (1962) developed a method to use Rayleigh-wave amplitudes. It is based upon the fact that in an isotropic half-space the Rayleigh-wave amplitudes decrease exponentially with depth, in other words that high-frequency content would be indicative of shallow depth. The method is faced with several difficulties, which we summarize as follows:

(1) Small shot depth and high quality factor (Q) of the medium have the same effect, namely enhanced high frequencies, which leads to an ambiguity in interpretations.

(2) Signal/noise ratio plays a significant role.

(3) Surface layering will lead to: (a) increased amount of high frequencies; (b) dispersion, i.e. a spreading of Rayleigh-wave energy over a longer time, with ensuing lower signal/noise; (c) higher modes, some of which may be more depth-sensitive than the fundamental Rayleigh mode.

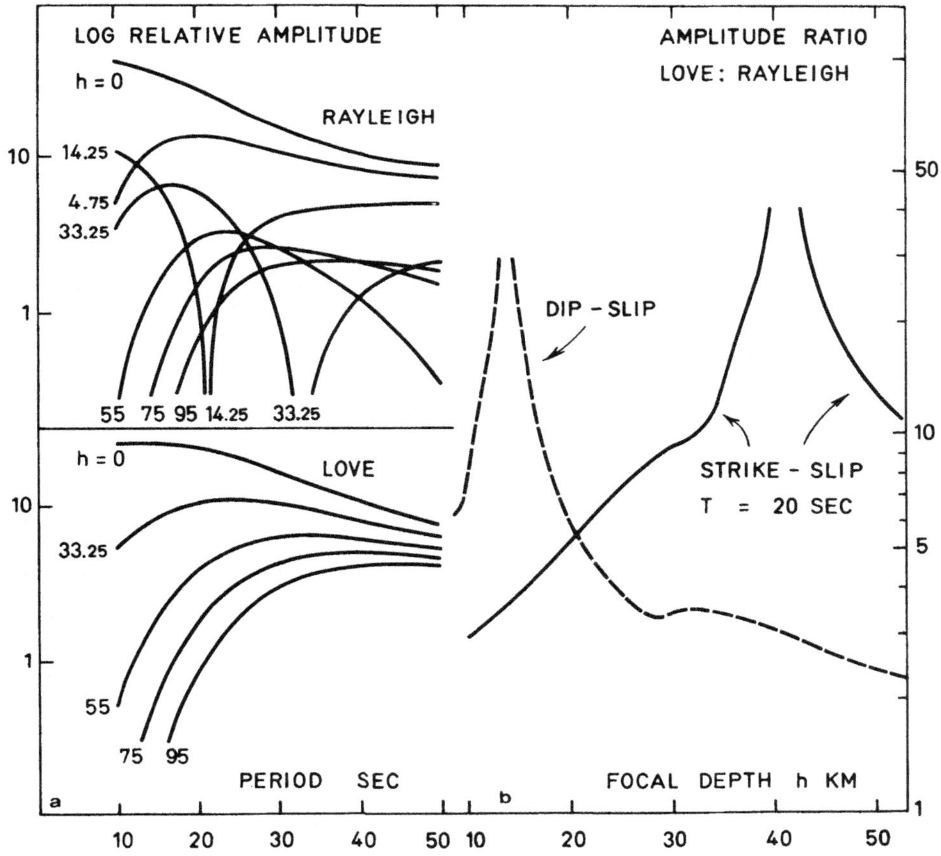

Fig.87. a) Examples of theoretical amplitude spectra of Rayleigh (vertical) and Love waves from a vertical strike-slip fault with a step time function in the Gutenberg continental model. After Tsai and Aki (1970b). b) Amplitude ratio Love/Rayleigh in dependence on focal depth. After Canitez and Toksöz (1971).

That increasing focal depth leads to relatively higher content of lower-frequency surface waves has also been emphasized by Marshall (1970).

Concerning focal-depth determination from surface-wave amplitude spectra, papers by Tsai and Aki (1970b, 1971) are very valuable, as they do not only develop the method and investigate its limitations, but also put it to test on shallow earthquakes. Starting from the normal-mode theory for Rayleigh and Love waves from a known dislocation source in a multi-layered medium, they demonstrate the focal-depth dependence for both phase and amplitude spectra within the period range 10–50 sec. See Fig.87a, which shows that Rayleigh-wave amplitude spectra are more reliable than Love-wave spectra for depth calculation. The method works if the depth is the dominant factor, and it is proposed that other spectral factors (crust and upper-mantle structure, uncertainty in fault-plane solutions, finiteness of source, source time function, path attenuation) can be ignored if held within specified limits. The method yields the depth to a few km from stations up to about 4000 km distance. It has significance also for discrimination between earthquakes and explosions. See also Tsai (1972). The method was applied to the San Fernando earthquake in 1971 by Canitez and Toksöz (1972).

However, the use of single spectra of either Rayleigh or Love waves will hardly lead to reliable depth estimates, because of uncontrolled influence from other effects besides the focal depth. Then, procedures based on combinations of Rayleigh- and Love-wave spectra will more likely lead to success. Ben-Menahem et al. (1968) and Harkrider (1970) emphasize the efficient use which can be made of spectral ratios of various mode-combinations of Rayleigh and Love waves for focal-depth determination. Harkrider (1970) finds the minima of the spectral ratio of fundamental-mode Rayleigh to Love waves as most promising for focal-depth determination by such methods, although even then independent information on source mechanism and earth model is necessary (Massé et al., 1973). In the ratio L/R (Love to Rayleigh) a number of source and path factors cancel and the ratio depends on frequency, radiation pattern, earth model, and focal depth. For given radiation pattern (determined e.g. from body waves) and earth model, the ratio L/R provides a measure of focal depth with an error of only a few kilometers (Canitez and Toksöz, 1971). See Fig.87b.

8.5 EARTHQUAKE PERIODICITY

A treatise on spectra and periodicities would hardly be complete without mentioning earthquake periodicities, even though this field has nowadays been mostly dismissed and abandoned. No doubt, it has played a certain role in the historical development of seismology. Several books and numerous papers, mostly of older date, have discussed such problems. A reference to Gutenberg and Richter (1954), which gives references to earlier works, may suffice. Frequently the underlying statistical data of earthquakes were incomplete or insufficient, also in some cases inadequate methods were used. All this naturally gave results of only questionable value and miscredited the field. Sometimes, significant periods were obtained for limited areas, whereas in global statistics periods were usually statistically insignificant.

Among newer attacks on this problem, we may mention Morgan et al. (1961), who, using a Fourier series expansion, meant to find a statistically significant annual period in the number of earthquakes. However, the data used by these authors were both incomplete and inconsistent. If number of earthquakes is used, it should preferably be a complete set above some specified magnitude. This need seems to be fairly well satisfied by the table by Duda (1965), listing earthquakes for magnitude $M \geq 7$ for the years 1897–1964. The monthly average number N of earthquakes for this set of data can be represented by the following formula, including only an annual term and adjusting each month to 30 days:

$$N = 1.53 + 0.10 \cos (t + 5^{\circ}.37) \qquad [66]$$

where t is counted from the beginning of the year. This formula is based on a homogeneous material. The number N reaches a minimum on June 25 (near aphelion) and a maximum on December 25 (near perihelion), i.e. opposite to Morgan et al. (1961). However, the most important point to note is that the annual variation is not statistically significant. And even if we would subdivide the material into different, smaller groups of shocks, it appears very unlikely that any significant periods would come out.

Power spectral methods, which could also be applied to earthquake series, have not improved the situation, even though these methods are not bound to preassigned periods as the Fourier series expansion method. For example, Lomnitz (1964) presented power spectra of numbers of earthquakes for the entire earth for the interval 1904–1952. The material was no doubt inhomogeneous, but this would hardly influence the conclusion that no significant periods existed. Similarly, Shlien and Toksöz (1970), investigating earthquake numbers in the interval 1961–1968, did not find any significant periods between 2 and 256 days in their power spectra. Nor did Vere-Jones and Davies (1966) find any significant periodicities in New Zealand earthquakes, using detailed statistical and spectral methods. Haubrich (1969) also found no evidence of any strong periodic component, possibly with some exception for a daily period, which, however, could be fortuitous, caused by a corresponding variation of detection threshold due to variation of background noise.

Still another remark has to be made on this kind of statistics. The mere number of earthquakes does not give the best representation of earthquake activity. This is much more reliably measured by the energy release or some suitable, related quantity. Therefore, any search for periodicities of seismicity should preferably be based also on such quantities and not only on numbers of shocks.

A closely related and equally controversial field concerns correlation of the number of earthquakes with various, supposedly releasing factors, such as earth tides (see section on microearthquakes 8.1.6), weather conditions (wind, rainfall, snowfall, atmospheric pressure), etc. A detailed discussion of earthquake statistics with numerous additional references is given by Utsu (1972).

Chapter 9

SPECTRAL STUDIES IN METEOROLOGY, OCEANOGRAPHY AND MICROSEISMOLOGY

In this chapter we shall make a combined study of spectral methods applied to meteorology, oceanography and microseismology (science about microseisms), because phenomena are closely related in these three fields. Partly the phenomena are genetically related (meteorological turbulence creating ocean surface waves, which in turn in some way or another create microseisms), partly the three phenomena exhibit time series with closely related properties. Meteorological turbulence, ocean surface waves, microseisms represent random (stochastic) phenomena, with infinite duration, though with varying intensity, as distinct from transient seismic signals from earthquakes and explosions. The similar nature of the three types of records also suggests that spectral methods will be closely related.

In studies in all three fields, we are very much concerned about scales of quantities involved, both in time and space. Quite naturally, this invites spectral analysis as the most efficient means to disentangle scales (frequencies, wavenumbers) involved and to study phenomena in relation to their scales. However, it is only relatively recently, within the last twenty years or so, that spectral methods have been applied extensively, excepting the long history of harmonic analysis of various meteorological elements. A paper by Rushton and Neumann (1957) contains applications of spectral analysis both to meteorology (turbulence) and to oceanography (ocean surface waves), including an extensive bibliography on earlier papers in these fields.

9.1 METEOROLOGY

9.1.1 Spectral versus traditional methods

Meteorology is one of those fields which, in its earlier development, has seen much more of harmonic analysis than most other sciences. As most meteorological elements are recorded versus time, the most usual procedure was to expand such observations into Fourier time series. Only one or two references to the meteorological literature may suffice to cover this point; see, for example, Conrad and Pollak (1950) and Von Hann and Süring (1940–1951, pp.81ff and 280ff). The purpose of such analyses was to extract all periods that a record could contain and to test them for their statistical significance (see section 5.1, also Stumpff,

1937). The book by Stumpff (1937) contains many examples from geophysics, especially meteorology, to which period analysis techniques of older date have been applied. However, when such data in addition were used for prediction purposes, the efforts almost always met with failure. As Panofsky (1955) says, the meteorological variables do not seem to be characterized by oscillations of particular periods, excepting the diurnal and annual variations.

Long time series of air temperature measurements were earlier mostly subjected only to harmonic analysis, but more recently power spectral density techniques have been applied also to such cases, one example being Roden (1966a). In his power spectra of air temperatures for western USA, only the annual and the semi-annual variation were found to be significant. This confirms a statement by Ward and Shapiro (1961a and b), who maintain that spectral techniques have not been able to reveal any more significant periods than known from traditional harmonic analysis. However, see also Landsberg et al. (1959). Harmonic analysis has for many years been extensively applied in all branches of climatology. On the other hand, Fourier integrals have been comparatively little used for climatological research up to the present time (Stringer, 1972, p.91).

Regular spectral analysis, using Fourier transforms, has led to great success in the study of a number of other meteorological phenomena during the last twenty years. The power spectra have thus been applied successfully to studies of such phenomena as wind velocities (both in frequency and wavenumber domains), turbulence, etc. The success of the power-spectrum method over the older Fourier series expansions depends on the fact that power spectra are continuous and represent a probability distribution and do not give discrete frequencies. Also cross-spectra have been applied with advantage for the same purposes often by displaying separately their real part (co-spectra) and their imaginary part (quadrature spectra).

The book by Panofsky and Brier (1958) gives a review of modern applications of statistics to meteorology, including one chapter on time series and spectral analysis. Meteorological applications of cross-spectrum analysis are reviewed by Panofsky (1967). Modern treatments of atmospheric turbulence, including application of spectral methods, are given by Panofsky and Deland (1959), Pasquill (1962a), and Lumley and Panofsky (1964). More recently, the papers presented at a conference in Stockholm in 1969 give a comprehensive review of spectral analysis in meteorology (Saxton, 1969), including 35 papers with references to most of the recent literature and 8 discussion reviews. Naturally, most of the papers concern atmospheric turbulence of all different scales. The significance of spectral analysis of such phenomena lies primarily in the fact that it provides a means to distinguish different scales from each other, both in time and space.

Turbulence studies of any medium, mostly atmosphere or ocean, lend themselves in a very natural way to the application of spectral concepts. The wind velocity components U, V, etc. can be written as the sum of average velocities

$\overline{U}$, $\overline{V}$, etc. and the turbulent deviations U', V', etc. (see textbooks in meteorology):

$$U = \overline{U} + U' \qquad \text{and} \qquad V = \overline{V} + V'$$

The turbulent or eddy stresses are proportional to expressions such as $\overline{U'V'}$, $\overline{U'^2}$ etc. Expanding $\overline{U'V'}$ we can write:

$$\overline{U'V'} = \overline{(U - \overline{U})(V - \overline{V})} = \overline{[U(t) - \overline{U}][V(t) - \overline{V}]}$$

Comparing with: $\overline{[U(t) - \overline{U}][V(t + \tau) - \overline{V}]} = C_{UV}(\tau)$

for a given point x, or [1]

$\overline{[U(x) - \overline{U}][V(x + \xi) - \overline{V}]} = C_{UV}(\xi)$

for a given time t

we see that in computing $\overline{U'V'}$ we are combining *simultaneous* values of U and V, whereas in computing the cross-correlation coefficient $C_{UV}(\tau)$, we keep one series $U(t)$ fixed in time but let the other $V(t + \tau)$ pass over it. This corresponds to the continuous nature of turbulence and in this way time variations of the correlation are introduced. The transform of $C_{UV}(\tau)$ yields the corresponding cross-power spectrum (section 3.3.4). The same consideration applies to the autocorrelation coefficient $C_{UU}(\tau)$ for calculation of power spectrum. Also, instead of varying time, we could vary any space coordinate x, i.e. at a fixed moment combining U at one point with V at a series of points along some profile. These considerations have equal validity to any kind of turbulence measurements, whether of momentum, heat, concentration of matter, etc.

From the relations [1] above we find immediately:

$$\overline{U'V'} = C_{UV}(0) = \frac{1}{2\pi}\int_{-\infty}^{\infty} E_{UV}(\omega)\,\mathrm{d}\omega = \int_{-\infty}^{\infty} E_{UV}(v)\,\mathrm{d}v$$

$$= \int_{-\infty}^{\infty} v\,E_{UV}(v)\,\mathrm{d}\ln v \qquad [2]$$

where $E_{UV}(v)$ is referred to as a *spectral tensor*. Thus, $\overline{U'V'}$ which is proportional to Reynolds eddy stress, is equal to the area under the spectrum $E_{UV}(v)$.

In meteorological research, the simultaneous behaviour of two (or more) parameters is often investigated. Earlier, this was usually done by calculating the over-all correlation coefficient between two variables, as defined in statistics, and by investigating its statistical significance. It may well happen that a positive correlation in one frequency range is more or less cancelled by a negative correlation in another frequency band, and as a result the over-all correlation would show up as insignificant. Cross-power spectra, on the other hand, reveal the frequency-dependence of correlations, and are thus a much more powerful technique. However, significance tests would need to be applied to cross-power spectra as well as to simple correlation coefficients. Panofsky and McCormick (1954) emphasize that the classical Reynolds definitions of turbulence do not specify what scales of motion are important, whereas power spectra and cross-spectra provide such information.

9.1.2 Spectral methods

With reference to section 4.6 we can list spectral methods in meteorology as follows:

(1) Indirect method or correlation-transform method (section 4.6.1). This has been the dominating method in meteorology, introduced essentially by Blackman and Tukey (1959) and further developed for meteorological applications by Muller (1966). Cross-power spectra are mostly split into co-spectra and quad-spectra. Variance bears a simple relation to power (section 5.1.1) which has been utilized in some calculations of power spectra.

(2) Direct method or periodogram method (section 4.6.2). This method is hardly used at all.

(3) Fast Fourier Transform method, FFT (section 4.6.3). Since its introduction around the middle of the 1960's, this method is gaining steadily increasing application also in meteorology. Examples of papers using FFT are Endlich et al., (1969), Madden and Julian (1971), and others.

Examples of application of various filters to meteorological time series can be found in papers by Craddock (1957) and Charnock (1957), used on temperature and wind records, respectively. It can be proved (see especially Jenkins and Watts, 1968), that correlation functions and power spectra for random observational series do not converge in any statistical sense to a definite value by increasing the observational interval T. In fact, tests show that beyond a certain T, fluctuations keep on existing indefinitely, in an undiminished degree. Therefore, once a minimum but sufficiently long time series T has been obtained for analysis, results will not be improved by prolonging T, in case of random data.

As a general rule, according to Blackman and Tukey (1959), it is not advisable to use lags exceeding 5 to 10% of the total record length T. According to some other suggestions, autocorrelation functions could with advantage be calculated for lags up to about 30% of the total number of observations (Jenkins, 1961b). Bendat (1958, pp.201–206) demonstrates that many observed properties of

TABLE XLVII

Atmospheric scales (after Fiedler and Panofsky, 1970)

Parameter	Microscale Turbulence scale	Mesoscale	Macroscale Synoptic scale
Period	< 1 hr	1 hr–48 hr	> 48 hr
Wavelength	< 20 km	20 km–500 km	> 500 km

atmospheric turbulence can be derived by assuming an autocorrelation function in the form of an exponentially decaying cosine function (Fig.22).

A critical study on calculation of meteorological spectra, especially from limited intervals of observations, was given by Kahn (1957). It is stated that spectra calculated for limited intervals are representative only for the interval used, but still problems remain about the reliability of the spectrum even for this interval. The spacing of observations within an observational interval is another matter of great significance (see, e.g. Ogura, 1957a and b). Shapiro and Ward (1963) say that there are considerable differences among spectral turbulence studies, probably due to the use of insufficient data. The significance of sampling and averaging observations has been emphasized in several papers. For derivations, with special regard to turbulence studies, see F. B. Smith (1962). Aliasing problems are discussed by Griffith et al. (1956). As emphasized by Wooldridge and Reiter (1970), there is considerable difficulty in comparing published spectra, due to variation both in computation and in presentation, including normalization.

Another important question concerns the influence on spectra of instruments and measuring techniques. Not many comparisons of this kind have been done. One example is given by Cornett and Brundidge (1970), who found good agreement between wind spectra determined from radar-tracked balloons and measurements on a near-by tower. Similar spectral comparisons of different measuring techniques are of general importance for any kind of geophysical observations.

In general, meteorological spectra cover a large range of frequencies and it is therefore customary to differ between different atmospheric scales (Table XLVII). From this it is clear that it is impossible to get an equal coverage of observations over a very extended range. The practical way to handle this is "patching", which means that different parts of a wide-range spectrum have to be derived from different sets of observations, with correspondingly different sampling rates. At the same time as this is a practical necessity, considerations of homogeneity become of great importance. "Patching" is not unique to the construction of meteorological spectra; in fact, it is a common procedure in the construction of any kind of wide-range geophysical spectra. Cf. section 6.5.4.

In meteorology, there is sometimes of interest to calculate seasonal mean spectra. In case of harmonic analysis, which is linear in the coefficients, this does not present any problem. But for power spectra we have to consider that these are quadratic in the coefficients, and then care has to be taken in calculation of seasonal means, as explained by Benton and Kahn (1958). Take as example the cross-power spectrum E_{12} of $F_1 = a_1 - ib_1$ and $F_2 = a_2 - ib_2$ (cf. section 3.3.4):

$$E_{12} = F_1^* F_2 = (a_1 + ib_1)(a_2 - ib_2) = (a_1a_2 + b_1b_2) - i(a_1b_2 - a_2b_1) \quad [3]$$

The co-spectrum (real part) of E_{12} is:

$$P_{12} = P(F_1,F_2) = \text{Re } E_{12} = a_1a_2 + b_1b_2 \quad [4]$$

We express each coefficient as the sum of its seasonal mean and the daily deviation from the mean:

$$\left.\begin{aligned} a_1 &= \bar{a}_1 + a_1' \qquad & b_1 &= \bar{b}_1 + b_1' \\ a_2 &= \bar{a}_2 + a_2' \qquad & b_2 &= \bar{b}_2 + b_2' \end{aligned}\right| \quad [5]$$

Introducing these into [4] we get:

$$P(F_1,F_2) = P(\bar{F}_1,\bar{F}_2) + P(\bar{F}_1,F_2') + P(F_1',\bar{F}_2) + P(F_1',F_2') \quad [6]$$

Averaging over one season, we obtain the following result, as time averages of F_1' and F_2' vanish:

$$\overline{P(F_1,F_2)} = P(\bar{F}_1,\bar{F}_2) + \overline{P(F_1',F_2')} \quad [7]$$

or in words: the seasonal mean of the spectrum = the spectrum of the seasonal mean + the seasonal mean of the perturbation spectrum.

Two-dimensional power-wavenumber spectral analysis has also found meteorological application, the first instance probably being an analysis of satellite cloud photographs (Leese and Epstein, 1963). The principle is the same as applied to ocean surface waves and in some gravity and geomagnetic studies (see Chapter 10) and refers back to fundamental principles laid down in Chapter 2. In the cloud study mentioned, power (indicating photographic brightness) is plotted as isolines in graphs with two perpendicular wavenumbers as axes. This method provides for more detailed and accurate analysis of cloud photographs than earlier visual examination. Analogously, Mak (1969) and Izawa (1972a and b) display power as isolines in the wavenumber-frequency domain.

Expansion in spherical harmonics is the most suitable approach in studies of phenomena of global extent. Besides some meteorological examples given in section 2.5.4, we note that spherical harmonics provide the natural mathematical apparatus for studies of atmospheric tides. Reviews are given by Siebert (1961) and Chapman and Malin (1970). As outlined by Siebert (1961), the calculation procedure is the following:

(1) Harmonic time expansions are made for pressure (and temperature) for as many stations as possible distributed around the earth.

(2) Amplitude and phases obtained under 1 are expressed for each wave component by spherical harmonic functions.

Combination of operations 1 and 2 leads to a representation in terms of sine and cosine functions of time or longitude and Legendre functions of latitude. The method as outlined here is equally well applicable to appropriate measurable quantities representing the oceanic tide and the solid earth tide, to which we shall return in later sections (9.2.3 and 10.1.2, respectively).

9.1.3 Spectral properties: coordinates, parameters, spurious minima

In recent studies of turbulence, using spectra, it has been customary to plot observations as non-dimensional logarithmic power spectra (section 5.4.1). The ordinate is chosen $= \omega E(\omega)/V_*^2$, as normalized logarithmic spectral power density ($E(\omega)$ = power, V_* = friction velocity). The abscissa is ln $(\omega z/V)$, with V = mean wind speed at height z. $\omega z/V$ is sometimes referred to as the *natural frequency*. Both coordinates are obviously non-dimensional. By this normalized representation, it is easier to compare different sets of observations. Moreover, the spectra are often expressed in analytical form, relating the ordinate to the abscissa, as defined here. Such equations then permit an immediate comparison of different observation series. See also section 5.4.1 and Table XXIV, where we investigated the properties of the logarithmic spectrum.

In case of turbulence, close connection between time and space properties may be expected, as Taylor (1938) emphasized for the case when turbulent velocities are low compared to the mean wind. Then, a time correlation could be rewritten as a space correlation by replacing time t by x/V (x = mean wind direction, V = mean wind speed). And a frequency spectrum could be changed into a wavenumber spectrum by replacing frequency ω by Vk, where k = wavenumber. This hypothesis implies equivalence of time and space averages, a property which goes under the general name of *ergodicity*; cf. *stationarity* which implies equivalence of different time samples, and *homogeneity* which implies equivalence between different space samples. Experimental tests by Panofsky and McCormick (1954), Panofsky et al. (1958), and Tsvang (1963) demonstrate that ergodicity holds with high accuracy, at least under the conditions of their experiments. Further information as to the validity of Taylor's hypothesis can be found in Panofsky and Deland (1959), Cramer (1959), Lappe and Davidson (1963), Lumley and Panofsky (1964). Restriction of the validity of the hypothesis to smaller-scale motions, with scales not exceeding 1 km, was announced by Chernikov et al. (1969) and Kao et al. (1970b).

Among spectral parameters (section 5.3) used in meteorological spectroscopy, we note especially the slope of power-wavenumber spectra, towards higher wavenumbers. A slope of k^{-3} (k = wavenumber) or $k^{-5/3}$ is often found and discussed for turbulence spectra. The dependence on frequency ω will be the same as on k, under the assumption of Taylor's hypothesis (Monin, 1967). This is a general property of meteorological spectra, at least within certain frequency ranges, while spectra usually become flatter towards lower frequencies and steeper towards higher frequencies. As for ocean surface waves (section 9.2), there is clear indication of prevailing higher slope $\sim\omega^{-3}$ in the high-frequency portion of the power spectra, as shown for example for zonal and meridional wind power spectra by Kao and Henderson (1970). They find a spectral maximum (i.e. zero slope) at a frequency around 0.01 cph (period about 4 days) for meridional wind spectra, explained as due to some periodicity in cyclonic activity, but not for zonal wind spectra.

The slope phenomenon is to be considered as an inertial decay or degeneration of turbulence, and then the same or similar slope manifests itself in any other property which depends upon the turbulence. Sometimes, a maximum (a "hump") is observed superposed on the general slope, and then this indicates another source of turbulent energy at some frequency (Roth, 1971). This general slope is a common property of several kinds of geophysical phenomena, and represents an energy transfer (or "cascading") from lower to higher frequencies or wavenumbers ("inertial subrange"). With reference to section 5.4, it is obvious that due consideration to the coordinates used has to be taken in slope determinations. Putting power $E(k) \sim k^{\gamma}$, the exponent γ is obtained as the straight-line slope in a graph of $\ln E(k)$ versus $\ln k$.

Among numerous meteorological papers dealing with spectral slope, we list the following examples of power spectral investigations, all with well agreeing slopes:

(1) Turbulent wind. MacCready (1962 a and b) studies the slope of turbulent wind power-frequency spectra, and Tsvang et al. (1963) study the slope for turbulent wind and temperature wavenumber spectra, finding a slope of $-5/3$ (cf. also Reiter and Foltz, 1967; Pinus and Šur, 1970; and Leith, 1971).

(2) Temperature. Panofsky (1969b) finds similar slope for the high-frequency portion of temperature spectra.

(3) Humidity. Elagina (1963) and Coantic and Leducq (1969) report the same slope for the humidity spectrum.

(4) Pressure. Power-frequency spectra of pressure were studied by Gossard (1960) and by Kimball and Lemon (1970), the latter finding a slope of -2 over a wide frequency range.

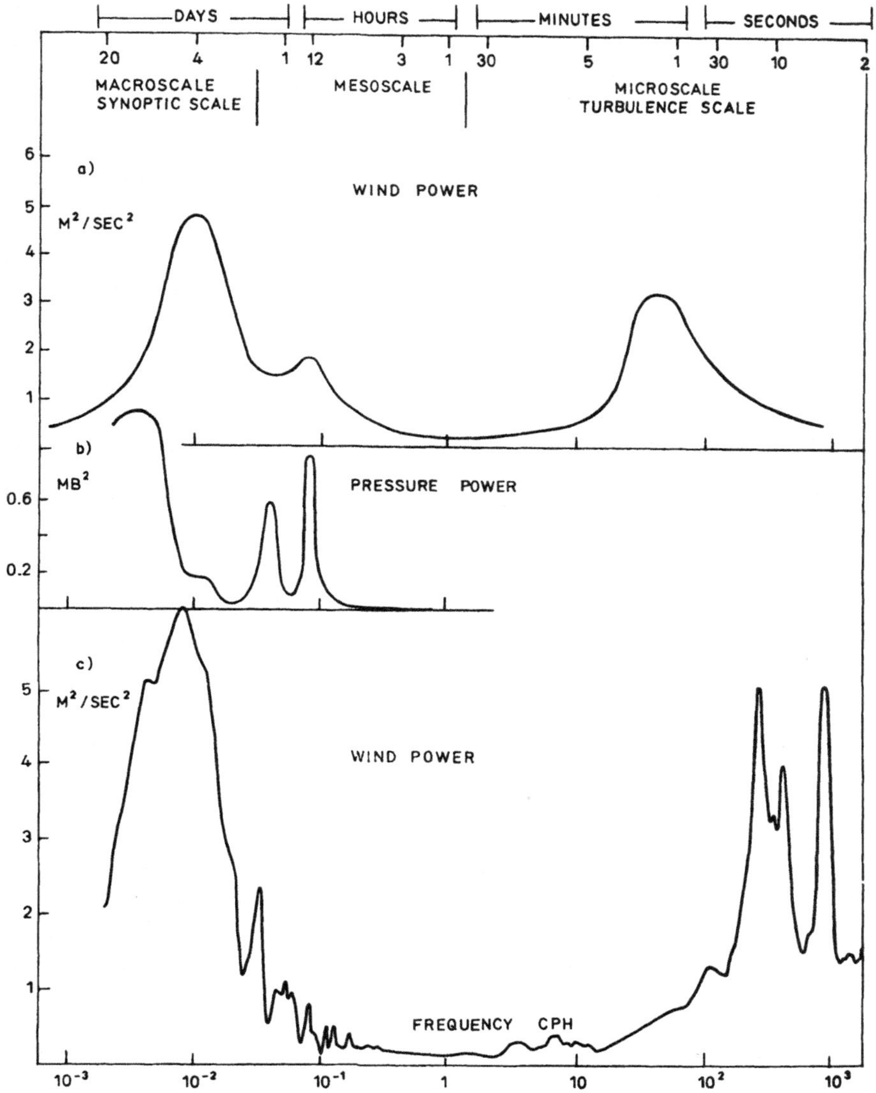

Fig.88. Examples of meteorological power spectra. a) Horizontal wind speed at about 100 m height at Brookhaven. After Van der Hoven (1957). b) Surface pressure at Palmyra Island. After Hwang (1970). c) Wind measured on buoys in the North Atlantic. After Millard (1971).

(5) Refractive index. Power-frequency spectra were studied by Thompson et al. (1960) and Jeske (1969), and power-wavenumber spectra by Gossard (1960); see also Bull and Neisser (1968).

(6) General. Hinze (1959, p.189 ff) gives a thorough discussion of the shape

of the complete turbulence spectrum. Garrat (1972) and Kaimal et al. (1972) find that spectra and co-spectra reduce to a family of curves which converge into single universal curves in the inertial subrange but spread out at lower frequencies.

Of special interest is a result of Panofsky and Van der Hoven (1955) and Van der Hoven (1957) that the horizontal wind speed spectrum exhibits a minimum around a period of 1 hour (or wavelength of about 10 km). The minimum separates two significant maxima at around 4 days and at around 1 min. See Fig.88. The minimum has no physical meaning in itself; it just results from the good separation of the two physically significant maxima. Similar situations often arise in seismology, for instance, when a P-wave and microseismic background show up as two distinct maxima at different frequencies, again separated by a minimum. Significant spectral minima may be important as information on destructive interference, in its turn containing valuable structural information. However, in the cases mentioned, we are concerned with spurious minima with no special physical significance. Cf. section 8.4.3.

For newer results about the minimum in the meteorological spectrum, just mentioned, consult Panofsky (1969a), Vinnichenko and Dutton (1969), and Vinnichenko (1970) with further references. Newer results are not quite conclusive concerning this minimum in the turbulence spectra. Whereas Panofsky (1969a) shows minima both at around 1/2 hour period and between the synoptic scale and the 1-year peak both for horizontal wind and temperature, others find no evidence for this. Methods for data sampling and data handling are considered to be responsible. The gaps appear to be less clear or absent at higher levels. A wind power spectrum, closely resembling the one by Van der Hoven (1957), has been derived from wind measurements on buoys at sea by Millard (1971). The high-frequency part of the spectrum (possibly due to buoy motions) may contaminate the low-frequency part by aliasing, if not carefully considered in the sampling process. A valuable discussion on atmospheric spectral gaps has been given by Fiedler and Panofsky (1970). Another instance of clear minima is exhibited by a power spectrum of atmospheric pressure, given by Hwang (1970).

9.1.4 Meteorological applications

The meteorological literature from the last twenty years contains a wealth of information on application of spectral analysis, too much to be all reviewed here. Table XLVIII gives a list of examples of observational studies in meteorology where spectral analysis has been applied. The cited papers usually contain extensive lists to further literature in this field. The list is representative enough to permit some general conclusions, both concerning methods and meteorological problems studied.

A large amount of information on *atmospheric turbulence* has been derived from spectral studies, such as the effects of stability, mean wind, height above

TABLE XLVIII

Examples of observational research in meteorology where spectral analysis has been applied

Reference	Problem (parameters studied)	Spectral method
Panofsky (1953)	variation of turbulence with height	spectra of kinetic energy, heat flux and stress
Panofsky and McCormick (1954)	turbulence at 100 m height	individual power spectra, co-spectra and quad-spectra of vertical and horizontal wind speeds
Estoque (1955)	meridional transfer of heat and momentum	cross-correlation and cross-power spectra between meridional wind and heat or momentum
Panofsky and Van der Hoven (1955)	horizontal wind at 91 m height	spectra and cross-spectra of velocity components
Chiu and Rib (1956)	laboratory investigation of turbulence	correlation and energy-wavenumber spectra
Griffith et al. (1956)	temperature fluctuation	power spectra
White and Cooley (1956)	meridional motion in mid-troposphere	wavenumber kinetic-energy spectra
Panofsky and Wolff (1957)	westerly winds at different latitudes	cross-spectra (co-spectra and quad-spectra) between zonal indices
Saltzman and Peixoto (1957)	wind over northern hemisphere	amplitude and phase wavenumber spectra (along latitudes) of zonal and meridional wind velocities
Van der Hoven (1957)	frequency of horizontal wind speed	power spectra
Angell (1958)	wind fluctuations (with time)	power and cross-power (variance and co-variance) of wind velocities
Benton and Kahn (1958)	large-scale turbulence	wavenumber power spectra (wind) and co-spectra (meridional transfer of momentum, heat and kinetic energy)
Bushnell and Huss (1958)	surface winds	power spectra, calculated from velocity variance
Businger and Suomi (1958)	vertical wind component	power–frequency spectra
Ely (1958)	surface winds	power spectra, calculated from wind velocity variances
Henry and Hess (1958)	large-scale turbulence	kinetic energy, momentum and heat flux as isolines with wavenumber (along latitude) and time as axes
Saltzman (1958)	large-scale turbulence	wavenumber spectra of kinetic energy, momentum transport, and transfer of energy between eddies and mean motion
Adel and Epstein (1959)	analysis of ozone parameters	power spectra of several ozone parameters

TABLE XLVIII (continued)

Reference	Problem (parameters studied)	Spectral method
Cramer (1959)	turbulent wind components at 2 m height in period range 2–100 sec	space and time correlations and power spectra
Landsberg et al. (1959)	climatological study of temperature and precipitation	power spectra (frequency)
Robinson (1959)	turbulence at 1 m height	power spectra and co-spectra of wind and temperature
Chiu (1960)	large-scale turbulence	power spectra (frequency) of heat and momentum transfer
Gossard (1960)	atmospheric pressure and refractive index	power spectra of pressure vs. frequency and of refractive index vs. wavenumber
Gurvich (1960)	vertical wind velocity near ground	power–frequency spectra
Panofsky and McCormick (1960)	vertical wind near surface	power spectra of vertical wind velocity in space and time
Rosenthal (1960)	large-scale turbulence in low latitudes	frequency spectra of power (variance) of zonal and meridional wind
Saltzman and Fleisher (1960)	conversion between potential and kinetic energy in a hemispheric field	amplitude–wavenumber spectra of pressure change and temperature; eddy potential energy vs. wave-number spectra
Shapiro and Ward (1960)	meridional wind	power spectra of kinetic energy (frequency and wavenumber)
Zwang (1960a)	temperature fluctuations in the surface layer	power–frequency spectra
Zwang (1960b)	temperature fluctuations at altitudes 100–1500 m	power–wavenumber spectra
Barrett (1961a)	general circulation	harmonic analysis of height of 500 mbar and 300 mbar surfaces vs. longitude; wave-energy spectra
Barrett (1961b)	general circulation	wavenumber spectra of flux of momentum and kinetic energy across latitudes
Davenport (1961)	turbulence near ground in strong wind	power and cross-power frequency and wavenumber spectra (via respective correlations)
Gurvich (1961)	turbulent flow of momentum near ground	power–wavenumber spectra
Harper (1961)	meridional circulation	power spectra of kinetic energy (frequency)
F. B. Smith (1961)	vertical wind	power–wavenumber spectra
Ward and Shapiro (1961a, b)	periodicities of several parameters	power (variance) spectra (frequency)

TABLE XLVIII (continued)

Reference	Problem (parameters studied)	Spectral method
Kahn (1962)	large-scale turbulence	power and cross-power spectra (wavenumber along latitude) of zonal wind, meridional wind, temperature
MacCready (1962a)	wind turbulence from sailplane	power–frequency spectra of longitudinal and vertical turbulence
Panofsky (1962)	turbulent energy in lowest 100 meters	power spectra
Pasquill (1962b)	vertical component of turbulent wind in neutral conditions near the ground	variance (power) spectra
Roden (1962)	temperature, cloudiness, wind over Atlantic	correlation and power spectra (frequency) of respective parameters
Saltzman and Fleisher (1962)	large-scale turbulence	spectra of kinetic wind energy (wavenumber along latitude)
Thompson (1962)	vertical wind	power–wavenumber spectra
Zubkovskii (1962)	horizontal wind velocity in the surface layer	power–frequency spectra
Elagina (1963)	humidity fluctuations near ground	power–frequency and –wavenumber spectra
Horn and Bryson (1963)	large-scale turbulence at 300, 500, 700 mbar and 25°, 45°, 65°N	power–wavenumber spectra
Leese and Epstein (1963)	analysis of satellite cloud photographs	two-dimensional power–wavenumber spectra of photographic brightness degree
Mantis (1963)	medium-scale turbulence	power spectra (frequency) of horizontal wind
Monin (1963)	mean zonal atmospheric motion	periodogram, autocorrelation, power spectra
Pfeffer and Zarichny (1963)	acoustic–gravity waves in the atmosphere	amplitude isolines with period and group velocity as coordinates
Pinus (1963)	horizontal wind velocity at 6–12 km height	power–wavenumber spectra
Reed et al. (1963)	sudden stratospheric warming	spectral form of energy equations
Shapiro and Ward (1963)	meridional wind at 500 mbar	kinetic energy wavenumber (along latitude) spectra
Tsvang (1963)	temperature fluctuations measured on tower and airplane	power–wavenumber spectra
Tsvang et al. (1963)	wind and temperature turbulence up to 300 m height	power–wavenumber spectra
Zubkovskii (1963)	vertical wind velocity at altitudes 50–2000 m	power–wavenumber spectra
Amelung (1964)	atmospheric pressure variations in time and space	power–frequency spectra at different localities

TABLE XLVIII (continued)

Reference	Problem (parameters studied)	Spectral method
Angell (1964)	vertical wind at different levels	spectra of wind inclination at individual levels, cross-spectra between levels
Dartt and Belmont (1964)	stratosphere zonal wind in the tropics	power–period spectra
Golitsyn (1964)	atmospheric pressure micro-pulsations in relation to wind	power–frequency spectra
Kao and Bullock (1964)	geostrophic wind	correlations and energy spectra
Kao and Woods (1964)	horizontal turbulent wind	power–wavenumber spectra
Murakami and Tomatsu (1964)	kinetic energy interaction between zonal flow and disturbances, and between disturbances	power–frequency spectra of inter-action terms (this would possibly be a field for application of bispectral analysis, section 3.6.4)
Saltzman and Teweles (1964)	exchange of kinetic energy between harmonic components	wavenumber kinetic-energy spectra of the wind
Berman (1965)	longitudinal wind	power spectra (normalized logarithmic spectra)
Eliasen and Machenhauer (1965)	spatial spectral distribution of kinetic energy over northern hemisphere	spherical-harmonic representation of stream function
Kaimal and Izumi (1965)	vertical wind velocities measured by sonic anemometers on tower	power–wavenumber spectra
Kao (1965)	large-scale turbulence	correlations and energy spectra
Panofsky and Singer (1965)	medium-scale turbulence	cross-spectra between wind components at various heights
Cagnetti and Giudici (1966)	vertical component of turbulent wind	power–frequency spectra
Chiu and Crutcher (1966)	meridional transport of angular momentum in troposphere and lower stratosphere	power–frequency spectra
Kao and Sands (1966)	zonal and meridional wind to 50 km height	power–wavenumber spectra
Noel (1966)	horizontal wind at 90–140 km height	power–wavenumber spectra
Payne and Lumley (1966)	wind-velocity fluctuations	wavenumber spectra
Reiter and Burns (1966)	turbulence in clear air	power–wavelength spectra
Roden (1966a)	long-term temperature variations in western USA	power–frequency spectra (0–6 cpy) for monthly mean, maximum and minimum temperatures 1821–1964
Roper (1966)	horizontal wind energy at 83–97 km height	power–period spectra (showing peaks at 24 and 12 hour periods)

TABLE XLVIII (continued)

Reference	Problem (parameters studied)	Spectral method
Ackerman (1967)	cloud structure from airplane observations	power–wavenumber spectra of temperature, liquid-water content, airspeed
Armendariz and Rachele (1967)	wind profiles from balloon data	power–frequency spectra by special filtering using truncated Fourier series
McCrory (1967)	pressure waves from explosions	amplitude spectra, displayed as isolines with group velocity and period as coordinates
Lappe et al. (1967)	vertical wind over land and ocean	power–wavenumber spectra
Reiter and Foltz (1967)	clear-air turbulence (CAT) from airplane observations	power–wavenumber spectra of wind speed
Weiler and Burling (1967)	turbulence over sea surface	power spectra and co-spectra of wind components
Wiin-Nielsen (1967)	atmospheric wind energy	Fourier series expansion and kinetic energy spectra
Bassanini et al. (1968)	geopotential of 500 mbar surface	power–frequency spectra
Busch and Panofsky (1968)	turbulence	power spectra (normalized logarithmic spectra)
Fichtl (1968) Fichtl and McVehil (1970)	wind turbulence measured on a 150-m meteorological tower	power–frequency spectra of longitudinal and lateral components of wind turbulence
Foote (1968)	constitution of clouds by pulsed-Doppler radar measurements	power–frequency spectra of returned power
Hsueh (1968)	intermediate-scale turbulence	energy–wavenumber spectra of wind
Kao and Al-Gain (1968)	diffusion in the atmosphere	correlations, power spectra and cross-spectra of wind components
Maruyama (1968) Yanai et al. (1968)	horizontal wind in tropical and sub-tropical Pacific at 0–30 km altitude	power–frequency spectra, coherence, phases
Panofsky and Mares (1968)	heat-flux and stress	co-spectra (normalized logarithmic spectra)
Eliasen and Machenhauer (1969)	world-wide circulation	spherical harmonic expansion of surfaces of constant pressure; quadrature spectra of wave motion
Endlich et al. (1969)	vertical wind profiles	power–wavenumber spectra of wind
Essenwanger and Reiter (1969)	relation between vertical wind shear and clear-air turbulence	power spectra
Kao and Powell (1969)	diffusion in the atmosphere	correlation, power spectra and co-spectra of wind components
Myrup (1969)	turbulence	power–frequency spectra
Revah (1969)	zonal wind at 80–110 km height (internal gravity waves)	power–frequency spectra

TABLE XLVIII (continued)

Reference	Problem (parameters studied)	Spectral method
Spizzichino (1969)	zonal wind at 80–110 km height (tides, gravity waves)	power–frequency spectra
Wallace and Chang (1969)	wave disturbances	cross-spectra (vs. frequency) of wind, temperature, humidity and surface pressure
Wendell (1969)	large-scale turbulence	kinetic energy in wavenumber–frequency space and in wavenumber space and in frequency space (similar methods as used in recent papers by Kao and collaborators)
Belmont and Dartt (1970)	periodicities of tropical stratospheric winds (at 50 mbar)	power–frequency spectra of zonal and meridional wind (peaks near 1, 3, and 15 day periods)
Bowne and Ball (1970)	comparison of rural and urban wind turbulence, measured on towers	power–frequency spectra of longitudinal, lateral and vertical wind components
Chang et al. (1970)	wind structure	power spectra, wind and humidity
Chattopadhyay (1970) Chattopadhyay and Rathor (1972)	ozone content in India	power–frequency spectra
Cornett and Brundidge (1970)	comparison of wind data by radar-tracked balloons with tower measurements	power–frequency spectra of wind speed and direction
Dyer (1970)	snowfall rates by optical measurements	autocorrelations and power–frequency spectra of snowfall rates in storms
Hwang (1970)	surface wind over a tropical island	power–frequency spectra of wind speed and pressure in the range 0.002–200 cph
Julian et al. (1970)	large-scale circulation	kinetic energy wavenumber (along latitude) spectra, including a discussion of methods in meteorological spectroscopy
Kao (1970)	temperature distribution in time and space	wavenumber–frequency and power (proportional to temperature squared) spectra of temperature, with the same methods as applied to wind by Kao and collaborators in other, recent papers
Kao and Henderson (1970)	particle dispersion in atmosphere by zonal and meridional winds	power spectra and co-spectra of zonal and meridional wind vs. frequency
Kao and Hill (1970)	diffusion in the atmosphere	correlation, power spectra and co-spectra of wind components

TABLE XLVIII (continued)

Reference	Problem (parameters studied)	Spectral method
Kao and Sagendorf (1970)	meridional heat transport	wavenumber–frequency spectra (same method as in other recent papers by Kao and collaborators)
Kao and Wendell (1970)	large-scale circulation	wavenumber–frequency spectra of wind, power spectra
Kao et al. (1970a)	large-scale circulation	wavenumber–frequency as well as frequency and wavenumber spectra of meridional transport of angular momentum
Kao et al. (1970b)	large-scale circulation	same methods as used by Kao and Wendell (1970)
Kimball and Lemon (1970)	air pressure and wind at soil surface	power–frequency spectra of pressure and wind
Lenschow (1970)	turbulence at 100–1000 m height from airplane measurements	power–wavenumber spectra of temperature and wind (vertical, longitudinal)
Mitsuta et al. (1970)	air-surface turbulent interaction	power–frequency and co-spectra of temperature, humidity, wind components
Miyake et al. (1970a)	turbulence over water	power spectra and co-spectra of wind and temperature (normalized logarithmic spectra)
Miyake et al. (1970b)	turbulent fluxes of momentum, heat, moisture, measured on airplane	power spectra and co-spectra vs. frequency
Nitta (1970a, b) Yanai and Murakami (1970a, b) Wallace (1971)	tropospheric waves in tropical Pacific	power-frequency spectra, coherence, power isolines with frequency and height as coordinates
Sitaraman (1970)	turbulence near surface	power spectra and co-spectra of wind (normalized logarithmic spectra)
S. D. Smith (1970)	wind turbulence over the sea	power spectra, co-spectra and quad-spectra of wind velocity components vs. wavenumber
S. D. Smith et al. (1970)	wind stress on ice sheet	power spectra, co-spectra, quad-spectra of wind components vs. frequency
Vinnichenko (1970)	kinetic energy of wind	frequency and wavenumber power spectra
Wooldridge and Reiter (1970)	large-scale circulation	power spectra and co-spectra of wind (frequency)
Axford (1971)	gravity waves in lower stratosphere	power spectra and cross-spectra (vs. wavelength) of wind and temperature

TABLE XLVIII (continued)

Reference	Problem (parameters studied)	Spectral method
Bhartendu (1971)	atmospheric potential gradient in relation to humidity, temperature, pressure, wind speed	frequency spectra of power, coherence, cross-power
Dickson (1971)	correlation between temperature and atmospheric circulation	power–frequency spectra of daily mean temperatures
Dutton (1971)	clear-air turbulence (CAT)	power–frequency and power–wavenumber spectra
Kao and Gebhard (1971)	clear-air turbulence (heat, momentum) in mid-stratosphere	frequency spectra of power and cross-power
Kao et al. (1971a)	meridional flux of heat at 500 mbar in southern hemisphere	power–wavenumber–frequency spectra
Kao et al. (1971b)	meridional flux of angular momentum at 500 mbar in southern hemisphere	power–wavenumber–frequency spectra
Madden and Julian (1971)	zonal wind oscillation	spectra and cross-spectra of zonal wind, temperature and pressure
Mantis and Pepin (1971)	vertical wind structure in troposphere and stratosphere	power–wavenumber spectra
McDonald et al. (1971)	wind turbulence at ground level in period range 1–100 sec	power, coherence, signal-to-noise vs. period
O'Neill and Ferguson (1971)	moisture flux in troposphere	power–frequency spectra of wind speed, humidity and moisture flux. Discussion on aliasing
Rao and Rao (1971)	meridional wind component over India	power–frequency spectra
Axford (1972)	turbulence (wind, temperature) measured on aircraft	power and cross-spectra vs. wavelengths from 250 m to 100 km
Brook (1972)	wind turbulence in a city	power–wavenumber spectra of wind (three components)
Garratt (1972)	wind and temperature over a water surface	power–frequency spectra
Izawa (1972b)	atmospheric disturbances over tropical Pacific	power–frequency–wavenumber spectra of pressure and wind
Kaimal et al. (1972)	wind and temperature fluctuations in the surface layer	power and cross-power frequency spectra
Martin (1972)	turbulence near ground	wavenumber–power spectra of temperature and humidity
McBean and Miyake (1972)	turbulent flux of momentum, heat, and moisture in 2 m surface layer	co-spectra, spectral correlation coefficients (coherence)
S. D. Smith (1972)	wind velocity and temperature over an ice sheet	power spectra, co-spectra, coherence vs. wavenumber
Thompson (1972)	vertical flux of heat and moisture over the sea to 200 m altitude	frequency spectra of power and cross-power
Hess and Clarke (1973)	kinetic energy (turbulent components U', V', W') in planetary boundary layer (balloon data)	power and cross-power frequency spectra of U', V', W'

ground and ground roughness, etc. A reference to Lumley and Panofsky (1964) will suffice, which gives a full account of results obtained up to 1964 and an extensive list of references, including also a number of Russian contributions. Summary discussions of turbulence spectra from Russian, Australian and American sources are given by Pinus et al. (1967) and Vinnichenko (1970). The fact that turbulence and wind studies have come to dominate spectral studies so far, may be appreciated also from the viewpoint that the wind, due to its advective property, is of fundamental importance in shaping the properties of the atmosphere. One practical aspect of the numerous studies of turbulence is to improve prediction. However, the atmospheric prediction problem is non-linear, which further adds to the complications. Cf. section 3.6.4.

As turbulence studies are of great significance also in other fields than meteorology, especially in various branches of technology, much material can be found in other journals, e.g. *Journal of Fluid Mechanics*, *Physics of Fluids*, and others, as well as in special books (Hinze, 1959). It is not our intention to review these here, as they mostly lie outside the realm of geophysics. Nor is it our intention to review results within meteorological turbulence, which would require a detailed study by themselves.

Spectral analysis has also stimulated new methods of acquiring data about the atmosphere. One recent and very promising approach concerns radio wave propagation through the atmosphere. The refractive index is a function of temperature, pressure and humidity. Refractivity power spectra are given by Thompson et al. (1960). Moreover, reflection and scattering phenomena depend on the motion of targets, i.e. on the wind structure. Observations of radio wave propagation are able to yield high resolution and thus inform about the fine structure of the atmosphere. The volume edited by Saxton (1969) contains several papers on this topic. The agreement between turbulence measurements by radar and on aircraft is very good, even in details of the spectrum (Chernikov et al., 1969). Radar-reflection techniques have made it possible to explore wind conditions also at high altitudes, of the order of 100 km or so. Zonal wind investigations presented as power-frequency spectra are given by Revah (1969) and Spizzichino (1969) from such measurements for the height interval of 80–110 km. Spizzichino (1969) demonstrates besides dominant winds also the semi-diurnal tide (found to be a rather regular oscillation), the diurnal tide (a very irregular oscillation, possibly due to interaction with other wind components) and internal gravity waves of various frequencies.

Of course, it could be envisaged that also other phenomena dependent on atmospheric structure in a multiple way would lend themselves to useful information on turbulence. Microseisms are such a phenomenon, which depend especially on the structure near the contact surface between atmosphere and ocean or land. However, because of the complexities of the whole system involved in microseism generation and propagation, they have not yet given such detailed information as radio waves.

But it is also emphasized that more traditional measurements of atmospheric elements need improvement for a better understanding of turbulence spectra (Pao, 1969). Especially, it is recommended to use instruments with a wide band of frequency response so as to cover a wide range of the spectrum; moreover, it is necessary that measuring platforms (aircraft, etc.) should not interact with atmospheric elements measured.

Another field of atmospheric studies concerns *propagation of pressure waves* (sound waves) from atmospheric explosions, volcanic eruptions, meteor impacts, thunder, and from earthquakes. New observational material was obtained from the large atmospheric nuclear explosions in the beginning of the 1960's. The problems of wave propagation in the atmosphere are in principle equivalent to those met with in seismology. For instance, amplitude spectra are displayed as isolines in graphs with group velocity and period as coordinates (see, e.g. Pfeffer and Zarichny, 1963; McCrory, 1967), i.e. a method well known from seismological studies of surface waves (section 7.2.2). Using this technique, Pfeffer and Zarichny (1963) were able to demonstrate the significance of a sound channel at around 85 km height in addition to the one at the tropopause. Power spectra of sound from thunder were studied by Few (1969), Bhartendu (1969), and Holmes et al. (1971). In spite of large variations from case to case, Bhartendu (1969) demonstrates persistent power density maxima at 52 and 96 cps. Mikumo (1968) deduced phase and amplitude spectra from microbarograph records of pressure waves originating from the Alaskan earthquake in 1964. Comparison of theoretical and observed barogram traces led to the conclusion that the pressure waves originated from rapid vertical ground displacements in the source area.

As in seismology, arrays (of microbarographs) have been employed for coherence and spectral studies. Pressure waves of meteorological origin, as recorded by an array of microbarographs, have been studied by spectral methods by Herron et al. (1969) and others. Mack and Flinn (1971) found larger coherence along the wave propagation than perpendicular to it, as deduced from a microbarograph array in Montana in the USA. Finally, as a general reference, the interested reader will find many papers on spectral studies of sound propagation in the atmosphere in some journals, especially the *Journal of the Acoustical Society of America*, too many to be reviewed here.

Atmospheric waves, especially tropospheric wave disturbances in the tropical Pacific, have been investigated by spectral means by several authors. The reader is referred to Wallace (1971), who, besides data, their analysis and an extensive review also presents a useful discussion on spectral methods with special application to such data. In addition to ordinary power-frequency spectra of wind components, coherence, etc, also displays are used of power and cross-power as isolines with frequency and height as coordinates (analogous to the power-isoline display with

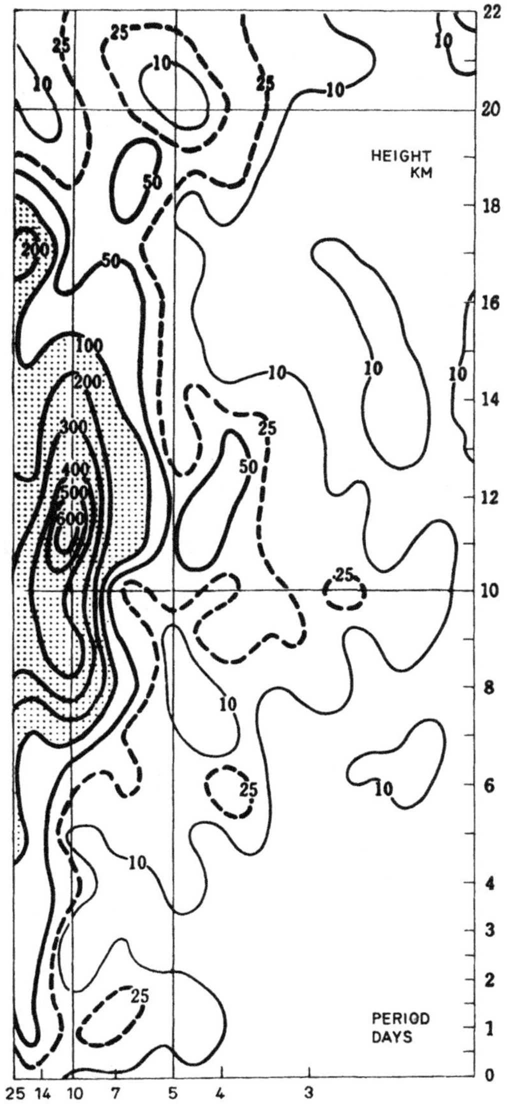

Fig.89. Height–frequency distribution of power spectral density ($m^2 \cdot sec^{-2} \cdot day$) for the meridional wind component at Christmas Island. After Yanai and Murakami (1970a).

frequency and time as coordinates, sections 1.6 and 7.2.2). One example is shown in Fig.89.

Another problem, to which power spectra have been applied, concerns the *constitution of clouds* (drop size distribution, liquid-water content, hailstone size, temperature, airspeed) as well as precipitation, as measured by vertically-pointing pulsed-Doppler radar installations (Foote, 1968; Battan and Theiss, 1972; Atlas et al., 1973), or by airplane observations (Ackerman, 1967). Especially the latter paper contains valuable considerations as to the calculation of power spectra for such observations. See also Rogers and Tripp (1964).

Paleoclimatological variations have been investigated by spectral methods, by application to phenomena which depend in a composite way on meteorological factors, especially temperature and precipitation. This concerns varved sediments (Agterberg and Banerjee, 1969) and tree rings. A general result is that most terrestrial variations are only weakly or not at all related to sunspot variations. This concerns precipitation data (Brier, 1961), varved sediment series, except for the 22-year period (Anderson, 1961; Anderson and Koopmans, 1963), and tree rings (Bryson and Dutton, 1961), whereas geomagnetic data show clear relations (Ward and Shapiro, 1961a). In a critical review of numerous applications of time series analysis to geophysical phenomena, Monin and Vulis (1971) conclude that there is no evidence for sun-influenced components of 11- and 22-year periods. Climatological data reveal clear temperature periods of 1.8–2.7 years, 5.6 years, 11 years and greater than 50 years, but none of 22 years, whereas precipitation shows no clear periods (Stringer, 1972, p.99).

Occasionally, also other meteorological problems have benefitted from spectral analysis (see for example Adel and Epstein, 1959; Landsberg et al., 1959; Rogers, 1963; Chattopadhyay, 1970; Chattopadhyay and Rathor, 1972). The spectral method has proved valuable not only in the study of observations, but also to suggest new pathways for *data collection and handling* (Peterson and Middleton, 1963; Morrissey and Muller, 1968) and it has had important repercussions within *theoretical meteorology* (Saltzman, 1957; Pedlosky, 1962; Dutton, 1963; Platzman, 1964; Kao, 1968; Merilees, 1968; Bretherton, 1969; Chiu, 1970).

There is no doubt that application of spectral analysis techniques to meteorological problems has meant an immense improvement in such studies. It has also to be recognized that meteorology faces spectral analysis with a four-dimensional pattern (x,y,z,t). Only more recent improvement and intensification of observational systems coupled with availability of large computers have made it possible to fully utilize the potentialities of spectral methods. The product has been in the form of numerous results, as witnessed by present-day meteorological journals. Just as one example, a recent number of the *Quarterly Journal of the Royal Meteorological Society* has four extensive papers, including spectra: Pasquill (1972) with a very useful review, Kaimal et al. (1972) with a great wealth of spectral information, Wyngaard and Coté (1972), and Garratt (1972).

9.2 OCEANOGRAPHY

In oceanography, spectral analysis of all kinds of sea-level fluctuations is of great significance. These are recorded versus time on tide-gauges or other equivalent apparatus, suitably tuned to different frequency ranges. In the order of generally increasing period, we may distinguish between the following phenomena in this connection:

(1) Ocean surface waves, from short capillary waves or ripples to long swell.

(2) Tsunamis or earthquake-generated ocean surface waves.

(3) Standing waves or free oscillations, called seiches, of limited sea basins of various sizes.

(4) Ocean tides.

(5) Sea-level fluctuations due to various meteorological influences, especially atmospheric pressure and wind, and other long-period effects.

In discussing the application of spectral methods in oceanography, we find it most convenient to deal with the various phenomena separately, in the same order as enumerated above.

9.2.1 Ocean surface waves

The ocean surface, as a boundary between water and air, represents a field of interaction between the two adjacent half-spaces (see Neumann and Pierson, 1966, pp.416–420). One example, which has attracted enormous interest, both from theoretical and observational sides, is the generation and propagation of ocean surface waves. Spectra of these waves reveal significant dependence on fetch and duration of the wind, and they have provided useful tools for the study of wave growth, propagation and decay, including wave prediction, as well as checks on theory by comparison of observed and theoretical spectra. See for example Phillips and Katz (1961), Moskowitz (1964), Pierson and Moskowitz (1964), Pierson (1964), Snyder and Cox (1966), Shonting (1968), Barnett (1968). Vetter and Bretschneider (1963) edited a volume on Ocean Wave Spectra, containing the reports presented at a conference at Easton, Maryland, in May, 1961. This book gives a very complete and authoritative account of ocean-wave research up to 1961, with special reference to application of spectral analysis. The book edited by Hill (1962), of fundamental importance to oceanographic studies and research, contains spectral studies of practically all phenomena connected with physical oceanography. The comprehensive textbooks by Kinsman (1965) and Neumann and Pierson (1966) contain detailed accounts of spectral calculations and results for ocean surface waves, including extensive bibliographies.

A thorough treatment of the situation up to the middle 1950's was given by Pierson (1955). See also Pierson and Marks (1952). The ocean surface elevation due to a single, harmonic progressive wave can be written as:

$$f_A(x,y,t) = A \cos\left[\frac{2\pi}{L}(x\cos\theta + y\sin\theta) - \frac{2\pi}{T}t + \varphi\right] \qquad [8]$$

where x,y = horizontal coordinates on the sea surface, θ = direction of wave propagation, measured from the positive x-axis, A = amplitude, L = wavelength, T = period, and φ = phase angle.

Since in case of deep water, the phase velocity $c = gT/2\pi$, where g = acceleration of gravity, then instead of $2\pi/L$ in [8] we can introduce ω^2/g. Applying the Fourier integral theorem to this three-dimensional case (cf. [15], Chapter 2), we can then write the surface elevation in the following form, considering propagation towards the half-plane of positive x:

$$f(x,y,t) = \int_{-\pi/2}^{\pi/2} \int_0^{\infty} a(\omega,\theta) \cos\left[\frac{\omega^2}{g}(x\cos\theta + y\sin\theta) - \omega t\right] d\omega\, d\theta$$

$$+ \int_{-\pi/2}^{\pi/2} \int_0^{\infty} b(\omega,\theta) \sin\left[\frac{\omega^2}{g}(x\cos\theta + y\sin\theta) - \omega t\right] d\omega\, d\theta \qquad [9]$$

Here, $a(\omega,\theta)$ and $b(\omega,\theta)$ are the Fourier coefficients, otherwise denoted a_n and b_n in this book. Pierson (1955) demonstrates the use of [9] partly in the case of certain given boundary conditions ($f(0,y,t)$ given), partly in the case of finite wave trains.

However, Pierson (1955) points out that the Fourier integral representation of the ocean surface is only of limited application, especially due to the extreme irregularity of the waves, and due to the impossibility to obtain the necessary observational material for formulating a useful Fourier integral representation. Nor can it be applied to shallow water as given in the form above. Therefore, recourse has been taken to statistical methods, usually describing the ocean surface as a stationary three-dimensional Gaussian process. In this statistical approach, the main emphasis is laid on the energy or power spectrum, which, when given, specifies all properties of interest of the ocean surface. For details, see Pierson (1955), who also discusses numerous oceanographic results obtained by this approach, among others those of practical importance in swell forecasting. See also Seiwell (1949), Rushton and Neumann (1957), Czepa and Schellenberger (1960), Cartwright (1962), Kinsman (1965). A similar approach will be most suitable also in other related fields with much complexity and variability, as dealt with in the whole of this chapter.

Just as in meteorology, power and cross-power spectra are generally used, often also coherence. Both frequency and wavenumber spectra are used. By and large, spectra of meteorological turbulence and ocean wave spectra show some similarities, and this is not accidental but has most likely genetic reasons. Moreover, both phenomena are to be considered as random processes.

In studies of ocean waves it is important to distinguish between the initial stage when waves grow up and the later fully developed sea. The initial stage represents a non-stationary and non-linear process, with close interaction between the atmosphere and the sea. An example of such studies, including display of temporal development of ocean-wave power spectra, is given by DeLeonibus and Simpson (1972). The wave development stage enters all efforts to forecast wave spectra. As demonstrated by Bunting (1970), such forecasts are feasible for up to 36-hour intervals for the North Atlantic, detailed wind information being the most critical factor. Another instance of spectral changes occurs for ocean waves when these proceed into shallow water. Predictions of such spectral changes are of great practical importance for harbour installations (Collins, 1972). Not only the growth of ocean waves but also their dissipation has been studied by spectral means (Hasselmann and Collins, 1968).

Wavenumber spectra require simultaneous recording of ocean waves at different points at suitable distances apart. An experiment of this kind is reported by Snodgrass et al. (1962) for the Californian coast, which among other things suggested the existence of standing wave patterns. No doubt, such research has also clear repercussions on explanation of microseisms (section 9.3). Wavenumber spectra were presented also by Scott (1969). A most extensive operation to obtain wavenumber spectra of ocean surface waves was made in 1953 in an area of the North Atlantic Ocean under the name of SWOP - Stereo Wave Observation Project (see Kinsman, 1965, pp.460–472). Simultaneous wave measurements at seven different places were made by Walden and Rubach (1967), who did not produce any power-wavenumber spectra but individual power-frequency spectra. These were studied in relation to water depth and meteorological conditions:

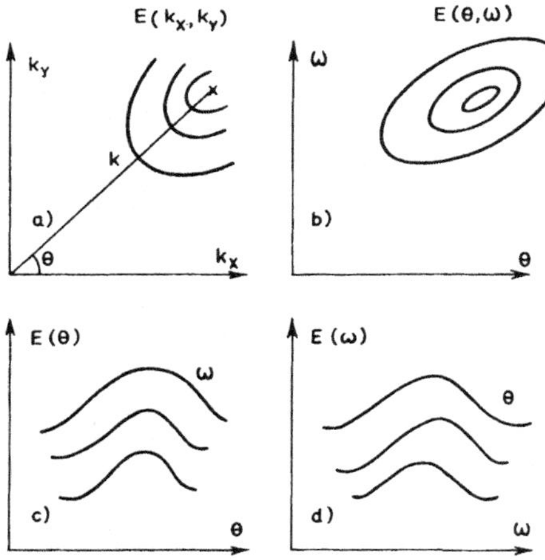

Fig.90. Different ways to display directional spectra of ocean surface waves. E = power density, θ = direction of arrival. a) Used by Barber (1963). c) Used by Longuet-Higgins et al. (1963), Rudnick (1969), Ewing (1969).

increasing water depth leads to increasing power and decreasing frequency for the spectral peak. Schule et al. (1971) deduced power-wavenumber spectra of ocean surface waves from airborne laser observations.

Directional spectra. Among spatial studies of ocean waves, also the distribution of wave propagation directions is of importance. From [9] we get the following expression for the spectral energy density $E(\omega,\theta)$ in the frequency-direction interval $\delta\omega\delta\theta$ (cf. section 4.6.2):

$$E(\omega,\theta)\,\delta\omega\,\delta\theta = \frac{1}{2}\sum_{\omega}^{\omega+\delta\omega}\sum_{\theta}^{\theta+\delta\theta}[a^2(\omega,\theta) + b^2(\omega,\theta)] \qquad [10]$$

$E(\omega,\theta)$ is termed the *directional energy spectrum* (Cartwright, 1962). Different ways to present such spectra are shown in Fig.90. Barber (1958, 1963) gives the principles of array design to explore the distribution of wave power with wave direction. The basic idea is that the cross-power of all elements in an array will depend on the direction of wave arrival, and this dependence will be different for different wavelengths. Properties, such as directional resolving power, are investigated for arrays of three detectors in a triangle and of four detectors in a star-configuration. Although primarily intended for ocean-wave research, these principles have general validity for any array, whether for seismic waves or radio waves. See also Kinsman (1965, pp.478–482).

Frequency-direction spectra can be determined in a number of ways: (1) from stereo-photographs of the sea surface; (2) from arrays of detectors; (3) from direct measurements of sea waves: (a) surface slope, (b) orbital velocity.

Nagata (1964), Bowden and White (1966) and Simpson (1969) combined

TABLE XLIX

Equivalence between frequency–direction spectra and wavenumber spectra $E_1(\omega,\theta) \leftrightarrows E_2(k_x, k_y)$ of ocean surface wave amplitudes

Case	ω	θ (from x-axis)	k_x	k_y
Deep water $L < 2h$	$g^{1/2}(k_x^2 + k_y^2)^{1/4}$	$\tan^{-1}\frac{k_y}{k_x}$	$\frac{\omega^2 \cos\theta}{g}$	$\frac{\omega^2 \sin\theta}{g}$
Shallow water $L \simeq h$	$[gh(k_x^2 + k_y^2)]^{1/2}$	$\tan^{-1}\frac{k_y}{k_x}$	$\frac{\omega \cos\theta}{(gh)^{1/2}}$	$\frac{\omega \sin\theta}{(gh)^{1/2}}$

L = wavelength, h = water depth, g = acceleration of gravity.

simultaneous orbital velocities (two horizontal components) and pressure of ocean waves for determination of directional spectra.

There is full equivalence between frequency–direction spectra and wavenumber spectra, which for ocean waves assumes particularly simple forms in case of deep water and of shallow water. A summary of transformation of coordinates in these two cases is given in Table XLIX. For information on fundamental properties of surface waves, the reader is referred to Lamb (1945).

Among *spectral parameters* studied, the slope of power spectra has attracted most attention, just as the case is with turbulence spectra in meteorology. The dependence of water-wave power density on frequency can be written as $E(\omega) \sim \omega^{\gamma}$ for ω-values beyond about twice the frequency of the major peak, and where the exponent γ usually has values around -5. Variations from this have been studied by Pierson (1959), due to non-linear effects, by Kinsman (1961) and by Garrett (1969). Comparison of the -5 law for ocean waves with the -3 law for atmospheric turbulence is no doubt of importance in any study of possible relations between these two phenomena. For spectra of internal gravity waves, Briscoe (1972) suggests two slopes: $k^{-5/3}$ for low wavenumbers and k^{-3} for high wavenumbers, the limit placed at $kh = 1$, where h is the thermocline depth. The k^{-3}-slope for high wavenumbers agrees with Dobson's (1970) stereo-photographic data.

Even though spectral slope has attracted much interest, it is clear that there must also be an increasing spectral part for lower frequencies. In this connection, it is of interest to note that Pierson and Moskowitz (1964) found the following formula for the power of the fully developed sea to represent observational data well (cf. Darbyshire, 1970a):

$$E(\omega) = \frac{\alpha g^2}{\omega^5} e^{-\beta g^4/V^4 \omega^4} \qquad [11]$$

Here α and β = dimensionless constants, g = acceleration of gravity, V = wind speed at 19.5 m above the sea surface. It is easily seen that [11] is dimensionally correct, as the left-hand side is power density, i.e. power (in this case square of length) per unit frequency interval.

Differentiating [11], we find for $E(\omega)$ = maximum that ω_{max} decreases with increasing V, in agreement with observations, or in formula:

$$\omega_{max} V = g\,(4\beta/5)^{1/4} \qquad \text{for } E(\omega) = \text{maximum} \qquad [12]$$

We find that for $\omega \geq 2\omega_{max}$ the exponential factor in [11] is ≥ 0.92, and therefore that $E(\omega)$ can be approximated by $\alpha g^2/\omega^5$ for so high frequencies, implying a spectral slope of -5, as discussed above. Other spectral properties may as well be calculated from [11]. Moreover, by virtue of the relations in section 3.3.3, any functional relation $E(\omega)$ corresponds to a certain relation $C_{11}(\tau)$ for the autocorrelation.

Equation [11] is also related to the question of stationarity, supposed to be true in spectral analysis of a random process. Equation [11] has to be taken as representing the power spectrum under stationary conditions, especially for a given value of the wind speed V. However, the wind speed is nearly always changing, i.e. $V = V(t)$. Then, we have to assume that the individual samples used for spectral estimation are taken so close together in time that the simultaneous V-variations can be neglected. Otherwise, non-stationary conditions would prevail. The situation is similar for any other random process, e.g. microseisms and atmospheric turbulence. Just as for atmospheric turbulence, explained above, it has been customary to display ocean-wave spectra in non-dimensional coordinates, for easier comparison of different sets of data. Frequently, the ordinate is of the shape $\ln\ [E(\omega)g^3/V^5]$ and the abscissa $\omega V/g$, both of which refer to non-dimensional quantities.

Since the 1950's a number of formulas corresponding to [11] have been derived by different authors. Useful summaries are given by Burling (1959) and Czepa and Schellenberger (1960). The exponent of ω in the first factor shows best agreement, most estimates being between 5 and 6. Burling (1959) finds that β is inversely dependent on fetch, which means that with increasing fetch the exponential factor will approach 1 and become constant. See also Kitaigorodski (1962), Kitaigorodskii and Strekalov (1962, 1963), Stewart (1967), DeLeonibus and Simpson (1972).

The application of spectral methods to ocean waves, including testing of theories, has prompted development of more efficient *methods for observations and recording* of such waves. Developments of this kind are reported by McIlwraith and Hays (1963), Barnett and Wilkerson (1967), Valenzuela et al. (1971), as well as in several papers in the book edited by Vetter and Bretschneider (1963). Observations of the vertical acceleration of the sea surface seem to offer good methods for extending spectra to higher frequencies (Garner, 1969). Besides observations from (stationary) weather ships, especially observations on oceanic platforms or towers in the open sea have been of great significance in clarifying the mutual

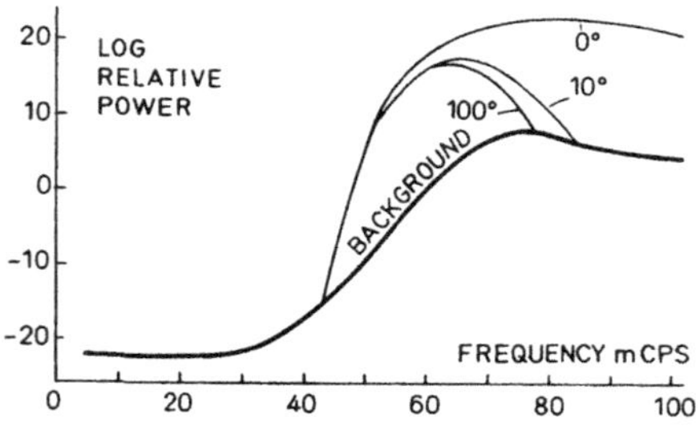

Fig.91. Typical ocean-wave power spectra, based on pressure measurements, at distances of 0°, 10°, and 100° from storm fetch and the average July background in the central Pacific. mCPS = millicycles/sec and 50 mCPS = 20 sec period. After Snodgrass et al. (1966).

interaction between sea and atmosphere (DeLeonibus, 1971), as well as laboratory research (Lai and Shemdin, 1971). Further information relevant to ocean-wave research can probably be derived from spectral studies of waves recorded on ice-sheets (Hunkins, 1962; LeSchack and Haubrich, 1964).

Ocean swell has been found to propagate practically over all oceans of the world from any given source, e.g. a cyclone. Snodgrass et al. (1966) report such a study using a profile of wave-recording stations across the Pacific from New Zealand to Alaska. Pressure detectors were used for the recording of ocean waves and readings were reduced to the surface, so as to yield surface-wave spectra. Averaged, idealized power spectra are shown in Fig.91 for the Pacific area.

Numerous practical applications of ocean-wave research have prompted development of spectral techniques. Practical applications include wave forecasting for various sea operations, design of ships and harbours, studies of beach erosion etc. Sea-wave observations have, however, to struggle with great scatter in observations. This is due to the fact that wave heights depend on numerous factors, such as wind speed and duration, fetch, atmospheric turbulence, etc. A recent method for ocean-wave prediction has been presented by Ewing (1971). His comparisons of predicted and observed frequency and direction spectra show satisfactory agreement.

9.2.2 Tsunamis

In an extensive investigation, Munk et al. (1959) report power spectra and cross-spectra for ocean surface waves, including tsunamis, in the period range of about 20 sec to 3 hours. This represents an intermediate range, between tides and shorter-period surface waves. In addition to reporting numerous observations in graphical form, the paper also gives a thorough discussion of the application of spectral techniques to ocean-wave records, including various sources or error and corrections for them. The spectra are calculated by first forming the correlation coefficients and then Fourier-transforming these. A similar intermediate period range—from 1/2 min to 12 hours—is treated by Munk (1962). This range in fact forms a relative minimum in the ocean-wave spectrum, between swell and tides, and is filled only occasionally and partially by various phenomena, among them tsunamis. This spectral gap is in a way a counterpart to the spectral gap of meteorological turbulence, centered at about the same periods. There may be a genetic connection between the two gaps.

Detailed analysis of tsunamis has a great practical importance in furnishing more reliable information for use by tsunami-warning systems. Active research is going on, particularly in the Pacific Ocean area. In summary, we can list the following factors which are of influence on observed tsunamis, especially their spectral structure:

(1) Source properties: focal depth (only shallow-focus earthquakes seem to

produce tsunamis), magnitude (larger magnitudes favouring lower frequencies), symmetry or lack of symmetry of the source (depending on ocean-bottom topography near the epicenter, in addition to source mechanism).

(2) Path properties: selective attenuation, favouring lower frequencies with increasing distance.

(3) Receiver properties: dimension and shape of water-bodies at observation point with ensuing resonance effect (seiche generation), which will enhance certain frequencies and suppress others.

(4) Contamination by other sea-level fluctuations: ocean surface waves (in a way corresponding to microseisms in a seismic record), tides (high tide will enhance the tsunami action, whereas low tide will prevent it).

Quite clearly, an observed tsunami spectrum depends on a sequence of factors which has a close analogy to the case we have encountered for seismic signals (Chapter 7 and 8). We shall now briefly discuss the various factors mentioned above.

Source properties. From measurements of dominant tsunami periods T (in minutes), formulas have been produced between log T and M with a striking numerical similarity to those reported in section 8.3.3 (Hatori, 1969b), suggesting an approximately constant ratio between tsunami periods and seismic wave periods. The literature, especially from Japan, contains much information to this effect (Takahasi and Aida, 1961, 1962, 1963; Hatori, 1967, 1968, 1969a, 1971).

Related to tsunamis are ocean waves generated by underwater volcanic eruptions, even though these seldom produce waves of noticeable amplitude at distant stations. A notable example is the Krakatoa volcanic eruption in 1883 in the Sunda Strait with gave rise to ocean waves recorded at many places around the Pacific Ocean and as far as the English Channel (by passage around Africa), in addition to earth-encircling atmospheric waves. As still another source, major underground nuclear explosions are able to generate resonant oscillations in nearby water basins (Olsen et al., 1972).

Receiver properties. Numerous observations testify to the effect that tsunamis generate seiches at the approach to bays and coasts, and that these seiches dominate the observations. The effect is quite similar to the recording of seismic waves through a seismograph, where the records are largely dominated by the seismograph response. We could likewise in the tsunami case describe the situation by the common filtering equation:

$$G_{st}(\omega) = F_t(\omega)\, H_s(\omega) \qquad [13]$$

where $G_{st}(\omega)$ = observed spectrum of tsunami t at station s, $F_t(\omega)$ = true tsunami spectrum, unaffected by local bay conditions, $H_s(\omega)$ = transfer function of the bay, the so-called harbour filter (cf. section 6.1.2).

Various operations are then possible by means of [13] with the aim at determining $F_t(\omega)$, either in absolute or in relative measure. Averaging over a number of different tsunamis, all with different spectra, we get:

$$\frac{1}{N_t}\sum_{t=1}^{N_t} G_{st}(\omega) = H_s(\omega)\underbrace{\frac{1}{N_t}\sum_{t=1}^{N_t} F_t(\omega)}_{\simeq \text{ constant}} \sim H_s(\omega) \qquad [14]$$

Knowing $H_s(\omega)$, either by the method just indicated or by other observations or theoretical calculations, it is an easy matter to obtain $F_t(\omega)$ from $G_{st}(\omega)/H_s(\omega)$. The function $F_t(\omega)$ gives the tsunami spectrum as uninfluenced by local conditions at the receiver (cf. Hatori, 1967). Also, taking the ratio between different tsunami spectra at the same station, we have:

$$G_{st'}(\omega)/G_{st''}(\omega) = F_{t'}(\omega)/F_{t''}(\omega) \qquad [15]$$

Cf. Takahasi and Aida (1963), Hatori (1968, 1969a and b).

There are numerous observations showing that the same tsunami produces different spectra at different harbours (depending on their local seiche conditions),

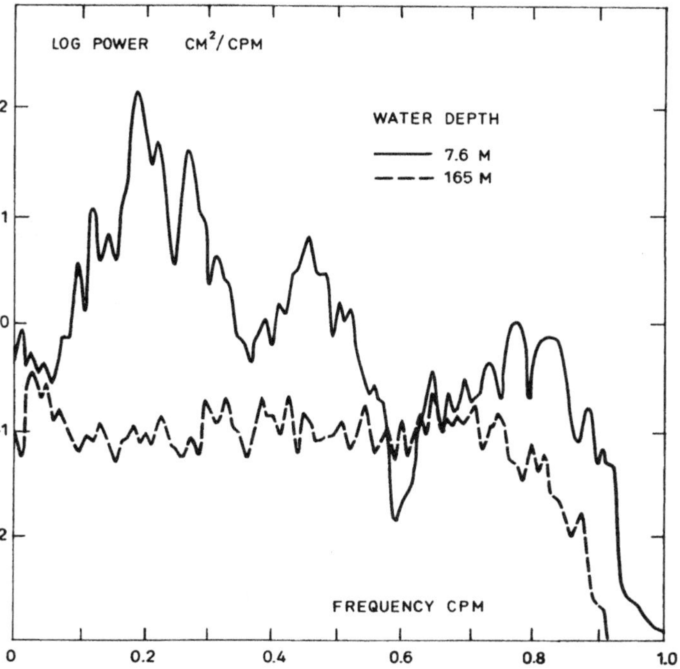

Fig.92. Power spectra of water-level oscillations in a bay (shallow water, 7.6 m deep) and on the open ocean (deeper water, 165 m). After Olsen and Hwang (1971).

while on the other hand, different tsunamis produce the same spectrum at a given harbour. See for example Takahasi and Aida (1961), Aida (1967), Munk (1962). From spectral analysis of tide records of the Chilean tsunami in May 1960, Hatori (1968) finds dominating periods of 60–80 min in north Japan but 40–50 min in southwest Japan. He emphasizes the similarity between different tsunami spectra for the same station, including an equal shaping of spectral peaks. The amplifying effect of bays is clearly demonstrated by Olsen and Hwang (1971) both from observations and theory, indicating power density amplifications of as much as 10^3 for selected frequencies (Fig.92).

However, spectra of tsunamis do not always suggest a clear-cut picture of local resonance (Higuchi, 1963; Loomis, 1966). For instance, evidence of tsunami-generated edge waves has been produced by Aida (1969) who studied Japanese records of tsunamis from the large Kurile Islands earthquakes in 1958 and 1963.

Analysis of tsunami records. Besides the analysis procedures just mentioned, also some other approaches have been suggested or tried. As shore-based tsunami observations to a greater or lesser extent are influenced by the local site, it would be desirable to have observations from the open sea, especially in order to correlate tsunami properties with earthquake source models in a more reliable way (Ben-Menahem and Rosenman, 1972). Alternatively, filtering of shore-based observations would be needed to better isolate source (and propagation) effects. An observed spectrum consists of the continuous ocean wave spectrum with superimposed peaks, due to the tsunami and to the seiche periods of the observation site. In a study of tsunamis at Hawaii, Loomis (1966) gives a detailed account of methods for calculation of spectra of tsunami records. By application of special band-pass filters to wave records, Royer and Reid (1971) were able to isolate tsunamis created by aftershocks of the Aleutian earthquake of March 9, 1957. Reflections of tsunami waves from different coast lines enter as disturbing factors. This latter effect has also been ascertained by Loomis (1966).

As distinct from ordinary ocean surface waves, tsunamis have an impulsive beginning, but as they may last up to one week, they will then mix with the ordinary surface waves. In a study of La Jolla records of the tsunami of the Chilean earthquake of May 22, 1960, Miller et al. (1962) found that the tsunami remained above background for a week, that its energy decayed as $E = E_0 \, e^{-at}$ with $a^{-1} \simeq$ 1/2 day in the frequency range 1–20 cycles per hour, and that the total tsunami energy was approximately $3 \cdot 10^{23}$ ergs (incidentally, this is about 20% of the total seismic wave energy of the earthquake, magnitude $M = 8.3$). A comparable decay constant was found for tidal waves by Munk et al. (1962).

9.2.3 Seiches and tides

Concerning *seiches* in lakes and bays, the literature has a wealth of information on pertinent harmonic analysis. One example, a classical one at least as far as

Sweden is concerned, is offered by the work of Bergsten (1926), who Fourier-analyzed the seiches of the Lake Vetter in southern Sweden. Newer techniques, including autocorrelation, power spectra and periodogram, were applied by Whittle (1954) in a study of seiche records from New Zealand. His paper contains also several statistical viewpoints on the methods used. In a period range intermediate between tides and swell, cross-power spectra between sea-level records from Argentine stations reveal partly peaks around 1 hour period, corresponding to seiches, partly a continuous spectrum, monotonically decreasing with increasing frequency (Inman et al., 1961). Power spectra of water levels in Lake Erie exhibit peaks at periods of 14, 9, 6 and 4 hours, due to longitudinal free oscillations, and in addition weaker peaks at 24 and 12 hours, due to tidal action (Platzman and Rao, 1964). Power spectra of seiche records in Japan reveal excitation of longer-period peaks by typhoons, cyclones and tsunamis (Hatori, 1969b). More recently, amplitude spectra, calculated by FFT method, have been applied to seiche data, which also clearly show the peaks corresponding to the seiche periods (Imasato, 1971).

Ocean tides have for many years been studied by harmonic analysis. These studies have been prompted for practical purposes, especially for tide prediction in harbours. This is also a field, where prediction has met with greater success, probably because the underlying causes are more regular in their behaviour. Complications arise mostly in shallow waters, due to local modification of the tides. By splitting the observed tides at Anchorage, Alaska, into 114 constituents, Zetler and Cummings (1967) have been able to make considerably improved predictions as compared to earlier efforts with fewer constituents.

Spectral analysis for tidal studies are also of significance. For example, from determinations of power spectra and coherence between simultaneous ocean tide records at several points, Wunsch et al. (1969) demonstrated the dominance of diurnal, semidiurnal and annual sea-level fluctuations. From tidal frequencies (about 1–3 cycles/day) the power spectra exhibit a gentle downward slope towards higher frequencies and a sharp rise towards lower frequencies (Munk and Bullard, 1963). For a discussion of spectral methods, see Zadro and Poretti (1972).

For *tidal prediction*, Munk and Cartwright (1966) presented a new technique, in which the ocean's response is evaluated for certain input functions, depending on gravitation and on radiant flux at the earth's surface. The residual record, i.e. the difference between observation and response, is due to irregular oscillations caused by wind, atmospheric pressure fluctuations, coast-line configurations, etc. Even though improvements in tidal prediction are only modest, this approach is more satisfactory from physical and theoretical points of view. The method is tested on wave records at Honolulu and at Newlyn (England). For further development, see Zetler et al. (1970).

In a somewhat related approach towards improved tidal prediction for the Thames estuary, Rossiter and Lennon (1968) subjected residuals (after subtraction

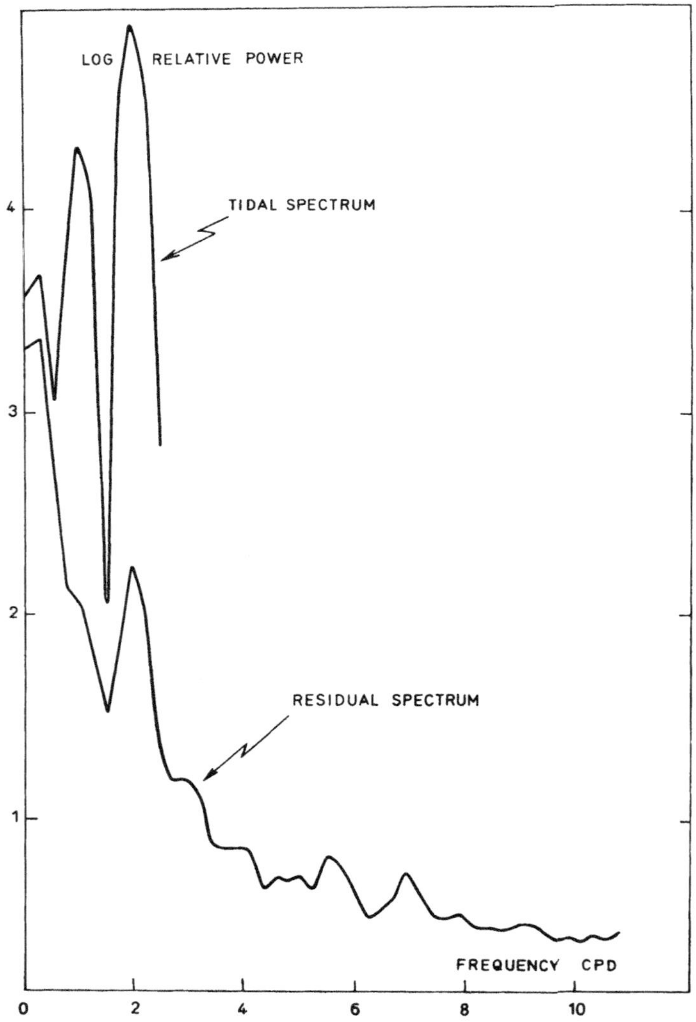

Fig.93. Power spectra of tides and residuals for Kochi, Japan. After Isozaki (1968).

of 60 harmonic constituents) to Fourier and power spectral analysis. See section 4.2.4. The same technique was applied to 53 points on the Japanese coasts by Isozaki (1968), who constructed power spectra of residuals between observed tides and calculated tides (using 30 harmonic constituents). See Fig.93. The residual peak at 2 cpd was quite persistent at all investigated stations, and at stations at bays or inland seas, additional higher-frequency peaks appeared, due to local effects superimposed on the astronomical tides. Similarly, Zetler and Lennon (1967) produced power spectra of residuals between observed and predicted tides, in an effort to compare different prediction schemes. The spectra of the residuals have the advantage to exhibit their frequency dependence. This method is naturally generally applicable to test any type of prediction, which can be put into spectral form, e.g. of ocean surface waves.

Related methods were also developed by Mosetti and Carrozzo (1971) and Mosetti and Manca (1972) for application to the Adriatic Sea. As in this case tides and seiches may appear with closely similar periods, narrow-band filtering is necessary to separate the two phenomena from each other. A method based on transfer functions was developed and applied to tidal prediction by Nowroozi (1970). Computations of the gravitational tide potential using high-precision data and new methods were presented by Cartwright and Tayler (1971).

Spectral comparison of tides and currents recorded at an ocean-bottom (3.9 km depth) installation off the coast of California, reveals that ocean currents at this location are produced primarily by tides (Nowroozi et al., 1968). The deep-sea location provides unique opportunities for studies of oceanic tidal spectra, free from coastal anomalies and non-linear effects caused by shallow depths (Nowroozi et al., 1966; Nowroozi, 1972a). The same installation has been used for simultaneous observations of oceanic tides and solid earth tides, displayed as amplitude–period spectra compared with theoretical spectra (Nowroozi et al., 1969).

9.2.4 Long-period sea-level fluctuations

Relations between ocean surface waves, sea level and meteorological parameters (atmospheric pressure and wind) have been investigated by means of power spectra, cross-spectra and coherence by Groves and Miyata (1967), Wunsch (1972a), and others. In research of this kind it is of great significance to make a correct selection of parameters to be combined, e.g., waves at one point may be more influenced by remote weather conditions, at least in some part of the spectrum. This is a fact of general validity in any research of such a comparative nature, even more so in microseismic research, when microseisms are combined with oceanographic and meteorological conditions. Similarly, because of the many influencing factors, caution is called for in interpretations of spectra, cross-spectra, coherence and phase between sea levels at different points (Hamon, 1962; Mysak, 1967a and b; Miyata and Groves, 1968; Askew et al., 1969; Longuet-Higgins, 1971). Examples of spectral studies of large-scale sea-level fluctuations are given by Groves (1956), who found a sea-level period of nearly four days for the central Pacific, explained by wind and wave effects, and by Shaw and Donn (1964) and Groves and Zetler (1964) with no very conclusive results. Using power spectral analysis of monthly means of sea level for the frequency range 0.0125–6 cycles per year (= periods of 80 years to 2 months), Haubrich and Munk (1959) were able to demonstrate the so-called pole tide with a period of 14 months, though it barely extends over noise level. A useful review, especially of long-period sea-level fluctuations together with genetic discussions, has been given by Donn et al. (1964).

Among factors of significance for long-period sea-level fluctuations, we note variations in atmospheric pressure, wind, current systems and shelf waves. The response of the sea level to varying atmospheric pressure is usually expressed in terms of a frequency-dependent barometric factor $H_{zp}(\omega)$ = the transfer function for the pressure-water level system (section 6.1.2). Multiplying the filtering equation between pressure $P(\omega)$ and pressure-induced sea-level oscillations $Z(\omega)$, i.e. $P(\omega)$. $H_{zp}(\omega) = Z(\omega)$, by $P^*(\omega)$, we get the corresponding relation between power densities:

$$E_p(\omega) \cdot H_{zp}(\omega) = E_{zp}(\omega) \qquad [16]$$

where $E_p(\omega)$ is the power spectrum of the input function (atmospheric pressure) and $E_{zp}(\omega)$ is the cross-power of sea level and barometric pressure (Roden, 1966b; Hamon, 1966; Mysak and Hamon, 1969). By having $E_{zp}(\omega)$ as the right-hand side and not just $E_z(\omega)$, we restrict the case to pressure-induced level fluctuations. Equation [16] may be split into its real and imaginary parts (Hamon, 1966).

Sea-level fluctuations due to varying barometric pressure (about 1.0 cm/mbar) have often been observed. Roden (1960) found for the west coast of North America a high coherence between sea-level fluctuations and atmospheric pressure for all frequencies investigated, from zero to six cycles per year. Power spectra of sea level show their largest values for low frequencies. Several observed deviations from these simple relations have been investigated theoretically by Mysak (1967a and b) for continental shelves and compared with observations. Hamon (1968) ascribes sea-level fluctuations near Australia to oceanic current systems. The importance of current systems for sea level is corroborated by Wyrtki and Graefe (1967).

Continental shelf waves may have amplitudes of centimeters, wavelengths of thousands of kilometers and periods of several days. For their detection, in sea-level records, spectral methods, often including well chosen filters, have been applied (Mooers and Smith, 1968; Mysak and Hamon, 1969). There appears to be a strong coupling to atmospheric pressure. Apparently abnormal effects of atmospheric pressure on sea level at Australian coastal stations were explained by influence of continental shelf waves (Hamon, 1966). See also Munk (1962).

9.2.5 Spectral studies of other oceanographic phenomena

Turbulence. Whereas turbulence studies dominate meteorological spectral analyses, sea-level fluctuations of all kinds dominate oceanographic spectral analyses. Turbulence no doubt exists also in the oceans, but lack of more abundant measurements, as compared with the atmosphere, has probably so far prevented more extensive application of spectral techniques. A review of oceanic turbulence studies by spectral means was given by Bowden (1962a). Examples of the method are provided by Bowden (1962b), who calculated auto- and cross-correlations and power-wavenumber spectra for turbulent velocity components measured near the sea bottom, and by Williams (1968), Stewart (1969), and Pao (1969), who gave power spectra for some temperature and velocity measurements in the ocean. They as well as Woods (1969) emphasize the mutual significance of combined turbulence studies for the ocean and the atmosphere. The wind structure just over the sea surface was investigated by S. D. Smith (1967, 1970), who displayed his measurements in terms of power spectra and cross-spectra of the different wind components. One problem is the difficulty to separate clearly between effects due to waves and those due to turbulence, except that waves dominate at low wavenumbers and turbulence dominates at high wavenumbers. There might be energy transfer between the two phenomena, and it may happen that energy measurements are, at least in part, ascribed to an incorrect phenomenon, when both coexist.

Quite generally, it may be stated that measurements of oceanic turbulence are quite analogous to those of atmospheric turbulence, as far as measured quantities and their three-dimensional distribution in space and in time concern, whereas naturally measuring techniques may differ. As an example of large-scale measurements of this type, we find Brekhovskikh et al. (1971) who present preliminary power spectra of currents and temperature at different depths in the tropical Atlantic. The diffusion of dye ejected into the ocean has provided a valuable means for studies of oceanic turbulence and circulation, to which also power-wavenumber spectra have been applied (Ichiye, 1967).

Currents. Measurements of velocities of water currents in oceans and lakes may also to advantage be handled by spectral methods. One example is Malone (1968), who applied frequency analysis of power and cross-power of current measurements in Lake Michigan. See also Ozmidov (1964). Vertical currents were investigated by spectral means by Voorhis (1968).

Kinetic energy density of intermediate scale (period below 10 hours, between tides and ocean surface waves) exhibits a slope of $-5/3$, i.e. quite in agreement with corresponding slopes for atmospheric turbulence, section 9.1.3 (note that this slope refers to power spectra, and that the corresponding slope for amplitude spectra is $-5/6$). This slope was reported by Webster (1969a) from current measurements in the ocean, by Krauss (1969) for ocean current components at different depths in the Baltic Sea, and by Cannon (1971) from measurements in an estuary.

Pochapsky (1966) found that internal waves cause most of the water movements at depth, using also power–frequency spectra of pressure, temperature and motion for around 2300 m depth in the North Atlantic. See also Cox (1962), Wunsch and Dahlen (1970), Kanari (1970), Schott (1971), and McWilliams (1972).

Lukasik and Grosch (1963) present power and cross-power spectra, coherences and phases for pressure and fluid velocity close to the ocean bottom, which are of special value in connection with studies of the boundary layer near the bottom.

Temperature. A lot of spectral studies of oceanic temperatures has been published. To achieve some order among these papers, we list them after the parameters measured. Denoting the temperature as $\bar{T} = \bar{T}(t,h)$, where t is time and h depth, we distinguish between the following special cases.

(1) $\bar{T} = \bar{T}(t,0)$, i.e. temperature–time fluctuation at the sea surface. From power spectra of sea-surface temperatures and low-frequency sea-level oscillations along the west coast of the USA, Roden (1963a, 1966 a and b) finds no other really significant peak than the one due to the annual variation. Cf. also Roden (1963c). Sea-surface temperatures in the East China Sea were analyzed by spectral methods by Moriyasu (1967) for the influence of advection and diffusion of different water masses and other effects. It has been emphasized above (section 9.1.1) that for meteorological phenomena, predictions based on simple harmonic analysis have failed. Spectra may provide better means for such purposes. Pierson (1959–60) has applied spectral methods to sea-surface temperatures and winds with a view to use the results for forecasting.

(2) $\bar{T} = \bar{T}(t,h_0)$, i.e. temperature–time fluctuations at different constant depths h_0. Hecht and White (1968) gave power spectra of temperature fluctuations to 300 m depth in the North Atlantic for the frequency range of 0.05 cph–43 cpm. They found a general power decrease with increasing frequency, though different in different frequency ranges, also superposed by peaks, indicating certain input of power. Cf. Roth (1971) concerning similar phenomena in meteorology. Hecht and White (1968) and Halpern (1971) found spectral peaks coinciding with the semidiurnal tide, the first-mentioned authors for the depth range 120–180 m. In a similar investigation of temperature fluctuations, Hecht and Hughes (1971) found in the presence of semidiurnal internal waves a cascading of power density with a general slope close to $-5/3$ towards higher frequencies (down to a period of around 3.6 sec), whereas in the absence of internal waves, the spectral shape was quite different. The internal waves no doubt represent an input of turbulent energy. Internal wave motion has been investigated by means of power–frequency spectra of temperature, density and circulation among others by Hollan (1966), demonstrating most internal unrest at periods of 1–10 minutes.

Power spectra of temperature close to the ocean bottom revealed tidal peaks in the Pacific but not in the North Atlantic, at least not for the localities investigated (Broek, 1969a and b). In Broek (1969a), methods are developed for suppressing aliasing effects and for determining the true frequency of each peak. In general, it appears as if only the semidiurnal and/or the annual variations dominate in deep-ocean temperature time series. In the category of paleoclimatological studies, we find Fourier series expansions of Atlantic Ocean temperatures for the last 300 000 years, determined from deep-sea cores, and correlated with astronomical data (Van den Heuvel, 1966).

(3) $\overline{T} = \overline{T}(t_0,h)$ i.e. temperature variation with depth h at given instants of time t_0. Spectral methods were applied by Roden (1964) and Cairns (1968) for investigation of time-variations of the vertical temperature distribution in the sea. Salinity–temperature–depth records were spectralized by Roden (1968). See also Roden and Groves (1960), Roden (1963b), and Lee and Cox (1966), the latter with a discussion of analysis of time series with missing data. For a study of turbulence and stratification in the ocean, Ichiye (1972) presented vertical wavenumber power spectra of temperature and salinity. McLeish (1970) found significant differences in horizontal wavenumber power spectra of temperature at the ocean surface and at slight depth.

(4) $\overline{T} = \overline{T}_0(t,h)$ = constant, i.e. depth–time relation for an isotherm surface T_0. Spectra of isotherm height variations were given by Darbyshire (1970b). The same type of observations, represented as power-frequency spectra, was used by Zalkan (1970) for studies of internal waves in the ocean off Baja California. From measurements of the depths of the 20° isotherm in the North Pacific, Wyrtki (1967) found from spectral peaks that eddies of a size of around 200 km exist, i.e. between ordinary turbulence and large-scale circulation. From temperature measurements off the Californian coast, Cairns and LaFond (1966) found, using power spectral techniques, that vertical oscillations of the thermocline were effected by winds and tides for periods exceeding 2 hours. (The term *thermocline* refers to the horizontal layer in which there is a rapid transition from warm water above to cold water below.) See also Wunsch (1972b).

Sound propagation through the ocean (so-called T-phases) has also been investigated by spectral methods. A useful review was given by Vigoureux and Hersey (1962), and application of filter theory to oceanic sound transmission problems can be found in the book by Tolstoy and Clay (1966). The reader can find much information on spectral studies of sound waves, propagated both through ocean and atmosphere, in addition to numerous technical applications, in special journals, notably the *Journal of the Acoustical Society of America.*

In graphs of frequency vs. time, T-phases exhibit the "Christmas-tree" shape (Northrop et al., 1960), implying considerable dispersion and where the "stem" corresponds to the beginning of the Airy phase. Spectral banding, i.e. persistence of certain frequency bands, derived from sonagram records of underwater volcanic eruptions, is interpreted as mainly due to interference effects in the water layer (Johnson and Norris, 1972), and would suggest that a profitable application of cepstral analysis (section 3.6.3) could be made. Hydrophone records of T-phases from explosions and one earthquake were spectralized by Milne (1959). Comparing power spectra of T-phases at Arctic (Tromsö) and equatorial (Trinidad) stations, Båth and Shahidi (1971) found a preponderance of lower frequencies at Trinidad as compared to Tromsö, explained by greater ocean depths (about 1.36 times) in the Trinidad case. Moreover, a clear spectral separation of T-phases and microseisms was found, inviting successful frequency filtering. For sound propagation in lakes, spectral methods have been useful in identifying various reflections, as from shore-lines etc. (DeNoyer et al., 1966). Spectra of both underwater and airborne volcanic sounds, investigated by Richards (1963), are related to the type of volcanic activity. An extensive review of ambient noise in the ocean, on the basis of pressure spectra in the range of 1 cps–100 kcps, has been given by Wenz (1962), including a detailed discussion of the origin of the noise within different spectral bands.

9.3 MICROSEISMOLOGY

The study of microseisms, the steady unrest of the ground, is a border-line field between meteorology, oceanography and seismology. Microseisms are no doubt of greatest concern to seismologists, but when their generation is to be explained, recourse must be taken to meteorological and oceanographic conditions. As a consequence, microseisms constitute a random process, like atmospheric turbulence and ocean surface waves. Therefore, spectral studies are closely related, both considering methods and results.

An enormous literature exists on microseisms, since around 100 years ago, but, as in other fields, it is only within the last twenty years that spectral techniques have been applied, when methods had been developed and computer facilities became available. A review of spectral studies of microseisms is given in Table L.

A main problem in microseismic research is to establish their origin. Whichever theory one may adhere to, spectral comparison of microseisms and of possible source and path properties will be of significance. It should be emphasized that the spectral analysis offers a powerful means to distinguish microseisms of different origin. Another problem concerns the wave modes involved in the microseismic motion. Both problems concerning origin and wave mode have to be studied in relation to the wave frequency—another point which emphasizes the significant role of spectral analysis. Therefore, we will find it most convenient to order our following study according to wave frequencies, just as we did in section 9.2 on ocean surface fluctuations.

9.3.1 Short-period microseisms ($T \leq 2$ sec about)

Spectral investigations of short-period microseisms, especially in the range of 0.5–5 cps, are of special importance because of their masking effect on short-period P-waves from distant events. Similar problems prevail in seismic prospection and crustal exploration, only that the frequencies involved are higher (Ivanova and Vasil'ev, 1964; Mikhota, 1968). From such viewpoints, short-period micro-

TABLE L

Review of observational spectral studies of microseisms

Reference	Period range (sec)	Location	Type of spectrum	Wave mode	Remark
Deacon (1949) and Darbyshire (1950)	1–12	surface	amplitude		confirmation of Longuet-Higgins' (1950) theory
Aki (1957)	0.07–1	surface	(autocorrelation)	Rayleigh and/or Love waves	
Moskvina and Shebalin (1958)	0.1–10	surface	amplitude		consistent minimum at 1 sec period
Akamatu (1961)	0.005–1	surface	power	Rayleigh and Love waves	tripartite station measurements
Kárník and Tobyáš (1961)	0.1–2	underground	amplitude		
Schneider (1961)	5–13	surface	amplitude	Love and Rayleigh waves	
Frantti et al. (1962)	0.03–2	surface	vertical amplitude		amplitude varies as v^{-2} for $v > 2$ cps
Haubrich and Iyer (1962)	2–170	surface	power	Rayleigh and/or Love waves	
Shima (1962)	0.08–0.5	surface and borehole	power		essentially traffic noise
Darbyshire (1963)	0.3–1	surface	power	Rayleigh and Love waves	confirmation of Longuet-Higgins' (1950) theory
Dinger (1963)	3–20	surface	power		confirmation of Longuet-Higgins' (1950) theory
Frantti (1963a)	0.01–5	surface	vertical particle velocity	Rayleigh waves	
Haubrich et al. (1963)	3–30	surface	vertical particle velocity	(Rayleigh waves)	

TABLE L (continued)

Reference	Period range (sec)	Location	Type of spectrum	Wave mode	Remark
Oliver and Page (1963)	7–9 14–18	surface	time-varying power	Rayleigh waves	sound spectrograph
Prentiss and Ewing (1963)	1–10	ocean bottom	vertical amplitude		signal/noise ratio for body waves comparable to land stations
Bradner and Dodds (1964)	0.1–8	ocean bottom and on land	power		see Fig.95
Douze (1964a, b, 1966)	0.2–8	deep well	amplitude, power, phase	fundamental- and higher-mode Rayleigh waves	
Isacks and Oliver (1964)	0.01–2	deep mine	particle velocity		
Milne and Clark (1964)	0.5–10	underwater, shallow-bottom, ice-covered	pressure, spectrograms	internal reflections	
Milne and Ganton (1964)	0.0001–2	ocean-bottom hydrophone under ice	pressure amplitude		mechanical activity associated with ice cover
Schneider and Backus (1964)	0.2–10	ocean bottom	power		signal/noise ratio slightly less on ocean bottom than on land
Schneider et al. (1964)	0.1–10	ocean bottom	power		signal/noise ratio on bottom equal to or less than on land
Vinnik and Pruchkina (1964)	0.2–2	surface	power		
Walker et al. (1964)	0.3–3 5–50	surface	time-varying power		continuously operating sound spectrograph

TABLE L (continued)

Reference	Period range (sec)	Location	Type of spectrum	Wave mode	Remark
Bradner et al. (1965)	0.1–10	ocean bottom	power	Rayleigh and Love waves	ocean bottom noise power one to five orders of magnitude higher than on land
De Bremaecker (1965)	3.3–20	surface	power		
Haubrich and MacKenzie (1965)	2–200	surface	power, cross-power		source identification
Robertson (1965)	0.2–10	surface	power		wind noise
Rykunov and Sedov (1965)	0.07–0.5	sea bottom	amplitude		sea-bottom noise approximately equal to land noise
Basham and Whitham (1966)	3–13	surface	amplitude		
Berckhemer and Akasche (1966)	0.1–3 10–100	surface	amplitude	surface waves (short period). Tilt etc. (long period)	relations to local wind investigated
Hinde and Gaunt (1966)	3–40	surface	power	Rayleigh and Love waves	instrumental development
Husebye and Jansson (1966)	0.4–5	surface (network)	power		
Kulhánek (1966)	0.3–1	mine	power		sharp maximum constantly near 2 cps
Latham and Sutton (1966)	1.5–30	ocean bottom	power	fundamental-mode Rayleigh waves	
Takano and Hagiwara (1966, 1968)	0.03–1	surface and deep well	amplitude, power		larger signal/noise ratio at depth
Zverev and Galkin (1966)	0.1–1	ocean bottom	amplitude		review paper

TABLE L (continued)

Reference	Period range (sec)	Location	Type of spectrum	Wave mode	Remark
Donn and Posmentier (1967)	3–20	surface	power		relation to microbaroms
Douze (1967)	0.3–6	deep hole and surface array	amplitude, power, phase, coherence	Rayleigh waves (fundamental and higher modes) and body waves	preponderance of body waves at shorter periods
Kulhánek (1967)	0.3–1 1–10	mine	power, amplitude		application of Wiener optimum noise filtering
Latham et al. (1967)	1–10	ocean bottom	pressure power	(Rayleigh waves)	interference effect (Longuet-Higgins, 1950) and direct coast effect
Savarensky et al. (1967)	2–10	surface	amplitude, phase	Rayleigh waves	phase-velocity dispersion for upper crustal structure study
Vinnik (1967a, b) Vinnik et al. (1967)	0.5–2 4–6	surface		fundamental- and higher-mode Rayleigh and Love waves, in addition to body waves	
Hirono et al. (1968, 1969), Suyehiro et al. (1970)	0.06–6	shallow holes	amplitude		
Latham and Nowroozi (1968)	5–10	ocean bottom	power	fundamental-mode Rayleigh waves	confirmation of Longuet-Higgins' (1950) theory
Sanford et al. (1968)	0.2–0.7	surface (alluvium)	power of particle velocity	fundamental- and higher-mode Rayleigh waves	train-generated microseisms
Abbas (1969)	5–11	surface	power	Rayleigh and Love waves	

TABLE L (continued)

Reference	Period range (sec)	Location	Type of spectrum	Wave mode	Remark
Capon (1969b)	20–40	surface (array)	power–wavenumber	fundamental-mode Rayleigh waves plus incoherent noise	
Darbyshire and Okeke (1969)	2–(100)	surface	power		primary and secondary microseisms
Der (1969)	0.1–10	deep well	power and cross-power	fundamental- and higher-mode Rayleigh and Love waves	vertical array
Haubrich and McCamy (1969)	2–25	surface (array)	power–wavenumber	Rayleigh, Love and body waves	
Lacoss et al. (1969)	~ 1–10	surface (array)	power–wavenumber	Rayleigh, Love and P-waves	see Fig.96
McGarr (1969)	4–10	surface, ocean bottom	vertical-component amplitude	(Rayleigh waves)	ocean-bottom microseisms 6–10 times larger for periods 5.4–7.0 sec
Bradner et al. (1970)	0.2–50	midwater, ocean bottom	power	leaky organ-pipe water oscillations	
Korhonen (1970, 1971)	3–25	surface	power		primary and secondary microseisms confirmed for Oulu, Finland
Bungum et al. (1971)	0.2–8	surface (array)	power		mainly frequency–wavenumber presentation
Dmitriyev (1971)	0.1–5	ocean bottom (shallow)	amplitude (pressure)		
Chandra and Cumming (1972)	2–60	surface	power	Rayleigh plus other waves	power displayed as isolines in period–azimuth graphs
Fix (1972)	0.1–2560	mine (130 m deep)	amplitude, power		longest period range investigated; also most quiet site

TABLE L (continued)

Reference	Period range (sec)	Location	Type of spectrum	Wave mode	Remark
Goforth et al. (1972)	0.05–5	surface	power		increased noise in 1–3 cps range over geothermal reservoirs
Iyer and Healy (1972)	1–7	surface (array)	power	surface and body waves	surface waves: 3–7 sec period body waves: 1–2 sec period
Kulhánek and Båth (1972)	1.2–5	surface	power		geographical distribution demonstrates significance also of smaller sea basins for short-period microseisms
Murphy et al. (1972)	10–130	surface	power		worldwide spectral minimum at 30–40 sec period
Savino et al. (1972a)	15–130	deep mine and surface	3-component amplitude		pronounced spectral minimum at 30–40 sec period, separating two different regimes
Savino et al. (1972b)	10–140	surface	amplitude, power		general appearance of minimum at 30–40 sec
Walzer (1972a)	3–9	surface	amplitude	Rayleigh waves	
Whorf (1972)	5–200	surface	power		quartz accelerometer. Spectral minimum at 30 sec, maxima at 18 and 9 sec

seisms have been investigated for their modal structure (Vinnik and Pruchkina, 1964), their frequency characteristics (Kulhánek, 1966), their source nature (Robertson, 1965; Sanford et al., 1968), their geographical distribution (Kulhánek and Båth, 1972), and their depth dependence (Takano and Hagiwara, 1966). The latter find decreasing signal and noise within a few hundred meter depth, but considerably larger noise decrease, with the consequence that signal/noise is larger at depth. From amplitude spectra of P- and S-waves from local earthquakes and of noise, Hirono et al. (1968, 1969) and Suyehiro et al. (1970) find an increase of signal/noise ratio by a factor of 3–4 for high frequencies (2–10 cps) at depths of 100–200 m in soft layers (sand, clay, silt) in Japan. Agreeing results were found in underground recordings of microearthquakes (section 8.1.6).

Frantti (1963a) investigated short-period microseisms (frequency range 0.2–100 cps) at numerous localities, especially in North America, and displayed his results as vertical-component particle-velocity spectra. He concludes that these microseisms originate close to the surface and consist essentially of surface waves, mostly of Rayleigh type. This agrees with findings for short-period microseisms by some seismologists, whereas Båth (1966a), who recorded the amplitude–depth variation for microseisms in the range of 8–33 cps, concluded that locally generated body waves constitute the major portion of the microseisms he studied. Using amplitude and power spectra of surface and deep-well microseisms of lower frequencies (0.2–5 cps) in different geological structures, Douze (1964a and b, 1966) concludes that these are mainly due to fundamental and especially 2nd-mode Rayleigh waves and that power decrease with depth depends on frequency, wind velocity, and type of ground. Power spectra of train-generated noise, investigated by Sanford et al. (1968), cover the range of 1.5–6 cps with peaks between 2 and 5 cps, explained as fundamental- and higher-mode Rayleigh waves.

Amplitude spectra of noise in the frequency range 2–15 cps on the bottom of the Black Sea reveal about the same noise level as on near-by continent for depths exceeding 1500 m, but more for shallower depth (Rykunov and Sedov, 1965). Higher noise level on the ocean floor (Pacific Ocean, depths 0.6–6 km) than on land (Aleutian Islands) in the frequency range 1–10 cps, is reported in a review paper by Zverev and Galkin (1966).

Using spectrograph recording, Milne and Clark (1964) studied ocean-bottom seismic noise in the period range 0.5–10 sec under Arctic sea ice. The line structure of the background (quasi-stationary) noise is explained in terms of multiple internal reflections of plane waves in a layered medium bounded at depth by the Mohorovičić discontinuity (cf. section 7.1.4), while ambient (transient) noise is ascribed to ice motion and cracking (Milne and Ganton, 1964).

9.3.2 Medium- and long-period microseisms (T over 2 sec)

A striking spectral feature in this range is the double-humped character, with

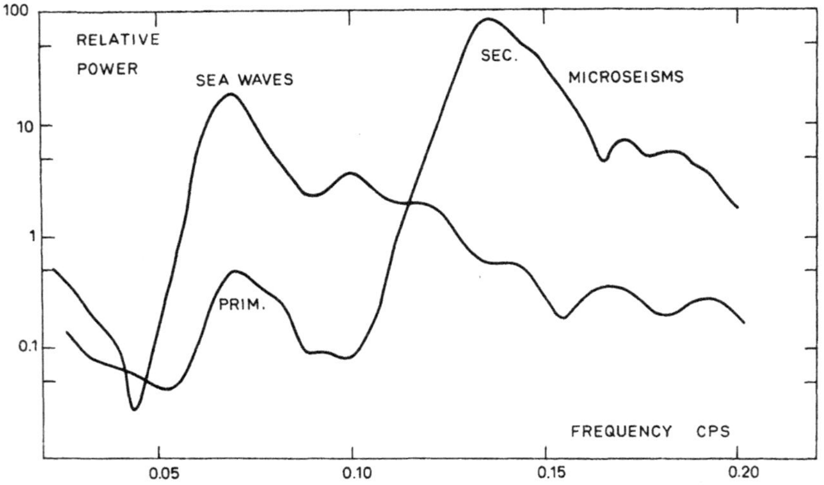

Fig.94. Primary (prim.) and secondary (sec.) microseisms and sea waves as power spectra. After Hinde and Gaunt (1966).

two distinct maxima. Microseisms with the same dominant period as ocean surface waves are sometimes referred to as *primary microseisms*, as distinct from *secondary microseisms* which have half this period (Fig.94). The double-humped spectra have attracted much attention (Oliver and Page, 1963; Haubrich et al., 1963; Haubrich and MacKenzie, 1965; Hinde and Gaunt, 1966; Darbyshire and Okeke, 1969; Korhonen, 1970, 1971). According to some observations, the two maxima derive from practically the same location (even though there may be different opinions about this location), but the question remains whether their mechanism of generation is the same. We could envisage the following alternatives: (1) they are both due to coast effects (surf), with fundamental and second harmonics; (2) the lower-frequency maximum (agreeing with ocean surface wave maxima) could be due to surf, whereas the higher-frequency maximum could be due to standing water-wave oscillations, in agreement with Longuet-Higgins' (1950) theory.

There seems to be fair agreement that primary microseisms derive from shallow water, and secondary from shallow and/or deep water. In Fig.94 (Hinde and Gaunt, 1966) the energy ratio between the peaks is about 180, and Haubrich et al. (1963) also find a ratio of about 100.

Deacon (1949) noted from spectra of microseisms at Kew and simultaneous ocean waves on the coast of Cornwall that the dominant periods have a ratio of 1/2, respectively. See also Darbyshire (1950, 1962) and Dinger (1963). This was considered as a strong support of Longuet-Higgins' (1950) theory, according to which microseisms originate from pressure on the sea bottom below standing sea waves. Relations between microseisms and ocean surface waves were investigated in the Pacific area by Snodgrass et al. (1966), who also found confirmation of Longuet-Higgins' theory for near-coast action. However, as many other investigators of microseisms, they had difficulty in assigning any particular area as origin of the microseisms, or as they state this, "microseisms may be associated with wave activity on beaches near the storm as well as on beaches near the station". A detailed observational and spectral analysis study would probably be able to distinguish between different source regions. Hinde and Gaunt (1966), using power spectra of microseisms in the frequency range 0.05–0.5 cps (Fig.94), recorded in Great Britain, and of ocean waves, also found good confirmation of Longuet-Higgins' (1950) theory, and that the microseisms consist of Rayleigh and Love waves.

The microseism studies have made much progress by a combined use of several different recording methods: in addition to traditional station recordings, more recently also ocean-bottom installations as well as surface arrays have contributed essentially.

Ocean-bottom recordings. A study of seismic records from different localities

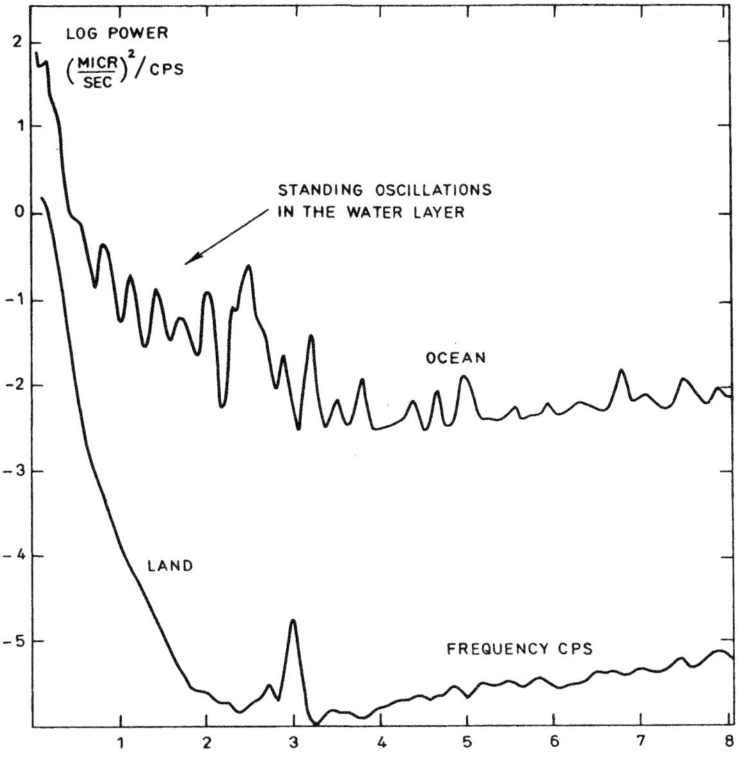

Fig.95. Seismic background power spectra (vertical component) on land and ocean bottom at Hawaii. After Bradner and Dodds (1964).

on the Pacific Ocean bottom (maximum around 6 km depth) is reported by Schneider et al. (1964). They found that both signals and noise are larger on the ocean bottom than on adjacent land stations with the effect that the signal/noise ratio in general is smaller on the ocean bottom than on land. The analysis is based on power spectra and coherence estimates. Similar results are reported by Schneider and Backus (1964) for 1.2 km depth off the Californian coast, while Prentiss and Ewing (1963) report similar signal/noise ratios as on land for several deep-ocean recordings in the Atlantic and the Arctic. Considerably larger noise on the Pacific Ocean bottom than on land was found by Bradner and Dodds (1964) and Latham and Nowroozi (1968), which indicates that a considerable part of the microseismic energy is generated at sea (Fig.95). The latter authors also confirmed Longuet-Higgins' theory, also that the observed microseisms consist of fundamental-mode Rayleigh waves travelling from sea to land (off coast of California). Theoretical comparison of their observations is offered by Abramovici (1968).

Bradner et al. (1965) constructed microseism spectra from records of ocean-bottom seismometers down to 5 km depth in the Pacific Ocean in an area roughly coinciding with the one used for swell studies by Snodgrass et al. (1966), also finding confirmation of Longuet-Higgins' theory. From records at 4.3 km ocean depth near Bermuda, Latham and Sutton (1966) found medium-period microseisms (displayed as power spectra) to have much larger amplitudes (due to different layering) than on Bermuda, but similar periods and similar time variations. They were fundamental-mode Rayleigh waves, generated by ocean-wave reflection at the island (in agreement with Longuet-Higgins' (1950) theory). Compared to Schneider et al. (1964), they found higher signal/noise ratio on the ocean bottom, at least for high-frequency signals from West Indies earthquakes.

Array recordings. Array stations have provided important data as to the modal structure of microseisms. From records at LASA in Montana, Lacoss et al. (1969) find the following results which may be typical of the interior of continents:

Frequency:	Wave structure:
> 0.3 cps	P-waves from distant oceanic sources
0.2–0.3 cps	higher-mode Rayleigh waves and P-waves
< 0.15 cps	fundamental-mode Rayleigh and Love waves

They express their data in terms of power density $E(k_x,k_y,\omega)$ as function of two-dimensional wavenumber (k_x,k_y) and frequency (ω). The power density is displayed as isolines in graphs with k_x and k_y as axes, one plot for each ω (this is the same display as shown in Fig.90a and used in some oceanographic work). See Fig.96. In such a graph, a vector from the origin to any point gives both the direction of wave arrival and the moveout velocity. In case of body waves, also an approximate distance determination is possible from the observed moveout velocity. However, as mentioned in section 2.5.5, Smart (1971) has warned against velocity determinations from such graphs. In a study of microseisms at the Nor-

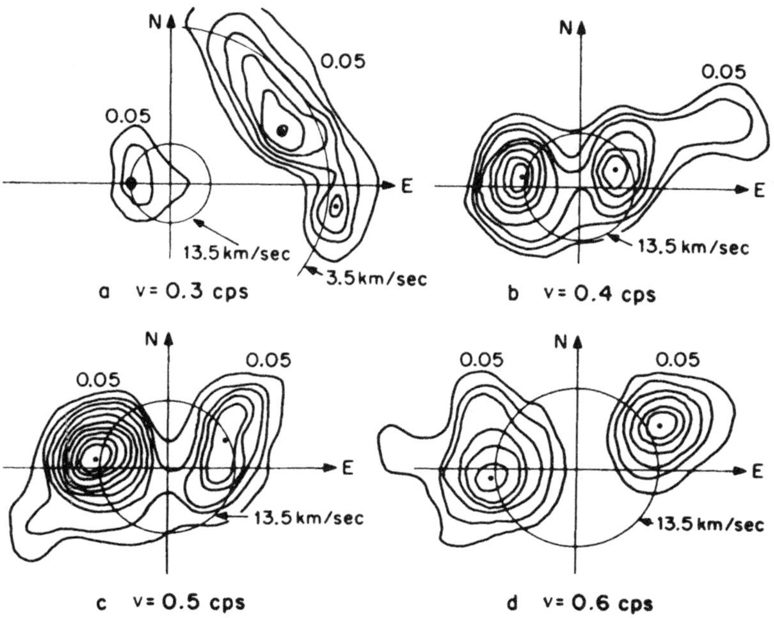

Fig.96. Power–wavenumber microseismic structure at LASA on December 2, 1965. After Lacoss et al. (1969).

wegian array NORSAR, Bungum et al. (1971) apply the same mode of presentation, and Vinnik (1967c) used the method for two Russian arrays. Results from LASA, closely agreeing with those of Lacoss et al. (1969) reported above, have been found also by Haubrich and McCamy (1969), who include a valuable description of spectral methods applied to microseismic array data.

In another investigation of LASA microseisms, Iyer and Healy (1972) find, besides ordinary surface-wave microseisms in the period range 3–7 sec, also body-wave microseisms in the range of 1.2 sec, believed to be generated locally. A detailed account of spectral relations between noise observed at a surface array and at depth is presented by Douze (1967).

The frequency range 0.05–0.025 cps (20–40 sec period) has been investigated with vertical-component LASA data by the same technique as above by Capon (1969b). He finds this noise to consist partly of fundamental-mode Rayleigh waves, believed to be caused by surf, partly of incoherent wave motion, which in some way is caused by local atmospheric pressure fluctuations. In case of surface waves, the frequency–wavenumber plots often show arc-like patterns. This is easy to understand as the wavenumber $k = \omega/c$ is constant for a constant ω (for given ω, the corresponding dispersion curve gives a certain value of the phase velocity c). Thus, $k_x^2 + k_y^2$ = constant, which is the equation for a circular arc, and whose angular extent depends on the spread in the arrival directions.

Analogous microseismic research, but using an ordinary network of stations, has been done by Chandra and Cumming (1972) in Canada, who display power and coherence as isolines with period and azimuth as coordinates. This is equivalent to the presentation $E(\omega,\theta)$ as explained already for ocean waves (Fig.90b). They find a coast effect to be the major source and that microseisms in the period range 4–7 sec consist mainly of Rayleigh waves, probably together with P-waves.

From comparative spectral studies of microbaroms (microbarograph records) and microseisms in the period range of 3–20 sec, Donn and Posmentier (1967) conclude that both phenomena have the same source, although microbaroms are modified by atmospheric conditions along the path.

9.3.3 Spectral shape

Summarizing the factors of influence on spectra of microseisms, we arrive at the following list:

(1) Source properties: ocean-wave periods (according to Longuet-Higgins' (1950) theory), intensity of source (larger intensity favouring longer periods).

(2) Path properties: selective attenuation, layering (including multiple-reflection propagation; cf. theory of Press and Ewing, 1948).

(3) Receiver properties: location (on surface or underground), surficial layering (which may enhance certain periods due to resonance and suppress others).

This list has no doubt a great similarity to the one presented above for tsunamis as well as for any kind of seismic waves. In general, all these properties enter the spectral problem, even though there are modifications in the various effects depending on the kind of wave studied.

A wide-band amplitude spectrum of microseisms was constructed on the basis of direct observations by Fix (1972), covering the largest period range so far (Fig.97). In addition to a general increase with increasing period, the microseismic spectrum shows the following typical features: a short-period maximum of local noise (A), two maxima corresponding to primary and secondary microseisms (C and B, respectively), and the pronounced minimum around 40 sec period (D). Berckhemer (1971), who constructed a similar spectrum by combining observations from different sources, demonstrated by showing also signal spectra for an $M = 7$ earthquake, that reception may be unfavourable near the microseism maximum at B but very favourable near the microseismic minimum at D.

There are two main factors that characterize the microseismic spectrum: partly a general slope, partly superposed maxima separated by spurious spectral minima.

It is a common experience that microseismic amplitudes and periods increase in parallelism (section 1.2.3). This is also confirmed by spectral analysis. The interesting thing is that such a dependence is common to several geophysical phenomena, as briefly summarized in Table LI, including geomagnetic spectra

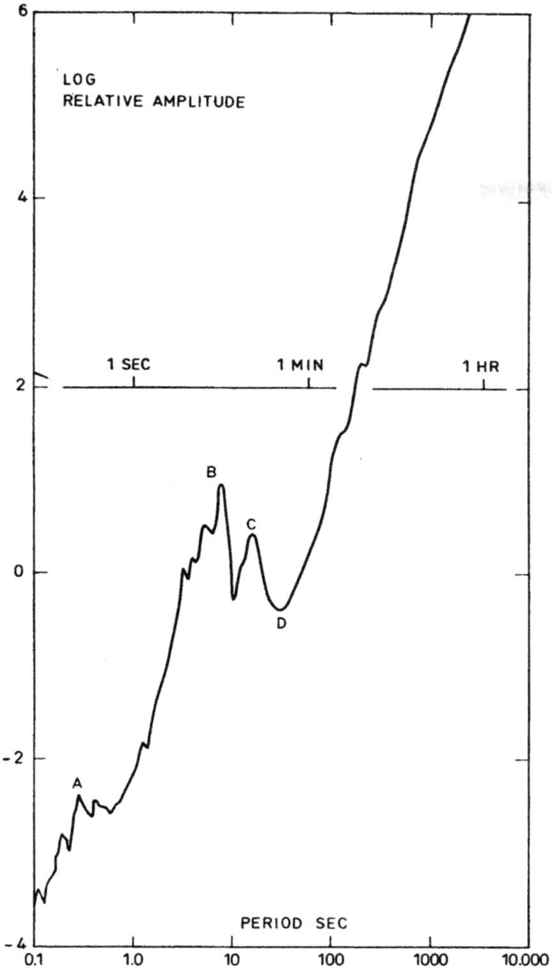

Fig.97. Horizontal microseismic amplitude spectral density. After Fix (1972).

(Chapter 10). For microseisms we would expect the slope to be distance-dependent, due to differential attenuation. This is in general agreement with observations, although it would be welcome with a more detailed spectral study of these phenomena.

As we have learnt in Chapters 7 and 8, spectra of seismic waves may exhibit a sequence of maxima and minima due to source properties (finite sources) and to structural properties (reverberations). In addition, minima may appear in a spectrum fortuitously, just between two different wave regimes due to different sources. Other related examples of double- and multiple-peaked spectra can be found in ocean wave spectra, which frequently exhibit minima or gaps between

TABLE LI

Power–frequency relations $E(\omega) \sim \omega^{\gamma}$

Phenomenon	γ	Reference
Meteorological turbulence	$-3, -5/3$	section 9.1.3
Ocean surface waves	-5	section 9.2.1
Ocean currents	$-5/3$	section 9.2.5
Ocean temperatures	$-5/3$	section 9.2.5
Microseisms	-4 (over 2 cps)	section 9.3.3
	-8 (below 2 cps)	
Ambient sea noise	-2 to -2.7	section 9.2.5
Geomagnetism	-2 to -2.3	section 10.1.1

low-frequency swell and more high-frequency locally generated surface waves (DeLeonibus and Simpson, 1972, and section 9.2). The microseismic spectrum offers several such examples, which we list here:

(1) Separation between short-period local noise (period around 0.5 sec) and more long-period microseisms of distant origin (5–6 sec period). A figure by Douze (1967) demonstrates this minimum very clearly. It was clearly demonstrated also by Moskvina and Shebalin (1958).

(2) Separation between primary and secondary microseisms, as already explained, with periods approximately equal to ocean surface waves and equal to half their periods, respectively (Fig.94).

(3) In recent extensions of microseismic studies to long periods, a clear minimum has shown up at 30–40 sec period (Fig.97). This is explained as a separation between microseisms due to swell (periods below 30 sec) and those generated by atmospheric effects and codas of earthquakes (periods over 40 sec). See further Savino et al. (1972a and b) and Whorf (1972). From a comparison of several stations around the world, Murphy et al. (1972) found clear indications that this minimum is a worldwide phenomenon. This microseismic minimum seems to offer a suitable "window" for more sensitive observations of surface waves and long-period body waves from weak events, and may therefore be advantageous for discrimination purposes.

Chapter 10

SPECTRAL STUDIES OF GRAVITY AND GEOMAGNETISM

Gravity and geomagnetism are closely related in the sense that both are potential fields, i.e., the respective forces are the space derivatives of quantities, termed potentials. By virtue of this similarity between the two fields of study, we can most conveniently discuss the spectral methods together, and make separation rather with regard to the independent variable used, such as frequency or wavenumber. Spectral studies of problems in gravity and geomagnetism represent by and large just another application of general principles, with the special amendments that such studies may involve. In a review article, Jacqmin and Pekar (1969) emphasize similarities and dissimilarities in the treatment of seismic and gravimetric problems. Especially in the frequency and wavenumber domains, the two fields have much in common, as far as computational techniques are concerned.

We shall limit our treatment to spectral methods, while for more detailed discussion of gravimetric or geomagnetic problems, the reader is referred to special books. For geomagnetism, there is a recent comprehensive and authoritative account to be found in the books edited by Matsushita and Campbell (1967). Also, we shall not enter upon problems of ionospheric research or any electromagnetic spectra, even though parts of the geomagnetic study have close relations to these fields.

10.1 TIME SERIES ANALYSES

In applying power spectral methods to the search for periods in magnetic as well as other geophysical data, there are in essence two methods which can be used:

(1) Get the power spectrum for the given time series and investigate maxima for statistical significance.

(2) Get the cross-power spectra between the given time series and a sequence of sinusoids with given amplitudes, frequencies and phases, and investigate maxima in the resulting cross-power spectra. The latter method is computationally more cumbersome in a general application, but could be used with advantage in the search for specified periods (cf. Nowroozi et al., 1969).

10.1.1 Geomagnetic frequency spectra

Geomagnetic fluctuations are probably unique among geophysical spectra by extending over such a wide range of periods or frequencies, from those of 10^4 cps (Heirtzler et al., 1960) to those of around 10^{-11} cps, i.e. a range of about 50 octaves. The latter corresponds to periods of several thousand years and is derived from paleomagnetic data (Yukutake, 1962). It is considered as primarily due to the westward drift of the geomagnetic field. For comparison, sea-level fluctuations extend approximately over the frequency range from 10^{-1} to 10^{-11} cps. Cf. section 1.7.

Time series analysis of geomagnetic variations has been done for many years. See for instance Chapman and Bartels (1940, 1951). It is only more recently, in the last ten to fifteen years, that power spectrum analysis has been applied to these investigations. In reviewing different spectral results, we distinguish for

TABLE LII

Period studies of geomagnetic elements, using spectral methods
A. Long-period variations

Reference	Period		
	years	months	days
Ward (1960) Ward and Shapiro (1961a)	11	6	27, 14, 9, 7, 5½, 4½
Eckhardt et al. (1963)		6	13.5
Currie (1966)	1	6	
Cehak and Pichler (1969b)			27
Edwards and Kurtz (1971) Džodenčuková (1972)			27 + harmonics
Bhargava (1972)	1	6	
Hauska (1972)			27, 13, 9, 6

B. Short-period variations (geomagnetic micropulsations)

Reference	Period range investigated	
	min	sec
Heirtzler et al. (1960)		0.0001–0.1
Santirocco and Parker (1963)		0.2–200
Campbell (1966, 1967)		0.2–600
Kleĭmenova (1965)		0.3–1
Ku et al. (1967) Jones (1969)		10–200
Prikner et al. (1972)		10–100
Sasai (1966)	6–	
Lambert and Caner (1965)	~ 10–30	
Best (1968)	20–120	

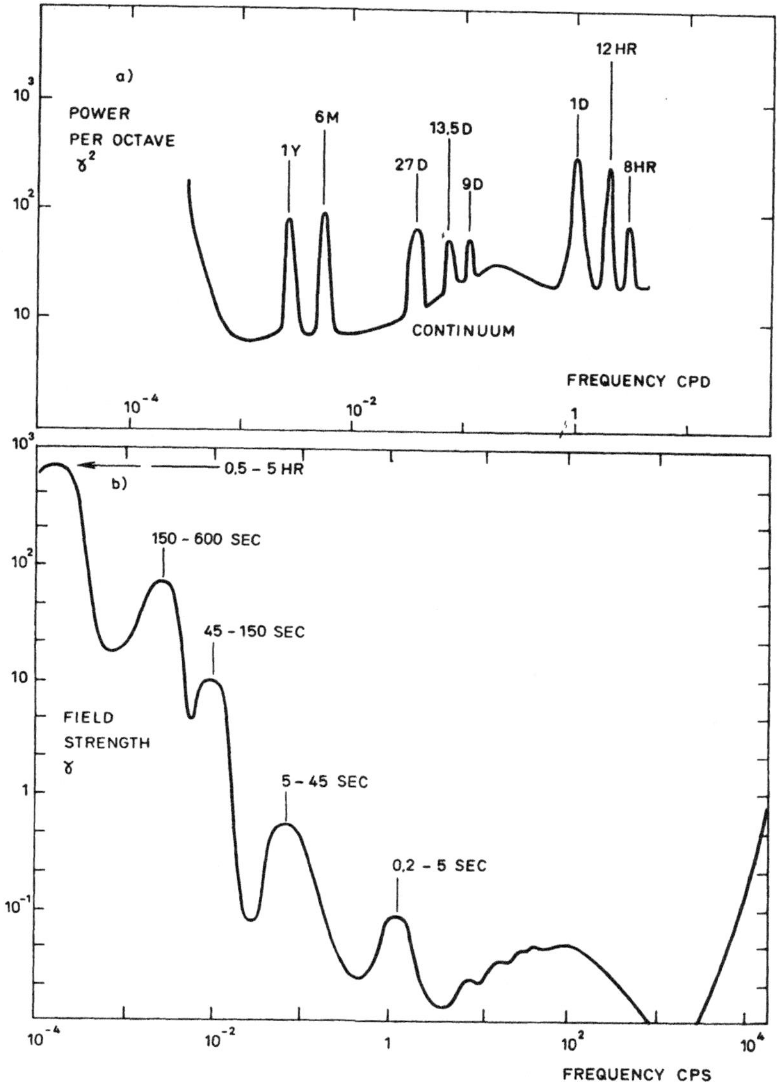

Fig.98. Geomagnetic frequency spectra. a) Power spectrum of the horizontal component. After Banks (1969). b) Geomagnetic pulsation spectrum. After Campbell (1966).

convenience between long-period variations (periods of days and more) and short-period variations (periods less than one day).

Long-period variations. Periods of geomagnetic elements, mostly found by power spectral analysis, are summarized in Table LII, to which we give the following comments (see also Fig.98). An outstanding period is 27 days and its higher harmonics (14, 9, 7, 5½, 4½ days). Currie's (1966) paper contains several useful considerations on methods in computing power spectra, especially for magnetic elements.

Hauska (1972), who found periods of 27 days and its harmonics, reports that cosmic-ray data show only the 27-day period (Stern, 1962), and he emphasizes that even though these periodicities are caused by solar and interplanetary effects, their exact reason is still largely unrevealed. Dynamo action in the ionosphere by corresponding variation in the air velocity is a probable explanation.

Geomagnetism is found to correlate well with sunspots, whereas meteorological variables do not. Application of cross-spectral analysis would probably have been of great interest. However, it appears as a justified question whether sunspot numbers really represent the kind of solar activity that is to be related with terrestrial phenomena. From investigations of power spectra of solar data and geomagnetism, using monthly mean values, Cehak and Pichler (1969a) suggest that the solar wind has a closer connection to variations in the geomagnetic field than sunspot numbers. For shorter intervals, Cehak and Pichler (1969b) find the 27-day period both in solar and magnetic data, whereas the magnetic data also exhibit large amplitudes at shorter periods of variable length. No significant lunar influence is found according to Rassbach et al. (1966) and Shapiro and Ward (1966), while Black (1970) has investigated lunar and solar variations of magnetic declination for the frequency range 0–4.5 cpd. He applies both autocorrelation power methods and FFT methods and gives much valuable detail on the spectral computations. The lunar and solar geomagnetic influences are explained as due to atmospheric tides. Geomagnetic lunar and solar daily variations within the same general frequency range were investigated by Gupta and Chapman (1969) both by harmonic analysis and by spectral (correlation-transform) analysis, with partial agreement. They found smaller amplitudes by harmonic analysis than by spectral analysis, but identical peak frequencies. Their paper is valuable not the least from the methodical point of view. Cf. also Coleman and Smith (1966).

Currie (1967, 1968) emphasizes that variations in the magnetic field with periods shorter than about 4 years are due to external influences, whereas those of longer periods than 4 years are due to internal influences (from the core). For the period range of about 4–40 years, the magnetic power spectrum $E_0(\omega)$ observed at the surface is related to the power spectrum $E_c(\omega)$ at the core–mantle boundary via the mantle power transfer function $|H_1(\omega)|^2$ as follows:

$$E_0(\omega) = |H_1(\omega)|^2 E_c(\omega) \qquad [1]$$

i.e. in a way familiar from seismological phenomena. This relation has been used in estimations of the lower-mantle conductivity.

For magnetic fields of external origin, on the other hand, external sources induce electric currents in the earth, which in their turn give rise to a magnetic field, depending upon the conductivity in the interior. A filtering equation, analogous to [1], holds also in this case (Sen, 1968):

$$E_i(\omega) = |H_2(\omega)|^2 E_e(\omega) \qquad [2]$$

where $E_i(\omega)$ = the output = magnetic power of internal origin, but generated by external factors, $E_e(\omega)$ = the input = magnetic power of external origin, $|H_2(\omega)|^2$ = the corresponding power transfer function depending upon conductivity in the earth. This latter quantity may be calculated from the two others and, compared with theoretical models, it yields valuable information about the earth's internal structure (Banks, 1969). See also Ku et al. (1970).

Short-period variations. A summary is given in Table LII of spectral investigations in the short-period range, mostly based on power spectra. It has been emphasized by Sugiura (1960) that an observer, moving in relation to the sun, will observe Doppler-shifted frequencies in harmonic analysis of a geomagnetic storm. In geomagnetic power–frequency spectra, Santirocco and Parker (1963) found the noise to show up as ripples on a generally decreasing power in the frequency range 0.005–5 cps (= period range 200–0.2 sec). Detrending and cepstral analysis would probably be able to exhibit the noise still better. The general power decrease with increasing frequency of 6–7 dB/octave is confirmed by Davidson (1964), Herron (1967), and Sen (1968). This general slope extends over a wide period range, from those of 0.0001–0.1 sec (Heirtzler et al., 1960) to those of 3.7–33 years (Currie, 1968). This "red-noise" phenomenon reminds of similar slopes encountered in meteorological and oceanographic wave spectra (section 9.3.3).

To this general period range belong geomagnetic pulsations, i.e. partly irregular transients of a few minutes duration, frequently in the shape of damped oscillations (occurring mostly at night), partly more regular beat-like oscillations (mostly in the daytime). Spectral amplitude and phase characteristics, especially of the transient type, have been investigated among others by Střeštík (1969 a and b, 1970, 1971), Prikner (1969a and b), Dobeš et al. (1970), Prikner et al. (1972), and by Střeštík et al. (1973).

The spectral analysis is generally extended only over the pulse duration. The average pulse shape in the time domain is constructed by first averaging spectra frequency-wise and then inverting the average spectrum, rather than by direct averaging in the time domain. Due their non-stationary character, time-dependent spectra are more apt to demonstrate the development of geomagnetic pulsations. Such spectral methods, showing intensity of the pulsation by isolines in graphs with period and time as coordinates, were constructed for both types of pulsations by Kato et al. (1966), by means of narrow band-pass filtering of magnetic tape records. Further detailed investigations of geomagnetic pulsations were made by Eleman (1967), also including some power spectra.

Lambert and Caner (1965) found that the coast effect in geomagnetic variations is due to conductivity differences of the land-ocean surface layers for periods less than 10–15 min and also of the upper mantle for periods larger than 30 min.

By means of power spectra of geomagnetic elements for periods of 6 min and over, Sasai (1966) in Japan tried to find connections with heat flow anomalies, using similar methods, and Whitham (1963) suggested a conducting layer near the base of the crust to explain anomalously low Z-components at Mould Bay in Canada. Similar observations and interpretations are reported for parts of Scotland by Osemeikhian and Everett (1968).

10.1.2 Tidal spectra

The earth's tides constitute a special branch of gravity effects. Earth tides, like oceanic tides, have for many years been subject to harmonic analysis. In his book, Melchior (1966) describes several different methods and computational systems for tidal harmonic analysis, but he also describes how computers more and more are taking over the calculations within the 1960's. Comparative tests show that different methods for tidal harmonic analysis generally yield results of good internal agreement, and the methods differ essentially only in computational aspects. It is characteristic that computational schemes for harmonic analysis by simple means have persisted longer in this field than in most other fields, even though application of spectral methods was advocated already by Jobert (1962). Moreover, we notice that the analysis is restricted to time series from single points, and more sophisticated techniques, like power and cross-power spectra, etc., are not applied to earth tides in Melchior's (1966) book. One contributing factor to this state of affairs is probably that harmonic analysis has largely proved to be more successful even for prediction purposes in this field than, for instance, in meteorology. The tides, especially those of the solid earth, represent a more deterministic phenomenon than atmospheric turbulence and ocean surface waves. Zetler (1964) also emphasizes the superiority of harmonic analysis to power spectrum analysis in the case of tides, because we are dealing with a limited number of unique frequencies rather than a continuum of energy, moreover the peaks have more or less constant phases and the signal/noise ratio is exceptionally high. Under such conditions, harmonic analysis is bound to give higher resolution than power spectra with their inherent smoothing.

Melchior's (1966) book contains a very extensive bibliography covering the field of earth tides (including theory, observations, harmonic analysis, results and relations to other observations) up to the year 1964. The reader will also find many papers on harmonic analysis of earth tides in *Bulletin d'informations, Marées terrestres* issued by the Permanent Commission on Earth Tides, International Association of Geodesy, under the editorship of P. Melchior in Brussels.

From the spectral methodical point of view, a number of papers are of special importance regarding application to earth-tide observations. For example, Barsenkov (1967 a and b) compared results of harmonic analysis and of spectral analysis of the same set of gravity data (investigated for tidal effects) and found

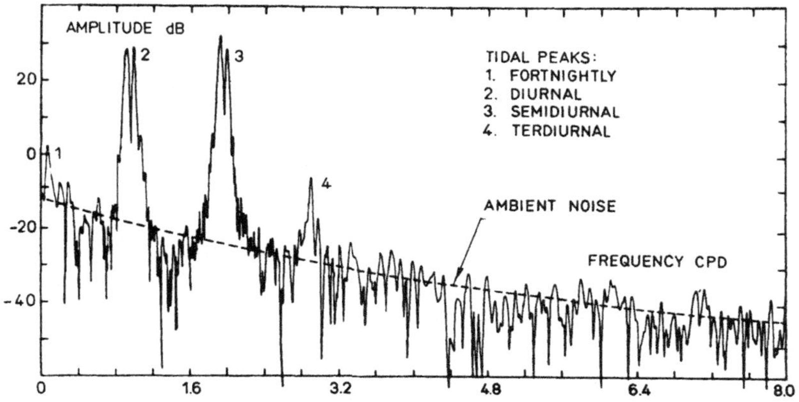

Fig.99. Amplitude–frequency tidal spectrum. After Prothero and Goodkind (1972).

good agreement. His spectral analysis is discussed by Melchior and Venedikov (1968). Later Barsenkov (1972) used spectral methods to isolate long-period tidal waves (periods up to 27 days) from gravimeter records, where reliable elimination of long-period drift becomes a problem. Likewise, Jobert (1963) compared results of spectral analysis with theoretical results for tidal data, also finding good agreement. Pan (1970, 1971) made a significant attempt to apply statistical methods (correlations, power spectra, filtering, etc.) to earth-tide observations, especially for error elimination. Harrison et al. (1963) applied both power spectrum analysis and direct Fourier transform analysis to a world-wide set of tidal gravity meter records from the IGY. The purpose of the power spectrum was to determine the general noise background and the Fourier analysis was concentrated to determination of amplitude and phase of the major tidal lines.

Among papers where spectral analysis has been applied to tides, we also note Kuo and Ewing (1966), who emphasize the great significance of ocean tidal loading in any observation of the earth tides. This was also studied by spectral methods by Blum and Hatzfeld (1970), and demonstrated by Bilham et al. (1972) from power and phase spectra of strainmeter records in England from differences between observed and theoretical tidal spectra. Other examples are Mikumo and Nakagawa (1968), Pil'nik (1970), and Prothero and Goodkind (1972), the latter presenting an amplitude–frequency tidal spectrum derived from a special gravimeter of high sensitivity (Fig.99).

It appears as if in the analysis of earth-tide records, different schools of thought have crystallized more than in other geophysical fields, partly concerning different methods of harmonic analysis, partly concerning harmonic versus spectral analysis, even though the general tendency now seems to be towards spectral methods.

10.1.3 Crustal movement

Closely related to tides is the study of secular crustal movements and their relation to gravity. Spatial and temporal spectral analysis has been applied also to such movements. One example is Mizoue (1967) who investigated both secular vertical movements and gravity anomalies along various levelling routes in Japan, USA and USSR. Tiltmeter and extensometer records contain generally a wide range of various phenomena, of which the most important are the following:

(1) Tides, both the solid earth tides, effects of varying loading due to oceanic tides in nearby oceans, and atmospheric tides.

(2) Meteorological effects: atmospheric pressure and its variations, temperature, rainfall, snow and ice covers, etc.

(3) Seismic phenomena: earthquakes and free oscillations of the earth.

(4) Slow crustal movement, due to tectonic and other forces.

(5) Instrumental disturbances.

As these various effects cover a wide frequency range, it is usually necessary to apply digital band-pass filtering to isolate different effects. A clear illustration of the efficiency of such filtering is given by T. Tanaka (1966), which was reproduced in Fig.57. By means of filtering and spectral techniques, T. Tanaka (1966, 1967 a and b, 1968, 1969) and Zadro (1961, 1966) could isolate and study various influences on ground movement from tiltmeter–extensometer records and pendulum–tidegauge–barograph records, respectively. In order to study phenomena in the earth's crust, a good instrumental coupling to the ground is essential, especially for such long periods that various extraneous effects may easily deteriorate the records. In order to ascertain a good coupling, S. W. Smith and Kind (1972) compared observed and calculated earth tide spectra.

Above, in section 9.2.4, we discussed sea-level fluctuation due to meteorological effects. Water-level fluctuation of longer periods may be caused by crustal deformation, and thus it forms another means to investigate such deformations. A paper by Dohler and Ku (1970) presents some pertinent power–frequency spectra for Canadian stations, and a paper by Caputo et al. (1972) for some Italian stations.

10.2 SPACE SERIES ANALYSES

10.2.1 Representation of two-dimensional fields

Data sampling. In subjecting a map of gravity or geomagnetism to computation of spectra or correlations, the usual procedure is to digitize the map, i.e. to get numerical samples from a certain grid covering the map. The sampling interval has to be carefully chosen, just as in dealing with sampling from a seismograph record, for instance. Considerations to be taken are the partly conflicting requirements imposed by aliasing, available computer core memory, and the type of structures to be studied.

Unequal spacing of data may present a certain problem in spectral studies. This does not so much concern time series, where generally continuous traces are available, alternatively digital recording with equally spaced digits is applied. On the other hand, in the space domain, information on gravity or geomagnetism is as a rule unequally spaced. A frequent procedure is then to determine values at equal spacing by interpolation, often from hand-drawn curves. This entails an uncertainty factor. Interpolation to get equally spaced data can also be made mathematically, by fitting polynomials to the given observations. But also such procedures are subject to some uncertainty. For such a study, see Naidu (1970b). It is therefore preferable to use original data for spectral analysis, if possible, even if unequally spaced. Examples of procedures where unequally spaced data can be used are the spherical harmonic analysis, as suggested by Fougere (1963), and the expansion into bicubic splines, suggested by Bhattacharyya (1969). Handling of unequally spaced time series by least-squares methods is discussed by Zetler et al. (1965). Cf. also section 4.2.5.

Another frequently encountered problem in handling spatially distributed data is the existence of "holes", i.e., areas where data are missing, within the region being investigated. Among methods to deal with this problem, we mention one due to Lewis and Dorman (1970) for the case of topography. Their procedure is as follows: (1) make a guess on likely values; (2) take the Fourier transform; (3) do low-pass filtering; (4) do inverse transformation; (5) use the values so derived. This procedure entails at least no discontinuities in the surface topography at the missing points.

Spherical harmonic expansion. For the mathematical representation of spatial distribution of gravity and geomagnetism, expansion in spherical harmonics is the natural method for spherical or hemispherical coverage (sections 2.5.1 and 2.5.4). For fundamentals of spherical harmonic expansions the reader is referred to Heiskanen and Vening Meinesz (1958) for gravity and to Chapman and Bartels (1940, 1951) for geomagnetism. Problems connected with such expansions are partly due to very uneven distribution of observations, partly to interdependence of observations, even at large separation. In recent time, observations of artificial satellite orbits have provided valuable data for spherical harmonic expansions of gravity (see, for example, King-Hele et al., 1963).

Matsushita and Maeda (1965a and b) expressed the magnetic potentials of the quiet solar and the lunar daily variations in spherical harmonics in terms of coefficients of the external and internal electric current systems. Their two papers contain extensive references to earlier works in these fields. Aeromagnetic measurements of total field intensity along world-encircling profiles are considered by Alldredge et al. (1963) for expansion both in Fourier series along profiles and for spherical harmonic analysis. For world charts with contour spacing of 1000 γ, spherical harmonic expansion to order and degree of at most 10 is sufficiently accurate. In the profile spectra, small wavelengths correspond to crustal sources and large wavelengths to core sources, whereas the mantle leads to a minimum centered around a wavelength of approximately 900 km.

Sinc-function expansion. For observations of smaller, local or regional extent, representation in a plane is often adequate and then several different methods, as listed in Table LIII, have been applied. Two-dimensional Fourier series expansion (cf. section 2.5.1) has been applied both to gravity and geomagnetism. Bhattacharyya (1965) emphasizes that this method permits high accuracy in calculation, for instance, of vertical derivatives and upward and downward continuation of the magnetic field, provided all possible Fourier coefficients are included.

TABLE LIII

Examples of spatial studies of gravity and geomagnetism[1]

Method	Gravity	Geomagnetism
Fourier series	Tsuboi and Fuchida (1937, 1938)	Nagata (1938)
	Tsuboi (1954)	Morelli and Mosetti (1961)
	Morelli and Mosetti (1961)	Alldredge et al. (1963)
	Morelli and Carrozzo (1963)	Bhattacharyya (1965)
	Henderson and Cordell (1971)	Hahn (1965)
Fourier-Bessel series	Tsuboi (1954)	
Sinc-functions	Tomoda and Aki (1955)	Oldham (1967)
	Tsuboi and Tomoda (1958)	Roy (1970)
	Oldham (1967)	
	Roy (1970)	
	Ku et al. (1971)	
Bicubic splines		Bhattacharyya (1969)
Spherical harmonics	Heiskanen and Vening Meinesz (1958)	Chapman and Bartels (1940, 1951)
	Kaula (1959, 1966a, b)	Fanselau and Kautzleben (1958)
		Mauersberger (1959)
		Alldredge et al. (1963)
		Fougere (1963, 1965)
		Matsushita and Maeda (1965a, b)
		Hurwitz et al. (1966)
		Yukutake and Tachiraka (1968)
		Malin and Pocock (1969)
		Yukutake (1971)
Autocorrelation	Mundt (1966)	Horton et al. (1964)
	Beryland (1971)	Mundt (1966)
	Milcoveanu (1971a, 1972)	Milcoveanu (1971b, 1972)

[1] Even though a separation is made here between gravity and geomagnetism, most methods are applicable to both fields.

TABLE LIII (continued)

Method	Gravity	Geomagnetism
Fourier spectra	Neidell (1965, 1966)	Horton et al. (1964)
Power spectra	Odegard and Berg (1965)	Neidell (1965, 1966)
	Naidu (1968)	Bhattacharyya (1966)
	Sharma and Geldart (1968)	Gudmundsson (1966, 1967)
	Shaw et al. (1969)	Herron (1966)
	Avasthi and Satyanarayana (1970)	Spector and Bhattacharyya (1966)
	Kanasewich and Agarwal (1970)	Caner et al. (1967)
	Lewis and Dorman (1970)	Simonenko and Roze (1967)
	Sharma et al. (1970)	Naidu (1968, 1969, 1970a, c)
	Cordell and Taylor (1971)	Kanasewich and Agarwal (1970)
	Davis (1971)	Lamden (1970)
	Ku et al. (1971)	Scollar (1970)
	Milcoveanu (1971a, 1972)	Spector and Grant (1970)
	Syberg (1972)	Agarwal and Kanasewich (1971)
		Bhattacharya (1971)
		Cordell and Taylor (1971)
		Milcoveanu (1971b, 1972)
		Mishra and Naidu (1971)
		Treitel et al. (1971)
		Green (1972)
		Syberg (1972)
Two-dimensional filtering	Dean (1958)	Dean (1958)
	Meskó (1965)	Hagiwara (1965)
	Darby and Davies (1967)	Lehmann (1970)
	Fuller (1967)	Scollar (1970)
	Agarwal and Lal (1972a, b)	
	Mihail and Nicolae (1972)	

The Fourier-Bessel series and the $(\sin x)/x$ series method have the advantage over the regular Fourier series method that their amplitudes become vanishingly small at some distance beyond the measuring interval, whereas the frequency concept is lost. Expansion of the gravitational anomaly $\Delta g(x,y)$ in two spatial coordinates as a sum of $(\sin x)/x \cdot (\sin y)/y$ terms takes the following form:

$$\Delta g(x,y) = \sum_{\xi,\eta} \Delta g(\xi,\eta) \frac{\sin(x-\xi)}{x-\xi} \frac{\sin(y-\eta)}{y-\eta} \qquad [3]$$

with measuring points (ξ,η) at mutual distances of π.

This expansion is shown by Tsuboi and Tomoda (1958) to be equivalent to the Fourier series expansion if a sufficiently large number of terms is included. As demonstrated by Oldham (1967) and Roy (1970), expansions of potential fields (both gravity and geomagnetic fields) as shown in [3] provide a convenient method for downward and especially for upward continuation of such fields.

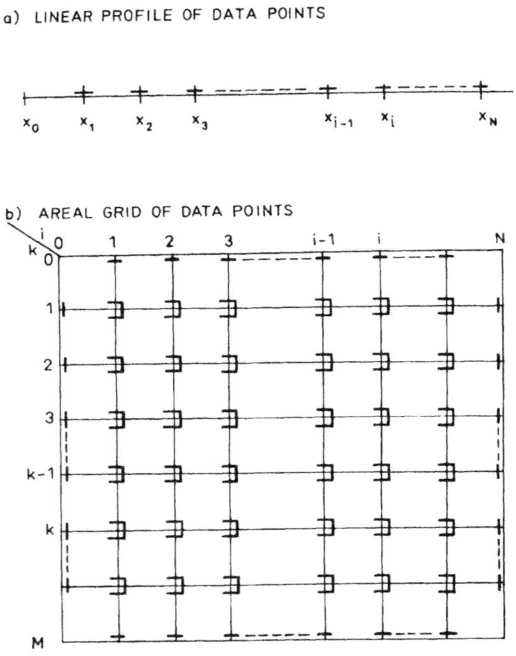

Fig.100. Scheme used for expansion in cubic splines. Heavy dashes mark where continuity conditions are used. In b) at interior corners, there are three independent continuity conditions for each continuous quantity.

Equation [3] expresses a two-dimensional convolution (Bracewell, 1965, p.243) in the space domain and corresponds to the sampling theorem (Papoulis, 1962, p.51). Sampling is equivalent to filtering. Cf. section 4.3.1. The convolution in [3] corresponds to multiplication in the wavenumber domain. By constructing the spectra corresponding to [3], it would be possible to extend this method such that the frequency concept would be recovered. The function $(\sin x)/x$ is equal to $\text{sinc}\,(x/\pi)$, whose spectrum is a rectangular function (example 3, Table V):

$$(\sin x)/x = \text{sinc}\,(x/\pi) \leftrightarrow \pi\, \Pi(k_x/2) \qquad [4]$$

The spectrum of [3] is the corresponding two-dimensional form, and it can be displayed as isolines in a wavenumber–wavenumber coordinate system. The techniques are quite analogous to those used in dealing with array-station data, which are also derived from sets of points, distributed according to some pattern over a plane.

Spline-function expansion. Another very promising method for spatial studies of potential field data (geomagnetism, gravity) involves piecewise expansion of observations in cubic polynomials (Bhattacharyya, 1969). For a linear profile

(Fig.100a), observations $f(x)$ are expressed in each interval $x_{i-1} \leq x \leq x_i$ as a cubic polynomial:

$$f_i(x) = c_0 + c_1(x - x_{i-1}) + c_2(x - x_{i-1})^2 + c_3(x - x_{i-1})^3$$
$$= \sum_{n=0}^{3} c_n^{(i)}(x - x_{i-1})^n \qquad [5]$$

Altogether this involves $4N$ unknowns, there being four coefficients c_0, c_1, c_2, c_3 to be determined for each interval. For their determination we have: (1) the following given quantities: $f(x_0)$, $f(x_1)$, $f(x_2)$, ..., $f(x_N)$, $f'(x_0)$, $f'(x_N)$, i.e. $N + 3$ data; and (2) the following continuity conditions: $f(x)$, $f'(x)$, $f''(x)$ are continuous for $x = 1, 2, ..., N - 1$, i.e. at all interior points, making a total of $3(N - 1)$ conditions.

Obviously, the number of given quantities plus conditions, i.e. $(N + 3) + 3(N - 1) = 4N =$ the number of unknowns. This method therefore leads to a unique representation of data by [5].

The total set of observations $f(x)$ can then be written as a sum over the polynomials for each interval:

$$f(x) = \sum_{i=0}^{N} f_i(x)[u(x - x_{i-1}) - u(x - x_i)]$$
$$= \sum_{i=0}^{N} \sum_{n=0}^{3} c_n^{(i)}(x - x_{i-1})^n [u(x - x_{i-1}) - u(x - x_i)] \qquad [6]$$

where $u(x - x_{i-1}) - u(x - x_i)$ is the difference between two unit-step functions and thus represents a rectangular data window.

The method can be immediately extended to a two-dimensional grid (Fig. 100b). For each rectangle $x_{i-1} \leq x \leq x_i$, $y_{k-1} \leq y \leq y_k$, data can be represented by a bicubic polynomial:

$$g_{ik}(x,y) = \sum_{m=0}^{3} \sum_{n=0}^{3} a_{mn}^{(ik)}(x - x_{i-1})^m (y - y_{k-1})^n \qquad [7]$$

This equation involves 16 coefficients for each rectangle, thus in all $16NM$ unknowns. For their determination we proceed in a way analogous to the one above:

(1) The quantities g, g_x, g_y, g_{xy} (subscripts indicating derivatives) are given at each corner, i.e., in all there are $4(N + 1)(M + 1)$ given quantities.

(2) The same quantities, g, g_x, g_y, g_{xy}, are continuous at adjacent corners: partly along the edges $2(N - 1) + 2(M - 1)$, partly at interior corners, where each one contributes by 3 conditions, in all $3(N - 1)(M - 1)$. Adding up for the

4 quantities, the total number of conditions becomes $8(N - 1) + 8(M - 1) + 12(N - 1)(M - 1)$.

The sum of the number of given quantities and the number of conditions is then easily seen to be $= 16NM$, i.e. equal to the total number of unknowns. Therefore, equation [7] then permits a unique representation of the given data.

As for the linear profile, the total set of two-dimensionally distributed data can be represented by a summation over the contributions [7] from each rectangle, by including appropriate rectangular windows:

$$g(x,y) = \sum_{i=0}^{N} \sum_{k=0}^{M} g_{ik}(x,y)[u(x - x_{i-1}) - u(x - x_i)] \cdot [u(y - y_{k-1}) - u(y - y_k)] \qquad [8]$$

Bhattacharyya (1969) has demonstrated the efficient use which bicubic splines permit in evaluating geomagnetic data, which can be summarized as follows:

(1) Bicubic splines permit an accurate and reliable representation of two-dimensional data and permit accurate calculation of various derivatives of the potential field data.

(2) Fourier transformation of [8] leads to amplitude and phase spectra which are more accurate for large wavelengths than spectra calculated from Fourier series (as verified by comparison with theoretical spectra for given bodies).

(3) Spline functions permit accurate interpolation of non-equispaced two-dimensional data.

The spline function method has no doubt even much wider application than those mentioned, practically to any one- or two-dimensional data presentation. The continuity conditions can be varied, only that their number suffices for a unique calculation of all coefficients. Development to higher dimensions would also be possible.

10.2.2 Spatial filtering

Derivatives of potential fields (gravity, geomagnetism) offer interesting applications of filtering processes (section 6.2.1). The dependence of gravity g at x,y on the vertical coordinate z can be written as the following Fourier transform:

$$g(x,y,z) = \frac{1}{4\pi^2} \int_{-\infty}^{\infty} \int_{-\infty}^{\infty} G(k_x,k_y)\, \mathrm{e}^{z\sqrt{k_x^2+k_y^2}}\, \mathrm{e}^{i(xk_x+yk_y)}\, \mathrm{d}k_x\, \mathrm{d}k_y \qquad [9]$$

where $\mathrm{e}^{z\sqrt{k_x^2+k_y^2}}$ is the filter transfer factor for vertical continuation of the field (Dean, 1958; Darby and Davies, 1967). Derivatives of gravity g are frequently needed in calculations. Taking the Nth derivative of g with respect to z we get:

$$\frac{\partial^N}{\partial z^N} g(x,y,z) = \frac{1}{4\pi^2} \int_{-\infty}^{\infty} \int_{-\infty}^{\infty} G(k_x,k_y)(k_x^2 + k_y^2)^{N/2} \, e^{z\sqrt{k_x{}^2+k_y{}^2}} \, e^{i(xk_x+yk_y)} \, dk_x \, dk_y \quad [10]$$

which, interpreted as a filtering equation, yields the following expression for the transfer factor of the Nth vertical derivative:

$$H_N(k_x,k_y,z) = (k_x^2 + k_y^2)^{N/2} \, e^{z\sqrt{k_x{}^2+k_y{}^2}} \quad [11]$$

The inverse transform of $H_N(k_x,k_y,z)$ is the impulse response $h_N(x,y,z)$:

$$h_N(x,y,z) = \frac{1}{4\pi^2} \int_{-\infty}^{\infty} \int_{-\infty}^{\infty} H_N(k_x,k_y,z) \, e^{i(xk_x+yk_y)} \, dk_x \, dk_y \quad [12]$$

Let $x \to m\Delta x$ and $y \to n\Delta y$ with m,n integers, and let the sampling intervals Δx and Δy be $= 1$. Furthermore, assume:

$$H(k_x,k_y,z) = 0 \qquad \text{for} \quad \left\{ \begin{matrix} |k_x| \geq \pi/\Delta x = \pi \\ \\ |k_y| \geq \pi/\Delta y = \pi \end{matrix} \right\} \quad \text{Nyquist warenumbers} \quad [13]$$

Also considering that $H_N(k_x,k_y,z)$ is an even function of k_x and k_y, we finally get from [12]:

$$h_N(m,n,z) = \frac{1}{\pi^2} \int_0^{\pi} \int_0^{\pi} H_N(k_x,k_y,z) \cos mk_x \cos nk_y \, dk_x \, dk_y \quad [14]$$

This agrees with equations derived in section 6.4.2, in connection with a general discussion of two-dimensional filtering. The advantage of this equation is that it yields values $h_N(m,n,z)$ by which a numerically given gravity field should be multiplied in order to achieve any operation involving any vertical derivative N of the gravity field. Special cases can be written down immediately, for instance the impulse response for the vertical gradient ($N = 1$) at $z = 1$:

$$h_1(m,n,1) = \frac{1}{\pi^2} \int_0^{\pi} \int_0^{\pi} \sqrt{k_x^2 + k_y^2} \, e^{\sqrt{k_x{}^2+k_y{}^2}} \cos mk_x \cos nk_y \, dk_x \, dk_y \quad [15]$$

In numerical calculations, formula [14] is used in its corresponding digital summation form.

In practical calculations, also a window function $w(m,n)$ may be applied to $h(m,n,z)$, yielding a modified impulse response function:

$$h'(m,n,z) = w(m,n)\, h(m,n,z) \qquad [16]$$

The same window functions as developed in section 4.4, are applicable to the space domain, with appropriate modifications. For example, the Hanning window will become as follows in the two-dimensional space domain:

$$w(m,n) = \begin{cases} \dfrac{1}{2}\left[1 + \cos\dfrac{\pi(m^2 + n^2)^{1/2}}{(X^2 + Y^2)^{1/2}}\right] & \text{for } \begin{cases} |m| \le X \\ \text{and} \\ |n| \le Y \end{cases} \\ 0 & \text{for } \begin{cases} |m| \ge X \\ \text{and/or} \\ |n| \ge Y \end{cases} \end{cases} \qquad [17]$$

The space window serves the same purpose as the time windows explained in Chapter 4.

Agarwal and Lal (1972a), whose paper was followed in this theoretical development, demonstrate that the Hanning window furnishes impulse responses in very good agreement with the theoretical values obtained from [14] or its inverse. In order to execute the various operations mentioned, Agarwal and Lal (1972b) derive coefficient sets which approximate theoretical curves very closely.

Lavin and Devane (1970) developed two-dimensional, circular symmetric wavenumber filters, applicable to both gravity and magnetic observations, and Davis (1971) used a filtering technique for the study of gravitational effects of faults. Ku et al. (1971) developed filtering methods for gravity, which are equally applicable in other areas of geophysics as well.

From a theoretical study, Bhattacharyya (1966) has shown that higher wavenumbers are caused by near-surface structures and lower wavenumbers by deeper structures, as a general result. This is also what one would expect intuitively, and it can be seen by considering the relation between the power spectra $E(k,0)$ on the surface and $E(k,z)$ at the depth z of the body, remembering that the E-functions are the squares of G in [9]:

$$E(k,0) = E(k,z)\, e^{-2zk} \qquad [18]$$

Therefore, for any given disturbing body, i.e. for any given source power function $E(k,z)$, the spectrum will be flatter, the smaller z is, i.e., there will be a higher proportion of higher wavenumbers than for larger z. This invites the application of filtering methods, high-pass and low-pass, to study shallow and deep structures,

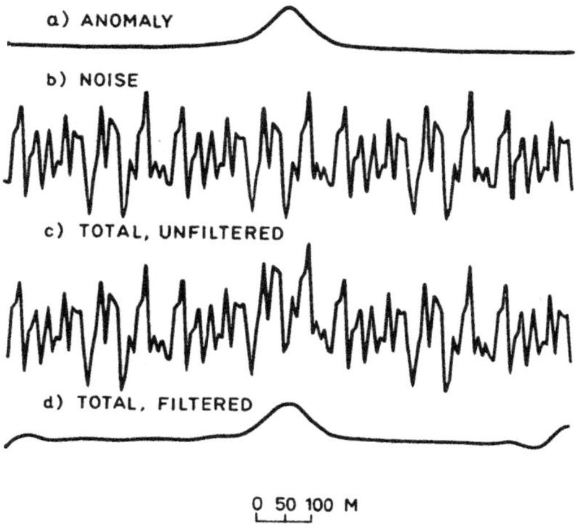

Fig.101. Example of spatial filtering of gravity record along a profile. a) Gravity anomaly from a sphere at 60 m depth. b) Noise. c) Input anomaly plus noise. d) Filtered output. The anomaly is hardly noticeable in c) but well seen in the filtered output d). After Fraser et al. (1966).

respectively (Dean, 1958; Bhattacharyya, 1967). A successful development of some such techniques can be found in a paper by Fraser et al. (1966). See Fig.101 and also Schaub (1961), Hagiwara (1967), and Avasthi and Satyanarayana (1970). It is important to note that this discussion assumes a disturbing body of known size and properties. If the body would have different properties and size, the scheme does not hold for intercomparison of cases. It should also be observed that a successful filtering as outlined above rests upon the assumption that the lower magnetic (or gravimetric) basement has a smooth surface which does not give rise to any high frequencies (Lehmann, 1970). Otherwise, such high frequencies will mix with those of shallower origin, making an effective filtering impossible.

10.2.3 Wavenumber spectra

The purpose of geophysical surveys of an area, such as in gravity and geomagnetism, is to deduce properties (depth, shape) of underground bodies causing observed anomalies. For any given body, it is possible to calculate theoretically the corresponding field on the surface. Direct comparisons in the space domain of observations and theoretical models have been done for many years. Spectra, both for theoretical models and for observed data, and their intercomparison have considerably improved such studies in recent years, by working in the wavenumber domain.

Theoretical models for which wavenumber spectra have been published are summarized in Table LIV. In addition, many papers, not reviewed here, give

TABLE LIV

Theoretical model spectra in gravity and geomagnetism

Reference	Type of body	Type of spectrum
(a) *Gravity*		
Odegard and Berg (1965)	sphere, cylinder, vertical fault	Fourier wavenumber spectra
Sharma and Geldart (1968)	inclined fault	Fourier wavenumber spectra
Sharma et al. (1970)	dike	Fourier wavenumber spectra
(b) *Geomagnetism*		
Spector and Bhattacharyya (1966)	poles, dipoles, lines of poles and of dipoles	power spectra and correlation functions
Bhattacharya (1971)	vertical dike	Fourier wavenumber spectrum
(c) *Gravity and geomagnetism*		
Naidu (1968)	randomly distributed sources (over an infinite, horizontal sheet, over a stack of sheets, over a semi-infinite medium)	power spectra
Milcoveanu (1971a, b)	cylinder, sphere, step, various types of faults, horst, graben, anticline, syncline, plate	Fourier wavenumber and power spectra, autocorrelations
Milcoveanu (1972)	inclined layer	ditto

space-domain solutions for gravity and magnetism caused by bodies of various shapes. Fourier transformation of such formulas will naturally yield the corresponding wavenumber spectra, and therefore such papers will prove useful also for spectral studies by providing at least a partial solution.

A few comments will be made to the examples in Table LIV. Naidu (1968) demonstrates that spectra of his potential fields on different levels are simply related to each other, also that the three component spectra are related such that knowledge of only one component is sufficient. A possible practical application of these results would be for the design of filters, by which random source effects could be eliminated from observations (corresponding to noise filtering in seismology), thus leaving non-random effects (corresponding to seismic signals) better accessible for further study. Naidu (1969, 1970a) developed the FFT method for calculation of two-dimensional power density spectra from aeromagnetic data and applied this method to data from Ontario. He emphasizes the significance of such spectra in connection with studies of surface geology. Similar methods were used by Agarwal and Kanasewich (1971). A paper by Green (1972) contains a useful discussion of procedures for obtaining magnetic wavenumber spectra.

Not only are gravimetric and geomagnetic fields wanted for various bodies, but also their derivatives (gradient, curvature, etc). By the derivation theorem, section 2.3.6, any derivative can be immediately expressed in the wavenumber domain, as soon as the wavenumber spectrum is given. Besides two-dimensional measurements, also those along linear profiles are of significance.

On the basis of assumptions about the properties of the disturbing body (shape, magnetization, density), it is possible to estimate its depth from the wavenumber spectra. Methods are based on the fact, mentioned in section 10.2.2, that shallow bodies lead to higher wavenumbers than deeper bodies. Therefore, the depth will be reflected in the slope of the power–wavenumber spectra, such that shallow bodies correspond to smaller slopes. This is understood by logarithmic differentiation of [18]:

$$\underbrace{\frac{\mathrm{d}\ln E(k,0)}{\mathrm{d}k}}_{\substack{\uparrow \\ \text{observed spectral} \\ \text{slope}}} = \underbrace{\frac{\mathrm{d}\ln E(k,z)}{\mathrm{d}k}}_{\substack{\uparrow \\ \text{calculated spectral} \\ \text{slope for assumed body}}} - \underbrace{2z}_{\substack{\uparrow \\ \text{depth to be} \\ \text{estimated}}} \qquad [19]$$

The method has an obvious similarity to the frequency-ratio method for attenuation measurements of seismic waves (section 7.4.1). Such methods, with special application to magnetic bodies, were discussed by Bhattacharya (1971) and Treitel et al. (1971). See also Gudmundsson (1966). Likewise, Mundt (1966) applied spatial autocorrelation along linear profiles to deduce depths to magnetic bodies or sources for gravity anomalies. Autocorrelation functions which decrease rapidly from their maximum are shown to correspond to shallower depths. This is understood by the fact that a shallow depth z corresponds to a flat, broad-band spectrum $E_z(k)$, which in turn corresponds to a narrow-band autocorrelation function, by the reciprocity of space and wavenumber domains (section 2.4.4).

Turning to the observational side, we note a number of relatively recent papers where spectral methods have been used to disentangle gravimetric and magnetic anomalies. For instance, Horton et al. (1964) studied observed aeromagnetic data from an area in Canada by linear power spectra and autocorrelations and areal (two-dimensional) autocorrelations. The method proved especially useful in determining structural trends and their spatial periodicities. Related studies over Japanese volcanoes were reported by Hagiwara (1965), applying the principle that upward and downward continuations of the geomagnetic field are equivalent to low-pass and high-pass two-dimensional spatial filtering, respectively (section 6.4.2). Spector and Grant (1970) interpret empirical power spectra of aeromagnetic data in terms of blocks of varying size, depth, and magnetization, using a theoretical development for this case. The spectra are expressed in radial wavenumbers (radial spectra), for which dominating wavenumbers and slopes depend on the depth to the blocks.

Neidell (1965, 1966) applies spectral analysis to sea-floor profiles of gravity, geomagnetism and sea-bottom topography. He emphasizes that the spacing between maxima in the obtained power–wavenumber spectra is the most important quantity to be determined, especially for testing certain models of sea-floor properties. This in fact suggests use of cepstrum analysis. The spectral methods will permit investigations of more complex geophysical theories. The application of magnetic spatial spectra seems particularly appropriate for investigation of the lineations which have been found on the ocean bottom on either side of ridge structures, and which relate to ocean-floor spreading and paleomagnetism (Luyden-dyk, 1969). A suggestive power–wavenumber spectrum from a profile roughly perpendicular to the Mid-Atlantic Ridge was presented already by Bullard and Mason (1963, p.187). By such analysis, an increase of wavelength of magnetic amplitude density has been found towards either side of the Mid-Atlantic Ridge (Heirtzler and Le Pichon, 1965). No significant coherence was found between trans-ridge spectra of magnetism and topography, even though both show a significant peak at 33 km wavelength. Gudmundsson (1967) displays magnetic measurements from the Carlsberg Ridge as two-dimensional wavenumber spectra (power isolines with wavenumbers k_x and k_y as axes).

10.3 COMBINED STUDIES

Most geophysical phenomena are interrelated to a greater or lesser extent. Connections extend across boundaries of different regimes of the earth, encompassing both the solid earth, the oceans, the atmosphere and the ionosphere. Moreover, they may extend across boundaries between different kinds of phenomena, such as those of mechanical nature and of electromagnetic nature. It is quite obvious that spectral methods offer powerful techniques to investigate and to disentangle such interrelationships. In this section we shall study spectral combinations of either the independent variables (time–space) or the dependent variables (gravity–geomagnetism) and combinations of either of these disciplines with some other fields.

10.3.1 Gravity–geomagnetism space studies

A joint analysis of two-dimensional wavenumber presentations of gravity and geomagnetism was given by Kanasewich and Agarwal (1970). From the two kinds of potential field data, they demonstrate calculations of J/ρ, i.e. the ratio of intensity of magnetization to the density, for the case that both kinds of anomalies derive from the same body. A two-dimensional FFT method is used for the calculations.

Another joint analysis of the spatial distributions of geomagnetism and gravity is indicated by Cordell and Taylor (1971). The magnetic anomaly Δm and the gravity anomaly Δg are given at a number of discrete points $j_1 = 0, 1, \ldots, N_1 - 1$, $j_2 = 0, 1, \ldots, N_2 - 1$ distributed over a field. We combine the two anomalies for each point into:

$$f_{j_1 j_2} = \Delta m_{j_1 j_2} + i\, \Delta g_{j_1 j_2} \qquad [20]$$

and form its discrete Fourier transform:

$$F_{k_1 k_2} = \frac{1}{N_1 N_2} \sum_{j_1=0}^{N_1-1} \sum_{j_2=0}^{N_2-1} f_{j_1 j_2} \exp\left[-2\pi i\left(\frac{k_1 j_1}{N_1} + \frac{k_2 j_2}{N_2}\right)\right] \qquad [21]$$

$k_1 = 0, 1, \ldots, N_1/2$; $k_2 = 0, 1, \ldots, N_2/2$ (Nyquist wavenumber)

Then it is easily shown that the real and imaginary parts of the separate Fourier terms for geomagnetism ($F^m_{k_1 k_2}$) and gravity ($F^g_{k_1 k_2}$) can be obtained from $F_{k_1 k_2}$ as expressed in the following compact form:

$$2\begin{bmatrix} \mathrm{Re} \\ \mathrm{Im} \\ \mathrm{Re} \\ \mathrm{Im} \end{bmatrix} \begin{matrix} F^m_{k_1 k_2} \\ \\ F^g_{k_1 k_2} \end{matrix} = \begin{bmatrix} \mathrm{Re} \\ \mathrm{Im} \\ \mathrm{Im} \\ -\mathrm{Re} \end{bmatrix} F_{k_1 k_2} + \begin{bmatrix} \mathrm{Re} \\ -\mathrm{Im} \\ \mathrm{Im} \\ \mathrm{Re} \end{bmatrix} F_{N_1 - k_1, N_2 - k_2} \qquad [22]$$

This kind of analysis has the advantage that both kinds of data are analyzed in one operation and the needs for computer core storage are minimized. Cordell and Taylor (1971) made a combined analysis of geomagnetic and gravity data in terms of magnetization and density, using FFT. Similar combined treatment of different geophysical fields (with common origin or otherwise related) will certainly prove useful.

10.3.2 Gravity–topography space studies

Gravity and topography show relatively high coherence, as expected. Quite generally, the relation between gravity anomalies and topography can be expressed by a convolution in the space domain and a corresponding multiplication in the wavenumber domain:

$$G(k) = \Gamma(k)\, H(k) \qquad [23]$$

where $G(k)$ is gravity anomaly, $\Gamma(k)$ topography and $H(k)$ the transfer function, k = wavenumber. Lewis and Dorman (1970) calculated $H(k)$ from two-dimensional transforms of gravity anomalies (Bouguer and free-air) and of topography, and were able to demonstrate that calculated $H(k)$ requires density differences down to depths of 400 km to explain the isostatic compensation. Cf. also Hagiwara (1967).

10.3.3 Geomagnetism time–space studies

Space studies presuppose the availability of arrays or networks of observing stations. But apart from this, several different techniques have been applied to combined space–time analyses of geomagnetic fields. The most important are the following equivalent methods, which differ essentially only in the order the different operations are undertaken.

(1) *Space–time analysis:* Space analysis (wavenumber spectra or spherical harmonic expansions) followed by time analysis (variation of spectral components with time). This procedure has been applied to secular variation of spherical harmonic coefficients. Such analyses generally encompass large intervals of time.

World-wide secular variations of geomagnetism are obtained from differences between coefficients of spherical harmonic expansions at successive epochs (Adam et al., 1963; Yukutake, 1971). Second differences yield the secular accelerations (Malin, 1969). Alternatively, Fourier series expansion may be applied to the coefficients of spherical harmonic expansions of archeomagnetic data at different epochs (Márton, 1970). Power spectra of geomagnetic secular variations (periods from less than 10 years up to 60 years) have been constructed by Slaucitajs (1964) for three observatories in the southern hemisphere.

(2) *Time–space analysis:* Time analysis (frequency spectra) followed by space analysis (mapping of spectral amplitude or power and phase for given frequencies). Such maps in fact represent partial Fourier transforms: if the given amplitude is $A(t,x,y)$, where x, y are the horizontal coordinates, the maps show $A(\omega,x,y)$, and correspondingly for the phase angle. This mode of presentation was applied by Lilley and Bennett (1972) for a geomagnetic array in Australia, demonstrating effects of coasts and of areas of high heat flow, and by Camfield et al. (1971) for an array in northwest USA and southwest Canada.

The geomagnetic depth-sounding method, which uses frequency spectra of the magnetic components at a network of stations, is able to reveal discontinuities in the electrical conductivity in the upper mantle, and is thus a useful complement to deep seismic sounding methods (Caner et al., 1967; Cochrane and Hyndman, 1970; Reitzel et al., 1970; Porath and Gough, 1971). On islands even as small as 10 km in diameter, complete reversals of the magnetic field between opposite sides have been observed for periods around 30 min. This so-called "island effect" is explained as due to electric currents induced in the surrounding sea water. It offers a suitable problem for combined spatial and temporal spectral analysis (cf. Sasai, 1967). Everett and Hyndman (1967) calculated transfer functions $H_N(\omega)$

and $H_E(\omega)$ between the vertical magnetic component $Z(\omega)$ and the two horizontal components $N(\omega)$ and $E(\omega)$ for a network of stations in Australia:

$$Z(\omega) = H_N(\omega) \cdot N(\omega) + H_E(\omega) \cdot E(\omega) \qquad [24]$$

The existence of a non-zero $Z(\omega)$ is mainly due to lateral inhomogeneities especially of temperature (electrical conductivity) or composition in the earth, and the coast effect is one example of such an inhomogeneity.

As an alternative to the mapping mentioned, combined studies of time and space variations of geomagnetic activity, by use of an array of stations and calculation techniques including coherence, phase- and cross-spectra, are of particular value (Lamden, 1970; Davidson and Heirtzler, 1968; Duffus et al., 1962a). The latter demonstrate a coherence over 0.9 for geomagnetic micropulsations of periods over 16 sec over a 10 km range. Duffus et al. (1962b) report more extensive measurements.

When geomagnetic pulsations can be recognized as propagating waves, determinations of relative phase shifts among a group of stations by spectral analysis permit calculation of phase-velocity dispersion. This is a general method for propagating waves, outlined in section 7.2.1. It has been applied to geomagnetic micropulsations by Herron (1966) for the period range of 15 sec to several hundred sec.

10.3.4 Geomagnetism–geoelectricity time–space studies

Closely related to the earth's magnetic field is its associated electric field, the magnetotelluric field being used as a common term. The purpose of magnetotelluric methods is to determine the electrical conductivity in the crust and upper mantle, which in turn may contribute to structural interpretations. Several measuring methods are used:

(1) Simultaneous measurements of time-varying magnetic and electric fields, two horizontal components of each (see for example Frolov, 1970; Jankowski and Pawliszyn, 1970).

(2) Measurement of time variations of the electric field at two stations some distance apart.

(3) Geomagnetic depth-sounding method, using three components of the magnetic field.

Jankowski and Pawliszyn (1970) advocate a combined use of magnetotelluric and geomagnetic methods for more reliable structural studies. Spectral methods have a given role in evaluating measurements from any of the above-mentioned items. For example, under item 1, amplitude (energy) ratios, phase differences and coherences between electric and magnetic fields, deduced from their respective spectra obtained at the same point, give useful information about conductivities and structural properties in the earth.

There are now quite numerous publications on spectral studies of magnetotelluric measurements, which we shall briefly summarize. Comparing electric and magnetic records at the same point, these generally display high similarity. Thus, Cantwell and Madden (1960) found a coherence of around 0.8–0.9 between electric and magnetic signals, H. W. Smith et al. (1961) and Jones et al. (1969) show closely similar power–frequency spectra of simultaneous measurements of earth current and of the magnetic field.

Analogously, the magnetotelluric field exhibits often great similarity at different stations. Thus, Hoffman and Horton (1966) found similar magnetotelluric power spectra and coherence at Tbilisi and at Tikhaya Bay in the Arctic, and Jones and Geldart (1968) found close relationships between power–density spectra of earth currents from simultaneous records at locations separated by 80 and 160 km.

From the structural interpretation point of view, there are also a number of spectral studies. Horton and Hoffman (1962) interpret peaks in their magnetotelluric power spectra as due to resonant frequencies. From magnetotelluric measurements in the period range 2–15000 sec, Caner and Auld (1968), using power–frequency spectra, find high conductivity at depths roughly coinciding with the upper-mantle low-velocity layer. Partly similar results were obtained by Caner et al. (1969). Related measurements for the Canadian Arctic are reported by Whitham and Andersen (1965), which however have not permitted any firm structural conclusions. Using power spectral methods (ratios between components and between stations), Srivastava and White (1971) demonstrate evidence of coast and island effects as well as differences in subsurface conductivity under ocean and continent, while Caner et al. (1971) and Caner (1971) demonstrate lateral conductivity variations under western Canada, constituting valuable contributions to the structural studies. Jones and Geldart (1967), using spectral analysis, found that the vertical component of earth currents in the frequency range 0.0012–0.12 cps has amplitudes and frequencies comparable with the horizontal component, contrary to some earlier theoretical considerations. Concerning ocean-floor magnetotelluric spectra for studies of conductivity and temperature in the crust and upper mantle, see, for example, Hermance (1969).

10.3.5 Geomagnetism and other time studies

Relations between phenomena of different kinds are generally dependent upon their scales, in time or space. Therefore, spectral analysis is particularly apt to clarify such relations. As one example we may mention the connection between geomagnetic fluctuations and the earth's toroidal oscillations after the Chilean 1960 earthquake, studied by Winch et al. (1963) by power spectral methods. The matter is somewhat controversial, as evidenced by Hirshberg et al. (1972) and Stacey and Westcott (1972).

Disturbances in the ionosphere, especially from nuclear explosions, produce sudden, pulse-like records on magnetometers with sufficient sensitivity and time resolution. Comparison of magnetic power spectra of the high-altitude nuclear explosion of July 9, 1962, with theory suggests a magnetohydrodynamic resonance mechanism in a wave-guide between 300 and 3000 km altitude (Kovach and Ben-Menahem, 1966). Cf. also section 10.1.1.

No significant connection between geomagnetic variations and the height gradient of the 100-mbar surface was found by London et al. (1959), using spectral and coherence techniques. On the other hand, a clear relation has been demonstrated between ocean surface waves and magnetic field variations, recorded by an ocean-bottom mounted magnetometer. Fraser (1966) found good agreement between observed and calculated magnetic power spectra under the supposition that the field is due to electric currents induced in the waves by their movement through the earth's magnetic field.

REFERENCES

Note. Depending upon the transliteration system used, some Russian author names appear with different spelling in different papers: Berson = Berzon, Gol'tsman = Holzmann, Gratsinskii = Gratsinsky, Gurvich = Gurwitsch, Khudzinski = Khudzinskii = Khudzinsky, Kitaigorodski = Kitaigorodskii, Savarenskii = Savarensky, Schaub = Shaub, Shur = Šur, Tsvang = Zwang. Also, Bhattacharyya = Bhattacharya, Korkman = Kogeus.

Abbas, M. J., 1969. Beobachtung der langperiodischen Mikroseismen an den Stationen Moxa, Collm und Berggiesshübel. *Gerlands Beitr. Geophys.*, 78: 353–368.

Abe, K., 1972a. Focal process of the South Sandwich Islands earthquake of May 26, 1964. *Phys. Earth Planet. Inter.*, 5: 110–122.

Abe, K., 1972b. Lithospheric normal faulting beneath the Aleutian trench. *Phys. Earth Planet. Inter.*, 5: 190–198.

Abe, K., 1972c. Mechanisms and tectonic implications of the 1966 and 1970 Peru earthquakes. *Phys. Earth Planet. Inter.*, 5: 367–379.

Abe, K., 1972d. Group velocities of oceanic Rayleigh and Love waves. *Phys. Earth Planet. Inter.*, 6: 391–396.

Abe, K., Satô, Y. and Frez, J., 1970. Free oscillations of the earth excited by the Kurile Islands earthquake 1963. *Bull. Earthquake Res. Inst.*, 48: 87–114.

Abramovici, F., 1968. Diagnostic diagrams and transfer functions for oceanic wave-guides. *Bull. Seismol. Soc. Am.*, 58: 427–456.

Abramovici, F., 1973. Numerical application of a technique for recovering the spectrum of a time function. *Geophys. J. R. Astron. Soc.*, 32: 65–78.

Ackerman, B., 1967. The nature of the meteorological fluctuations in clouds. *J. Appl. Meteorol.*, 6: 61–71.

Adam, N. V., Ben'kova, N. P., Orlov, V. P., Osipov, N. K. and Tyurmina, L. O., 1963. Spherical analysis of the main geomagnetic field and secular variations. *Geomagn. Aeron.*, 3: 271–285.

Adams, W. M. and Allen, D. C., 1961a. Reading seismograms with digital computers. *Bull. Seismol. Soc. Am.*, 51: 61–67.

Adams, W. M. and Allen, D. C., 1961b. Seismic decoupling for explosions in spherical underground cavities. *Geophysics*, 26: 772–799.

Adams, W. M. and Carder, D. S., 1960. Seismic decoupling for explosions in spherical cavities. *Geofis. Pura Appl.*, 47: 17–29.

Adel, A. and Epstein, E. S., 1959. Power-spectrum analysis of atmospheric ozone parameters. *J. Meteorol.*, 16: 548–555.

Agarwal, B. N. P. and Lal, T., 1972a. Calculation of the vertical gradient of the gravity field using the Fourier transform. *Geophys. Prospect.*, 20: 448–458.

Agarwal, B. N. P. and Lal, T., 1972b. Application of frequency analysis in two-dimensional gravity interpretation. *Geoexploration*, 10: 91–100.

Agarwal, R. G. and Kanasewich, E. R., 1971. Automatic trend analysis and interpretation of potential field data. *Geophysics*, 36: 339–348.

Agterberg, F. P. and Banerjee, I., 1969. Stochastic model for the deposition of varves in glacial Lake Barlow-Ojibway, Ont., Canada. *Can. J. Earth Sci.*, 6: 625–652.

Aida, I., 1967. Water level oscillations on the continental shelf in the vicinity of Miyagi-Enoshima. *Bull. Earthquake Res. Inst.*, 45: 61–78.

Aida, I., 1969. On the edge waves of the Iturup tsunami. *Bull. Earthquake Res. Inst.*, 47: 43–54.

Aitken, A. C., 1962. *Statistical Mathematics*. Oliver and Boyd, 8th ed., 153 pp.

Akamatu, K., 1961. On microseisms in frequency range from 1 c/s to 200 c/s. *Bull. Earthquake Res. Inst.*, 39: 23–75.

Aki, K., 1956. Correlogram analyses of seismograms by means of a simple automatic computer. *J. Phys. Earth*, 4: 71–79.

Aki, K., 1957. Space and time spectra of stationary stochastic waves, with special reference to microtremors. *Bull. Earthquake Res. Inst.*, 35: 415–456.

Aki, K., 1960a. The use of Love waves for the study of earthquake mechanism. *J. Geophys. Res.*, 65: 323–331.

Aki, K., 1960b. Study of earthquake mechanism by a method of phase equalization applied to Rayleigh and Love waves. *J. Geophys. Res.*, 65: 729–740.

Aki, K., 1960c. Interpretation of source functions of circum-Pacific earthquakes obtained from long-period Rayleigh waves. *J. Geophys. Res.*, 65: 2405–2417.

Aki, K., 1960d. Further study of the mechanism of circum-Pacific earthquakes from Rayleigh waves. *J. Geophys. Res.*, 65: 4165–4172.

Aki, K., 1962. Accuracy of the Rayleigh wave method for studying the earthquake mechanism. *Bull. Earthquake Res. Inst.*, 40: 91–105.

Aki, K., 1964a. Study of Love and Rayleigh waves from earthquakes with fault plane solutions or with known faulting (three parts). *Bull. Seismol. Soc. Am.*, 54: 511–570.

Aki, K., 1964b. A note on surface waves from the Hardhat nuclear explosion. *J. Geophys. Res.*, 69: 1131–1134.

Aki, K., 1966a. Generation and propagation of G-waves from the Niigata earthquake of June 16, 1964. Part 1. A statistical analysis. *Bull. Earthquake Res. Inst.*, 44: 23–72.

Aki, K., 1966b. Generation and propagation of G-waves from the Niigata earthquake of June 16, 1964. Part 2. Estimation of earthquake moment, released energy, and stress-strain drop from the G-wave spectrum *Bull. Earthquake Res. Inst.*, 44: 73–88.

Aki, K., 1967. Scaling law of seismic spectrum. *J. Geophys. Res.*, 72: 1217–1231.

Aki, K., 1968. Seismic displacements near a fault. *J. Geophys. Res.*, 73: 5359–5376.

Aki, K., 1969. Analysis of the seismic coda of local earthquakes as scattered waves. *J. Geophys. Res.*, 74: 615–631.

Aki, K., 1972. Scaling law of earthquake source time-function. *Geophys. J. R. Astron. Soc.*, 31: 3–25.

Aki, K. and Kaminuma, K., 1963. Phase velocity of Love waves in Japan. Part 1. Love waves from the Aleutian shock of March 9, 1957. *Bull. Earthquake Res. Inst.*, 41: 243–259.

Aki, K. and Nordquist, J. M., 1961. Automatic computation of impulse response seismograms of Rayleigh waves for mixed paths. *Bull. Seismol. Soc. Am.*, 51: 29–34.

Aki, K. and Tsai, Y.-B., 1972. Mechanism of Love-wave excitation by explosive sources. *J. Geophys. Res.*, 77: 1452–1475.

Aki, K. and Tsujiura, M., 1959. Correlational study of near earthquake waves. *Bull. Earthquake Res. Inst.*, 37: 207–232.

Aki, K., Tsujiura, M., Hori, M. and Gotô, K., 1958. Spectral study of near earthquake waves (1). *Bull. Earthquake Res. Inst.*, 36: 71–98.

Aki, K., Matumoto, H., Tsujiura, M. and Maruyama, T., 1965. A digital, tele-recorded, long-period seismograph system. *Bull. Earthquake Res. Inst.*, 43: 381–397.

Alavi, A. S. and Jenkins, G. M., 1965. An example of digital filtering. *Appl. Stat., R. Stat. Soc.*, C 14: 70–74.

Alexander, S. S. and Phinney, R. A., 1966. A study of the core–mantle boundary using P-waves diffracted by the earth's core. *J. Geophys. Res.*, 71: 5943–5958.

Allam, A. M., 1970. An investigation into the nature of microtremors through experimental studies of seismic waves. *Bull. Int. Inst. Seismol. Earthquake Eng.*, 7: 1–59.

Allam, A. and Shima, E., 1967. An investigation into the nature of microtremors. *Bull. Earthquake Res. Inst.*, 45: 43–59.

Alldredge, L. R., Van Voorhis, G. D. and Davis, T. M., 1963. A magnetic profile around the world. *J. Geophys. Res.*, 68: 3679–3692.

Allsopp, D. F., Burke, M. D. and Cumming, G. L., 1972. A digital seismic recording system. *Bull. Seismol. Soc. Am.*, 62: 1641–1647.

Al-Sadi, H. N., 1973. Dependence of the P-wave amplitude spectrum on focal depth. *Pure Appl. Geophys.*, 104: 439–452.

Alsop, L. E., 1964a. Spheroidal free periods of the earth observed at eight stations around the world. *Bull. Seismol. Soc. Am.*, 54: 755–776.

Alsop, L. E., 1964b. Excitation of free oscillations of the earth by the Kurile Islands earthquake of 13 October 1963. *Bull. Seismol. Soc. Am.*, 54: 1341–1348.

Alsop, L. E. and Brune, J. N., 1965. Observation of free oscillations excited by a deep earthquake. *J. Geophys. Res.*, 70: 6165–6174.

Alsop, L. E., Sutton, G. H. and Ewing, M., 1961a. Free oscillations of the earth observed on strain and pendulum seismographs. *J. Geophys. Res.*, 66: 631–641.

Alsop, L. E., Sutton, G. H. and Ewing, M., 1961b. Measurement of Q for very long period free oscillations. *J. Geophys. Res.*, 66: 2911–2915.

Alterman, Z. and Loewenthal, D., 1972. Computer generated seismograms. In: B. A. Bolt (Editor), *Methods in Computational Physics*, Vol. 12. Acad. Press, pp. 35–164.

Alterman, Z., Jarosch, H. and Pekeris, C. L., 1959. Oscillations of the earth. *Proc. R. Soc. London, Ser. A*, 252: 80–95.

Amelung, U., 1964. Das quadratische Spektrum des Luftdruckjahresganges über Europa. *Meteorol. Rundsch.*, 17: 49–54.

An, V. A., 1965. On the possibility of using analog-digital conversion for recording micro-variations in the earth's electromagnetic field. *Izv. Acad. Sci. USSR, Phys. Solid Earth*, 9: 647–649 (Engl. ed.).

Anderson, D. L., 1967. The anelasticity of the mantle. *Geophys. J. R. Astron Soc.*, 14: 135–164.

Anderson, D. L. and Kovach, R. L., 1964. Attenuation in the mantle and rigidity of the core from multiply reflected core phases. *Proc. Natl. Acad. Sci.*, 51: 168–172.

Anderson, R. Y., 1961. Solar-terrestrial climatic patterns in varved sediments. *Ann. N. Y. Acad. Sci.*, 95: 424–439.

Anderson, R. Y. and Koopmans, L. H., 1963. Harmonic analysis of varve time series. *J. Geophys. Res.*, 68: 877–893.

Angell, J. K., 1958. Lagrangian wind fluctuations at 300 mbar derived from transosonde data. *J. Meteorol.*, 15: 522–530.

Angell, J. K., 1964. Correlations in the vertical component of the wind at heights of 600, 1600 and 2600 ft. at Cardington. *Q. J. R. Meteorol. Soc.*, 90: 307–312.

Anstey, N. A., 1964. Correlation techniques—a review. *Geophys. Prospect.*, 12: 355–382.

Anstey, N. A., 1966. The sectional auto-correlogram and the sectional retro-correlogram. Part I: The sectional auto-correlogram. *Geophys. Prospect.*, 14: 389–411.

Anstey, N. A. and Newman, P., 1966. The sectional auto-correlogram and the sectional retro-correlogram. Part II: The sectional retro-correlogram. *Geophys. Prospect.*, 14: 411–426.

Archambeau, C. B., 1972. The theory of stress wave radiation from explosions in prestressed media. *Geophys. J. R. Astron. Soc.*, 29: 329–366.

Archambeau, C. B. and Flinn, E. A., 1965. Automated analysis of seismic radiation for source characteristics. *Proc. IEEE*, 53: 1876–1884.

Archambeau, C. B., Flinn, E. A. and Lambert, D. G., 1966. Detection, analysis, and interpretation of teleseismic signals. 1. Compressional phases from the Salmon event. *J. Geophys. Res.*, 71: 3483–3501.

Archambeau, C. B., Flinn, E. A. and Lambert, D. G., 1969. Fine structure of the upper mantle. *J. Geophys. Res.*, 74: 5825–5865.

Arkani-Hamed, J. and Toksöz, M. N., 1968. Analysis and correlation of geophysical data. *Suppl. Nuovo Cimento, Ser. I*, 6: 22–66.

Armendariz, M. and Rachele, H., 1967. Determination of a representative wind profile from balloon data. *J. Geophys. Res.*, 72: 2997–3006.

Arsac, J., 1966. *Fourier Transforms and the Theory of Distributions.* Prentice-Hall, 318 pp.

Asada, T. and Takano, K., 1963. Attenuation of short period P-waves in the mantle. *J. Phys. Earth*, 11: 25–34. Also in *VESIAC Spec. Rep.*, 4410-52-X, 1964: 41–53.

Askew, R., Miyata, M. and Groves, G. W., 1969. On the observed low coherence between sea-level records on the opposite sides of Oahu Island. *J. Geophys. Res.*, 74: 7058–7062.

Atlas, D., Srivastava, R. C. and Sekhon, R. S., 1973. Doppler radar characteristics of precipitation at vertical incidence. *Rev. Geophys. Space Phys.*, 11: 1–35.

Avasthi, D. N. and Satyanarayana, M., 1970. Inferences regarding the upper mantle from the gravity data in Punjab and Ganga plains. *Proc. 2nd Symp. UMP, Geophys. Res. Board and Natl. Geophys. Res. Inst. (Hyderabad)*, 11: 25–37.

Axford, D. N., 1971. Spectral analysis of an aircraft observation of gravity waves. *Q. J. R. Meteorol. Soc.*, 97: 313–321.

Axford, D. N., 1972. A case study of a high-level Canberra flight on 11 September 1968. *Q. J. R. Meteorol. Soc.*, 98: 420–430.

Backus, G. and Gilbert, F., 1961. The rotational splitting of the free oscillations of the earth. *Proc. Natl. Acad. Sci.*, 47: 362–371.

Backus, M. M., 1959. Water reverberations—their nature and elimination. *Geophysics*, 24: 233–261.

Backus, M. M., 1966. Teleseismic signal extraction. *Proc. R. Soc. London, Ser. A*, 290: 343–367.

Baer, L. and Withee, G. W., 1971. A methodology for defining an operational synoptic temporal oceanic sampling system. *J. Appl. Meteorol.*, 10: 1053–1065.

Bakun, W. H., 1971. Crustal model parameters from P-wave spectra. *Bull. Seismol. Soc. Am.*, 61: 913–935.

Bakun, W. H. and Eisenberg, A., 1970. Fourier integrals and quadrature-introduced aliasing. *Bull. Seismol. Soc. Am.*, 60: 1291–1296.

Bakun, W. H. and Johnson, L. R., 1971. Short-period spectral discriminants for explosions. *Geophys. J. R. Astron. Soc.*, 22: 139–152.

Balakina, L. M., Vvedenskaya, A. V. and Kolesnikov, Yu. A., 1966. Study of the boundary of the earth's core from spectral analysis of seismic waves. *Izv. Acad. Sci. USSR, Phys. Solid Earth*, 8: 492–499 (Engl. ed.).

Bancroft, A. M., 1966. Seismic spectra and detection probabilities from explosions in Lake Superior. *Am. Geophys. Union, Geophys. Monogr.*, 10: 234–240.

Banks, R. J., 1969. Geomagnetic variations and the electrical conductivity of the upper mantle. *Geophys. J. R. Astron. Soc.*, 17: 457–487.

Barazangi, M. and Isacks, B., 1971. Lateral variations of seismic-wave attenuation in the upper mantle above the inclined earthquake zone of the Tonga Island arc: deep anomaly in the upper mantle. *J. Geophys. Res.*, 76: 8493–8516.

Barber, N. F., 1958. Optimum arrays for direction finding. *N.Z. J. Sci.*, 1: 35–51.

Barber, N. F., 1961. *Experimental Correlograms and Fourier Transforms*. Pergamon Press, 136 pp.

Barber, N. F., 1963. The directional resolving power of an array of wave detectors. In: R. C. Vetter and C. L. Bretschneider (Editors), *Ocean Wave Spectra*. Prentice-Hall, pp. 137–150.

Barber, N. F., 1966. Fourier methods in geophysics. In: S. K. Runcorn (Editor), *Methods and Techniques in Geophysics*, Vol. 2. Interscience, pp. 123–204.

Barnett, T. P., 1968. On the generation, dissipation, and prediction of ocean wind waves. *J. Geophys. Res.*, 73: 513–529.

Barnett, T. P. and Wilkerson, J. C., 1967. On the generation of ocean wind waves as inferred from airborne radar measurements of fetch-limited spectra. *J. Mar. Res.*, 25: 292–328.

Barrett, E. W., 1961a. Some applications of harmonic analysis to the study of the general circulation. I. Harmonic analysis of some daily and five-day mean hemispheric contour charts. *Beitr. Phys. Atmos.*, 33: 280–332.

Barrett, E. W., 1961b. Some applications of harmonic analysis to the study of the general circulation. II. Calculations of some product-spectra and related quantities. *Beitr. Phys. Atmos.*, 33: 333–355.

Barsenkov, S. N., 1967a. A spectral analysis of tidal variations in the force of gravity at Talgar. *Izv. Acad. Sci. USSR, Phys. Solid Earth*, 3: 170–174 (Engl. ed.).

Barsenkov, S. N., 1967b. Calculation of third-order tides from gravimetric observations. *Izv. Acad. Sci. USSR, Phys. Solid Earth*, 5: 284–286 (Engl. ed.).

Barsenkov, S. N., 1972. Identification of long-period waves in tidal variations of gravity. *Izv. Acad. Sci. USSR, Phys. Solid Earth*, 6: 365–368 (Engl. ed.).

Basham, P. W. and Ellis, R. M., 1969. The composition of P-codas using magnetic tape seismograms. *Bull. Seismol. Soc. Am.*, 59: 473–486.

Basham, P. W. and Whitham, K., 1966. Microseismic noise on Canadian seismograph records in 1962 and station capabilities. *Publ. Dom. Obs. Ottawa*, 32: 123–135.

Basham, P. W., Weichert, D. H. and Anglin, F. M., 1970. An analysis of the "Benham" aftershock sequence using Canadian recordings. *J. Geophys. Res.*, 75: 1545–1556.

Bassanini, P., Riccucci, A. and Todaro, C., 1968. Analisi delle serie temporali del geopotenziale a 500 mb su Roma. *Atti Conv. Ann., Assoc. Geofis. Ital.*, 17: 335–343.

Båth, M., 1949. *An Investigation of the Uppsala Microseisms*. Thesis Uppsala Univ., Almqvist and Wiksell, 168 pp.

Båth, M., 1953. Microseismic period spectra and related problems in the Scandinavian area. *Natl. Res. Council (Wash.), Publ.*, 306: 56–64.

Båth, M., 1962. Seismic records of explosions—especially nuclear explosions. Part III. *Res. Inst. Natl. Def. (Stockh.), Rep. A* 4270–4721, 116 pp.

Båth, M., 1966a. Underground measurements of short-period noise. *Ann. Geofis.*, 19: 107–117.

Båth, M., 1966b. Earthquake energy and magnitude. In: L. H. Ahrens, F. Press, S. K. Runcorn and H. C. Urey (Editors), *Physics and Chemistry of the Earth*, Vol. 7. Pergamon Press, pp.115–165.

Båth, M., 1968. *Mathematical Aspects of Seismology*. Elsevier, 415 pp.

Båth, M. and Burman, S., 1972. Walsh spectroscopy of Rayleigh waves caused by underground detonations. In: R. W. Zeek and A. E. Showalter (Editors), *Applications of Walsh Functions. 1972 Proc., Naval Res. Lab.*, AD-744 650: 48–63.

Båth, M. and Crampin, S., 1965. Higher modes of seismic surface waves—relations to channel waves. *Geophys. J. R. Astron. Soc.*, 9: 309–321.

Båth, M. and López Arroyo, A., 1962. Attenuation and dispersion of G-waves. *J. Geophys. Res.*, 67: 1933–1942.

Båth, M. and Shahidi, M., 1971. T-phases from Atlantic earthquakes. *Pure Appl. Geophys.*, 92: 74–114.

Battan, L. J. and Theiss, J. B., 1972. Observed Doppler spectra of hail. *J. Appl. Meteorol.*, 11: 1001–1007.

Beaudet, P. R., 1970. Synthesis of nonstationary seismic signals. *Bull. Seismol. Soc. Am.*, 60: 1615–1624.

Beaudet, P. R. and Wolfson, S. J., 1970. Digital filters for response spectra. *Bull. Seismol. Soc. Am.*, 60: 1001–1013.

Behrens, J., Dresen, L. and Hinz, E., 1969. Modellseismische Untersuchungen der dynamischen Parameter von Kopfwelle und Reflexion im überkritischen Bereich. *Z. Geophys.*, 35: 43–68.

Belmont, A. D. and Dartt, D. G., 1970. The variability of tropical stratospheric winds. *J. Geophys. Res.*, 75: 3133–3145.

Belotelov, V. L. and Kondorskaya, N. V., 1964. The spectra of body waves from Kamchatka earthquakes. *Izv. Acad. Sci. USSR, Geophys. Ser.*, 4: 285–289 (Engl. ed.).

Belotelov, V. L. and Rykunov, L. N., 1963. A device for digitizing seismograms. *Izv. Acad. Sci. USSR, Geophys. Ser.*, 3: 293–294 (Engl. ed.).

Bendat, J. S., 1958. *Principles and Applications of Random Noise Theory.* Wiley, 431 pp.

Bendat, J. S. and Piersol, A. G., 1971. *Random Data: Analysis and Measurement Procedures.* Wiley-Interscience, 407 pp.

Benioff, H., Harrison, J. C., LaCoste, L., Munk, W. H. and Slichter, L. B., 1959. Searching for the earth's free oscillations. *J. Geophys. Res.*, 64: 1334–1337.

Benioff, H., Press, F. and Smith, S., 1961. Excitation of the free oscillations of the earth by earthquakes. *J. Geophys. Res.*, 66: 605–619.

Ben-Menahem, A., 1961. Radiation of seismic surface-waves from finite moving sources. *Bull. Seismol. Soc. Am.*, 51: 401–435.

Ben-Menahem, A., 1962. Radiation of seismic body waves from a finite moving source in the earth. *J. Geophys. Res.*, 67: 345–350.

Ben-Menahem, A., 1964. Spectral response of an elastic sphere to dipolar point-sources. *Bull. Seismol. Soc. Am.*, 54: 1323–1340.

Ben-Menahem, A., 1965. Observed attenuation and *Q*-values of seismic surface waves in the upper mantle. *J. Geophys. Res.*, 70: 4641–4651.

Ben-Menahem, A., 1971. The force system of the Chilean earthquake of 1960 May 22. *Geophys. J. R. Astron. Soc.*, 25: 407–417.

Ben-Menahem, A. and Aboodi, E., 1971. Tectonic patterns in the northern Red Sea region. *J. Geophys. Res.*, 76: 2674–2689.

Ben-Menahem, A. and Harkrider, D. G., 1964. Radiation patterns of seismic surface waves for buried dipolar point sources in a flat stratified earth. *J. Geophys. Res.*, 69: 2605–2620.

Ben-Menahem, A. and Rosenman, M., 1972. Amplitude patterns of tsunami waves from submarine earthquakes. *J. Geophys. Res.*, 77: 3097–3128.

Ben-Menahem, A. and Singh, S. J., 1972. Computation of models of elastic dislocations in the earth. In: B. A. Bolt (Editor), *Methods in Computational Physics*, Vol. 12. Academic Press, pp.299–375.

Ben-Menahem, A. and Toksöz, M. N., 1962. Source mechanism from spectra of long-period seismic surface waves. 1. The Mongolian earthquake of December 4, 1957. *J. Geophys. Res.*, 67: 1943–1955.

Ben-Menahem, A. and Toksöz, M. N., 1963a. Source mechanism from spectra of long-period seismic surface waves. 2. The Kamchatka earthquake of November 4, 1952. *J. Geophys. Res.*, 68: 5207–5222.

Ben-Menahem, A. and Toksöz, M. N., 1963b. Source mechanism from spectra of long-period seismic surface waves. 3. The Alaska earthquake of July 10, 1958. *Bull. Seismol. Soc. Am.*, 53: 905–919.

Ben-Menahem, A., Smith, S. W. and Teng, T.-L., 1965. A procedure for source studies from spectrums of long-period seismic body waves. *Bull. Seismol. Soc. Am.*, 55: 203–235.

Ben-Menahem, A., Jarosch, H. and Rosenman, M., 1968. Large-scale processing of seismic data in search of regional and global stress patterns. *Bull. Seismol. Soc. Am.*, 58: 1899–1932.

Ben-Menahem, A., Rosenman, M. and Harkrider, D. G., 1970. Fast evaluation of source parameters from isolated surface-wave signals. Part I. Universal tables. *Bull. Seismol. Soc. Am.*, 60: 1337–1387.

Ben-Menahem, A., Israel, M. and Levité, U., 1971. Theory and computation of amplitudes of terrestrial line spectra. *Geophys. J. R. Astron. Soc.*, 25: 307–406.

Ben-Menahem, A., Rosenman, M. and Israel, M., 1972. Source mechanism of the Alaskan earthquake of 1964 from amplitudes of free oscillations and surface waves. *Phys. Earth Planet. Inter.*, 5: 1–29.

Benton, G. S. and Kahn, A. B., 1958. Spectra of large-scale atmospheric flow at 300 millibars. *J. Meteorol.*, 15: 404–410.

Berckhemer, H., 1962. Die Ausdehnung der Bruchfläche im Erdbebenherd und ihr Einfluss auf das seismische Wellenspektrum. *Gerlands Beitr. Geophys.*, 71: 5–26.

Berckhemer, H., 1971. The concept of wide band seismometry. *Obs. R. Belg. Comm. A13, Sér. Géophys.*, 101: 214–220.

Berckhemer, H. and Akasche, B., 1966. Seismic ground noise and wind. *Ber. Inst. Meteorol. Geophys. Univ. Frankfurt/Main*, 12: 1–26.

Berckhemer, H. and Jacob, K. H., 1968. Investigation of the dynamical process in earthquake foci by analyzing the pulse shape of body waves. *Ber. Inst. Meteorol. Geophys. Univ. Frankfurt/Main*, 13: 1–85.

Berckhemer, H., Müller, St. and Sellevoll, M., 1961 Die Krustenstruktur in Südwestdeutschland aus Phasen-Geschwindigkeitsmessungen an Rayleigh-Wellen. *Z. Geophys.*, 27: 151–163.

Berg, J. W., Jr. and Papageorge, G. E., 1964. Elastic displacement of primary waves from explosive sources. *Bull. Seismol. Soc. Am.*, 54: 947–959.

Bergsten, F., 1926. Seiches of Lake Vetter. *Geogr. Ann. (Stockh.)*, 8: 1–73.

Berman, S., 1965. Estimating the longitudinal wind spectrum near the ground. *Q. J. R. Meteorol. Soc.*, 91: 302–317.

Berson, I. S., 1959. Some spectral characteristics of waves reflected from thin layers. *Izv. Acad. Sci. USSR, Geophys. Ser.*, 5: 453–461 (Engl. ed.).

Beryland, N. G., 1971. On the possibilities of autocorrelation analysis in studies of the structure of a gravity field. *Izv. Acad. Sci. USSR, Phys. Solid Earth*, 1: 42–48 (Engl. ed.).

Berzon, I. S., 1961. Methods and some results of interpretation of data of seismic waves' spectrum analysis. *Int. Assoc. Seismol. Phys. Earth's Inter.*, A 21: 65–77.

Berzon, I. S., 1965. The determination of a model of a thinly layered medium by the simultaneous use of amplitude- and phase-spectrum characteristics of the layer. *Izv. Acad. Sci. USSR, Phys. Solid Earth*, 6: 363–367 (Engl. ed.).

Berzon, I. S. and Ratnikova, L. I., 1971. Approximate calculations of reflected seismic waves in finely stratified multilayered media. *Izv. Acad. Sci. USSR, Phys. Solid Earth*, 10: 730–738 (Engl. ed.).

Best, A., 1968. Eine Analyse quasiperiodischer Variationen des Magnetfeldes während geomagnetischer Stürme. *Pure Appl. Geophys.*, 69: 193–204.

Bhargava, B. N., 1972. A two-component model of the annual line in the spectrum of the geomagnetic field. *Ann. Géophys.*, 28: 357–361.

Bhartendu, 1969. Audio-frequency pressure variations from lightning discharges. *J. Atmos. Terr. Phys.*, 31: 743–747.

Bhartendu, 1971. Relation of the atmospheric potential gradient with meteorological elements—cross-power spectral analysis. *Pure Appl. Geophys.*, 88: 210–227.

Bhattacharya, B., 1971. Analysis of a vertical dike, infinitely deep, striking north by Fourier transform. *Pure Appl. Geophys.*, 89: 134–138.

Bhattacharyya, B. K., 1965. Two-dimensional harmonic analysis as a tool for magnetic interpretation. *Geophysics*, 30: 829–857.

Bhattacharyya, B. K., 1966. Continuous spectrum of the total-magnetic-field anomaly due to a rectangular prismatic body. *Geophysics*, 31: 97–121.

Bhattacharyya, B. K., 1967. Some general properties of potential fields in space and frequency domain: a review. *Geoexploration*, 5: 127–143.

Bhattacharyya, B. K., 1969. Bicubic spline interpolation as a method for treatment of potential field data. *Geophysics*, 34: 402–423.

Bilham, R., Evans, R., King, G., Lawson, A. and McKenzie, D., 1972. Earth strain tides observed in Yorkshire, England with a simple wire strainmeter. *Geophys. J. R. Astron. Soc.*, 29: 473–485.

Bingham, Ch., Godfrey, M. D. and Tukey, J. W., 1967. Modern techniques of power spectrum estimation. *IEEE Trans. Audio Electroacoustics*, AU-15: 56–66.

Black, D. I., 1970. Lunar and solar magnetic variations at Abinger: their detection and estimation by spectral analysis via Fourier transforms. *Philos. Trans. R. Soc. London, Ser. A*, 268: 233–263.

Black, D. I. and Scollar, I., 1969. Spatial filtering in the wave-vector domain. *Geophysics*, 34: 916–923.

Blackman, R. B. and Tukey, J. W., 1959. *The Measurement of Power Spectra*. Dover Publications, 190 pp.

Bloch, S. and Hales, A. L., 1968. New techniques for the determination of surface wave phase velocities. *Bull. Seismol. Soc. Am.*, 58: 1021–1034.

Bloch, S., Hales, A. L. and Landisman, M., 1969. Velocities in the crust and upper mantle of southern Africa from multi-mode surface wave dispersion. *Bull. Seismol. Soc. Am.*, 59: 1599–1629.

Block, B. and Dratler, J., 1972. A review of tidal, earth normal mode and seismic data obtained with quartz torsion accelerometers. *Geophys. J. R. Astron. Soc.*, 31: 239–269.

Blum, P. A. and Gaulon, R., 1971. Détection et traitement des ondes sismiques de très basses fréquences. *Ann. Géophys.*, 27: 123–140.

Blum, P. A. and Hatzfeld, D., 1970. Etude régionale de l'influence océanique sur l'inclinaison—Premiers résultats à la station De Moulis. *Obs. R. Belg. Comm. (Géophys.)*, A9: 102–105.

Blume, J. A., 1970. An engineering intensity scale for earthquakes and other ground motion. *Bull. Seismol. Soc. Am.*, 60: 217–229.

Bodoky, T., Korvin, G., Liptai, I. and Sipos, J., 1972. An analysis of the initial seismic pulse near underground explosions. *Geofiz. Közl. (Budapest)*, 20, 3–4: 7–27.

Bogert, B. P., 1961a. Seismic data collection, reduction, and digitization. *Bull. Seismol. Soc. Am.*, 51: 515–525.

Bogert, B. P., 1961b. An observation of free oscillations of the earth. *J. Geophys. Res.*, 66: 643–646.

Bogert, B. P., Healy, M. J. R. and Tukey, J. W., 1963. The quefrency alanysis of time series for echoes: cepstrum, pseudo-autocovariance, cross-cepstrum and saphe cracking. In: M. Rosenblatt (Editor), *Time Series Analysis*. Wiley, pp.209–243.

Bois, P., 1972. Analyse sequentielle. *Geophys. Prospect.*, 20: 497–513.

Bollinger, G. A., 1968. Determination of earthquake fault parameters from long-period P-waves. *J. Geophys. Res.*, 73: 785–807.

Bollinger, G. A., 1970. Fault length and fracture velocity for the Kyushu, Japan, earthquake of October 3, 1963. *J. Geophys. Res.*, 75: 955–964.

Bolt, B. A., 1964. Recent information on the earth's interior from studies of mantle waves and eigenvibrations. In: L. H. Ahrens, F. Press, S. K. Runcorn and H. C. Urey (Editors), *Physics and Chemistry of the Earth*, Vol. 5. Pergamon Press, pp.55–119.

Bolt, B. A. and Marussi, A., 1962. Eigenvibrations of the earth observed at Trieste. *Geophys. J. R. Astron. Soc.*, 6: 299–311. Revised (1963) in *Geophys. J. R. Astron. Soc.*, 7: 510–512.

Bolt, B. A., Niazi, M. and Sommerville, M. R., 1970. Diffracted ScS and the shear velocity at the core boundary. *Geophys. J. R. Astron. Soc.*, 19: 299–305.

Bonjer, K.-P. and Fuchs, K., 1970. Crustal structure in southwest Germany from spectral transfer ratios of longperiod bodywaves. *Int. Upper Mantle Proj., Sci. Rep.* 27: 198–202.

Bonjer, K.-P., Fuchs, K. and Wohlenberg, J., 1970. Crustal structure of the East African rift system from spectral response ratios of long-period body waves. *Z. Geophys.*, 36: 287–297.

Boore, D. M. and Toksöz, M. N., 1969. Rayleigh wave particle motion and crustal structure. *Bull. Seismol. Soc. Am.*, 59: 331–346.

Boore, D. M., Larner, K. L. and Aki, K., 1971. Comparison of two independent methods for the solution of wave-scattering problems: response of a sedimentary basin to vertically incident SH waves. *J. Geophys. Res.*, 76: 558–569.

Borcherdt, R. D., 1970. Effects of local geology on ground motion near San Francisco Bay. *Bull. Seismol. Soc. Am.*, 60: 29–61.

Bowden, K. F., 1962a Turbulence. In: M. N. Hill (Editor), *The Sea*, Vol. 1. Interscience, pp.802–825.

Bowden, K. F., 1962b. Measurements of turbulence near the sea bed in a tidal current. *J. Geophys. Res.*, 67: 3181–3186.

Bowden, K. F. and White, R. A., 1966. Measurements of the orbital velocities of sea waves and their use in determining the directional spectrum. *Geophys. J. R. Astron. Soc.*, 12: 33–54.

Bowne, N. E. and Ball, J. T., 1970. Observational comparison of rural and urban boundary layer turbulence. *J. Appl. Meteorol.*, 9: 862–873.

Boyarskiy, E. A. and Kogan, M. G., 1968. Linear smoothing of measurements made with marine gravimeters. *Izv. Acad. Sci. USSR, Phys. Solid Earth*, 10: 640–643 (Engl. ed.).

Bozhko, G. N. and Starovoit, O. E., 1968. Phase velocities of Rayleigh waves on the Russian platform. In: E. Bisztricsány (Editor), *Proc. 8th Assem. Eur. Seismol. Comm.* Akad. Kiadó, Budapest, pp.339–343.

Bracewell, R., 1965. *The Fourier Transform and Its Applications.* McGraw-Hill, 381 pp.

Bradner, H. and Dodds, J. G., 1964. Comparative seismic noise on the ocean bottom and on land. *J. Geophys. Res.*, 69: 4339–4348.

Bradner, H., Dodds, J. G. and Foulks, R. E., 1965. Investigation of microseism sources with ocean-bottom seismometers. *Geophysics*, 30: 511–526.

Bradner, H., De Jerphanion, L. G. and Langlois, R., 1970. Ocean microseism measurements with a neutral bouyancy free-floating midwater seismometer. *Bull. Seismol. Soc. Am.*, 60: 1139–1150.

Brekhovskikh, L. M., Fedorov, K. N., Fomin, L. M., Koshlyakov, M. N. and Yampolsky, A. D., 1971. Large-scale multi-buoy experiments in the tropical Atlantic. *Deep-Sea Res.*, 18: 1189–1206.

Bretherton, F. P., 1969. Momentum transport by gravity waves. *Q. J. R. Meteorol. Soc.*, 95: 213–243.

Brier, G. W., 1961. Some statistical aspects of long-term fluctuations in solar and atmospheric phenomena. *Ann. N. Y. Acad. Sci.*, 95: 173–187.

Brier, G. W., 1968. Long-range prediction of the zonal westerlies and some problems in data analysis. *Rev. Geophys.*, 6: 525–551.

Brillinger, D. R., 1968. Estimation of the cross-spectrum of a stationary bivariate Gaussian process from its zeros. *J. R. Stat. Soc.*, B 30: 145–159.

Briscoe, M. G., 1972. A note on internal gravity wave spectra. *J. Geophys. Res.*, 77: 3278–3280.

Broek, H. W., 1969a. Determination of the aliasing band corresponding to each spectrum peak in ocean-bottom temperature spectra. *J. Geophys. Res.*, 74: 5439–5448.

Broek, H. W., 1969b. Fluctuations in bottom temperature at 2000 meter depth off the Blake Plateau. *J. Geophys. Res.*, 74: 5449–5452.

Brook, R. R., 1972. The measurement of turbulence in a city environment. *J. Appl. Meteorol.*, 11: 443–450.

Brooks, J. A., 1969. Rayleigh waves in southern New Guinea. I. Higher mode group velocities. *Bull. Seismol. Soc. Am.*, 59: 945–958.

Brown, R. J., 1973. Slowness and azimuth at the Uppsala array. Part 1: array calibration and event location. *Pure Appl. Geophys.*, 105: 759–769.

Brune, J. N., 1960. Radiation pattern of Rayleigh waves from the southeast Alaska earthquake of July 10, 1958. *Publ. Dom. Obs. Ottawa*, 24: 373–383.

Brune, J. N., 1962. Attenuation of dispersed wave trains. *Bull. Seismol. Soc. Am.*, 52: 109–112.

Brune, J. N., 1965. The Sa phase from the Hindu Kush earthquake of July 6, 1962. *Pure Appl. Geophys.*, 62: 81–95.

Brune, J. N., 1970. Tectonic stress and the spectra of seismic shear waves from earthquakes. *J. Geophys. Res.*, 75: 4997–5009.

Brune, J. N., 1971. Correction to "Tectonic stress and the spectra of seismic shear waves from earthquakes". *J. Geophys. Res.*, 76: 5502.

Brune, J. N. and King, C.-Y., 1967. Excitation of mantle Rayleigh waves of period 100 seconds as a function of magnitude. *Bull. Seismol. Soc. Am.*, 57: 1355–1365.

Brune, J. N. and Pomeroy, P. W., 1963. Surface wave radiation patterns for underground nuclear explosions and small-magnitude earthquakes. *J. Geophys. Res.*, 68: 5005–5028.

Brune, J. N., Nafe, J. E. and Oliver, J., 1960. A simplified method for the analysis and synthesis of dispersed wave trains. *J. Geophys. Res.*, 65: 287–304.

Brune, J. N., Nafe, J. E. and Alsop, L. E., 1961a. The polar phase shift of surface waves on a sphere. *Bull. Seismol. Soc. Am.*, 51: 247–257.

Brune, J. N., Benioff, H. and Ewing, M., 1961b. Long-period surface waves from the Chilean earthquake of May 22, 1960, recorded on linear strain seismographs. *J. Geophys. Res.*, 66: 2895–2910.

Bryson, R. A. and Dutton, J. A., 1961. Some aspects of the variance spectra of tree rings and varves. *Ann. N. Y. Acad. Sci.*, 95: 580–604.

Buchbinder, G. G. R., 1968. Amplitude spectra of PcP and P phases. *Bull. Seismol. Soc. Am.*, 58: 1797–1819.

Buchheim, W. and Smith, S. W., 1961. The earth's free oscillations observed on earth tide instruments at Tiefenort, East Germany. *J. Geophys. Res.*, 66: 3608–3610.

Bufe, C. G. and Willis, D. E., 1969. High frequency teleseismic energy from Aleutian sources. *Bull. Seismol. Soc. Am.*, 59: 2061–2070.

Bulakh, Ye. G., 1970. Elimination of the regional background in the interpretation of gravitational and magnetic anomalies. *Izv. Acad. Sci. USSR, Phys. Solid Earth*, 2: 127–129 (Engl. ed.).

Bull, G. and Neisser, J., 1968. Untersuchungen der atmosphärischen Feinstruktur mit Hilfe von Ausbreitungsmessungen im Mikrowellenbereich. *Gerlands Beitr. Geophys.*, 77: 394–410.

Bullard, E. C., 1967. The removal of trend from magnetic surveys. *Earth Planet. Sci. Lett.*, 2: 293–300.

Bullard, E. C. and Mason, R. G., 1963. The magnetic field over the oceans. In: M. N. Hill (Editor), *The Sea*, Vol. 3. Interscience, pp.175–217.

Bungum, H., Rygg, E. and Bruland, L., 1971. Short-period seismic noise structure at the Norwegian seismic array. *Bull. Seismol. Soc. Am.*, 61: 357–373.

Bunting, D. C., 1970. Evaluating forecasts of ocean-wave spectra. *J. Geophys. Res.*, 75: 4131–4143.

Burke, M. D., Kanasewich, E. R., Malinsky, J. D. and Montalbetti, J. F., 1970. A wide-band digital seismograph system. *Bull. Seismol. Soc. Am.*, 60: 1417–1426.

Burkhardt, H., 1964. Some physical aspects of seismic scaling laws for underwater explosions. *Geophys. Prospect.*, 12: 192–214.

Burkhardt, H., Rosenbach, O. and Vees, R., 1968. Der seismische Impuls bei Unterwassersprengungen in verschiedenen Registrierentfernungen. In: E. Bisztricsány (Editor), *Proc. 8th Assem. Eur. Seismol. Comm.* Akad. Kiadó, Budapest, pp.33–42.

Burling, R. W., 1959. The spectrum of waves at short fetches. *Dtsche Hydrogr. Z.*, 12: 45–64 and 96–117.

Busch, N. E. and Panofsky, H. A., 1968. Recent spectra of atmospheric turbulence. *Q. J. R. Meteorol. Soc.*, 94: 132–148.

Bushnell, R. H. and Huss, P. O., 1958. A power spectrum of surface winds. *J. Meteorol.*, 15: 180–183.

Businger, J. A. and Suomi, V. E., 1958. Variance spectra of the vertical wind component derived from observations with the sonic anemometer at O'Neill, Nebraska in 1953. *Arch. Meteorol. Geophys. Bioklimatol.*, A 10: 415–425.

Bustamante, J. I., 1964. Response spectra of earthquakes on very soft clay. *Bull. Seismol. Soc. Am.*, 54: 855–866.

Cagnetti, P. and Giudici, G., 1966. Determinazione dello spettro della componente verticale del vento in località Casaccia. *Atti Conv. Ann., Assoc. Geofis. Ital.*, 15: 147–165.

Cairns, J. L., 1968. Thermocline strength fluctuations in coastal waters. *J. Geophys. Res.*, 73: 2591–2595.

Cairns, J. L. and LaFond, E. C., 1966. Periodic motions of the seasonal thermocline along the southern California coast. *J. Geophys. Res.*, 71: 3903–3915.

Camfield, P. A., Gough, D. I. and Porath, H., 1971. Magnetometer array studies in the northwestern United States and southwestern Canada. *Geophys. J. R. Astron. Soc.*, 22: 201–221.

Campbell, G. A. and Foster, R. M., 1948. *Fourier Integrals for Practical Applications*. Van Nostrand, 177 pp.

Campbell, W. H., 1966. A review of the equatorial studies of rapid fluctuations in the earth's magnetic field. *Ann. Géophys.*, 22: 492–501.

Campbell, W. H., 1967. Geomagnetic pulsations. In: S. Matsushita and W. H. Campbell (Editors), *Physics of Geomagnetic Phenomena*. Acad. Press, *Int. Geophys. Ser.*, 11: 821–909.

Caner, B., 1971. Quantitative interpretation of geomagnetic depth-sounding data in western Canada. *J. Geophys. Res.*, 76: 7202–7216.

Caner, B. and Auld, D. R., 1968. Magneto-telluric determination of upper-mantle conductivity structure at Victoria, British Columbia. *Can. J. Earth Sci.*, 5: 1209–1220.

Caner, B., Cannon, W. H. and Livingstone, C. E., 1967. Geomagnetic depth sounding and upper-mantle structure in the Cordillera region of western North America. *J. Geophys. Res.*, 72: 6335–6351.

Caner, B., Camfield, P. A., Andersen, F. and Niblett, E. R., 1969. A large-scale magnetotelluric survey in Western Canada. *Can. J. Earth Sci.*, 6: 1245–1261.

Caner, B., Auld, D. R., Dragert, H. and Camfield, P. A., 1971. Geomagnetic depth-sounding and crustal structure in western Canada. *J. Geophys. Res.*, 76: 7181–7201.

Canitez, N., 1972. Source mechanism and rupture propagation in the Mudurnu Valley, Turkey, earthquake of July 22, 1967. *Pure Appl. Geophys.* 93: 116–124.

Canitez, N. and Toksöz, M. N., 1971. Focal mechanism and source depth of earthquakes from body- and surface-wave data. *Bull. Seismol. Soc. Am.*, 61: 1369–1379.

Canitez, N. and Toksöz, M. N., 1972. Static and dynamic study of earthquake source mechanism: San Fernando earthquake. *J. Geophys. Res.*, 77: 2583–2594.

Cannon, G. A., 1971. Statistical characteristics of velocity fluctuations at intermediate scales in a coastal plain estuary. *J. Geophys. Res.*, 76: 5852–5858.

Cantwell, T. and Madden, T. R., 1960. Preliminary report on crustal magnetotelluric measurements. *J. Geophys. Res.*, 65: 4202–4205.

Capon, J., 1969a. High-resolution frequency-wavenumber spectrum analysis. *Proc. IEEE*, 57: 1408–1418.

Capon, J., 1969b. Investigation of long-period noise at the Large Aperture Seismic Array. *J. Geophys. Res.*, 74: 3182–3194.

Capon, J., 1970. Analysis of Rayleigh-wave multipath propagation at LASA. *Bull. Seismol. Soc. Am.*, 60: 1701–1731.

Capon, J., 1972. Long-period signal processing results for LASA, NORSAR and ALPA. *Geophys. J. R. Astron. Soc.*, 31: 279–296.

Capon, J. and Evernden, J. F., 1971. Detection of interfering Rayleigh waves at LASA. *Bull. Seismol. Soc. Am.*, 61: 807–849.

Capon, J., Greenfield, R. J. and Kolker, R. J., 1967. Multidimensional maximum-likelihood processing of a Large Aperture Seismic Array. *Proc. IEEE*, 55: 192–211.

Capon, J., Greenfield, R. J. and Lacoss, R. T., 1969. Long-period signal processing results for the Large Aperture Seismic Array. *Geophysics*, 34: 305–329.

Caputo, M., Pieri, L. and Rossi Tesi, F., 1972. Land subsidence in Venice and Porto Corsini. *Ann. Geofis.*, 25: 55–61.

Carrozzo, M. T. and Mosetti, F., 1966. Coefficients and tables for two-dimensional periodal analysis. *Boll. Geofis. Teor. Appl.*, 8: 264–285.

Cartwright, D. E., 1962. Waves, analysis and statistics. In: M. N. Hill (Editor), *The Sea*, Vol. 1. Interscience, pp.567–589.

Cartwright, D. E. and Tayler, R. J., 1971. New computations of the tide-generating potential. *Geophys. J. R. Astron. Soc.*, 23: 45–73.

Casten, U., Chowdhury, K. R. and Gutdeutsch, R., 1969. Anlage, Durchführung und Analyse von Hydrophon- und Geophonregistrierungen bei seismischen Messungen im Okerstausee. *Hamb. Geophys. Einzelschriften*, H. 12, 137 pp.

Cehak, K. and Pichler, H., 1969a. Die Persistenz der solaren und geomagnetischen Aktivität (I. Mitteilung). *Arch. Meteorol. Geophys. Bioklimatol.*, A 18: 147–165.

Cehak, K. and Pichler, H., 1969b. Die Persistenz der solaren und geomagnetischen Aktivität (II. Mitteilung). *Arch. Meteorol. Geophys. Bioklimatol.*, A 18: 365–376.

Chan, S. H. and Leong, L. S., 1972. Analysis of least-squares smoothing operators in the frequency domain. *Geophys. Prospect.*, 20: 892–900.

Chander, R. and Brune, J. N., 1965. Radiation pattern of mantle Rayleigh waves and the source mechanism of the Hindu Kush earthquake of July 6, 1962. *Bull. Seismol. Soc. Am.*, 55: 805–819.

Chander, R., Alsop, L. E. and Oliver, J., 1968. On the synthesis of shear-coupled PL-waves. *Bull. Seismol. Soc. Am.*, 58: 1849–1877.

Chandra, N. N. and Cumming, G. L., 1972. Rotated power spectra of microseisms. *Can. J. Earth Sci.*, 9: 325–338.

Chandra, U., 1970a. Analysis of body-wave spectra for earthquake energy determination. *Bull. Seismol. Soc. Am.*, 60: 539–563.

Chandra, U., 1970b. Comparison of focal mechanism solutions obtained from P- and S-wave data. *J. Geophys. Res.*, 75: 3411–3420.

Chandra, U., 1970c. Stationary phase approximation in focal mechanism determination. *Bull. Seismol. Soc. Am.*, 60: 1221–1229.

Chang, C.-P., Morris, V. F. and Wallace, J. M., 1970. A statistical study of easterly waves in the western Pacific: July–December 1964. *J. Atmos. Sci.*, 27: 195–201.

Chang, M.-S., 1969. Mass transport in deep-water long-crested random gravity waves. *J. Geophys. Res.*, 74: 1515–1536.

Chapman, S. and Bartels, J., 1940, 1951. *Geomagnetism*, Vol. 2. Clarendon Press, pp.543–1049.

Chapman, S. and Malin, S. R. C., 1970. Atmospheric tides, thermal and gravitational: nomenclature, notation and new results. *J. Atmos. Sci.*, 27: 707–710.

Charnock, H., 1957. Notes on the specification of atmospheric turbulence. *J. R. Stat. Soc.*, A 120: 398–408.

Chattopadhyay, J., 1970. Power spectrum analysis of atmospheric ozone content over north India. *Pure Appl. Geophys.*, 83: 111–119.

Chattopadhyay, J. and Rathor, H. S., 1972. Power spectrum analysis of atmospheric ozone during the summer. *Pure Appl. Geophys.*, 95: 186–193.

Chernikov, A. A., Mel'nichuk, Yu. V., Pinus, N. Z., Shmeter, S. M. and Vinnichenko, N. K., 1969. Investigations of the turbulence in convective atmosphere using radar and aircraft. *Radio Sci.*, 4: 1257–1259.

Chiu, W., 1960. The spectra of large-scale turbulent transfer of momentum and heat. *J. Meteorol.*, 17: 435–441.

Chiu, W.-C., 1967. On the interpretation of the energy spectrum. *Am. J. Phys.*, 35: 642–648.

Chiu, W.-C., 1970. On the spectral equations and the statistical energy spectrum of atmospheric motions in the frequency domain. *Tellus*, 22: 608–619.

Chiu, W.-C. and Crutcher, H. L., 1966. The spectrums of angular momentum transfer in the atmosphere. *J. Geophys. Res.*, 71: 1017–1032.

Chiu, W.-C. and Rib, L. N., 1956. The rate of dissipation of energy and the energy spectrum in a low-speed turbulent jet. *Trans. Am. Geophys. Union*, 37: 13–26.

Chopra, A. K., 1966. The importance of the vertical component of earthquake motions. *Bull. Seismol. Soc. Am.*, 56: 1163–1175.

Choudhury, M. A., 1972. P-wave attenuation in the mantle. *Z. Geophys.*, 38: 447–453.

Choudhury, M. A. and Dorel, J., 1973. Spectral ratio of short-period ScP and ScS phases in relation to the attenuation in the mantle beneath the Tasman Sea and the Antarctic region. *J. Geophys. Res.*, 78: 462–469.

Cisternas, A., 1961. Crustal structure of the Andes from Rayleigh wave dispersion. *Bull. Seismol. Soc. Am.*, 51: 381–388.

Clarke, G. K. C., 1969. Optimum second-derivative and downward-continuation filters. *Geophysics*, 34: 424–437.

Cleary, J. R. and Peaslee, D. C., 1962. Amplitude perturbations in Rayleigh waves. *J. Geophys. Res.*, 67: 4741–4749.

Clewell, D. H. and Simon, R. F., 1950. Seismic wave propagation. *Geophysics*, 15: 51–60.

Cloud, W. K. and Carder, D. S., 1969. Ground effects from the Boxcar and Benham nuclear explosions. *Bull. Seismol. Soc. Am.*, 59: 2371–2381.

Cloud, W. K. and Hudson, D. E., 1961. A simplified instrument for recording strong motion earthquakes. *Bull. Seismol. Soc. Am.*, 51: 159–174.

Cloud, W. K. and Perez, V., 1967. Accelerograms—Parkfield earthquake. *Bull. Seismol. Soc. Am.*, 57: 1179–1192.

Clough, R. W., 1962. Earthquake analysis by response spectrum superposition. *Bull. Seismol. Soc. Am.*, 52: 647–660.

Clowes, R. M. and Kanasewich, E. R., 1970. Seismic attenuation and the nature of reflecting horizons within the crust. *J. Geophys. Res.*, 75: 6693–6705.

Clowes, R. M., Kanasewich, E. R. and Cumming, G. L., 1968. Deep crustal seismic reflections at near-vertical incidence. *Geophysics*, 33: 441–451.

Coantic, M. and Leducq, D., 1969. Turbulent fluctuations of humidity and their measurement. *Radio Sci.*, 4: 1169–1174.

Cochran, W. T., Cooley, J. W., Favin, D. L., Helms, H. D., Kaenel, R. A., Lang, W. W., Maling, G. C., Jr., Nelson, D. E., Rader, C. M. and Welch, P. D., 1967. What is the Fast Fourier Transform? *IEEE Trans. Audio Electroacoustics*, AU-15: 45–55.

Cochrane, N. A. and Hyndman, R. D., 1970. A new analysis of geomagnetic depth-sounding data from western Canada. *Can. J. Earth Sci.*, 7: 1208–1218.

Cohen, T. J., 1970. Source-depth determinations using spectral, pseudo-autocorrelation and cepstral analysis. *Geophys. J. R. Astron. Soc.*, 20: 223–231.

Coleman, P. J., Jr. and Smith, E. J., 1966. An interpretation of the subsidiary peaks at periods near 27 days in the power spectra of C_1 and K_p. *J. Geophys. Res.*, 71: 4685–4686.

Collins, J. I., 1972. Prediction of shallow-water spectra. *J. Geophys. Res.*, 77: 2693–2707.

Connes, J., Blum, P. A. and Jobert, G. and N., 1962. Observations des oscillations propres de la terre. *Ann. Géophys.*, 18: 260–268.

Conrad, V. and Pollak, L. W., 1950. *Methods in Climatology*. Harvard Univ. Press, 2nd ed., 459 pp.

Console, R. and Peronaci, F., 1971. Studio degli accelerogrammi ottenuti nei recenti eventi sismici dell'Italia centrale. *Ann. Geofis.*, 24: 497–514.

Coode, A. M., 1966. An analysis of major tectonic features. *Geophys. J. R. Astron. Soc.*, 12: 55–66.

Coode, A. M., 1967. The spherical harmonic analysis of major tectonic features. In: S. K. Runcorn (Editor), *Mantles of the Earth and Terrestrial Planets*. Interscience, pp.489–498.

Cook, E. E. and Taner, M. T., 1969. Velocity spectra and their use in stratigraphic and lithologic differentiation. *Geophys. Prospect.*, 17: 433–448.

Cooley, J. W. and Tukey, J. W., 1965. An algorithm for the machine calculation of complex Fourier series. *Math. Comput.*, 19: 297–301.

Cooley, J. W., Lewis, P. A. W. and Welch, P. D., 1967a. Historical notes on the Fast Fourier Transform. *IEEE Trans. Audio Electroacoustics*, AU-15: 76–79.

Cooley, J. W., Lewis, P. A. W. and Welch, P. D., 1967b. Application of the Fast Fourier Transform to computation of Fourier integrals, Fourier series, and convolution integrals. *IEEE Trans. Audio Electroacoustics*, AU-15: 79–84.

Cordell, L. and Taylor, P. T., 1971. Investigation of magnetization and density of a North Atlantic seamount using Poisson's theorem. *Geophysics*, 36: 919–937.

Cornett, J. S. and Brundidge, K. C., 1970. A comparison of the wind spectra calculated from data obtained by independent sampling methods. *Mon. Weather Rev.*, 98: 233–237.

Cox, C. S., 1962. Internal waves, Part II. In: M. N. Hill (Editor), *The Sea*, Vol. 1. Interscience, pp.752–763.

Craddock, J. M., 1957. An analysis of the slower temperature variations at Kew Observatory by means of mutually exclusive band pass filters. *J. R. Stat. Soc.*, A 120: 387–397.

Cramer, H. E., 1959. Measurements of turbulence structure near the ground within the frequency range from 0.5 to 0.01 cycles sec^{-1}. *Adv. Geophys.*, 6: 75–96.

Crampin, S. and Båth, M., 1965. Higher modes of seismic surface waves: Mode separation. *Geophys. J. R. Astron. Soc.*, 10: 81–92.

Crawford, F. S., Jr., 1965. *Waves. Berkeley Physics Course*, Vol. 3. McGraw-Hill, 600 pp.

Currie, R. G., 1966. The geomagnetic spectrum—40 days to 5.5 years. *J. Geophys. Res.*, 71: 4579–4598.

Currie, R. G., 1967. Magnetic shielding properties of the earth's mantle. *J. Geophys. Res.*, 72: 2623–2633.

Currie, R. G., 1968. Geomagnetic spectrum of internal origin and lower mantle conductivity. *J. Geophys. Res.*, 73: 2779–2786.

Czepa, O. and Schellenberger, G., 1960. Methoden und Ergebnisse der statistischen Seegangsanalyse. *Gerlands Beitr. Geophys.*, 69: 206–239.

Darby, E. K. and Davies, E. B., 1967. The analysis and design of two-dimensional filters for two-dimensional data. *Geophys. Prospect.*, 15: 383–406.

Darbyshire, J., 1950. Identification of microseismic activity with sea waves. *Proc. R. Soc. London, Ser. A*, 202: 439–448.

Darbyshire, J., 1962. Microseisms. In: M. N. Hill (Editor), *The Sea*, Vol. 1. Interscience, pp. 700–719.

Darbyshire, J., 1963. A study of microseisms in South Africa. *Geophys. J. R. Astron. Soc.*, 8: 165–175.

Darbyshire, J., 1970a. Wave measurements with a radar altimeter over the Irish Sea. *Deep-Sea Res.*, 17: 893–901.

Darbyshire, J., 1970b. Time variation of the depth of a high salinity layer in the Celtic Sea. *Deep-Sea Res.*, 17: 903–911.

Darbyshire, J. and Okeke, E. O., 1969. A study of primary and secondary microseisms recorded in Anglesey. *Geophys. J. R. Astron. Soc.*, 17: 63–92.

Dartt, D. G. and Belmont, A. D., 1964. Periodic features of the 50-millibar zonal winds in the tropics. *J. Geophys. Res.*, 69: 2887–2893.

Dash, B. P. and Obaidullah, K. A., 1970. Determination of signal and noise statistics using correlation theory. *Geophysics*, 35: 24–32.

Davenport, A. G., 1961. The spectrum of horizontal gustiness near the ground in high winds. *Q. J. R. Meteorol. Soc.*, 87: 194–211.

Davenport, W. B., Jr. and Root, W. L., 1958. *An Introduction to the Theory of Random Signals and Noise*. McGraw-Hill, 393 pp.

Davidson, M. J., 1964. Average diurnal characteristics of geomagnetic power spectrums in the period range 4.5 to 1000 seconds. *J. Geophys. Res.*, 69: 5116–5119.

Davidson, M. J. and Heirtzler, J. R., 1968. Spatial coherence of geomagnetic rapid variations. *J. Geophys. Res.*, 73: 2143–2162.

Davies, D., Kelly, E. J. and Filson, J. R., 1971. Vespa process for analysis of seismic signals. *Nature, Phys. Sci.*, 232(26): 8–13.

Davies, J. B. and Smith, S. W., 1968. Source parameters of earthquakes, and discrimination between earthquakes and nuclear explosions. *Bull. Seismol. Soc. Am.*, 58: 1503–1517.

Davis, L. L. and West, L. R., 1973. Observed effects of topography on ground motion. *Bull. Seismol. Soc. Am.*, 63: 283–298.

Davis, T. M., 1971. A filtering technique for interpreting the gravity anomaly generated by a two-dimensional fault. *Geophysics*, 36: 554–570.

Deacon, G. E. R., 1949. Recent studies of waves and swell. *Ann. N. Y. Acad. Sci.*, 51: 475–482.

Dean, W. C., 1958. Frequency analysis for gravity and magnetic interpretation. *Geophysics*, 23: 97–127.

Dean, W. C., 1964. Seismological applications of Laguerre expansions. *Bull. Seismol. Soc. Am.*, 54: 395–407.

De Bremaecker, J. Cl., 1964. Detection of small arrivals. *Bull. Seismol. Soc. Am.*, 54: 2141–2163.

De Bremaecker, J. Cl., 1965. Microseisms from hurricane "Hilda". *Science*, 148: 1725–1727.

De Bremaecker, J. Cl., Donoho, P. and Michel, J. G., 1962. A direct digitizing seismograph. *Bull. Seismol. Soc. Am.*, 52: 661–672.

De Bremaecker, J. Cl., Sitton, G. A., Rusk, S. K., Graham, M. H. and Schutz, T. C., 1963. The Rice digital seismograph system. *J. Geophys. Res.*, 68: 5029–5034.

DeLeonibus, P. S., 1971. Momentum flux and wave spectra observations from an ocean tower. *J. Geophys. Res.*, 76: 6506–6527.

DeLeonibus, P. S. and Simpson, L. S., 1972. Case study of duration-limited wave spectra observed at an open ocean tower. *J. Geophys. Res.*, 77: 4555–4569.

Denham, D., 1968. Thickness of the earth's crust in Papua New Guinea and the British Solomon Islands. *Aust. J. Sci.*, 30: 277.

DeNoyer, J., 1964. High-frequency microearthquakes recorded at Quetta, Pakistan. *Bull. Seismol. Soc. Am.*, 54: 2133–2139.

DeNoyer, J., Willis, D. E. and Wilson, J. T., 1962. Observed asymmetry of amplitudes from a high explosive source. *Bull. Seismol. Soc. Am.*, 52: 133–137.

DeNoyer, J. M., Frantti, G. E. and Willis, D. E., 1966. Short note on underwater sound measurements from the Lake Superior experiment. *Am. Geophys. Union, Geophys. Monogr.*, 10: 241–248.

Der, Z. A., 1969. Surface wave components in microseisms. *Bull. Seismol. Soc. Am.*, 59: 665–672.

Der, Z., Massé, R. and Landisman, M., 1970. Effects of observational errors on the resolution of surface waves at intermediate distances. *J. Geophys. Res.*, 75: 3399–3409.

Derr, J. S., 1969a. Free oscillation observations through 1968. *Bull. Seismol. Soc. Am.*, 59: 2079–2099.

Derr, J. S., 1969b. Internal structure of the earth inferred from free oscillations. *J. Geophys. Res.*, 74: 5202–5220.

Derr, J. S., 1970. Discrimination of earthquakes and explosions by the Rayleigh-wave spectral ratio. *Bull. Seismol. Soc. Am.*, 60: 1653–1668.

Dewart, G. and Toksöz, M. N., 1965. Crustal structure in east Antarctica from surface wave dispersion. *Geophys. J. R. Astron. Soc.*, 10: 127–139.

Dickson, R. R., 1971. On the relationship of variance spectra of temperature to the large-scale atmospheric circulation. *J. Appl. Meteorol.*, 10: 186–193.

Dinger, J. E., 1963. Comparison of ocean-wave and microseism spectrums as recorded at Barbados, West Indies. *J. Geophys. Res.*, 68: 3465–3471.

Dmitriyev, V. A., 1971. On the spectral composition of underwater seismic noise in coastal regions of the sea. *Izv. Acad. Sci. USSR, Phys. Solid Earth*, 4: 278–281 (Engl. ed.).

Dobeš, K., Prikner, K. and Střeštík, J., 1970. Processings of records of rapid variations of the geomagnetic field by computer. *Trav. Inst. Géophys. Acad. Tchécoslov. Sci.*, 18: 335–348.

Dobeš, K., Střeštík, J. and Prikner, K., 1971. Numerical calculation of frequency-time displays using amplitude-time records. *Studia Geophys. Geodaet.*, 15: 331–339.

Dobrin, M. B., Ingalls, A. L. and Long, J. A., 1965. Velocity and frequency filtering of seismic data using laser light. *Geophysics*, 30: 1144–1178.

Dobson, E. B., 1970. Measurement of the fine-scale structure of the sea. *J. Geophys. Res.*, 75: 2853–2856.

Dohler, G. C. and Ku, L. F., 1970. Presentation and assessment of tides and water level records for geophysical investigations. *Can. J. Earth Sci.*, 7: 607–625.

Donn, W. L. and Posmentier, E. S., 1967. Infrasonic waves from the marine storm of April 7, 1966. *J. Geophys. Res.*, 72: 2053–2061.

Donn, W. L., Pattullo, J. G. and Shaw, D. M., 1964. Sea-level fluctuations and long waves. In: H. Odishaw (Editor), *Research in Geophysics*, Vol. 2. M.I.T. Press, pp.243–269.

Dorman L.-R. M., 1968. Anelasticity and the spectra of body waves. *J. Geophys. Res.*, 73: 3877–3883.

Douglas, A., Hudson, J. A. and Blamey, C., 1972. A quantitative evaluation of seismic signals at teleseismic distances. III. Computed P- and Rayleigh wave seismograms. *Geophys. J. R. Astron. Soc.*, 28: 385–410.

Douglas, B. M. and Ryall, A., 1972. Spectral characteristics and stress drop for microearthquakes near Fairview Peak, Nevada. *J. Geophys. Res.*, 77: 351–359.

Douglas, B. M., Ryall, A. and Williams, R., 1970. Spectral characteristics of central Nevada microearthquakes. *Bull. Seismol. Soc. Am.*, 60: 1547–1559.

Douze, E. J., 1964a. Signal and noise in deep wells. *Geophysics*, 29: 721–732.

Douze, E. J., 1964b. Rayleigh waves in short-period seismic noise. *Bull. Seismol. Soc. Am.*, 54: 1197–1212.

Douze, E. J., 1966. Noise attenuation in shallow holes. *Bull. Seismol. Soc. Am.*, 56: 619–632.

Douze, E. J., 1967. Short-period seismic noise. *Bull. Seismol. Soc. Am.*, 57: 55–81.

Dragasevic, T., 1970. On some characteristics of waves generated by explosion and earthquake. *Boll. Geofis. Teor. Appl.*, 12: 208–224.

Dratler, J., Farrell, W. E., Block, B. and Gilbert, F., 1971. High-*Q* overtone modes of the earth. *Geophys. J. R. Astron. Soc.*, 23: 399–410.

Duclaux, F., 1971. Etude des spectres du signal sismique initial lors d'un tir nucléaire souterrain.

Determination du coefficient d'atténuation dans le granite. *Pure Appl. Geophys.*, 85: 75–89.

Duclaux, F., Albaret, A., Ferrieux, H. and Perrier, M., 1969. Etudes séismiques effectuées à l'occasion des tirs nucléaires souterrains français. *Ann. Géophys.*, 25: 681–692.

Duda, S. J., 1965. Secular seismic energy release in the circum-Pacific belt. *Tectonophysics*, 2: 409–452.

Duffus, H. J., Shand, J. A. and Wright, C. S., 1962a. Short-range spatial coherence of geomagnetic micropulsations. *Can. J. Phys.*, 40: 218–225.

Duffus, H. J., Kinnear, J. K., Shand, J. A. and Wright, C. S., 1962b. Spatial variations in geomagnetic micropulsations. *Can. J. Phys.*, 40: 1133–1152.

Duke, C. M., Luco, J. E., Carriveau, A. R., Hradilek, P. J., Lastrico, R. and Ostrom, D., 1970. Strong earthquake motion and site conditions: Hollywood. *Bull. Seismol. Soc. Am.*, 60: 1271–1289.

Dutton, J. A., 1962. Space and time response of airborne radiation sensors for the measurement of ground variables. *J. Geophys. Res.*, 67: 195–205.

Dutton, J. A., 1963. The rate of change of the kinetic energy spectrum of flow in a compressible fluid. *J. Atmos. Sci.*, 20: 107–114.

Dutton, J. A., 1971. Clear-air turbulence, aviation, and atmospheric science. *Rev. Geophys. Space Phys.*, 9: 613–657.

Dyer, R. M., 1970. Persistence in snowfall intensities measured at the ground. *J. Appl. Meteorol.*, 9: 29–34.

Dziewonski, A. M. and Hales, A. L., 1972. Numerical analysis of dispersed seismic waves. In: B. A. Bolt (Editor), *Methods in Computational Physics*, Vol. 11. Academic Press, pp.39–85.

Dziewonski, A. and Landisman, M., 1970. Great circle Rayleigh and Love wave dispersion from 100 to 900 seconds. *Geophys. J. R. Astron. Soc.*, 19: 37–91.

Dziewonski, A., Landisman, M., Bloch, S., Satô, Y. and Asano, S., 1968. Progress report on recent improvements in the analysis of surface wave observations. *J. Phys. Earth*, 16 (Spec. Issue): 1–26.

Dziewonski, A., Bloch, S. and Landisman, M., 1969. A technique for the analysis of transient seismic signals. *Bull. Seismol. Soc. Am.*, 59: 427–444.

Dziewonski, A., Mills, J. and Bloch, S., 1972. Residual dispersion measurement—a new method of surface-wave analysis. *Bull. Seismol. Soc. Am.*, 62: 129–139.

Džodenčuková, A., 1972. Spectrum of the geomagnetic field in central Europe. *Studia Geophys. Geodaet.*, 16: 202–208.

Eckhardt, D., Larner, K. and Madden, T., 1963. Long-period magnetic fluctuations and mantle electrical conductivity estimates. *J. Geophys. Res.*, 68: 6279–6286.

Edwards, R. N. and Kurtz, R. D., 1971. 27-day recurrence phenomena in the geomagnetic field at Alert. *Can. J. Earth Sci.*, 8: 1382–1387.

Elagina, L. G., 1963. Measurement of the frequency spectra of pulsations in absolute humidity in the near-ground layer of the atmosphere. *Izv. Acad. Sci. USSR, Geophys. Ser.*, 12: 1133–1137 (Engl. ed.).

Eleman, F., 1967. Studies of giant pulsations, continuous pulsations, and pulsation trains in the geomagnetic field. *Ark. Geofys. (Stockh)*, 5: 231–282.

Eliasen, E. and Machenhauer, B., 1965. A study of the fluctuations of the atmospheric planetary flow patterns represented by spherical harmonics. *Tellus*, 17: 220–238.

Eliasen, E. and Machenhauer, B., 1969. On the observed large-scale atmospheric wave motions. *Tellus*, 21: 149–166.

Ellis, R. M. and Basham, P. W., 1968. Crustal characteristics from short-period P-waves. *Bull. Seismol. Soc. Am.*, 58: 1681–1700.

Ellsaesser, H. W., 1966. Expansion of hemispheric meteorological data in antisymmetric surface spherical harmonic (Laplace) series. *J. Appl. Meteorol.*, 5: 263–276.

Ely, R. P., Jr., 1958. Spectral analysis of the *u*-component of wind velocity at three meters. *J. Meteorol.*, 15: 196–201.

Embree, P., Burg, J. P. and Backus, M. M., 1963. Wide-band velocity filtering—the Pie-Slice process. *Geophysics*, 28: 948–974.

Endlich, R. M., Singleton, R. C. and Kaufman, J. W., 1969. Spectral analysis of detailed vertical wind speed profiles. *J. Atmos. Sci.*, 26: 1030–1041.

Erdélyi, A. (Editor), 1954. *Tables of Integral Transforms*, Vol. 1. Bateman Manuscript Project, McGraw-Hill, 391 pp.

Espinosa, A. F., 1969. Ground amplification study at two sites near Bakersfield, California. *Earthquake Notes*, 40(3): 3–20.

Espinosa, A. F., Sutton, G. H. and Miller, H. J., 1962. A transient technique for seismograph calibration. *Bull. Seismol. Soc. Am.*, 52: 767–779.

Essenwanger, O. and Reiter, E. R., 1969. Power spectrum, structure function, vertical wind shear, and turbulence in troposphere and stratosphere. *Arch. Meteorol. Geophys. Bioklimatol.*, A 18: 17–24.

Estoque, M. A., 1955. The spectrum of large-scale turbulent transfer of momentum and heat. *Tellus*, 7: 177–185.

Everett, J. E. and Hyndman, R. D., 1967. Geomagnetic variations and electrical conductivity structure in southwestern Australia. *Phys. Earth Planet. Inter.*, 1: 24–34.

Evernden, J. F., Best, W. J., Pomeroy, P. W., McEvilly, T. V., Savino, J. M. and Sykes, L. R., 1971. Discrimination between small-magnitude earthquakes and explosions. *J. Geophys. Res.*, 76: 8042–8055.

Ewing, J. A., 1969. Some measurements of the directional wave spectrum. *J. Mar. Res.*, 27: 163–171.

Ewing, J. A., 1971. A numerical wave prediction method for the North Atlantic Ocean. *Dtsche Hydrogr. Z.*, 24: 241–261.

Ewing, M., Mueller, S., Landisman, M. and Satô, Y., 1959. Transient analysis of earthquake and explosion arrivals. *Geofis. Pura Appl.*, 44: 83–118.

Ewing, M., Mueller, S., Landisman, M. and Satô, Y., 1961a. Transient phenomena in explosive sound. In: L. Cremer (Editor), *Proc. 3rd Int. Congr. Acoustics, 1959*, I: 274–276.

Ewing, M., Mueller, S., Landisman, M. and Satô, Y., 1961b. Dispersive transients in earthquake signals. In: L. Cremer (Editor), *Proc. 3rd Int. Congr. Acoustics, 1959*, I: 426–428.

Fail, J. P. and Grau, G., 1963. Les filtres en éventail. *Geophys. Prospect.*, 11: 131–163.

Fairbridge, R. W. (Editor), 1961. Solar variations, climatic change, and related geophysical problems. *Ann. N. Y. Acad. Sci.*, 95: 1–740.

Fanselau, G. and Kautzleben, H., 1958. Die analytische Darstellung des geomagnetischen Feldes. *Geofis. Pura Appl.*, 41: 33–72.

Fara, H. D. and Scheidegger, A. E., 1961. Statistical geometry of porous media. *J. Geophys. Res.*, 66: 3279–3284.

Fedotov, S. A. and Boldyrev, S. A., 1969. Frequency dependence of the body-wave absorption in the crust and the upper mantle of Kuril-Island chain. *Izv. Acad. Sci. USSR, Phys. Solid Earth*, 9: 553–562 (Engl. ed.).

Fernandez, L. M., S. J., 1967. Master curves for the response of layered systems to compressional seismic waves. *Bull. Seismol. Soc. Am.*, 57: 515–543.

Fernandez, L. M. and Careaga, J., 1968. The thickness of the crust in central United States and La Paz, Bolivia, from the spectrum of longitudinal seismic waves. *Bull. Seismol. Soc. Am.*, 58: 711–741.

Few, A. A., 1969. Power spectrum of thunder. *J. Geophys. Res.*, 74: 6926–6934.

Fichtl, G. H., 1968. Characteristics of turbulence observed at the NASA 150-m meteorological tower. *J. Appl. Meteorol.*, 7: 838–844.

Fichtl, G. H. and McVehil, G. E., 1970. Longitudinal and lateral spectra of turbulence in the atmospheric boundary layer at the Kennedy Space Center. *J. Appl. Meteorol.*, 9: 51–63.

Fiedler, F. and Panofsky, H. A., 1970. Atmospheric scales and spectral gaps. *Bull. Am. Meteorol. Soc.*, 51: 1114–1119.

Filson, J. and Frasier, C. W., 1972. Multisite estimation of explosive source parameters. *J. Geophys. Res.*, 77: 2045–2061.

Filson, J. and McEvilly, T. V., 1967. Love wave spectra and the mechanism of the 1966 Parkfield sequence. *Bull. Seismol. Soc. Am.*, 57: 1245–1258.

Finetti, I., Nicolich, R. and Sancin, S., 1971. Review on the basic theoretical assumptions in seismic digital filtering. *Geophys. Prospect.*, 19: 292–320.

Fisher, R. A., 1929. Test of significance in harmonic analysis. *Proc. R. Soc. London, Ser.* A, 125: 54–59.

Fix, J. E., 1972. Ambient earth motion in the period range from 0.1 to 2560 sec. *Bull. Seismol. Soc. Am.*, 62: 1753–1760.

Fofonoff, N. P., 1969. Spectral characteristics of internal waves in the ocean. *Deep-Sea Res.*, 16 (Suppl.): 59–71.

Foote, G. B., 1968. Variance spectrum analysis of Doppler radar observations in continuous precipitation. *J. Appl. Meteorol.*, 7: 459–464.

Foster, M. R. and Guinzy, N. J., 1967. The coefficient of coherence: its estimation and use in geophysical data processing. *Geophysics*, 32: 602–616.

Fougere, P. F., 1963. Spherical harmonic analysis. 1. A new method and its verification. *J. Geophys. Res.*, 68: 1131–1139.

Fougere, P. F., 1965. Spherical harmonic analysis. 2. A new model derived from magnetic observatory data for epoch 1960.0. *J. Geophys. Res.*, 70: 2171–2179.

Frantti, G. E., 1963a. The nature of high-frequency earth noise spectra. *Geophysics*, 28: 547–562.

Frantti, G. E., 1963b. Spectral energy density for quarry explosions. *Bull. Seismol. Soc. Am.*, 53: 989–996.

Frantti, G. E., 1963c. Energy spectra for underground explosions and earthquakes. *Bull. Seismol. Soc. Am.*, 53: 997–1005.

Frantti, G. E., 1965. Attenuation of Pn from offshore Maine explosions. *Bull. Seismol. Soc. Am.*, 55: 417–423.

Frantti, G. E., Willis, D. E. and Wilson, J. T., 1962. The spectrum of seismic noise. *Bull. Seismol. Soc. Am.*, 52: 113–121.

Fraser, D. C., 1966. The magnetic fields of ocean waves. *Geophys. J. R. Astron. Soc.*, 11: 507–517.

Fraser, D. C., Fuller, B. D. and Ward, S. H., 1966. Some numerical techniques for application in mining exploration. *Geophysics*, 31: 1066–1077.

Frasier, C. W. and Filson, J., 1972. A direct measurement of the earth's short-period attenuation along a teleseismic ray path. *J. Geophys. Res.*, 77: 3782–3787.

Frolov, B. K., 1970. Some properties of the correlation and autocorrelation functions of the magnetotelluric process and experiments on the determination of impedance. *Izv. Acad. Sci. USSR, Phys. Solid Earth*, 8: 524–527 (Engl. ed.).

Fukao, Y., 1972. Source process of a large deep-focus earthquake and its tectonic implications—the western Brazil earthquake of 1963. *Phys. Earth Planet. Inter.*, 5: 61–76.

Fuller, B. D., 1967. Two-dimensional frequency analysis and design of grid operators. *Mining Geophys., Soc. Explor. Geophys.*, 2: 658–708.

Furumoto, A. S., 1972. Source mechanism study by Rayleigh wave analysis. In: Comm. Alaska Earthquake, Div. Earth Sci., Natl. Res. Council (Editors), *The Great Alaska Earthquake of 1964. Seismology and Geodesy*. Natl. Acad. Sci., pp.259–264.

Garner, D. M., 1969. Vertical surface acceleration in a wind-generated sea. *Dtsche Hydrogr. Z.*, 22: 163–168.

Garratt, J. R., 1972. Studies of turbulence in the surface layer over water (Lough Neagh). Part II. Production and dissipation of velocity and temperature fluctuations. *Q. J. R. Meteorol. Soc.*, 98: 642–657.

Garrett, J., 1969. Some new observations on the equilibrium region of the wind-wave spectrum. *J. Mar. Res.*, 27: 273–277.

Gaulon, R., 1971. Observations des oscillations propres sphéroïdales et toroïdales. I. Mode fondamental pour des ordres compris entre 2 et 51. *Ann. Géophys.*, 27: 141–149.

Gaulon, R., 1972a. Observations des vibrations propres sphéroïdales et toroïdales. II. Ordres supérieurs à 51. *Ann. Géophys.*, 28: 225–239.

Gaulon, R., 1972b. Observations des vibrations propres sphéroïdales et toroïdales. III. Observations des harmoniques. *Ann. Géophys.*, 28: 241–246.

Gentleman, W. M. and Sande, G., 1966. Fast Fourier transforms—for fun and profit. *Proc. Fall Joint Comput. Conf. AFIPS, Washington, D.C., Spartan*, 29: 563–578.

Ghosh, A. K. and Scheidegger, A. E., 1971. A study of natural wiggly lines in hydrology. *J. Hydrol.*, 13: 101–126.

Gilbert, F. and Backus, G., 1965. The rotational splitting of the free oscillations of the earth, 2. *Rev. Geophys.*, 3: 1–9.

Gilbert, F. and MacDonald, G. J. F., 1960. Free oscillations of the earth. I. Toroidal oscillations. *J. Geophys. Res.*, 65: 675–693.

Glassman, J. A., 1970. A generalization of the Fast Fourier Transform. *IEEE Trans. Comput.*, C-19: 105–116.

Glover, P. and Alexander, S. S., 1969. Lateral variations in crustal structure beneath the Montana LASA. *J. Geophys. Res.*, 74: 505–531.

Godfrey, M. D., 1965. An exploratory study of the bi-spectrum of economic time series. *Appl. Stat., R. Stat. Soc.*, C 14: 48–69.

Godin, G., 1967. The analysis of current observations. *Int. Hydrogr. Rev.*, 44: 149–165.

Goertzel, G., 1960. Fourier analysis. In: A. Ralston and H. S. Wilf (Editors), *Mathematical Methods for Digital Computers.* Wiley, pp.258–262.

Goforth, T. T., Douze, E. J. and Sorrells, G. G., 1972. Seismic noise measurements in a geothermal area. *Geophys. Prospect.*, 20: 76–82.

Gold, B. and Rader, C. M., 1969. *Digital Processing of Signals.* McGraw-Hill, 269 pp.

Gold, L. W., 1964. Analysis of annual variations in ground temperature at an Ottawa site. *Can. J. Earth Sci.*, 1: 146–157.

Golitsyn, G. S., 1964. On the time spectrum of micropulsations in atmospheric pressure. *Izv. Acad. Sci. USSR, Geophys. Ser.*, 8: 761–763 (Engl. ed.).

Gol'tsman (Holzmann), F. M., 1957. Application of linear systems to the filtration of compound oscillations. *Izv. Acad. Sci. USSR, Geophys. Ser.*, 5: 42–52 (Engl. ed.).

Goodman, N. R., 1961. Some comments on spectral analysis of time series. *Technometrics*, 3: 221–228.

Goodman, N. R. and Dubman, M. R., 1969. Theory of time-varying spectral analysis and complex Wishart matrix processes. In: P. R. Krishnaiah (Editor), *Multivariate Analysis*, II. Acad. Press, pp.351–366.

Gossard, E. E., 1960. Spectra of atmospheric scalars. *J. Geophys. Res.*, 65: 3339–3351.

Gostev, M. A. and Fedotov, S. A., 1964. Spectral characteristics of foreshocks and aftershocks of the catastrophic earthquake of 6 November 1958. *Izv. Acad. Sci. USSR, Geophys. Ser.*, 5: 405–411 (Engl. ed.).

Gratsinskii, V. G., 1962. On the spectrum of a segment of a sinusoidal curve. *Izv. Acad. Sci. USSR, Geophys. Ser.*, 11: 967–969 (Engl. ed.).

Gratsinsky, V. G., 1962. Distortions of seismic pulse spectra during analysis. *Izv. Acad. Sci. USSR, Geophys. Ser.*, 3: 233–239 (Engl. ed.).

Grau, G. (Editor), 1966. *Le filtrage en sismique*. I. L'Institut Français du Pétrole, Technip, 234 pp.

Green, A. G., 1972. Magnetic profile analysis. *Geophys. J. R. Astron. Soc.*, 30: 393–403.

Green, P. E., Jr., Kelly, E. J., Jr. and Levin, M. J., 1966. A comparison of seismic array processing methods. *Geophys. J. R. Astron. Soc.*, 11: 67–84.

Griffith, H. L., Panofsky, H. A. and Van der Hoven, I., 1956. Power spectrum analysis over large ranges of frequency. *J. Meteorol.*, 13: 279–282.

Grossling, B. F., 1959. Seismic waves from the underground atomic explosion in Nevada. *Bull. Seismol. Soc. Am.*, 49: 11–32.

Groves, G. W., 1955. Numerical filters for discrimination against tidal periodicities. *Trans. Am. Geophys. Union*, 36: 1073–1084.

Groves, G. W., 1956. Periodic variation of sea level induced by equatorial waves in the easterlies. *Deep-Sea Res.*, 3: 248–252.

Groves, G. W. and Hannan, E. J., 1968. Time series regression of sea level on weather. *Rev. Geophys.*, 6: 129–174.

Groves, G. W. and Miyata, M., 1967. On weather-induced long waves in the equatorial Pacific. *J. Mar. Res.*, 25: 115–128.

Groves, G. W. and Zetler, B. D., 1964. The cross-spectrum of sea level at San Francisco and Honolulu. *J. Mar. Res.*, 22: 269–275.

Gubbins, D., Scollar, I. and Wisskirchen, P., 1971. Two-dimensional digital filtering with Haar and Walsh transforms. *Ann. Géophys.*, 27: 85–104.

Gudmundsson, G., 1966. Interpretation of one-dimensional magnetic anomalies by use of the Fourier-transform. *Geophys. J. R. Astron. Soc.*, 12: 87–97.

Gudmundsson, G., 1967. Spectral analysis of magnetic surveys. *Geophys. J. R. Astron. Soc.*, 13: 325–337.

Guha, S. K., 1970. The effect of focal depth on the spectra of P-waves. I. Theoretical formulation. *Bull. Seismol. Soc. Am.*, 60: 1437–1456.

Guha, S. K. and Stauder, W., 1970. The effect of focal depth on the spectra of P-waves. II. Observational studies. *Bull. Seismol. Soc. Am.*, 60: 1457–1477.

Gupta, J. C. and Chapman, S., 1969. Lunar daily harmonic geomagnetic variation as indicated by spectral analysis. *J. Atmos. Terr. Phys.*, 31: 233–252.

Gurbuz, B. M., 1970. A study of the earth's crust and upper mantle using travel times and spectrum characteristics of body waves. *Bull. Seismol. Soc. Am.*, 60: 1921–1935.

Gurvich, A. S., 1960. Frequency spectra and functions of distribution of probabilities of vertical wind velocity components. *Izv. Acad. Sci. USSR, Geophys. Ser.*, 7: 695–703 (Engl. ed.).

Gurvich, A. S., 1961. On the spectral composition of turbulent flow of momentum. *Izv. Acad. Sci. USSR, Geophys. Ser.*, 10: 1031–1032 (Engl. ed.).

Gurvich, I. I., 1966a. Spectra of waves from a spherical emitter in a homogeneous absorbing medium. *Izv. Acad. Sci. USSR, Phys. Solid Earth*, 5: 304–309 (Engl. ed.).

Gurvich, I. I., 1966b. Determination of the spectrum of a seismic impulse of an explosion close to the focus, according to experimental data. *Izv. Acad. Sci. USSR, Phys. Solid Earth*, 11: 688–692 (Engl. ed.).

Gurvich, I. I., 1967. The dependence of seismic wave spectra in an absorbing medium on weight of charge. *Izv. Acad. Sci. USSR, Phys. Solid Earth*, 1: 44–49 (Engl. ed.).

Gurwitsch, I. I., 1970. *Seismische Erkundung*. Geest and Portig, 699 pp.

Gutenberg, B., 1957. Spectrum of P and S in records of distant earthquakes. *Z. Geophys.*, 23: 316–319.

Gutenberg, B., 1958. Attenuation of seismic waves in the earth's mantle. *Bull. Seismol. Soc. Am.*, 48: 269–282.

Gutenberg, B. and Richter, C. F., 1954. *Seismicity of the Earth*. Princeton Univ. Press, 310 pp.

Gutenberg, B. and Richter, C. F., 1956. Earthquake magnitude, intensity, energy, and acceleration, 2. *Bull. Seismol. Soc. Am.*, 46: 105–145.

Haase, K. H., 1961. Eine neue Methode der Fourier-Analyse und -Synthese. In: L. Cremer (Editor), *Proc. 3rd Int. Congr. Acoustics, 1959*, II: 727–729.

Haddon, R. A. W. and Bullen, K. E., 1969. An earth model incorporating free earth oscillation data. *Phys. Earth Planet. Inter.*, 2: 35–49.

Hagiwara, Y., 1965. Analysis of the results of the aeromagnetic surveys over volcanoes in Japan (I). *Bull. Earthquake Res. Inst.*, 43: 529–547.

Hagiwara, Y., 1967. Analyses of gravity values in Japan. *Bull. Earthquake Res. Inst.*, 45: 1091–1228.

Hahn, A., 1965. Two applications of Fourier's analysis for the interpretation of geomagnetic anomalies. *J. Geomagn. Geoelectr.*, 17: 195–225.

Hales, A. L. and Roberts, J. L., 1970. The travel times of S and SKS. *Bull. Seismol. Soc. Am.*, 60: 461–489.

Hall, F., 1950. Communication theory applied to meteorological measurements. *J. Meteorol.*, 7: 121–129.

Halpern, D., 1971. Semidiurnal internal tides in Massachusetts Bay. *J. Geophys. Res.*, 76: 6573–6584.

Hamming, R. W., 1962. *Numerical Methods for Scientists and Engineers.* McGraw-Hill, 411 pp.

Hamon, B. V., 1962. The spectrums of mean sea level at Sydney, Coff's Harbour, and Lord Howe Island. *J. Geophys. Res.*, 67: 5147–5155.

Hamon, B. V., 1966. Continental shelf waves and the effects of atmospheric pressure and wind stress on sea level. *J. Geophys. Res.*, 71: 2883–2893.

Hamon, B. V., 1968. Spectrum of sea level at Lord Howe Island in relation to circulation. *J. Geophys. Res.*, 73: 6925–6927.

Hamon, B. V. and Hannan, E. J., 1963. Estimating relations between time series. *J. Geophys. Res.*, 68: 6033–6041.

Hanks, T. C. and Thatcher, W., 1972. A graphical representation of seismic source parameters. *J. Geophys. Res.*, 77: 4393–4405.

Hanks, T. C. and Wyss, M., 1972. The use of body-wave spectra in the determination of seismic-source parameters. *Bull. Seismol. Soc. Am.*, 62: 561–589.

Hannan, E. J., 1966. Spectral analysis for geophysical data. *Geophys. J. R. Astron. Soc.*, 11: 225–236.

Hannon, W. J., 1964. An application of the Haskell-Thomson matrix method to the synthesis of the surface motion due to dilatational waves. *Bull. Seismol. Soc. Am.*, 54: 2067–2079.

Hannon, W. J. and Kovach, R. L., 1966. Velocity filtering of seismic core phases. *Bull. Seismol. Soc. Am.*, 56: 441–454.

Harkrider, D. G., 1964. Theoretical and observed acoustic-gravity waves from explosive sources in the atmosphere. *J. Geophys. Res.*, 69: 5295–5321.

Harkrider, D. G., 1970. Surface waves in multilayered elastic media. Part II. Higher mode spectra and spectral ratios from point sources in plane layered earth models. *Bull. Seismol. Soc. Am.*, 60: 1937–1987.

Harkrider, D. G. and Anderson, D. L., 1966. Surface wave energy from point sources in plane layered earth models. *J. Geophys. Res.*, 71: 2967–2980.

Harmuth, H. F., 1972. *Transmission of Information by Orthogonal Functions.* Springer Verlag, 2nd ed., 393 pp.

Harper, B. P., 1961. Energy spectra of 500-mbar meridional circulation indices. *J. Meteorol.*, 18: 487–493.

Harris, B. (Editor), 1967. *Spectral Analysis of Time Series.* Wiley, 319 pp.

Harrison, J. C., Ness, N. F., Longman, I. M., Forbes, R. F. S., Kraut, E. A. and Slichter, L. B.,

1963. Earth-tide observations made during the International Geophysical Year. *J. Geophys. Res.*, 68: 1497–1516.

Hasegawa, H. S., 1969. A study of the effects of the Yellowknife crustal structure upon the P coda of teleseismic events. *Geophys. J. R. Astron. Soc.*, 18: 159–175.

Hasegawa, H. S., 1970. Short-period P-coda characteristics in the eastern Canadian shield. *Bull. Seismol. Soc. Am.*, 60: 839–858.

Hasegawa, H. S., 1971a. Analysis of teleseismic signals from underground nuclear explosions originating in four geological environments. *Geophys. J. R. Astron. Soc.*, 24: 365–381.

Hasegawa, H. S., 1971b. Crustal transfer ratios of short- and long-period body waves recorded at Yellowknife. *Bull. Seismol. Soc. Am.*, 61: 1303–1320.

Hasegawa, H. S., 1972. Analysis of amplitude spectra of P-waves from earthquakes and underground explosions. *J. Geophys. Res.*, 77: 3081–3096.

Haskell, N. A., 1953. The dispersion of surface waves on multilayered media. *Bull. Seismol. Soc. Am.*, 43: 17–34.

Haskell, N. A., 1960. Crustal reflection of plane SH-waves. *J. Geophys. Res.*, 65: 4147–4150.

Haskell, N. A., 1962. Crustal reflection of plane P- and SV-waves. *J. Geophys. Res.*, 67: 4751–4767.

Haskell, N. A., 1964. Total energy and energy spectral density of elastic wave radiation from propagating faults. *Bull. Seismol. Soc. Am.*, 54: 1811–1841.

Haskell, N. A., 1966a. Total energy and energy spectral density of elastic wave radiation from propagating faults. Part II. A statistical source model. *Bull. Seismol. Soc. Am.*, 56: 125–140.

Haskell, N. A., 1966b. The leakage attenuation of continental crustal P-waves. *J. Geophys. Res.*, 71: 3955–3967.

Hasselmann, K. and Collins, J. I., 1968. Spectral dissipation of finite-depth gravity waves due to turbulent bottom friction. *J. Mar. Res.*, 26: 1–12.

Hasselmann, K., Munk, W. and MacDonald, G., 1963. Bispectra of ocean waves. In: M. Rosenblatt (Editor), *Time Series Analysis*. Wiley, pp.125–139.

Hatherton, T., 1960. Microseisms at Scott Base. *Geophys. J. R. Astron. Soc.*, 3: 381–405.

Hatori, T., 1967. The wave form of tsunami on the continental shelf. *Bull. Earthquake Res. Inst.*, 45: 79–90.

Hatori, T., 1968. Study on distant tsunamis along the coast of Japan. Part 2, Tsunamis of South American origin. *Bull. Earthquake Res. Inst.*, 46: 345–359.

Hatori, T., 1969a. A study of the wave sources of the Hiuganada tsunamis. *Bull. Earthquake Res. Inst.*, 47: 55–63.

Hatori, T., 1969b. Analyses of oceanic long-period waves at Hachijo Island. *Bull. Earthquake Res. Inst.*, 47: 863–874.

Hatori, T., 1971. Tsunami sources in Hokkaido and southern Kuril regions. *Bull. Earthquake Res. Inst.*, 49: 63–75.

Hattori, S., 1972. Investigation of seismic waves generated by small explosions. *Bull. Int. Inst. Seismol. Earthquake Eng.*, 9: 27–105.

Haubrich, R. A., 1965. Earth noise, 5 to 500 millicycles per second. 1. Spectral stationarity, normality, and nonlinearity. *J. Geophys. Res.*, 70: 1415–1427.

Haubrich, R. A., 1968. Array design. *Bull. Seismol. Soc. Am.*, 58: 977–991.

Haubrich, R. A., 1969. Spectra of earthquake time series. *EOS, Trans. Am. Geophys. Union*, 50: 409–410.

Haubrich, R. A. and Iyer, H. M., 1962. A digital seismograph system for measuring earth noise. *Bull. Seismol. Soc. Am.*, 52: 87–93.

Haubrich, R. A. and MacKenzie, G. S., 1965. Earth noise, 5 to 500 millicycles per second. 2. Reaction of the earth to oceans and atmosphere. *J. Geophys. Res.*, 70: 1429–1440.

Haubrich, R. A. and McCamy, K., 1969. Microseisms: Coastal and pelagic sources. *Rev. Geophys.*, 7: 539–571.

Haubrich, R. A., Munk, W. H. and Snodgrass, F. E., 1963. Comparative spectra of microseisms and swell. *Bull. Seismol. Soc. Am.*, 53: 27–37.

Haubrich, R., Jr. and Munk, W., 1959. The pole tide. *J. Geophys. Res.*, 64: 2373–2388.

Hauska, H., 1972. Geomagnetic activity—Periodic variations and connection to solar and interplanetary magnetic fields. *Acta Univ. Ups., Abstr. Upps. Diss. Fac. Sci.*, No. 216: 9 pp.

Hays, W. W., 1969. Amplitude and frequency characteristics of elastic wave types generated by the underground nuclear detonation, Boxcar. *Bull. Seismol. Soc. Am.*, 59: 2283–2293.

Healy, J. H., King, C.-Y. and O'Neill, M. E., 1971. Source parameters of the Salmon and Sterling nuclear explosions from seismic measurements. *J. Geophys. Res.*, 76: 3344–3355.

Hecht, A. and Hughes, P., 1971. Observations of temperature fluctuations in the upper layers of the Bay of Biscay. *Deep-Sea Res.*, 18: 663–684.

Hecht, A. and White, R. A., 1968. Temperature fluctuations in the upper layer of the ocean. *Deep-Sea Res.*, 15: 339–353.

Heirtzler, J. R. and Le Pichon, X., 1965. Crustal structure of the mid-ocean ridges. 3. Magnetic anomalies over the mid-Atlantic ridge. *J. Geophys. Res.*, 70: 4013–4033.

Heirtzler, J. R., Nichols, D. L. and Santirocco, R. A., 1960. Some observations of the geomagnetic fluctuation spectrum at audio frequencies. *J. Geophys. Res.*, 65: 2345–2347.

Heiskanen, W. A. and Vening Meinesz, F. A., 1958. *The Earth and Its Gravity Field.* McGraw-Hill, 470 pp.

Helmberger, D. and Wiggins, R. A., 1971. Upper-mantle structure of midwestern United States. *J. Geophys. Res.*, 76: 3229–3245.

Henderson, R. G. and Cordell, L., 1971. Reduction of unevenly spaced potential field data to a horizontal plane by means of finite harmonic series. *Geophysics*, 36: 856–866.

Henry, R. M. and Hess, S. L., 1958. A study of the large-scale spectra of some meteorological parameters. *J. Meteorol.*, 15: 397–403.

Herbst, R. F., Werth, G. C. and Springer, D. L., 1961. Use of large cavities to reduce seismic waves from underground explosions. *J. Geophys. Res.*, 66: 959–978.

Hermance, J. F., 1969. Resolution of ocean floor magnetotelluric data. *J. Geophys. Res.*, 74: 5527–5532.

Herron, T. J., 1966. Phase characteristics of geomagnetic micropulsations. *J. Geophys. Res.*, 71: 871–889.

Herron, T. J., 1967. An average geomagnetic power spectrum for the period range 4.5 to 12900 seconds. *J. Geophys. Res.*, 72: 759–761.

Herron, T. J., Tolstoy, I. and Kraft, D. W., 1969. Atmospheric pressure background fluctuations in the mesoscale range. *J. Geophys. Res.*, 74: 1321–1329.

Hess, G. D. and Clarke, R. H., 1973. Time spectra and cross-spectra of kinetic energy in the planetary boundary layer. *Q. J. R. Meteorol. Soc.*, 99: 130–153.

Higuchi, H., 1963. On the behavior of Chilean tsunami in the Seto inland sea. *Geophys. Pap. dedicated to Prof. K. Sassa (Kyoto)*. Publ. Office, Geophys. Inst., Kyoto, pp.49–57.

Hill, M. N. (Editor), 1962. *The Sea. Vol. 1, Physical Oceanography.* Interscience, 864 pp.

Hinde, B. J. and Gaunt, D. I., 1966. Some new techniques for recording and analysing microseisms. *Proc. R. Soc. London, Ser. A*, 290: 297–317.

Hinich, M. J., 1967. Estimation of exponential power spectra. *J. Acoust. Soc. Am.*, 42: 422–427.

Hinich, M. J. and Clay, C. S., 1968. The application of the discrete Fourier transform in the estimation of power spectra, coherence, and bispectra of geophysical data. *Rev. Geophys.*, 6: 347–363.

Hinze, J. O., 1959. *Turbulence.* McGraw-Hill, 586 pp.

Hirasawa, T., 1965. Source mechanism of the Niigata earthquake of June 16, 1964, as derived from body waves. *J. Phys. Earth*, 13: 35–66.

Hirasawa, T. and Berry, M. J., 1971. Reflected and head waves from a linear transition layer in a fluid medium. *Bull. Seismol. Soc. Am.*, 61: 1–25.

Hirasawa, T. and Nagata, T., 1966. Spectral analysis of geomagnetic pulsations from 0.5 to 100 sec in period for the quiet sun condition. *Pure Appl. Geophys.*, 65: 102–124.

Hirasawa, T. and Stauder, W., S. J., 1964. Spectral analysis of body waves from the earthquake of February 18, 1956. *Bull. Seismol. Soc. Am.*, 54: 2017–2055.

Hirasawa, T. and Stauder, W., S. J., 1965. On the seismic body waves from a finite moving source. *Bull. Seismol. Soc. Am.*, 55: 237–262.

Hirono, T., Suyehiro, S., Furuta, M. and Koide, K., 1968. Noise attenuation in shallow holes (I). *Pap. Meteorol. Geophys.*, 19: 323–339.

Hirono, T., Suyehiro, S., Furuta, M. and Sato, K., 1969. Noise attenuation in shallow holes (II). Improvement of signal to noise ratio. *Pap. Meteorol. Geophys.*, 20: 189–206.

Hirshberg, J., Currie, R. G. and Breiner, S., 1972. Long-period geomagnetic fluctuations after the earthquake. In: Comm. Alaska Earthquake, Div. Earth Sci., Natl. Res. Council (Editors), *The Great Alaska Earthquake of 1964. Seismology and Geodesy*. Natl. Acad. Sci., pp.520–522.

Hoffman, A. A. J. and Horton, C. W., 1966. An analysis of some magnetotelluric records from Tikhaya Bay, U.S.S.R. *J. Geophys. Res.*, 71: 4047–4052.

Hollan, E., 1966. Das Spektrum der internen Bewegungsvorgänge der westlichen Ostsee im Periodenbereich von 0.3 bis 60 Minuten. *Dtsche Hydrogr. Z.*, 19: 193–218 and 285–298.

Holloway, J. L., Jr., 1958. Smoothing and filtering of time series and space fields. *Adv. Geophys.*, 4: 351–389.

Holmes, C. R., Brook, M., Krehbiel, P. and McCrory, R., 1971. On the power spectrum and mechanism of thunder. *J. Geophys. Res.*, 76: 2106–2115.

Holzmann, F. M., 1960a. The frequency theory of interference systems. *Izv. Acad. Sci. USSR, Geophys. Ser.*, 1: 3–11 and 2: 142–148 (Engl. ed.).

Holzmann, F. M., 1960b. On the experimental analysis of interferences and of the reliability of the results of the grouping of signals. *Izv. Acad. Sci. USSR, Geophys. Ser.*, 12: 1140–1146 (Engl. ed.).

Horai, K.-I. and Simmons, G., 1968. Seismic travel time anomaly due to anomalous heat flow and density. *J. Geophys. Res.*, 73: 7577–7588.

Horai, K. and Simmons, G., 1969. Spherical harmonic analysis of terrestrial heat flow. *Earth Planet. Sci. Lett.*, 6: 386–394.

Horn, L. H. and Bryson, R. A., 1963. An analysis of the geostrophic kinetic energy spectrum of large-scale atmospheric turbulence. *J. Geophys. Res.*, 68: 1059–1064.

Horn, W., 1960. The harmonic analysis, according to the least-square rule, of tide observations upon which an unknown drift is superposed. *Boll. Geofis. Teor. Appl.*, 2: 218–222.

Horton, C. W. and Hoffman, A. A. J., 1962. Power spectrum analysis of the telluric field at Tbilisi, U.S.S.R., for periods from 2.4 to 60 minutes. *J. Geophys. Res.*, 67: 3369–3371.

Horton, C. W., Hempkins, W. B. and Hoffman, A. A. J., 1964. A statistical analysis of some aeromagnetic maps from the northwestern Canadian shield. *Geophysics*, 29: 582–601.

Housner, G. W., 1955. Properties of strong ground motion earthquakes. *Bull. Seismol. Soc. Am.*, 45: 197–218.

Housner, G. W. and Trifunac, M. P., 1967. Analysis of accelerograms—Parkfield earthquake. *Bull. Seismol. Soc. Am.*, 57: 1193–1220.

Housner, G. W., Martel, R. R. and Alford, J. L., 1953. Spectrum analysis of strong-motion earthquakes. *Bull. Seismol. Soc. Am.*, 43: 97–119.

Howell, B. F., Jr., 1966a. Simple digitizer for paper seismograms. *Bull. Seismol. Soc. Am.*, 56: 605–608.

Howell, B. F., Jr., 1966b. Lake Superior seismic experiment: frequency spectra and absorption. *Am. Geophys. Union, Geophys. Monogr.*, 10: 227–233.

Howell, B. F., Jr. and Baybrook, T. G., 1967. Scale-model study of refraction along an irregular interface. *Bull. Seismol. Soc. Am.*, 57: 443–446.

Howell, B. F., Jr., Andrews, A. B. and Huber, R. E., 1959. Photomechanical method of frequency analysis of seismic pulses. *Geophysics*, 24: 692–705.

Howell, B. F., Jr., Lavin, P. M., Watson, R. J., Cheng, Y. Y. and Lin, J. L., 1967. Method for recognizing repeated pulse sequences in a seismogram. *J. Geophys. Res.*, 72: 3225–3232.

Howell, B. F., Jr., Lundquist, G. M. and Yiu, S. K., 1970. Integrated and frequency-band magnitude, two alternative measures of the size of an earthquake. *Bull. Seismol. Soc. Am.*, 60: 917–937.

Hsueh, Y., 1968. Mesoscale turbulence spectra over the Indian Ocean. *J. Atmos. Sci.*, 25: 1052–1057.

Huang, Y. T., 1966. Spectral analysis of digitized seismic data. *Bull. Seismol. Soc. Am.*, 56: 425–440.

Hudleston, P. J., 1973. Fold morphology and some geometrical implications of theories of fold development. *Tectonophysics*, 16: 1–46.

Hudson, D. E., 1962. Some problems in the application of spectrum techniques to strong-motion earthquake analysis. *Bull. Seismol. Soc. Am.*, 52: 417–430.

Hudson, D. E., 1972. Local distribution of strong earthquake ground motions. *Bull. Seismol. Soc. Am.*, 62: 1765–1786.

Hudson, D. E. and Housner, G. W., 1958. An analysis of strong-motion accelerometer data from the San Francisco earthquake of March 22, 1957. *Bull. Seismol. Soc. Am.*, 48: 253–268.

Hudson, D. E., Alford, J. L. and Iwan, W. D., 1961. Ground accelerations caused by large quarry blasts. *Bull. Seismol. Soc. Am.*, 51: 191–202.

Hunkins, K., 1962. Waves on the Arctic Ocean. *J. Geophys. Res.*, 67: 2477–2489.

Hunkins, K., 1967. Inertial oscillations of Fletcher's ice island (T-3). *J. Geophys. Res.*, 72: 1165–1174.

Hurwitz, L., Knapp, D. G., Nelson, J. H. and Watson, D. E., 1966. Mathematical model of the geomagnetic field for 1965. *J. Geophys. Res.*, 71: 2373–2383.

Husebye, E. S. and Jansson, B., 1966. Application of array data processing techniques to the Swedish seismograph stations. *Pure Appl. Geophys.*, 63: 82–104.

Hwang, H. J., 1970. Power density spectrum of surface wind speed on Palmyra Island. *Mon. Weather Rev.*, 98: 70–74.

Ibrahim, A. K., 1969. Determination of crustal thickness from spectral behavior of SH-waves. *Bull. Seismol. Soc. Am.*, 59: 1247–1258.

Ibrahim, A. K., 1971a. Effects of a rigid core on the reflection and transmission coefficients from a multi-layered core–mantle boundary. *Pure Appl. Geophys.*, 91: 95–113.

Ibrahim, A. K., 1971b. The amplitude ratio PcP/P and the core–mantle boundary. *Pure Appl. Geophys.*, 91: 114–133.

Ichikawa, M. and Basham, P. W., 1965. Variations in short-period records from Canadian seismograph stations. *Can. J. Earth Sci.*, 2: 510–542.

Ichiye, T., 1967. Upper ocean boundary-layer flow determined by dye diffusion. In: K. F. Bowden, F. N. Frenkiel and I. Tani (Editors), *Boundary Layers and Turbulence. The Physics of Fluids Suppl.* American Institute of Physics, pp.270–277.

Ichiye, T., 1972. Power spectra of temperature and salinity fluctuations in the slope water off Cape Hatteras. *Pure Appl. Geophys.*, 96: 205–216.

Imasato, N., 1971. Study of seiche in Lake Biwa-ko (II)—On a numerical experiment by nonlinear two-dimensional model. *Contrib. Geophys. Inst., Kyoto Univ.*, 11: 77–90.

Inman, D., Munk, W. and Balay, M., 1961. Spectra of low-frequency ocean waves along the Argentine shelf. *Deep-Sea Res.*, 8: 155–164.

Inston, H. H., Marshall, P. D. and Blamey, C., 1971. Optimization of filter bandwidth in spectral analysis of wavetrains. *Geophys. J. R. Astron. Soc.*, 23: 243–250.

Irikura, K., Matsuo, K. and Yoshikawa, S., 1971. An analysis of strong motion accelerograms near the epicenter. *Bull. Disaster Prev. Res. Inst. (Kyoto)*, 20(4): 267–288.

Isacks, B. and Oliver, J., 1964. Seismic waves with frequencies from 1 to 100 cycles per second recorded in a deep mine in northern New Jersey. *Bull. Seismol. Soc. Am.*, 54: 1941–1979.

Isozaki, I., 1968. An investigation on the variations of sea level due to meteorological disturbances on the coast of the Japanese islands (I). On the accuracy of tide predictions. *Pap. Meteorol. Geophys.*, 19: 401–426.

Israelson, H., 1971. Spectral content of teleseismic P-waves recorded at the Hagfors Observatory. *Geophys. J. R. Astron. Soc.*, 25: 89–95.

Ivanova, T. G. and Vasil'ev, Yu. I., 1964. Selection of optimum apparatus characteristics when recording head waves from the crystalline basement. *Izv. Acad. Sci. USSR, Geophys. Ser.*, 5: 383–391 (Engl. ed.).

Iyengar, R. N. and Iyengar, K. T. S. R., 1969. A nonstationary random process model for earthquake accelerograms. *Bull. Seismol. Soc. Am.*, 59: 1163–1188.

Iyer, H. M., 1964. A frequency–velocity–energy diagram for the study of dispersive surface waves. *Bull. Seismol. Soc. Am.*, 54: 183–190.

Iyer, H. M., 1968. Determination of frequency–wavenumber spectra using seismic arrays. *Geophys. J. R. Astron. Soc.*, 16: 97–117.

Iyer, H. M., 1971. Variation of apparent velocity of teleseismic P-waves across the Large-Aperture Seismic Array, Montana. *J. Geophys. Res.*, 76: 8554–8567.

Iyer, H. M. and Healy, J. H., 1972. Evidence for the existence of locally-generated body waves in the short-period noise at the Large Aperture Seismic Array, Montana. *Bull. Seismol. Soc. Am.*, 62: 13–29.

Izawa, T., 1972a. Some considerations on the continuous space–time spectral analysis of atmospheric disturbances. *Pap. Meteorol. Geophys.*, 23: 33–71.

Izawa, T., 1972b. Statistical aspects of lower atmospheric disturbances delineated from conventional and satellite data over the tropical Pacific. *Pap. Meteorol. Geophys.*, 23: 73–120.

Jackson, D. D. and Anderson, D. L., 1970. Physical mechanisms of seismic-wave attenuation. *Rev. Geophys. Space Phys.*, 8: 1–63.

Jackson, P. L., 1964. Time-varying spectra through optical diffraction scanning. *Bull. Seismol. Soc. Am.*, 54: 485–500.

Jackson, P. L., 1965. Analysis of variable-density seismograms by means of optical diffraction. *Geophysics*, 30: 5–23.

Jackson, P. L., 1967. Truncations and phase relationships of sinusoids. *J. Geophys. Res.*, 72: 1400–1403.

Jackson, P. L., 1968. Sectional correlograms and convolutions by a simple optical method. *Geophysics*, 33: 747–754.

Jacob, K. H. and Hamada, K., 1972. The upper mantle beneath the Aleutian Island arc from pure-path Rayleigh-wave dispersion data. *Bull. Seismol. Soc. Am.*, 62: 1439–1453.

Jacqmin, A. and Pekar, L., 1969. Réflexions sur les applications de la transformée de Fourier en sismique et en gravimétrie. *Geophys. Prospect.*, 17: 294–326.

Jaeger, R. M. and Schuring, D. J., 1966. Spectrum analysis of terrain of Mare Cognitum. *J. Geophys. Res.*, 71: 2023–2028.

James, D. E. and Linde, A. T., 1971. A source of major error in the digital analysis of World Wide Standard Station seismograms. *Bull. Seismol. Soc. Am.*, 61: 723–728.

Jankowski, J. and Pawliszyn, J., 1970. Combined utilization of natural variations in the earth's electromagnetic field for investigation of the deep structure. *Publ. Inst. Geophys. Polish Acad. Sci.*, 34: 121–128.

Jansson, B. and Husebye, E. S., 1968. Application of array data processing techniques to a network of ordinary seismograph stations. *Pure Appl. Geophys.*, 69: 80–99.

Jeans, J. H., 1923. The propagation of earthquake waves. *Proc. R. Soc. London, Ser. A*, 102: 554–574.

Jeffreys, H., 1964. Note on Fourier analysis. *Bull. Seismol. Soc. Am.*, 54: 1441–1444.

Jenkins, G. M., 1961a. General considerations in the analysis of spectra. *Technometrics*, 3: 133–166.

Jenkins, G. M., 1961b. Comments on the discussions of Messrs. Tukey and Goodman. *Technometrics*, 3: 229–232.

Jenkins, G. M., 1965. A survey of spectral analysis. *Appl. Stat., R. Stat. Soc.*, C 14: 2–32.

Jenkins, G. M. and Watts, D. G., 1968. *Spectral Analysis and Its Applications.* Holden-Day, 525 pp.

Jennison, R. C., 1961. *Fourier Transforms and Convolutions for the Experimentalist.* Pergamon Press, 120 pp.

Jenschke, V. A. and Penzien, J., 1964. Ground motion accelerogram analysis including dynamical instrumental correction. *Bull. Seismol. Soc. Am.*, 54: 2087–2098.

Jensen, O. G. and Ellis, R. M., 1970. Wave propagation in a horizontally-layered crust in terms of linear system theory. *Can. J. Earth Sci.*, 7: 1185–1193.

Jensen, O. G. and Ellis, R. M., 1971. Generation of synthetic seismograms using linear systems theory. *Can. J. Earth Sci.*, 8: 1409–1422.

Jeske, H., 1969. Die Feinstruktur des Brechungsindexfeldes in Höhen zwischen 50 und 2400 m über See. *Z. Geophys.*, 35: 529–550.

Jobert, G., 1962. Remarques sur l'analyse spectrale des variations de la pesanteur. *Marées Terr., Bull. d'Inform.*, 30: 740–742.

Jobert, G., 1963. Comparaison des résultats de l'analyse spectrale des marées terrestres avec les résultats théoriques. *Marées Terr., Bull. d'Inform.*, 33: 1013–1016.

Johnson, G. G. and Vand, V., 1967. Application of a Fourier data smoothing technique to the meteoritic crater Ries Kessel. *J. Geophys. Res.*, 72: 1741–1750.

Johnson, L. R., 1969. Array measurements of P-velocities in the lower mantle. *Bull. Seismol. Soc. Am.*, 59: 973–1008.

Johnson, R. H. and Norris, R. A., 1972. Significance of spectral banding in hydroacoustic signals from submarine volcanic eruptions: Myojin 1970. *J. Geophys. Res.*, 77: 4461–4469.

Jones, F. W., 1969. Periodic variations in micropulsation dominant period. *Phys. Earth Planet. Inter.*, 2: 50–54.

Jones, F. W. and Geldart, L. P., 1967. Vertical telluric currents. *Earth Planet. Sci. Lett.*, 2: 69–74.

Jones, F. W. and Geldart, L. P., 1968. Spectral analysis of vertical and horizontal earth current components at separated locations. *Can. J. Earth Sci.*, 5: 1512–1517.

Jones, F. W., Ku, C. C. and Geldart, L. P., 1969. Earth currents and magnetic field variations in the period range 10–200 s. *Geophys. J. R. Astron. Soc.*, 17: 15–38.

Jones, R. H., 1965. A reappraisal of the periodogram in spectral analysis. *Technometrics*, 7: 531–542.

Jones, R. H., 1972. Aliasing with unequally spaced observations. *J. Appl. Meteorol.*, 11: 245–254.

Julian, P. R., Washington, W. M., Hembree, L. and Ridley, C., 1970. On the spectral distribution of large-scale atmospheric kinetic energy. *J. Atmos. Sci.*, 27: 376–387.

Kahn, A. B., 1957. A generalization of average-correlation methods of spectrum analysis. *J. Meteorol.*, 14: 9–17.

Kahn, A. B., 1962. Large-scale atmospheric spectra at 200 mbar. *J. Atmos. Sci.*, 19: 150–158.

Kaimal, J. C. and Izumi, Y., 1965. Vertical velocity fluctuations in a nocturnal low-level jet. *J. Appl. Meteorol.*, 4: 576–584.

Kaimal, J. C., Wyngaard, J. C., Izumi, Y. and Coté, O. R., 1972. Spectral characteristics of surface-layer turbulence. *Q. J. R. Meteorol. Soc.*, 98: 563–589.

Kampé de Fériet, J. and Frenkiel, F. N., 1962. Correlations and spectra for non-stationary random functions. *Math. Comput.*, 16: 1–21.

Kanai, K., 1957. Semi-empirical formula for the seismic characteristics of the ground. *Bull. Earthquake Res. Inst.*, 35: 309–325.

Kanai, K., 1958. A study of strong earthquake motions. *Bull. Earthquake Res. Inst.*, 36: 295–309.

Kanai, K., 1961. An empirical formula for the spectrum of strong earthquake motions. *Bull. Earthquake Res. Inst.*, 39: 85–95.

Kanai, K., 1962a. On the spectrum of strong earthquake motions. *Bull. Earthquake Res. Inst.*, 40: 71–90.

Kanai, K., 1962b. On the predominant period of earthquake motions. *Bull. Earthquake Res. Inst.*, 40: 855–860.

Kanai, K. and Yoshizawa, S., 1958. The amplitude and the period of earthquake motions (II). *Bull. Earthquake Res. Inst.*, 36: 275–293.

Kanai, K., Tanaka, T. and Yoshizawa, S., 1959. Comparative studies of earthquake motions on the ground and underground. (Multiple reflection problem). *Bull. Earthquake Res. Inst.*, 37: 53–87.

Kanai, K., Yoshizawa, S. and Suzuki, T., 1963. An empirical formula for the spectrum of strong earthquake motions. II. *Bull. Earthquake Res. Inst.*, 41: 261–270.

Kanai, K., Tanaka, T. and Yoshizawa, S., 1965. On microtremors. IX. (Multiple reflection problem). *Bull. Earthquake Res. Inst.*, 43: 577–588.

Kanai, K., Tanaka, T., Yoshizawa, S., Morishita, T., Osada, K. and Suzuki, T., 1966a. Comparative studies of earthquake motions on the ground and underground. II. *Bull. Earthquake Res. Inst.*, 44: 609–643.

Kanai, K., Tanaka, T., Morishita, T. and Osada, K., 1966b. Observation of microtremors. XI. (Matsushiro earthquake swarm area). *Bull. Earthquake Res. Inst.*, 44: 1297–1333.

Kanamori, H., 1967a. Spectrum of P and PcP in relation to the mantle–core boundary and attenuation in the mantle. *J. Geophys. Res.*, 72: 559–571.

Kanamori, H., 1967b. Spectrum of short-period core phases in relation to the attenuation in the mantle. *J. Geophys. Res.*, 72: 2181–2186.

Kanamori, H., 1970a. Synthesis of long-period surface waves and its applications to earthquake source studies—Kurile Islands earthquake of October 13, 1963. *J. Geophys. Res.*, 75: 5011–5027.

Kanamori, H., 1970b. The Alaska earthquake of 1964: Radiation of long-period surface waves and source mechanism. *J. Geophys. Res.*, 75: 5029–5040.

Kanamori, H., 1970c. Velocity and *Q* of mantle waves. *Phys. Earth Planet. Inter.*, 2: 259–275.

Kanamori, H., 1971a. Focal mechanism of the Tokachi-Oki earthquake of May 16, 1968: Contortion of the lithosphere at a junction of two trenches. *Tectonophysics*, 12: 1–13.

Kanamori, H., 1971b. Seismological evidence for a lithosphere normal faulting—the Sanriku earthquake of 1933. *Phys. Earth Planet. Inter.*, 4: 289–300.

Kanamori, H. and Abe, K., 1968. Deep structure of island arcs as revealed by surface waves. *Bull. Earthquake Res. Inst.*, 46: 1001–1025.

Kanari, S., 1970. Internal waves in Lake Biwa (I). *Bull. Disaster Prev. Res. Inst. (Kyoto)*, 19(3): 19–26.

Kanasewich, E. R. and Agarwal, R. G., 1970. Analysis of combined gravity and magnetic fields in wave number domain. *J. Geophys. Res.*, 75: 5702–5712.

Kao, S.-K., 1965. Some aspects of the large-scale turbulence and diffusion in the atmosphere. *Q. J. R. Meteorol. Soc.*, 91: 10–17.

Kao, S.-K., 1968. Governing equations and spectra for atmospheric motion and transports in frequency, wave-number space. *J. Atmos. Sci.*, 25: 32–38.

Kao, S.-K., 1970. Wavenumber–frequency spectra of temperature in the free atmosphere. *J. Atmos. Sci.*, 27: 1000–1007.

Kao, S.-K. and Al-Gain, A. A., 1968. Large-scale dispersion of clusters of particles in the atmosphere. *J. Atmos. Sci.*, 25: 214–221.

Kao, S.-K. and Bullock, W. S., 1964. Lagrangian and Eulerian correlations and energy spectra of geostrophic velocities. *Q. J. R. Meteorol. Soc.*, 90: 166–174.

Kao, S.-K. and Gebhard, J. B., 1971. An analysis of heat-, momentum-transports, and spectra for clear air turbulence in mid-stratosphere. *Pure Appl. Geophys.*, 88: 180–185.

Kao, S.-K. and Henderson, D., 1970. Large-scale dispersion of clusters of particles in various flow patterns. *J. Geophys. Res.*, 75: 3104–3113.

Kao, S.-K. and Hill, W. R., 1970. Characteristics of the large-scale dispersion of particles in the southern hemisphere. *J. Atmos. Sci.*, 27: 126–132.

Kao, S.-K. and Powell, D. C., 1969. Large-scale dispersion of clusters of particles in the atmosphere. II. Stratosphere. *J. Atmos. Sci.*, 26: 734–740.

Kao, S.-K. and Sagendorf, J. F., 1970. The large-scale meridional transport of sensible heat in wavenumber–frequency space. *Tellus*, 22: 172–185.

Kao, S.-K. and Sands, E. E., 1966. Energy spectrums, mean and eddy kinetic energies of the atmosphere between surface and 50 kilometers. *J. Geophys. Res.*, 71: 5213–5219.

Kao, S.-K. and Wendell, L. L., 1970. The kinetic energy of the large-scale atmospheric motion in wavenumber–frequency space: I. Northern hemisphere. *J. Atmos Sci.*, 27: 359–375.

Kao, S.-K. and Woods, H. D., 1964. Energy spectra of meso-scale turbulence along and across the jet stream. *J. Atmos. Sci.*, 21: 513–519.

Kao, S.-K., Tsay, C. Y. and Wendell, L. L., 1970a. The meridional transport of angular momentum in wavenumber–frequency space. *J. Atmos. Sci.*, 27: 614–626.

Kao, S.-K., Jenne, R. L. and Sagendorf, J. F., 1970b. The kinetic energy of large-scale atmospheric motion in wavenumber–frequency space: II. Mid-troposphere of the southern hemisphere. *J. Atmos. Sci.*, 27: 1008–1020.

Kao, S.-K., Jenne, R. L. and Sagendorf, J. F., 1971a. Wavenumber–frequency spectra of the meridional transport of sensible heat in the mid-troposphere of the southern hemisphere. *Pure Appl. Geophys.*, 86: 159–170.

Kao, S.-K., Jenne, R. L. and Sagendorf, J. F., 1971b. Spectral characteristics of the meridional transport of angular momentum in the mid-troposphere of the southern hemisphere. *Pure Appl. Geophys.*, 86: 171–183.

Karapetian, N. K., 1964. The determination of the energy of earthquakes with an account of the frequency spectrum of seismic vibrations. *Publ. Bur. Cent. Int. Séism.*, A23: 55–61.

Kárník, V. and Tobyáš, V., 1961. Underground measurements of the seismic noise level. *Studia Geophys. Geodaet.*, 5: 231–236.

Kasahara, K., 1957. The nature of seismic origins as inferred from seismological and geodetic observations (1). *Bull. Earthquake Res. Inst.*, 35: 473–532.

Kasahara, K., 1958. The nature of seismic origins as inferred from seismological and geodetic observations (2). *Bull. Earthquake Res. Inst.*, 36: 21–53.

Kasahara, K., 1960. An attempt to detect azimuth effect on spectral structures of seismic waves (The Alaskan earthquake of April 7, 1958). *Bull. Earthquake Res. Inst.*, 38: 207–218.

Kato, Y., Aoyama, I., Toyama, F. and Morioka, A., 1966. Frequency analysis of the geomagnetic pulsations. *Sci. Rep. Tôhoku Univ., Geophys.*, 18: 65–76.

Kats, S. A., 1963. A method of calculating complex spectra of impulse functions by means of an amplitude analyzer. *Izv. Acad. Sci. USSR, Geophys. Ser.*, 10: 932–937 (Engl. ed.).

Kaula, W. M., 1959. Statistical and harmonic analysis of gravity. *J. Geophys. Res.*, 64: 2401–2421.

Kaula, W. M., 1966a. Tests and combination of satellite determinations of the gravity field with gravimetry. *J. Geophys. Res.*, 71: 5303–5314.

Kaula, W. M., 1966b. Global harmonic and statistical analysis of gravimetry. *Am. Geophys. Union, Geophys. Monogr.*, 9: 58–67.

Kaula, W. M., 1967. Theory of statistical analysis of data distributed over a sphere. *Rev. Geophys.*, 5: 83–107.

Kawamura, M., 1970. Short-period geomagnetic micropulsations with period of about 1 second in the middle and low latitudes. *Geophys. Mag.*, 35: 1–54.

Keay, C. S. L., Ellyett, C. D. and Brown, T. A., 1966. Absence of unusual periodicities in radar meteor rates. *J. Geophys. Res.*, 71: 1409–1411.

Keilis-Borok, V. I., 1961. Differences in the spectra of surface waves from earthquakes and from explosions. *Tr. Inst. Fiz. Zemli*, 15(182): 88–101. Also in English translation (1962): *Seismic Effects of Underground Explosions.* Consultants Bureau, pp.71–82.

Keilis-Borok, V. I. and Yanovskaya, T. B., 1962. Dependence of the spectrum of surface waves on the depth of the focus within the earth's crust. *Izv. Acad. Sci. USSR, Geophys. Ser.*, 11: 955–959 (Engl. ed.).

Kemp, W. C. and Eger, D. T., 1967. The relationships among sequences with applications to geological data. *J. Geophys. Res.*, 72: 739–751.

Kenyon, K. E., 1968. Wave–wave interactions of surface and internal waves. *J. Mar. Res.*, 26: 208–231.

Kersta, L. G., 1948. Amplitude cross-section representation with the sound spectrograph. *J. Acoust. Soc. Am.*, 20: 796–801.

Kertz, W., 1966. Filterverfahren in der Geophysik. *Gerlands Beitr. Geophys.*, 75: 1–33.

Key, F. A., 1967. Signal-generated noise recorded at the Eskdalemuir seismometer array station. *Bull. Seismol. Soc. Am.*, 57: 27–37.

Key, F. A., 1968. Some observations and analyses of signal-generated noise. *Geophys. J. R. Astron. Soc.*, 15: 377–392.

Kharkevich, A. A., 1960. *Spectra and Analysis.* Consultants Bureau, 222 pp.

Khattri, K., 1972. Body wave directivity functions for two-dimensional fault model and kinematic parameters of a deep focus earthquake. *J. Geophys. Res.*, 77: 2062–2071.

Khattri, K., 1973. Earthquake focal mechanism studies—a review. *Earth-Sci. Rev.*, 9: 19–63.

Khattri, K. N., 1969a. Determination of earthquake fault plane, fault area, and rupture velocity from the spectra of long period P-waves and the amplitude of SH-waves. *Bull. Seismol. Soc. Am.*, 59: 615–630.

Khattri, K. N., 1969b. Focal mechanism of the Brazil deep focus earthquake of November 3, 1965, from the amplitude spectra of isolated P-waves. *Bull. Seismol. Soc. Am.*, 59: 691–704.

Khudzinski, L. L., 1961. On the determination of parameters of layers of intermediate thickness from the spectra of reflected waves. *Izv. Acad. Sci. USSR, Geophys. Ser.*, 5: 439–444 (Engl. ed.).

Khudzinskii, L. L., 1966. On the determination of some parameters of homogeneous layers by means of their spectral phase characteristics. *Izv. Acad. Sci. USSR, Phys. Solid Earth*, 5: 319–324 (Engl. ed.).

Khudzinskii, L. L. and Melamud, A. Ya., 1957. Frequency analyser for seismic vibrations. *Izv. Acad. Sci. USSR, Geophys. Ser.*, 9: 24–45 (Engl. ed.).

Khudzinsky, L. L., 1962. Determination of some of the spectral features of layered media. *Izv. Acad. Sci. USSR, Geophys. Ser.* 3: 195–203 (Engl. ed.).

Kimball, B. A. and Lemon, E. R., 1970. Spectra of air pressure fluctuations at the soil surface. *J. Geophys. Res.*. 75: 6771–6777.

King, C.-Y., Bakun, W. H. and Murdock, J. N., 1972. Source parameters of nuclear explosions MILROW and LONGSHOT from teleseismic P-waves. *Geophys. J. R. Astron. Soc.*, 31: 27–44.

King-Hele, D. G., Cook, G. E. and Rees, J. M., 1963. Determination of the even harmonics in the earth's gravitational potential. *Geophys. J. R. Astron. Soc.*, 8: 119–145.

Kinsman, B., 1961. Some evidence on the effect of nonlinearity on the position of the equilibrium range in wind-wave spectra. *J. Geophys. Res.*, 66: 2411–2415.

Kinsman, B., 1965. *Wind Waves.* Prentice-Hall, 676 pp.

Kishimoto, Y., 1964. Investigation on the origin mechanism of earthquakes by the Fourier analysis of seismic body waves (I). *Bull. Disaster Prev. Res. Inst. (Kyoto)*, 13: 1–37.

Kisslinger, C., 1960. Motion at an explosive source as deduced from surface waves. *Earthquake Notes*, 31(1-2): 5–17.

Kitaigorodski, S. A., 1962. Applications of the theory of similarity to the analysis of wind-generated wave motion as a stochastic process. *Izv. Acad. Sci. USSR, Geophys. Ser.*, 1: 73–80 (Engl. ed.).

Kitaigorodskii, S. A. and Strekalov, S. S., 1962. Contribution to an analysis of the spectra of wind-caused wave action, I. *Izv. Acad. Sci. USSR, Geophys. Ser.*, 9: 765–769 (Engl. ed.).

Kitaigorodskii, S. A. and Strekalov, S. S., 1963. On the analysis of spectra of wind-generated wave motion, II. *Izv. Acad. Sci. USSR, Geophys. Ser.*, 8: 754–760 (Engl. ed.).

Kleĭmenova, N. G., 1965. Certain results of observations of the natural electromagnetic field in the 1–20 cps band at the polar stations of Tiksi and Lovozero. *Izv. Acad. Sci. USSR, Phys. Solid Earth*, 3: 207–211 (Engl. ed.).

Knopoff, L. and Gilbert, F., 1959. Radiation from a strike-slip fault. *Bull. Seismol. Soc. Am.*, 49: 163–178.

Knopoff, L. and Pilant, W. L., 1966. A proposal for a seismological study of the Gulf of California. *Geofis. Inter. (Mexico)*, 6: 1–21.

Knopoff, L., Mueller, S. and Pilant, W. L., 1966. Structure of the crust and upper mantle in the Alps from the phase velocity of Rayleigh waves. *Bull. Seismol. Soc. Am.*, 56: 1009–1044.

Knopoff, L., Berry, M. J. and Schwab, F. A., 1967. Tripartite phase velocity observations in laterally heterogeneous regions. *J. Geophys. Res.*, 72: 2595–2601.

Ko, H.-Y. and Scott, R. F., 1967. Deconvolution techniques for linear systems. *Bull. Seismol. Soc. Am.*, 57: 1393–1408.

Koenig, W. and Ruppel, A. E., 1948. Quantitative amplitude representation in sound spectrograms. *J. Acoust. Soc. Am.*, 20: 787–795.

Koenig, W., Dunn, H. K. and Lacy, L. Y., 1946. The sound spectrograph. *J. Acoust. Soc. Am.*, 18: 19–49.

Kogan, S. D., 1972. A study of the dynamics of a longitudinal wave reflected from the earth's core. *Izv. Acad. Sci. USSR, Phys. Solid Earth*, 6: 339–349 (Engl. ed.).

Kogeus (Korkman), K., 1968. A synthesis of short-period P-wave records from distant explosion sources. *Bull. Seismol. Soc. Am.*, 58: 663–680.

Kolsky, H., 1953. *Stress Waves in Solids.* Clarendon Press, 211 pp.

Kondorskaya, N. V., Pavlov, L. N., Pustovitenko, A. N. and Solov'ev, S. L., 1967. Some results of a study of the longitudinal-wave spectra from earthquakes in the Far East. *Izv. Acad. Sci. USSR, Phys. Solid Earth*, 1: 36–43 (Engl. ed.).

Korhonen, H., 1970. Spectral composition of microseisms recorded at seismograph station Oulu. *UGGI, Monogr.*, 31: 33–52.

Korhonen, H., 1971. Types of storm microseism spectra at Oulu. *Nordia*, No. 3: 71 pp.

Korkman, K., 1968. Aftershock P-wave spectra and dynamic features of the Aleutian Islands earthquake sequence of February 4, 1965. *Tectonophysics*, 5: 245–266.

Kosarev, G. L., 1971. A study of the structure of the earth's crust beneath a seismic station from the spectra of longitudinal seismic waves. *Izv. Acad. Sci. USSR, Phys. Solid Earth*, 7: 476–481 (Engl. ed.).

Kosminskaya, I. P., 1971. *Deep Seismic Sounding of the Earth's Crust and Upper Mantle*. Consultants Bureau, 184 pp.

Kosminskaya, I. P. and Zverev, S. M., 1968. Abilities of explosion seismology in oceanic and continental crust and mantle studies. *Can. J. Earth Sci.*, 5: 1091–1100.

Kovach, R. L., 1967. Attenuation of seismic body waves in the mantle. *Geophys. J. R. Astron. Soc.*, 14: 165–170.

Kovach, R. L. and Anderson, D. L., 1964. Attenuation of shear waves in the upper and lower mantle. *Bull. Seismol. Soc. Am.*, 54: 1855–1864.

Kovach, R. L. and Ben-Menahem, A., 1966. Analysis of geomagnetic micropulsations due to high-altitude nuclear explosions. *J. Geophys. Res.*, 71: 1427–1433.

Kovach, R. L., Lehner, F. and Miller, R., 1963. Experimental ground amplitudes from small surface explosions. *Geophysics*, 28: 793–798.

Krauss, W., 1969. Typical features of internal wave spectra. In: M. Sears (Editor), *Progress in Oceanography*, Vol. 5. Pergamon Press, pp.95–101.

Krishna, J., Chandrasekaran, A. R. and Saini, S. S., 1969. Analysis of Koyna accelerogram of December 11, 1967. *Bull. Seismol. Soc. Am.*, 59: 1719–1731.

Ku, C. C., Jones, F. W. and Geldart, L. P., 1967. Analogue spectral analysis of geomagnetic micropulsations in the range 10 to 200 seconds. *Can. J. Phys.*, 45: 3743–3751.

Ku, C. C., Geldart, L. P. and Jones, F. W., 1970. The response method for the analysis of geomagnetic micropulsations. *Phys. Earth Planet. Inter.*, 2: 294–299.

Ku, C. C., Telford, W. M. and Lim, S. H., 1971. The use of linear filtering in gravity problems. *Geophysics*, 36: 1174–1203.

Kubotera, A., 1963. Volcanic micro tremor of the second kind. *Geophys. Pap. dedicated to Prof. K. Sassa (Kyoto)*. Publ. Office, Geophys. Inst., Kyoto, pp.199–209.

Kudo, K. and Shima, E., 1970. Attenuation of shear waves in soil. *Bull. Earthquake Res. Inst.*, 48: 145–158.

Kudo, K., Allam, A. M., Onda, I. and Shima, E., 1970. Attenuation of Love waves in soil layers. *Bull. Earthquake Res. Inst.*, 48: 159–170.

Kulhánek, O., 1966. The spectrum of short-period seismic noise. *Studia Geophys. Geodaet.*, 10: 472–475.

Kulhánek, O., 1967. Seismic noise filtering using digital computers. *Trav. Inst. Géophys. Acad. Tchécoslov. Sci.*, 273: 255–286.

Kulhánek, O., 1968. Seismic noise discrete filtering. *Pure Appl. Geophys.*, 69: 5–11.

Kulhánek, O., 1971. P-wave amplitude spectra of Nevada underground nuclear explosions. *Pure Appl. Geophys.*, 88: 121–136.

Kulhánek, O., 1973a. Source parameters of some presumed Semipalatinsk underground nuclear explosions. *Pure Appl. Geophys.*, 102: 51–66.

Kulhánek, O., 1973b. Signal and noise coherence determination for the Uppsala seismograph array station. *Pure Appl. Geophys.*, 109: 1653–1671.

Kulhánek, O. and Båth, M., 1972. Power spectra and geographical distribution of short-period microseisms in Sweden. *Pure Appl. Geophys.*, 94: 148–171.

Kulhánek, O. and Klíma, K., 1970. The reliable frequency band for amplitude spectra corrections. *Geophys. J. R. Astron. Soc.*, 21: 235–242.

Kuo, J. T. and Ewing, M., 1966. Spatial variations of tidal gravity. *Am. Geophys. Union, Geophys. Monogr.*, 10: 595–610.

Kurita, T., 1966. Attenuation of long-period P-waves and Q in the mantle. *J. Phys. Earth*, 14: 1–14.

Kurita, T., 1968. Attenuation of short-period P-waves and Q in the mantle. *J. Phys. Earth*, 16: 61–78.

Kurita, T., 1969a. Spectral analysis of seismic waves. Part 1. Data windows for the analysis of transient waves. *Spec. Contrib. Geophys. Inst., Kyoto Univ.*, 9: 97–122.

Kurita, T., 1969b. Crustal and upper-mantle structure in Japan from amplitude and phase spectra of long-period P-waves. Part 1. Central mountain area. *J. Phys. Earth*, 17: 13–41.

Kurita, T., 1969c. Crustal and upper-mantle structure in Japan from amplitude and phase spectra of long-period P-waves. Part 2. Kanto Plain. *Spec. Contrib. Geophys. Inst., Kyoto Univ.*, 9: 137–166.

Kurita, T., 1970. Crustal and upper-mantle structure in Japan from amplitude and phase spectra of long-period P-waves. Part 3. Chugoku Region. *J. Phys. Earth*, 18: 53–78.

Kurita, T., 1971. Inferences of a layered structure from S-wave spectra. Part 2. Study of the structure in selected regions of Japan. *J. Phys. Earth*, 19: 111–142.

Kurita, T., 1973. A procedure for elucidating fine structure of the crust and upper mantle from seismological data. *Bull. Seismol. Soc. Am.*, 63: 189–209.

Kurita, T. and Mikumo, T., 1971. Inferences of a layered structure from S-wave spectra. Part 1. Theoretical considerations of S-wave spectrum method. *J. Phys. Earth*, 19: 93–110.

Lacoss, R. T., 1971. Data adaptive spectral analysis methods. *Geophysics*, 36: 661–675.

Lacoss, R. T., Kelly, E. J. and Toksöz, M. N., 1969. Estimation of seismic noise structure using arrays. *Geophysics*, 34: 21–38.

Lai, R. J. and Shemdin, O. H., 1971. Laboratory investigation of air turbulence above simple water waves. *J. Geophys. Res.*, 76: 7334–7350.

Lamb, H., 1945. *Hydrodynamics*. Dover Publ., 6th ed., 738 pp.

Lambert, A. and Caner, B., 1965. Geomagnetic "depth-sounding" and the coast effect in western Canada. *Can. J. Earth Sci.*, 2: 485–509.

Lambert, D. G., Flinn, E. A. and Archambeau, C. B., 1972. A comparative study of the elastic wave radiation from earthquakes and underground explosions. *Geophys. J. R. Astron. Soc.*, 29: 403–432.

Lamden, R. J., 1970. A Fourier covariance analysis of the long period magnetic field variations in the British Isles during 1958. *Geophys. J. R. Astron. Soc.*, 20: 177–189.

Lampard, D. G., 1954. Generalization of the Wiener-Khintchine theorem to nonstationary processes. *J. Appl. Phys.*, 25: 802–803.

Lanczos, C., 1966. *Discourse on Fourier Series*. Oliver and Boyd, 255 pp.

Landisman, M., Mueller, S., Bolt, B. and Ewing, M., 1962. Transient analysis of seismic core phases. *Geofis. Pura Appl.*, 52: 41–52.

Landisman, M., Dziewonski, A. and Satô, Y., 1969. Recent improvements in the analysis of surface wave observations. *Geophys. J. R. Astron. Soc.*, 17: 369–403.

Landisman, M., Usami, T., Satô, Y. and Massé, R., 1970. Contributions of theoretical seismograms to the study of modes, rays, and the earth. *Rev. Geophys. Space Phys.*, 8: 533–589.

Landsberg, H. E., Mitchell, J. M., Jr. and Crutcher, H. L., 1959. Power spectrum analysis of climatological data for Woodstock College, Maryland. *Mon. Weather Rev.*, 87: 283–298.

Laning, J. H., Jr. and Battin, R. H., 1956. *Random Processes in Automatic Control*. McGraw-Hill, 434 pp.

Lansraux, G. and Delisle, C., 1962. Diffraction itérée de la figure (sin X/X) par un réseau infini. *Can. J. Phys.*, 40: 1113–1132.

Lappe, U. O. and Davidson, B., 1963. On the range of validity of Taylor's hypothesis and the Kolmogoroff spectral law. *J. Atmos. Sci.*, 20: 569–576.

Lappe, U. O., Kirwan, A. D., Jr. and Adelfang, S. I., 1967. Some aspects of boundary-layer turbulence over land and ocean. In: K. F. Bowden, F. N. Frenkiel and I. Tani (Editors), *Boundary Layers and Turbulence. The Physics of Fluids Suppl.* American Institute of Physics, pp.206–208.

Laster, S. J. and Linville, A. F., 1966. Application of multichannel filtering to the separation of dispersive modes of propagation. *J. Geophys. Res.*, 71: 1669–1701.

Lastrico, R. M., Duke, C. M. and Ohta, Y., 1972. Effects of site and propagation path on recorded strong earthquake motions. *Bull. Seismol. Soc. Am.*, 62: 933–954.

Latham, G. V. and Nowroozi, A. A., 1968. Waves, weather, and ocean bottom microseisms. *J. Geophys. Res.*, 73: 3945–3956.

Latham, G. V. and Sutton, G. H., 1966. Seismic measurements on the ocean floor. 1. Bermuda area. *J. Geophys. Res.*, 71: 2545–2573.

Latham, G. V., Anderson, R. S. and Ewing, M., 1967. Pressure variations produced at the ocean bottom by hurricanes. *J. Geophys. Res.*, 72: 5693–5704.

Lavin, P. M. and Devane, J. F., S. J., 1970. Direct design of two-dimensional digital wavenumber filters. *Geophysics*, 35: 1073–1078.

Leblanc, G., S. J., 1967. Truncated crustal transfer functions and fine crustal structure determination. *Bull. Seismol. Soc. Am.*, 57: 719–733.

Leblanc, G. and Howell, B. F., Jr., 1967. Spectral study of short-period P-waves. *Can. J. Earth Sci.*, 4: 1049–1063.

Lee, W. H. K., 1963. Heat flow data analysis. *Rev. Geophys.*, 1: 449–479.

Lee, W. H. K. and Cox, C. S., 1966. Time variation of ocean temperatures and its relation to internal waves and oceanic heat flow measurements. *J. Geophys. Res.*, 71: 2101–2111.

Lee, W. H. K. and Kaula, W. M., 1967. A spherical harmonic analysis of the earth's topography. *J. Geophys. Res.*, 72: 753–758. Correction in *J. Geophys. Res.* (1973), 78: 478–481.

Lee, W. H. K. and MacDonald, G. J. F., 1963. The global variation of terrestrial heat flow. *J. Geophys. Res.*, 68: 6481–6492.

Leese, J. A. and Epstein, E. S., 1963. Application of two-dimensional spectral analysis to the quantification of satellite cloud photographs. *J. Appl. Meteorol.*, 2: 629–644.

Leese, J. A., Novak, C. S. and Clark, B. B., 1971. An automated technique for obtaining cloud motion from geosynchronous satellite data using cross correlation. *J. Appl. Meteorol.*, 10: 118–132.

Lehmann, H. J., 1970. Examples for the separation of fields of magnetic sources in different depths by the harmonic analysis method. *Boll. Geofis. Teor. Appl.*, 12: 97–117.

Leith, C. E., 1971. Atmospheric predictability and two-dimensional turbulence. *J. Atmos. Sci.*, 28: 145–161.

Lenschow, D. H., 1970. Airplane measurements of planetary boundary layer structure. *J. Appl. Meteorol.*, 9: 874–884.

LeSchack, L. A. and Haubrich, R. A., 1964. Observations of waves on an ice-covered ocean. *J. Geophys. Res.*, 69: 3815–3821.

Levshin, A. L., Neigaus, M. G. and Sabitova, T. M., 1965. Spectra of Love waves and the depth of the normal source. *Geophys. J. R. Astron. Soc.*, 9: 253–259.

Levshin, A. L., Pisarenko, V. F. and Pogrebinsky, G. A., 1972. On a frequency–time analysis of oscillations. *Ann. Géophys.*, 28: 211–218.

Lewis, B. T. R. and Dorman, L. M., 1970. Experimental isostasy. 2. An isostatic model for the U.S.A. derived from gravity and topographic data. *J. Geophys. Res.*, 75: 3367–3386.

Liebermann, R. C. and Pomeroy, P. W., 1969. Relative excitation of surface waves by earthquakes and underground explosions. *J. Geophys. Res.*, 74: 1575–1590.

Lighthill, M. J., 1958. *Introduction to Fourier Analysis and Generalized Functions.* Cambridge Univ. Press, 79 pp.

Lilley, F. E. M. and Bennett, D. J., 1972. An array experiment with magnetic variometers near the coasts of southeast Australia. *Geophys. J. R. Astron. Soc.*, 29: 49–64.

Linde, A. T. and Sacks, I. S., 1971. Errors in the spectral analysis of long-period seismic body waves. *J. Geophys. Res.*, 76: 3326–3336.

Linde, A. T. and Sacks, I. S., 1972. Dimensions, energy, and stress release for South American deep earthquakes. *J. Geophys. Res.*, 77: 1439–1451.

Lindsey, J. P., 1960. Elimination of seismic ghost reflections by means of a linear filter. *Geophysics*, 25: 130–140.

Linville, A. F. and Laster, S. J., 1966. Numerical experiments in the estimation of frequency–wavenumber spectra of seismic events using linear arrays. *Bull. Seismol. Soc. Am.*, 56: 1337–1355.

Linville, A. F. and Laster, S. J., 1967. Near-surface dispersion studies at Tonto Forest Seismological Observatory. *Bull. Seismol. Soc. Am.*, 57: 311–340.

Liou, M.-L., 1964. Numerical techniques of Fourier transforms with applications. In: W. R. Perkins and P. M. Hinojosa (Editors), *Proc. 2nd Ann. Allerton Conf. on Circuit and System Theory*. Dept. Electrical Eng. and Coordinated Sci. Labor., Univ. of Illinois, pp.114–134.

Liu, S. C., 1969. Autocorrelation and power spectral density functions of the Parkfield earthquake of June 27, 1966. *Bull. Seismol. Soc. Am.*, 59: 1475–1493.

Liu, S. C., 1970. Evolutionary power spectral density of strong-motion earthquakes. *Bull. Seismol. Soc. Am.*, 60: 891–900.

Liu, S. C. and Jhaveri, D. P., 1969. Spectral and correlation analysis of ground-motion accelerograms. *Bull. Seismol. Soc. Am.*, 59: 1517–1534.

Lomnicki, Z. A. and Zaremba, S. K., 1957. On the estimation of autocorrelation in time series. *Ann. Math. Stat.*, 28: 140–158.

Lomnitz, C., 1964. Estimation problems in earthquake series. *Tectonophysics*, 2: 193–203.

London, J., Ruff, I. and Tick, L. J., 1959. The relationship between geomagnetic variations and the circulation at 100 mbar. *J. Geophys. Res.*, 64: 1827–1833.

Long, L. T. and Berg, J. W., Jr., 1969. Transmission and attenuation of the primary seismic wave, 100 to 600 km. *Bull. Seismol. Soc. Am.*, 59: 131–146.

Longuet-Higgins, M. S., 1950. A theory of the origin of microseisms. *Philos. Trans. R. Soc. London, Ser. A*, 243: 1–35.

Longuet-Higgins, M. S., 1971. On the spectrum of sea level at Oahu. *J. Geophys. Res.*, 76: 3517–3522.

Longuet-Higgins, M. S., Cartwright, D. E. and Smith, N. D., 1963. Observations of the directional spectrum of sea waves using the motions of a floating buoy. In: R. C. Vetter and C. L. Bretschneider (Editors), *Ocean Wave Spectra*. Prentice-Hall, pp.111–132.

Loomis, H. G., 1966. Spectral analysis of tsunami records from stations in the Hawaiian Islands. *Bull. Seismol. Soc. Am.*, 56: 697–713.

López Arroyo, A., 1968. Ondas superficiales de largo periodo registradas en Toledo y Málaga. *Rev. Geofis.*, 27: 1–26.

López Arroyo, A. and Udías, A., 1972. Aftershock sequence and focal parameters of the February 28, 1969 earthquake of the Azores–Gibraltar fracture zone. *Bull. Seismol. Soc. Am.*, 62: 699–719.

Lossovskiy, E. K., 1968. Synthetic seismograms and the possible spectral attenuation of multiply reflected waves. *Izv. Acad. Sci. USSR, Phys. Solid Earth*, 6: 374–378 (Engl. ed.).

Loynes, R. M., 1968. On the concept of the spectrum for non-stationary processes. *J. R. Stat. Soc.*, B 30: 1–20.

Lukasik, S. J. and Grosch, C. E., 1963. Pressure–velocity correlations in ocean swell. *J. Geophys. Res.*, 68: 5689–5699.

Lumley, J. L. and Panofsky, H. A., 1964. *The Structure of Atmospheric Turbulence*. Interscience, 239 pp.

Luydendyk, B. P., 1969. Origin of short-wavelength magnetic lineations observed near the ocean bottom. *J. Geophys. Res.*, 74: 4869–4881.

Lynch, R. D., 1969. Response spectra for Pahute Mesa nuclear events. *Bull. Seismol. Soc. Am.*, 59: 2295–2309.

Lysmer, J., Seed, H. B. and Schnabel, P. B., 1971. Influence of base-rock characteristics on ground response. *Bull. Seismol. Soc. Am.*, 61: 1213–1231.

Maasha, N. and Molnar, P., 1972. Earthquake fault parameters and tectonics in Africa. *J. Geophys. Res.*, 77: 5731–5743.

MacCready, P. B., Jr., 1962a. Turbulence measurements by sailplane. *J. Geophys. Res.*, 67: 1041–1050.

MacCready, P. B., Jr., 1962b. The inertial subrange of atmospheric turbulence. *J. Geophys. Res.*, 67: 1051–1059.

MacDonald, G. J. F., 1965. The bispectra of atmospheric pressure records. *Proc. IBM Sci. Comp. Symp. Stat., 1963*, pp.247–264.

MacDonald, G. J. F. and Ness, N. F., 1961. A study of the free oscillations of the earth. *J. Geophys. Res.*, 66: 1865–1911.

Mack, H., 1969. Nature of short-period P-wave signal variations at LASA. *J. Geophys. Res.*, 74: 3161–3170.

Mack, H. and Flinn, E. A., 1971. Analysis of the spatial coherence of short-period acoustic-gravity waves in the atmosphere. *Geophys. J. R. Astron. Soc.*, 26: 255–269.

Mack, H. and Smart, E., 1972. Frequency domain processing of digital microbarograph array data. *J. Geophys. Res.*, 77: 488–490.

Madden, R. A. and Julian, P. R., 1971. Detection of a 40–50 day oscillation in the zonal wind in the tropical Pacific. *J. Atmos. Sci.*, 28: 702–708.

Mak, M.-K., 1969. Laterally driven stochastic motions in the tropics. *J. Atmos. Sci.*, 26: 41–64.

Malin, S. R. C., 1969. Geomagnetic secular variation and its changes, 1942.5 to 1962.5. *Geophys. J. R. Astron. Soc.*, 17: 415–441.

Malin, S. R. C. and Chapman, S., 1970. The determination of lunar daily geophysical variations by the Chapman-Miller method. *Geophys. J. R. Astron. Soc.*, 19: 15–35.

Malin, S. R. C. and Pocock, S. B., 1969. Geomagnetic spherical harmonic analysis. *Pure Appl. Geophys.*, 75: 117–132.

Malone, F. D., 1968. An analysis of current measurements in Lake Michigan. *J. Geophys. Res.*, 73: 7065–7081.

Manchee, E. B., 1972. Short-period seismic discrimination. *Nature*, 239: 152–153.

Manchee, E. B. and Hasegawa, H. S., 1973. Seismic spectra of Yucca Flat underground explosions observed at Yellowknife, Northwest Territories. *Can. J. Earth Sci.*, 10: 421–427.

Mangiarotty, R. A. and Turner, B. A., 1967. Wave radiation Doppler effect correction for motion of a source, observer and the surrounding medium. *J. Sound Vibr.*, 6: 110–116.

Manley, R. G., 1945. *Waveform Analysis*. Chapman and Hall, 275 pp.

Mantis, H. T., 1963. The structure of winds of the upper troposphere at mesoscale. *J. Atmos. Sci.*, 20: 94–106.

Mantis, H. T. and Pepin, T. J., 1971. Vertical temperature structure of the free atmosphere at mesoscale. *J. Geophys. Res.*, 76: 8621–8628.

Manzoni, G., 1967. Theoretical evaluation of the perturbation on power spectra due to random errors in the spacing of the sampling instants. *Boll. Geofis. Teor. Appl.*, 9: 248–252.

Marshall, P. D., 1970. Aspects of the spectral differences between earthquakes and underground explosions. *Geophys. J. R. Astron. Soc.*, 20: 397–416.

Marshall, P. D. and Burton, P. W., 1971. The source-layering function of underground explosions and earthquakes—an application of a "common path" method. *Geophys. J. R. Astron. Soc.*, 24: 533–537.

Martin, H. C., 1972. Humidity and temperature microstructure near the ground. *Q. J. R. Meteorol. Soc.*, 98: 440–446.

Márton, P., 1970. Secular variation of the geomagnetic virtual dipole field during the last 2000 years as inferred from the spherical harmonic analysis of the available archeomagnetic data. *Pure Appl. Geophys.*, 81: 163–176.

Maruyama, T., 1968. Time sequence of power spectra of disturbances in the equatorial lower stratosphere in relation to the quasi-biennial oscillation. *J. Meteorol. Soc. Japan*, 46: 327–341.

Massé, R. P., Lambert, D. G. and Harkrider, D. G., 1973. Precision of the determination of focal depth from the spectral ratio of Love/Rayleigh surface waves. *Bull. Seismol. Soc. Am.*, 63: 59–100.

Matsushita, S. and Campbell, W. H. (Editors), 1967. *Physics of Geomagnetic Phenomena*, 2 vol. Acad. Press, 1398 pp.

Matsushita, S. and Maeda, H., 1965a. On the geomagnetic solar quiet daily variation field during the IGY. *J. Geophys. Res.*, 70: 2535–2558.

Matsushita, S. and Maeda, H., 1965b. On the geomagnetic lunar daily variation field. *J. Geophys. Res.*, 70: 2559–2578.

Matumoto, T., 1958. Calibration of an electromagnetic seismograph by means of the frequency analysis. *Bull. Earthquake Res. Inst.*, 36: 55–64.

Matumoto, T., 1959. On the spectral structure of earthquake waves. Its influence on magnitude scale. *Bull. Earthquake Res. Inst.*, 37: 265–277.

Matumoto, T., 1960. On the spectral structure of earthquake waves. The relation between magnitude and predominant period. *Bull. Earthquake Res. Inst.*, 38: 13–27.

Matumoto, T., 1971. Seismic body waves observed in the vicinity of Mount Katmai, Alaska, and evidence for the existence of molten chambers. *Geol. Soc. Am. Bull.*, 82: 2905–2920.

Mauersberger, P., 1959. Mathematische Beschreibung und statistische Untersuchung des Hauptfeldes und der Säkularvariation. In: G. Fanselau (Editor), *Geomagnetismus und Aeronomie* III. VEB Deutscher Verlag der Wissenschaften, pp.93–213.

McBean, G. A. and Miyake, M., 1972. Turbulent transfer mechanisms in the atmospheric surface layer. *Q. J. R. Meteorol. Soc.*, 98: 383–398.

McCamy, K., 1967. An investigation and application of the crustal transfer ratio as a diagnostic for explosion seismology. *Bull. Seismol. Soc. Am.*, 57: 1409–1428.

McCamy, K. and Meyer, R. P., 1964. A correlation method of apparent velocity measurement. *J. Geophys. Res.*, 69: 691–699.

McConnell, R. K., Jr., 1968. Viscosity of the mantle from relaxation time spectra of isostatic adjustment. *J. Geophys. Res.*, 73: 7089–7105.

McCrory, R. A., 1967. Atmospheric pressure waves from nuclear explosions. *J. Atmos. Sci.*, 24: 443–447.

McDonald, J. A., Douze, E. J. and Herrin, E., 1971. The structure of atmospheric turbulence and its application to the design of pipe arrays. *Geophys. J. R. Astron. Soc.*, 26: 99–109.

McEvilly, T. V., 1964. Central U.S. crust–upper mantle structure from Love and Rayleigh wave phase velocity inversion. *Bull. Seismol. Soc. Am.*, 54: 1997–2015.

McEvilly, T. V. and Peppin, W. A., 1972. Source characteristics of earthquakes, explosions and afterevents. *Geophys. J. R. Astron. Soc.*, 31: 67–82.

McGarr, A., 1969. Amplitude variations of Rayleigh waves—propagation across a continental margin. *Bull. Seismol. Soc. Am.*, 59: 1281–1305.

McGarr, A., Hofmann, R. B. and Hair, G. D., 1964. A moving-time-window signal-spectra process. *Geophysics*, 29: 212–220.

McIlwraith, C. G. and Hays, S. D., 1963. Ocean-wave measurements by sonar. *J. Mar. Res.*, 21: 94–101.

McIvor, I. K., 1964. Methods of spectral analysis of seismic data. *Bull. Seismol. Soc. Am.*, 54: 1213–1232.

McLeish, W., 1970. Spatial spectra of ocean surface temperature. *J. Geophys. Res.*, 75: 6872–6877.

McWilliams, J., 1972. Observations of kinetic energy correspondences in the internal wave field. *Deep-Sea Res.*, 19: 793–811.

Medwin, H., Clay, C. S., Berkson, J. M. and Jaggard, D. L., 1970. Traveling correlation function of the heights of wind-blown water waves. *J. Geophys. Res.*, 75: 4519–4524.

Melchior, P., 1966. *The Earth Tides*. Pergamon Press, 458 pp.

Melchior, P. and Venedikov, A., 1968. Derivation of the wave M_3 ($8^h.279$) from the periodic tidal deformations of the earth. *Phys. Earth Planet. Inter.*, 1: 363–372.

Mendiguren, J. A., 1971. Focal mechanism of a shock in the middle of the Nazca plate. *J. Geophys. Res.*, 76: 3861–3879.

Merilees, P. E., 1968. The equations of motion in spectral form. *J. Atmos. Sci*, 25: 736–743.

Merkel, R. H. and Alexander, S. S., 1969. Use of correlation analysis to interpret continental margin ECOOE refraction data. *J. Geophys. Res.*. 74: 2683–2697.

Meskó, A., 1965. Some notes concerning the frequency analysis for gravity interpretation. *Geophys. Prospect.*, 13: 475–488.

Mihail, I. and Nicolae, M., 1972. Separation of cumulated effects in the Bouguer map. *Rev. Roum. Géol. Géophys. Géogr., Sér. Géophys.*, 16: 193–199.

Mikhaĭlova, N. G., Pariĭskiĭ, B. S. and Saks, M. V., 1966. The spectral characteristics of multiple transition layers ("layer packets"). *Izv. Acad. Sci. USSR, Phys. Solid Earth*, 1: 6–12 (Engl. ed.).

Mikhota, G. G., 1968. Deep-wave spectra and optimal characteristics of a DSS apparatus for the conditions of western Uzbekistan. *Izv. Acad. Sci. USSR, Phys. Solid Earth*, 1: 40–46 (Engl. ed.).

Mikumo, T., 1968. Atmospheric pressure waves and tectonic deformation associated with the Alaskan earthquake of March 28, 1964. *J. Geophys. Res.*, 73: 2009–2025.

Mikumo, T., 1969. Long-period P waveforms and the source mechanism of intermediate earthquakes. *J. Phys. Earth*, 17: 169–192.

Mikumo, T., 1971. Source process of deep and intermediate earthquakes as inferred from long-period P and S waveforms. *J. Phys. Earth*, 19: 1–19.

Mikumo, T., 1972. Focal process of deep and intermediate earthquakes around Japan as inferred from long-period P and S waveforms. *Phys. Earth Planet. Inter.*, 6: 293–299.

Mikumo, T. and Kurita, T., 1968. *Q*-distribution for long-period P-waves in the mantle. *J. Phys. Earth*, 16: 11–29.

Mikumo, T. and Kurita, T., 1971. Inferences of a layered structure from S-wave spectra. Part 3. SH and SV waves and some related problems. *J. Phys. Earth*, 19: 243–257.

Mikumo, T. and Nakagawa, I., 1968. Some problems on the analysis of the earth tides. *J. Phys. Earth*, 16: 87–95.

Milcoveanu, D., 1971a. Expresiile mărimilor spectrale relative la modelele utilizate în interpretarea gravimetrică. *Stud. Cerc. Geol. Geofiz. Geograf. (Bucuresti), Ser. Geofiz.*, 9: 39–71.

Milcoveanu, D., 1971b. Expresiile mărimilor spectrale relative la modelele utilizate în interpretarea magnetometrică. *Stud. Cerc. Geol. Geofiz. Geograf. (Bucuresti), Ser. Geofiz.*, 9: 297–322.

Milcoveanu, D., 1972. Analiza spectrală a modelului strait înclinat bidimensional semiinfinit. *Stud. Cerc. Geol. Geofiz. Geograf. (Bucuresti), Ser. Geofiz.*, 10: 185–205.

Millard, R. C., Jr., 1971. Wind measurements from buoys: A sampling scheme. *J. Geophys. Res.*, 76: 5819–5828.

Miller, G. R., Munk, W. H. and Snodgrass, F. E., 1962. Long-period waves over California's continental borderland. Part II. Tsunamis. *J. Mar. Res.*, 20: 31–41.

Miller, W. F., 1963. The Caltech digital seismograph. *J. Geophys. Res.*, 68: 841–847.

Milne, A. R., 1959. Comparison of spectra of an earthquake T-phase with similar signals from nuclear explosions. *Bull. Seismol. Soc Am.*, 49: 317–329.

Milne, A. R. and Clark, S. R., 1964. Resonances in seismic noise under Arctic sea-ice. *Bull. Seismol. Soc. Am.*, 54: 1797–1809.

Milne, A. R. and Ganton, J. H., 1964. Ambient noise under Arctic-Sea ice. *J. Acoust. Soc. Am.*, 36: 855-863.

Mino, K., Onoguchi, T. and Mikumo, T., 1968. Focal mechanism of earthquakes on island arcs in the southwest Pacific region. *Bull. Disaster Prev. Res. Inst. (Kyoto)*, 18(2): 78–96.

Mishra, D. C. and Naidu, P. S., 1971. Two-dimensional power spectral analysis of aeromagnetic fields. *Bull. Natl. Geophys. Res. Inst. (Hyderabad)*, 9: 49–55.

Mitchell, B. J., 1973. Radiation and attenuation of Rayleigh waves from the southeastern Missouri earthquake of October 21, 1965. *J. Geophys. Res.*, 78: 886–899.

Mitsuta, Y., Hanafusa, T. and Maitani, T., 1970. Experimental studies of turbulent transfer processes in the boundary layer over bare soil. *Bull. Disaster Prev. Res. Inst. (Kyoto)*, 19(4): 45–58.

Miyake, M., Stewart, R. W. and Burling, R. W., 1970a. Spectra and cospectra of turbulence over water. *Q. J. R. Meteorol. Soc.*, 96: 138–143.

Miyake, M., Donelan, M. and Mitsuta, Y., 1970b. Airborne measurement of turbulent fluxes. *J. Geophys. Res.*, 75: 4506–4518.

Miyata, M. and Groves, G. W., 1968. Note on sea level observations at two nearby stations. *J. Geophys. Res.*, 73: 3965–3967.

Mizoue, M., 1967. Modes of secular vertical movements of the earth's crust. Part 1. *Bull. Earthquake Res. Inst.*, 45: 1019–1090.

Mohammadioun, B., 1965a. Structure de la croûte et spectres d'énergie des ondes longitudinales P. *C. R. Acad. Sci. Paris*, 261: 3181–3184.

Mohammadioun, B., 1965b. Structure du manteau supérieur et spectres d'énergie des ondes longitudinales. *C. R. Acad. Sci. Paris*, 261: 4472–4474.

Mohammadioun, B., 1966. Structure du manteau inférieur et spectres d'énergie des ondes longitudinales. *C. R. Acad. Sci. Paris*, 262: 156–159.

Molnar, P., 1971. P-wave spectra from underground nuclear explosions. *Geophys. J. R. Astron. Soc.*, 23: 273–286.

Molnar, P. and Wyss, M., 1972. Moments, source dimensions and stress drops of shallow-focus earthquakes in the Tonga–Kermadec arc. *Phys. Earth Planet. Inter.*, 6: 263–278.

Molnar, P., Savino, J., Sykes, L. R., Liebermann, R. C., Hade, G. and Pomeroy, P. W., 1969. Small earthquakes and explosions in western North America recorded by new high gain, long period seismographs. *Nature*, 224: 1268–1273.

Molotova, L. V., 1964. The relationship between the frequency spectra of seismic vibrations and explosion conditions. *Izv. Acad. Sci. USSR, Geophys. Ser.*, 12: 1059–1072 (Engl. ed.).

Molotova, L. V., 1966. Velocity dispersion of body waves in terrigenous rocks. *Izv. Acad. Sci. USSR, Phys. Solid Earth*, 8: 500–506 (Engl. ed.).

Monin, A. S., 1963. Stationary and periodic time series in the general circulation of the atmosphere. In: M. Rosenblatt (Editor), *Time Series Analysis*. Wiley, pp. 144–151.

Monin, A. S., 1967. Turbulence in the atmospheric boundary layer. In: K. F. Bowden, F. N. Frenkiel and I. Tani (Editors), *Boundary Layers and Turbulence. The Physics of Fluids Suppl.* American Institute of Physics, pp.31–37.

Monin, A. S. and Vulis, I. L., 1971. On the spectra of long-period oscillations of geophysical parameters. *Tellus*, 23: 337–345.

Monroe, A. J., 1962. *Digital Processes for Sampled Data Systems*. Wiley, 490 pp.

Montalbetti, J. F. and Kanasewich, E. R., 1970. Enhancement of teleseismic body phases with a polarization filter. *Geophys. J. R. Astron. Soc.*, 21: 119–129.

Mooers, C. N. K. and Smith, R. L., 1968. Continental shelf waves off Oregon. *J. Geophys. Res.*, 73: 549–557.

Mooney, H. M., 1970. Upper-mantle inhomogeneity beneath New Zealand: seismic evidence. *J. Geophys. Res.*, 75: 285–309.

Morelli, C. and Carrozzo, M. T., 1963. Calculation of the anomalous gravity gradient in elevation from Bouguer's anomalies. *Boll. Geofis. Teor. Appl.*, 5: 308–336.

Morelli, C. and Mosetti, F., 1961. Su un tipo di interpretazione semianalitica dei risultati dei rilievi geofisici. *Boll. Geofis. Teor. Appl.*, 3: 61–76.

Morgan, W. J., Stoner, J. O. and Dicke, R. H., 1961. Periodicity of earthquakes and the invariance of the gravitational constant. *J. Geophys. Res.*, 66: 3831–3843.

Moriyasu, S., 1967. On the anomaly of the sea surface temperature in the East China Sea (I). *Oceanogr. Mag.*, 19: 201–220.

Morrissey, E. G. and Muller, F. B., 1968. Spectral aspects of upper wind measurement systems. *Can. Meteorol. Mem.*, 26: 1–35.

Mosetti, F. and Carrozzo, M. T., 1971. Some considerations on a method for tides and seiches analysis. *Boll. Geofis. Teor. Appl.*, 13: 76–94.

Mosetti, F. and Manca, B., 1972. New methods of tides and seiches analysis. Application to the Adriatic Sea. *Boll. Geofis. Teor. Appl.*, 14: 105–127.

Moskowitz, L., 1964. Estimates of the power spectrums for fully developed seas for wind speeds of 20 to 40 knots. *J. Geophys. Res.*, 69: 5161–5179.

Moskvina, A. G., 1971. Possible determination of certain characteristic quantities of an earthquake focus from body-wave spectra. *Izv. Acad. Sci. USSR, Phys. Solid Earth*, 11: 754–761 (Engl. ed.).

Moskvina, A. G. and Shebalin, N. V., 1958. Seismograph frequency characteristics at the station "Pulkovo". *Izv. Acad. Sci. USSR, Geophys. Ser.*, 11: 801–803 (Engl. ed.).

Mueller, R. A., 1969. Seismic energy efficiency of underground nuclear detonations. *Bull. Seismol. Soc. Am.*, 59: 2311–2323.

Mueller, R. A. and Murphy, J. R., 1971. Seismic characteristics of underground nuclear detonations. Part 1. Seismic spectrum scaling. *Bull. Seismol. Soc. Am.*, 61: 1675–1692.

Muller, F. B., 1966. Notes on the meteorological application of power spectrum analysis. *Can. Meteorol. Mem.*, 24: 1–84.

Mundt, W., 1966. Autokorrelationsanalyse für das geomagnetische Feld und die Bouguer-Schwere im Gebiet der DDR. *Gerlands Beitr. Geophys.*, 75: 351–370.

Munk, W. H., 1962. Long ocean waves. In: M. N. Hill (Editor), *The Sea*, Vol. 1. Interscience, pp.647–663.

Munk, W. H. and Bullard, E. C., 1963. Patching the long-wave spectrum across the tides. *J. Geophys. Res.*, 68: 3627–3634.

Munk, W. H. and Cartwright, D. E., 1966. Tidal spectroscopy and prediction. *Philos. Trans. R. Soc. London, Ser. A*, 259: 533–581.

Munk, W. and Hassan, El S. M., 1961. Atmospheric excitation of the earth's wobble. *Geophys. J. R. Astron. Soc.*, 4: 339–358.

Munk, W. H., Snodgrass, F. E. and Tucker, M. J., 1959. Spectra of low-frequency ocean waves. *Bull. Scripps Inst. Oceanogr., Univ. Calif.*, 7: 283–362.

Munk, W. H., Miller, G. R. and Snodgrass, F. E., 1962. Long-period waves over California's continental borderland. Part III. The decay of tsunamis and the dissipation of tidal energy. *J. Mar. Res.*, 20: 119–120.

Munk, W. H., Zetler, B. and Groves, G. W., 1965. Tidal cusps. *Geophys. J. R. Astron. Soc.*, 10: 211–219.

Munkelt, K., 1959. Formeln zur harmonischen Analyse von Gezeitenerscheinungen denen ein unbekannter Gang überlagert ist. *Dtsche Hydrogr. Z.*, 12: 189–195.

Munuera, J. M., 1969. Magnitud sísmica determinada por ondas sísmicas. *Rev. Geofis.*, 28: 245–276.

Murakami, T. and Tomatsu, K., 1964. The spectrum analysis of the energy interaction terms in the atmosphere. *J. Meteorol. Soc. Japan*, 42: 14–25.

Murphy, A. J., Savino, J., Rynn, J. M. W., Choy, G. L. and McCamy, K., 1972. Observations of long-period (10–100 sec) seismic noise at several worldwide locations. *J. Geophys. Res.*, 77: 5042–5049.

Murphy, B. L., 1972. Variation of Rayleigh-wave amplitude with yield and height of burst for intermediate-altitude nuclear detonations. *J. Geophys. Res.*, 77: 808–817.

Murphy, J. R. and Lahoud, J. A., 1969. Analysis of seismic peak amplitudes from underground nuclear explosions. *Bull. Seismol. Soc. Am.*, 59: 2325–2341.

Murphy, J. R., Davis, A. H. and Weaver, N. L., 1971. Amplification of seismic body waves by low-velocity surface layers. *Bull. Seismol. Soc. Am.*, 61: 109–145.

Myrup, L. O., 1969. Turbulence spectra in stable and convective layers in the free atmosphere. *Tellus*, 21: 341–354.

Mysak, L. A., 1967a. On the theory of continental shelf waves. *J. Mar. Res.*, 25: 205–227.

Mysak, L. A., 1967b. On the very low frequency spectrum of the sea level on a continental shelf. *J. Geophys. Res.*, 72: 3043–3047.

Mysak, L. A. and Hamon, B. V., 1969. Low-frequency sea level behavior and continental shelf waves off North Carolina. *J. Geophys. Res.*, 74: 1397–1405.

Nafe, J. E. and Brune, J. N., 1960. Observations of phase velocity for Rayleigh waves in the period range 100 to 400 seconds. *Bull. Seismol. Soc. Am.*, 50: 427–439.

Nagamune, T., 1971. Source regions of great earthquakes. *Geophys. Mag.*, 35: 333–399.

Nagata, T., 1938. Magnetic anomalies and the corresponding subterranean mass distribution. *Bull. Earthquake Res. Inst.*, 16: 550–577.

Nagata, Y., 1964. The statistical properties of orbital wave motions and their application for the measurement of directional wave spectra. *J. Oceanogr. Soc. Japan*, 19: 169–181.

Naidu, P., 1968. Spectrum of the potential field due to randomly distributed sources. *Geophysics*, 33: 337–345.

Naidu, P. S., 1969. Estimation of spectrum and cross-spectrum of aeromagnetic field using Fast Digital Fourier Transform (FDFT) techniques. *Geophys. Prospect.*, 17: 344–361.

Naidu, P. S., 1970a. Fourier transform of large-scale aeromagnetic field using a modified version of Fast Fourier Transform. *Pure Appl. Geophys.*, 81: 17–25.

Naidu, P. S., 1970b. A statistical study of the interpolation of randomly spaced geophysical data. *Geoexploration*, 8: 61–70.

Naidu, P. S., 1970c. Statistical structure of aeromagnetic field. *Geophysics*, 35: 279–292.

Naidu, P. S., 1971. Statistical structure of geomagnetic field reversals. *J. Geophys. Res.*, 76: 2649–2662.

Nakamura, Y., 1964. Model experiments of refracted arrivals from a linear transition layer. *Bull. Seismol. Soc. Am.*, 54: 1–8.

Nakamura, Y., 1968. Head waves from a transition layer. *Bull. Seismol. Soc. Am.*, 58: 963–976.

Nakamura, Y. and Howell, B. F., Jr., 1964. Maine seismic experiment: Frequency spectra of refraction arrivals and the nature of the Mohorovičić discontinuity. *Bull. Seismol. Soc. Am.*, 54: 9–18.

Nakhamkin, S. A., 1969. Fan filtration. *Izv. Acad. Sci. USSR, Phys. Solid Earth*, 11: 686–691 (Engl. ed.).

Nash, D. M., Jr. and Barnes, T. G., 1968. Geoacoustic applications of electroseismic sources. *J. Acoust. Soc. Am.*, 44: 1671–1674.

Neidell, N. S., 1965. A geophysical application of spectral analysis. *Appl. Stat., R. Stat. Soc.*, C 14: 75–88.

Neidell, N. S., 1966. Spectral studies of marine geophysical profiles. *Geophysics*, 31: 122–134.

Ness, N. F., Harrison, J. C. and Slichter, L. B., 1961. Observations of the free oscillations of the earth. *J. Geophys. Res.*, 66: 621–629.

Neumann, G. and Pierson, W. J., Jr., 1966. *Principles of Physical Oceanography*. Prentice-Hall, 545 pp.

Niazi, M., 1969. Source dynamics of the Dasht-e Bayãz earthquake of August 31, 1968. *Bull. Seismol. Soc. Am.*, 59: 1843–1861.

Niazi, M., 1971. Seismic dissipation in deep seismic zones from the spectral ratio of pP/P. *J. Geophys. Res.*, 76: 3337–3343.

Nigam, N. C. and Jennings, P. C., 1969. Calculation of response spectra from strong-motion earthquake records. *Bull. Seismol. Soc. Am.*, 59: 909–922.

Nitta, T., 1970a. Statistical study of tropospheric wave disturbances in the tropical Pacific region. *J. Meteorol. Soc. Japan*, 48: 47–60.

Nitta, T., 1970b. On the role of transient eddies in the tropical troposphere. *J. Meteorol. Soc. Japan*, 48: 348–359.

Noel, T. M., 1966. Estimates of spectral energy density applied to vertical profiles of the horizontal wind in the lower ionosphere. *J. Geophys. Res.*, 71: 5749–5752.

Noll, A. M., 1964. Short-time spectrum and "cepstrum" techniques for vocal-pitch detection. *J. Acoust. Soc. Am.*, 36: 296–302.

Noll, A. M., 1967. Cepstrum pitch determination. *J. Acoust. Soc. Am.*, 41: 293–309.

Noponen, I., 1966. Surface wave phase velocities in Finland. *Bull. Seismol. Soc. Am.*, 56: 1093–1104.

Noponen, I., 1969. Wave velocities in crust and upper 300 kilometers of mantle in western and central Japan. *Bull. Int. Inst. Seismol. Earthquake Eng.*, 6: 11–37.

Northrop, J., Blaik, M. and Tolstoy, I., 1960. Spectrum analysis of T-phases from the Agadir earthquake, February 29, 1960, 23h 40m 12s GCT, 30 °N, 9 °W (USCGS). *J. Geophys. Res.*, 65: 4223–4224.

Nowroozi, A. A., 1965. Eigenvibrations of the earth after the Alaskan earthquake. *J. Geophys. Res.*, 70: 5145–5156.

Nowroozi, A. A., 1966. Terrestrial spectroscopy following the Rat Island earthquake. *Bull. Seismol. Soc. Am.*, 56: 1269–1288.

Nowroozi, A. A., 1967. Table for Fisher's test of significance in harmonic analysis. *Geophys. J. R. Astron. Soc.*, 12: 517–520.

Nowroozi, A. A., 1968. Measurement of *Q*-values from the free oscillations of the earth. *J. Geophys. Res.*, 73: 1407–1415.

Nowroozi, A. A., 1970. Spectrum and prediction of tides off the coast of California. *Ann. Géophys.*, 26: 259–271.

Nowroozi, A. A., 1972a. Long-term measurements of pelagic tidal height off the coast of northern California. *J. Geophys. Res.*, 77: 434–443.

Nowroozi, A. A., 1972b. Characteristic periods of fundamental and overtone oscillations of the earth following a deep-focus earthquake. *Bull. Seismol. Soc. Am.*, 62: 247–274.

Nowroozi, A. A. and Alsop, L. E., 1968. Torsional free periods of the earth observed at six stations around the earth. *Suppl. Nuovo Cimento, Ser. I*, 6: 133–146.

Nowroozi, A. A., Sutton, G. and Auld, B., 1966. Oceanic tides recorded on the sea floor. *Ann. Géophys.*, 22: 512–517.

Nowroozi, A. A., Ewing, M., Nafe, J. E. and Fleigel, M., 1968. Deep ocean current and its correlation with the ocean tide off the coast of northern California. *J. Geophys. Res.*, 73: 1921–1932.

Nowroozi, A. A., Kuo, J. and Ewing, M., 1969. Solid earth and oceanic tides recorded on the ocean floor off the coast of northern California. *J. Geophys. Res.*, 74: 605–614.

Nyquist, H., 1928. Certain topics in telegraph transmission theory. *Trans. AIEE*, 47: 617–644.

O'Brien, P. N. S., 1965. Seismic observations 20 km from explosions in a lake. *Boll. Geofis. Teor. Appl.*, 7: 144–164.

O'Brien, P. N. S., 1967a. Analysis of a small number of seismic records along an E–W Alpine profile. *Boll. Geofis. Teor. Appl.*, 9: 22–65.

O'Brien, P. N. S., 1967b. Quantitative discussion on seismic amplitudes produced by explosions in Lake Superior. *J. Geophys. Res.*, 72: 2569–2575.

Odaka, T. and Usami, T., 1970. Theoretical seismograms and earthquake mechanism. Part III. Azimuthal variation of the initial phase of surface waves. *Bull. Earthquake Res. Inst.*, 48: 669–689.

Odegard, M. E. and Berg, J. W., Jr., 1965. Gravity interpretation using the Fourier integral. *Geophysics*, 30: 424–438.

Officer, C. B., 1958. *Sound Transmission*. McGraw-Hill, 284 pp.

Ogura, Y., 1957a. Spectrum modification due to the use of finite differences. *J. Meteorol.*, 14: 77–80.

Ogura, Y., 1957b. The influence of finite observation intervals on the measurement of turbulent diffusion parameters. *J. Meteorol.*, 14: 176–181.

Ogura, Y., 1959. Diffusion from a continuous source in relation to a finite observation interval. *Adv. Geophys.*, 6: 149–158.

Okada, H., Suzuki, S. and Asano, S., 1970. Anomalous underground structure in the Matsushiro earthquake swarm area as derived from a fan shooting technique. *Bull. Earthquake Res. Inst.*, 48: 811–833.

Okano, K. and Hirano, I., 1971. Seismic wave attenuation in the vicinity of Kyoto. *Bull. Disaster Prev. Res. Inst. (Kyoto)*, 21(1): 99–108.

Oldham, C. H. G., 1967. The (sin x)/x· (sin y)/y method for continuation of potential fields. *Mining Geophys., Soc. Explor. Geophys.*, 2: 591–605.

Oliver, J. and Page, R., 1963. Concurrent storms of long and ultralong period microseisms. *Bull. Seismol. Soc. Am.*, 53: 15–26.

Olsen, K. and Hwang, L.-S., 1971. Oscillations in a bay of arbitrary shape and variable depth. *J. Geophys. Res.*, 76: 5048–5064.

Olsen, K. H., Stewart, J. N., McNeil, J. E. and Vitousek, M. J., 1972. Long-period water-wave measurements for the MILROW and CANNIKIN nuclear explosions. *Bull. Seismol. Soc. Am.*, 62: 1559–1578.

O'Neill, A. D. J. and Ferguson, H. L., 1971. A spectral investigation of horizontal moisture flux in the troposphere. *J. Appl. Meteorol.*, 10: 14–22.

Osemeikhian, J. E. A. and Everett, J. E., 1968. Anomalous magnetic variations in southwestern Scotland. *Geophys. J. R. Astron. Soc.*, 15: 361–366.

Otnes, R. K. and Enochson, L., 1972. *Digital Time Series Analysis*. Wiley, 467 pp.

Ōtsuka, M., 1963. Some considerations on the wave forms of ScS phases. *Geophys. Pap. Dedicated to Prof. K. Sassa (Kyoto)*. Publ. Office, Geophys. Inst., Kyoto, pp.415–425.

Ozmidov, R. V., 1964. On the large-scale characteristics of horizontal ocean current velocities. *Izv. Acad. Sci. USSR, Geophys. Ser.*, 11: 1032–1037 (Engl. ed.).

Page, C. H., 1952. Instantaneous power spectra. *J. Appl. Phys.*, 23: 103–106.

Pan, C., 1970. The gravitational factor from mid-continental body tides and its statistical analysis. *Tectonophysics*, 9: 15–46.

Pan, C., 1971. Fundamental statistical techniques for the analysis of earth tides. *Bull. Seismol. Soc. Am.*, 61: 203–215.

Panofsky, H. A., 1953. The variation of the turbulence spectrum with height under superadiabatic conditions. *Q. J. R. Meteorol. Soc.*, 79: 150–153.

Panofsky, H. A., 1955. Meteorological applications of power-spectrum analysis. *Bull. Am. Meteorol. Soc.*, 36: 163–166.

Panofsky, H. A., 1962. The budget of turbulent energy in the lowest 100 meters. *J. Geophys. Res.*, 67: 3161–3165.

Panofsky, H. A., 1967. Meteorological applications of cross-spectrum analysis. In: B. Harris (Editor), *Spectral Analysis of Time Series*. Wiley, pp.109–132.

Panofsky, H. A., 1969a. Spectra of atmospheric variables in the boundary layer. *Radio Sci.*, 4: 1101–1109.

Panofsky, H. A., 1969b. The spectrum of temperature. *Radio Sci.*, 4: 1143–1146.

Panofsky, H. A. and Brier, G. W., 1958. *Some Applications of Statistics to Meteorology*. Pa. State Univ., Coll. Mineral Industr., University Park, Pa., 224 pp.

Panofsky, H. A. and Deland, R. J., 1959. One-dimensional spectra of atmospheric turbulence in the lowest 100 metres. *Adv. Geophys.*, 6: 41–62.

Panofsky, H. A. and Mares, E., 1968. Recent measurements of cospectra for heat-flux and stress. *Q. J. R. Meteorol. Soc.*, 94: 581–585.

Panofsky, H. A. and McCormick, R. A., 1954. Properties of spectra of atmospheric turbulence at 100 meters. *Q. J. R. Meteorol. Soc.*, 80: 546–564.

Panofsky, H. A. and McCormick, R. A., 1960. The spectrum of vertical velocity near the surface. *Q. J. R. Meteorol. Soc.*, 86: 495–503.

Panofsky, H. A. and Singer, I. A., 1965. Vertical structure of turbulence. *Q. J. R. Meteorol. Soc.*, 91: 339–344.

Panofsky, H. A. and Van der Hoven, I., 1955. Spectra and cross-spectra of velocity components in the mesometeorological range. *Q. J. R. Meteorol. Soc.*, 81: 603–606.

Panofsky, H. A. and Wolff, P., 1957. Spectrum and cross-spectrum analysis of hemispheric westerly index. *Tellus*, 9: 195–200.

Panofsky, H. A., Cramer, H. E. and Rao, V. R. K., 1958. The relation between Eulerian time and space spectra. *Q. J. R. Meteorol. Soc.*, 84: 270–273.

Panter, P. F., 1965. *Modulation, Noise, and Spectral Analysis*. McGraw-Hill, 759 pp.

Pao, Y.-H., 1969. Spectra of internal waves and turbulence in stratified fluids. *Radio Sci.*, 4: 1315–1320.

Papoulis, A., 1962. *The Fourier Integral and Its Applications*. McGraw-Hill, 318 pp.

Parkinson, W. D., 1971. An analysis of the geomagnetic diurnal variation during the International Geophysical Year. *Gerlands Beitr. Geophys.*, 80: 199–232.

Parks, J. K., 1960. A comparison of power spectra of ocean waves obtained by an analog and a digital method. *J. Geophys. Res.*, 65: 1557–1563.

Pasquill, F., 1962a. *Atmospheric Diffusion*. Van Nostrand, 297 pp.

Pasquill, F., 1962b. Recent broad-band spectral measurements of turbulence in the lower atmosphere. *J. Geophys. Res.*, 67: 3025–3028.

Pasquill, F., 1972. Some aspects of boundary layer description. *Q. J. R. Meteorol. Soc.*, 98: 469–494.

Passechnik, I. P., 1968. On determining frequency dependence of absorption coefficient of longitudinal seismic waves propagating in the earth's mantle. In: E. Bisztricsány (Editor), *Proc. Eighth Assem. Eur. Seismol. Comm.* Akad. Kiadó, Budapest, pp.183–188.

Payne, F. R. and Lumley, J. L., 1966. One-dimensional spectra derived from an airborne hot-wire anemometer. *Q. J. R. Meteorol. Soc.*, 92: 397–401.

Payo, G., 1969. Atenuación de ondas sísmicas. *Rev. Geofís.*, 28: 277–315.

Payo, G., 1970. Structure of the crust and upper mantle in the Iberian shield by means of a long period triangular array. *Geophys. J. R. Astron. Soc.*, 20: 493–508.

Pease, C. B., 1967. A note on the spectrum analysis of transients and the loudness of sonic bangs. *J. Sound Vibr.*, 6: 310–314.

Pedlosky, J., 1962. Spectral considerations in two-dimensional incompressible flow. *Tellus*, 14: 125–132.

Pekeris, C. L., Alterman, Z. and Jarosch, H., 1961a. Comparison of theoretical with observed values of the periods of free oscillation of the earth. *Proc. Natl. Acad. Sci.*, 47: 91–98.

Pekeris, C. L., Alterman, Z. and Jarosch, H., 1961b. Rotational multiplets in the spectrum of the earth. *Phys. Rev.*, 122: 1692–1700.

Penzien, J., 1965. Applications of random vibration theory in earthquake engineering. *Bull. Int. Inst. Seismol. Earthquake Eng.*, 2: 47–69.

Petersen, D. P. and Middleton, D., 1963. On representative observations. *Tellus*, 15: 387–405.

Pfeffer, R. L. and Zarichny, J., 1963. Acoustic-gravity wave propagation in an atmosphere with two sound channels. *Geofis. Pura Appl.*, 55: 175–199.

Phillips, O. M. and Katz, E. J., 1961. The low-frequency components of the spectrum of wind-generated waves. *J. Mar. Res.*, 19: 57–69.

Phinney, R. A., 1964. Structure of the earth's crust from spectral behavior of long-period body waves. *J. Geophys. Res.*, 69: 2997–3017.

Phinney, R. A. and Alexander, S. S., 1966. P-wave diffraction theory and the structure of the core–mantle boundary. *J. Geophys. Res.*, 71: 5959–5975.

Phinney, R. A. and Cathles, L. M., 1969. Diffraction of P by the core: a study of long-period amplitudes near the edge of the shadow. *J. Geophys. Res.*, 74: 1556–1574.

Phinney, R. A. and Smith, S. W., 1963. Processing of seismic data from an automatic digital recorder. *Bull. Seismol. Soc. Am.*, 53: 549–562.

Pho, H. T., 1971. Etude comparative des spectres d'énergie de la composante verticale de l'onde P observée en deux stations voisines. *Ann. Géophys.*, 27: 303–310.

Pierson, W. J., Jr., 1955. Wind generated gravity waves. *Adv. Geophys.*, 2: 93–178.

Pierson, W. J., Jr., 1959. A note on the growth of the spectrum of wind-generated gravity waves as determined by non-linear considerations. *J. Geophys. Res.*, 64: 1007–1011.

Pierson, W. J., Jr., 1959–1960. On the use of time series concepts and spectral and cross-spectral analyses in the study of long-range forecasting problems. *J. Mar. Res.*, 18: 112–132.

Pierson, W. J., Jr., 1964. The interpretation of wave spectrums in terms of the wind profile instead of the wind measured at a constant height. *J. Geophys. Res.*, 69: 5191–5203.

Pierson, W. J. and Marks, W., 1952. The power spectrum analysis of ocean-wave records. *Trans. Am. Geophys. Union.*, 33: 834–844.

Pierson, W. J., Jr. and Moskowitz, L., 1964. A proposed spectral form for fully developed wind seas based on the similarity theory of S. A. Kitaigorodskii. *J. Geophys. Res.*, 69: 5181–5190.

Pilant, W. L. and Knopoff, L., 1964. Observations of multiple seismic events. *Bull. Seismol. Soc. Am.*, 54: 19–39.

Pil'nik, G. P., 1970. Astronomical observations of earth tides. *Izv. Acad. Sci. USSR, Phys. Solid Earth*, 3: 145–151 (Engl. ed.).

Pinus, N. Z., 1963. Statistical characteristics of the horizontal component of the wind velocity at heights of 6–12 km. *Izv. Acad. Sci. USSR, Geophys. Ser.*, 1: 105–107 (Engl. ed.).

Pinus, N. Z. and Šur, G. N., 1970. Einige Resultate experimenteller Untersuchungen der Turbulenz in der Troposphäre und Stratosphäre. *Gerlands Beitr. Geophys.*, 79: 363–378.

Pinus, N. Z., Reiter, E. R., Shur, G. N. and Vinnichenko, N. K., 1967. Power spectra of turbulence in the free atmosphere. *Tellus*, 19: 206–213.

Pisarenko, V. F., 1972. On the estimation of spectra by means of non-linear functions of the covariance matrix. *Geophys. J. R. Astron. Soc.*, 28: 511–531.

Platzman, G. W., 1964. An exact integral of complete spectral equations for unsteady one-dimensional flow. *Tellus*, 16: 422–431.

Platzman, G. W. and Rao, D. B., 1964. Spectra of Lake Erie water levels. *J. Geophys. Res.*, 69: 2525–2535.

Plutchok, R. and Broome, P., 1969. Modeling of seismic signals from large underwater explosions to predict the spectra and covariance functions. *Bull. Seismol. Soc. Am.*, 59: 1137–1147.

Pochapsky, T. E., 1966. Measurements of deep water movements with instrumented neutrally buoyant floats. *J. Geophys. Res.*, 71: 2491–2504.

Pollack, H. N., 1963a. Effect of delay time and number of delays on the spectra of ripple-fired shots. *Earthquake Notes*, 34(1): 1–12.

Pollack, H. N., 1963b. An experimental study of source motion synthesis from first arrivals. *Bull. Seismol. Soc. Am.*, 53: 955–963.

Pomeroy, P. W., 1963. Long-period seismic waves from large, near-surface nuclear explosions. *Bull. Seismol. Soc. Am.*, 53: 109–149.

Pomeroy, P. W. and Sutton, G. H., 1960. The use of galvanometers as band-rejection filters in electromagnetic seismographs. *Bull. Seismol. Soc. Am.*, 50: 135–151.

Popov, V. V. and Chernyavkina, M. K., 1960. Several results of observations of deformations of the earth's surface at the geophysical station "Yalta". *Izv. Acad. Sci. USSR, Geophys. Ser.*, 7: 671–676 (Engl. ed.).

Porath, H. and Gough, D. I., 1971. Mantle conductive structures in the western United States from magnetometer array studies. *Geophys. J. R. Astron. Soc.*, 22: 261–275.

Porcello, L. J., Heerema, C. E. and Massey, N. G., 1969. Optical processing of planetary radar data: preliminary results. *J. Geophys. Res.*, 74: 1111–1115.

Posmentier, E. S. and Herrmann, R. W., 1971. Cophase: an ad hoc array processor. *J. Geophys. Res.*, 76: 2194–2201.

Power, D. V., 1969. Analysis of earth motions and seismic sources by power spectral density. *Bull. Seismol. Soc. Am.*, 59: 1071–1091.

Prentiss, D. D. and Ewing, J. I., 1963. The seismic motion of the deep ocean floor. *Bull. Seismol. Soc. Am.*, 53: 765–781.

Press, F., 1956a. Rigidity of the earth's core. *Science*, 124: 1204.

Press, F., 1956b. Determination of crustal structure from phase velocity of Rayleigh waves. Part I: Southern California. *Bull. Geol. Soc. Am.*, 67: 1647–1658.

Press, F., 1964. Long-period waves and free oscillations of the earth. In: H. Odishaw (Editor), *Research in Geophysics*, Vol. 2. M.I.T. Press, pp.1–26.

Press, F. and Ewing, M., 1948. A theory of microseisms with geologic applications. *Trans. Am. Geophys. Union*, 29: 163–174.

Press, F., Ben-Menahem, A. and Toksöz, M. N., 1961. Experimental determination of earthquake fault length and rupture velocity. *J. Geophys. Res.*, 66: 3471–3485.

Press, F., Dewart, G. and Gilman, R., 1963. A study of diagnostic techniques for identifying earthquakes. *J. Geophys. Res.*, 68: 2909–2928.

Presti, A. J., 1966. High-speed sound spectrograph. *J. Acoust. Soc. Am.*, 40: 628–634.

Priestley, M. B., 1965a. The role of bandwidth in spectral analysis. *Appl. Stat., R. Stat. Soc.*, C 14: 33–47.

Priestley, M. B., 1965b. Evolutionary spectra and non-stationary processes. *J. R. Stat. Soc.*, B 27: 204–229.

Priestley, M. B., 1967. Power spectral analysis of non-stationary random processes. *J. Sound Vibr.*, 6: 86–97.

Priestley, M. B., 1971. Time-dependent spectral analysis and its application in prediction and control. *J. Sound Vibr.*, 17: 517–534.

Prikner, K., 1969a. Rapid changes in the Fourier spectra of Pi1 pulsations during magnetically disturbed periods. Part I. Frequency analysis. *Studia Geophys. Geodaet.*, 13: 276–292.

Prikner, K., 1969b. Rapid changes in the Fourier spectra of Pi1 pulsations during magnetically disturbed periods. Part II. Amplitude analysis. *Studia Geophys. Geodaet.*, 13: 444–456.

Prikner, K., Střeštík, J. and Dobeš, K., 1972. Frequency analysis of geomagnetic beating-type pulsations in the Pc3 range. *Studia Geophys. Geodaet.*, 16: 262–270.

Prothero, W. A. and Goodkind, J. M., 1972. Earth-tide measurements with the superconducting gravimeter. *J. Geophys. Res.*, 77: 926–937.

Proverbio, E. and Quesada, V., 1972. Long-term components in polar motion. *Ann. Geofis.*, 25: 37–54.

Raitt, R. W., 1969. Anisotropy of the upper mantle. *Am. Geophys. Union, Geophys. Monogr.*, 13: 250–256.

Ralston, A. and Wilf, H. S. (Editors), 1960. *Mathematical Methods for Digital Computers*. Wiley, 293 pp.

Randall, M. J. and Knopoff, L., 1970. The mechanism at the focus of deep earthquakes. *J. Geophys. Res.*, 75: 4965–4976.

Rao, V. B. and Rao, S. T., 1971. A theoretical and synoptic study of western disturbances. *Pure Appl. Geophys.*, 90: 193–208.

Rassbach, M. E., Dessler, A. J. and Cameron, A. G. W., 1966. The lunar period, the solar period, and K_p. *J. Geophys. Res.*, 71: 4141–4146.

Ratnikova, L. I. and Levshin, A. L., 1967. Calculation of the spectral characteristics of thin-layered media. *Izv. Acad. Sci. USSR, Phys. Solid Earth*, 2: 96–102 (Engl. ed.).

Reed, J. W., 1971. Low-frequency periodicities in Panama rainfall runoff. *J. Appl. Meteorol.*, 10: 666–673.

Reed, R. J., Wolfe, J. L. and Nishimoto, H., 1963. A spectral analysis of the energetics of the stratospheric sudden warming of early 1957. *J. Atmos. Sci.*, 20: 256–275.

Reiter, E. R. and Burns, A., 1966. The structure of clear-air turbulence derived from "Topcat" aircraft measurements. *J. Atmos. Sci.*, 23: 206–212.

Reiter, E. R. and Foltz, H. P., 1967. The prediction of clear-air turbulence over mountainous terrain. *J. Appl. Meteorol.*, 6: 549–556.

Reitzel, J. S., Gough, D. I., Porath, H. and Anderson, C. W., III, 1970. Geomagnetic deep sounding and upper-mantle structure in the western United States. *Geophys. J. R. Astron. Soc.*, 19: 213–235.

Revah, I., 1969. Etude des vents de petite échelle observés au moyen des trainées météoriques. *Ann. Géophys.*, 25: 1–45.

Richards, A. F., 1963. Volcanic sounds: investigation and analysis. *J. Geophys. Res.*, 68: 919–928.

Richter, C. F., 1958. *Elementary Seismology*. Freeman, 768 pp.

Ricker, N., 1940. The form and nature of seismic waves and the structure of seismograms. *Geophysics*, 5: 348–366.

Ricker, N., 1943. Further developments in the wavelet theory of seismogram structure. *Bull. Seismol. Soc. Am.*, 33: 197–228.

Ricker, N., 1944. Wavelet functions and their polynomials. *Geophysics*, 9: 314–323.

Ricker, N., 1953. The form and laws of propagation of seismic wavelets. *Geophysics*, 18: 10–36.

Rinner, K., 1960. Einfluss des Ganges auf die Ergebnisse der harmonischen Analyse. *Boll. Geofis. Teor. Appl.*, 2(5): 223–234.

Riznichenko, Yu. V., 1970. The seismic risk problem viewed under a new angle. *Izv. Acad. Sci. USSR, Phys. Solid Earth*, 4: 227–236 (Engl. ed.).

Robertson, H., 1965. Physical and topographic factors as related to short-period wind noise. *Bull. Seismol. Soc. Am.*, 55: 863–877.

Robinson, E. A., 1966. Collection of Fortran II programs for filtering and spectral analysis of single channel time series. *Geophys. Prospect.*, 14 (Suppl. 1): 1–52.

Robinson, E. A., 1967a. *Statistical Communication and Detection with special reference to Digital Data Processing of Radar and Seismic Signals*. Griffin, 362 pp.

Robinson, E. A., 1967b. *Multichannel Time Series Analysis with Digital Computer Programs*. Holden-Day, 298 pp.

Robinson, E. A. and Treitel, S., 1967. Principles of digital Wiener filtering. *Geophys. Prospect.*, 15: 311–333.

Robinson, G. D., 1959. Vertical motion and the transfer of heat and momentum near the ground. *Adv. Geophys.*, 6: 259–267.

Roden, G. I., 1960. On the nonseasonal variations in sea level along the west coast of North America. *J. Geophys. Res.*, 65: 2809–2826.

Roden, G. I., 1962. On sea-surface temperature, cloudiness and wind variations in the tropical Atlantic. *J. Atmos. Sci.*, 19: 66–80.

Roden, G. I., 1963a. On statistical estimation of monthly extreme sea-surface temperatures along the west coast of the United States. *J. Mar. Res.*, 21: 172–190.

Roden, G. I., 1963b. On sea level, temperature, and salinity variations in the central, tropical Pacific and on Pacific Ocean islands. *J. Geophys. Res.*, 68: 455–472.

Roden, G. I., 1963c. Sea level variations at Panama. *J. Geophys. Res.*, 68: 5701–5710.

Roden, G. I., 1964. Shallow temperature inversions in the Pacific Ocean. *J. Geophys. Res.*, 69: 2899–2914.

Roden, G. I., 1966a. A modern statistical analysis and documentation of historical temperature records in California, Oregon and Washington, 1821–1964. *J. Appl. Meteorol.*, 5: 3–24.

Roden, G. I., 1966b. Low-frequency sea level oscillations along the Pacific coast of North America. *J. Geophys. Res.*, 71: 4755–4776.

Roden, G. I., 1968. Spectral analysis and interpretation of salinity–temperature–depth records. *J. Geophys. Res.*, 73: 535–539.

Roden, G. I. and Groves, G. W., 1960. On the statistical prediction of ocean temperatures. *J. Geophys. Res.*, 65: 249–263.

Rodriguez-Portugal, C. and Udías, A., 1972. Estudio del mecanismo y determinacion de los parametros dinamicos del foco del terremoto de Azores del 17 de mayo de 1964. *Rev. Geofís.*, 31: 63–86.

Rogers, A. M., Jr. and Kisslinger, C., 1972. The effect of a dipping layer on P-wave transmission. *Bull. Seismol. Soc. Am.*, 62: 301–324.

Rogers, R. R., 1963. Radar measurement of velocities of meteorological scatterers. *J. Atmos. Sci.*, 20: 170–174.

Rogers, R. R. and Tripp, B. R., 1964. Some radar measurements of turbulence in snow. *J. Appl. Meteorol.*, 3: 603–610.

Roper, R. G., 1966. The semidiurnal tide in the lower thermosphere. *J. Geophys. Res.*, 71: 5746–5748.

Rosenblatt, M. and Van Ness, J. W., 1965. Estimation of the bispectrum. *Ann. Math. Stat.*, 36: 1120–1136.

Rosenthal, S. L., 1960. Some estimates of the power spectra of large-scale disturbances in low latitudes. *J. Meteorol.*, 17: 259–263.

Rossiter, J. R., 1962. Long-term variations in sea-level. In: M. N. Hill (Editor), *The Sea*, Vol. 1. Interscience, pp.590–610.

Rossiter, J. R., 1967. An analysis of annual sea level variations in European waters. *Geophys. J. R. Astron. Soc.*, 12: 259–299.

Rossiter, J. R. and Lennon, G. W., 1968. An intensive analysis of shallow water tides. *Geophys. J. R. Astron. Soc.*, 16: 275–293.

Roth, R., 1971. Turbulence spectra with two separated regions of production. *J. Appl. Meteorol.*, 10: 430–432.

Roy, A., 1970. Gravity and magnetic interpretation on uneven topography by (sin X)/X method of continuation. *Geoexploration*, 8: 37–40.

Royer, T. C. and Reid, R. O., 1971. The detection of secondary tsunamis. *Tellus*, 23: 136–142.

Rudnick, P., 1956. The spectrum of the variation in latitude. *Trans. Am. Geophys. Union*, 37: 137–142.

Rudnick, P., 1969. Wave directions from a large spar buoy. *J. Mar. Res.*, 27: 7–21.

Rushton, S. and Neumann, J., 1957. Some applications of time series analysis to atmospheric turbulence and oceanography. *J. R. Stat. Soc.*, A 120: 409–425.

Ryall, A., VanWormer, J. D. and Jones, A. E., 1968. Triggering of micro-earthquakes by earth tides, and other features of the Truckee, California, earthquake sequence of September, 1966. *Bull. Seismol. Soc. Am.*, 58: 215–248.

Rygg, E., 1971. The dispersive effect of a medium expressed as a convolutional filter. *Seismol. Obs., Bergen, Sci. Rep.*, 7: 1–28.

Rykunov, L. N., 1961. A correlation method of determining the velocities of microseisms. *Izv. Acad. Sci. USSR, Geophys. Ser.*, 7: 686–687 (Engl. ed.).

Rykunov, L. N. and Sedov, V. V., 1965. Seismic noise in the 2- to 15-cps frequency range on the bottom of the Black Sea. *Izv. Acad. Sci. USSR, Phys. Solid Earth*, 7: 443–448 (Engl. ed.).

Sacks, I. S., 1972. The *Q* at the base of the mantle. *Carnegie Inst. Year Book*, 71: 325–327.

Sacks, S., 1966. Diffracted wave studies of the earth's core. 1. Amplitudes, core size, and rigidity. *J. Geophys. Res.*, 71: 1173–1181.

Saito, T., 1960. Period analysis of geomagnetic pulsations by a sona-graph method. *Sci. Rep. Tôhoku Univ., Geophys.*, 12: 105–113.

Saltzman, B., 1957. Equations governing the energetics of the larger scales of atmospheric turbulence in the domain of wave number. *J. Meteorol.*, 14: 513–523.

Saltzman, B., 1958. Some hemispheric spectral statistics. *J. Meteorol.*, 15: 259–263.

Saltzman, B. and Fleisher, A., 1960. The modes of release of available potential energy in the atmosphere. *J. Geophys. Res.*, 65: 1215–1222.

Saltzman, B. and Fleisher, A., 1962. Spectral statistics of the wind at 500 mbar. *J. Atmos. Sci.*, 19: 195–204.

Saltzman, B. and Peixoto, J. P., 1957. Harmonic analysis of the mean northern-hemisphere wind field for the year 1950. *Q. J. R. Meteorol. Soc.*, 83: 360–364.

Saltzman, B. and Teweles, S., 1964. Further statistics on the exchange of kinetic energy between harmonic components of the atmospheric flow. *Tellus*, 16: 432–435.

Sanford, A. R., Carapetian, A. G. and Long, L. T., 1968. High-frequency microseisms from a known source. *Bull. Seismol. Soc. Am.*, 58: 325–338.

Santirocco, R. A. and Parker, D. G., 1963. The polarization and power spectrums of Pc micropulsations in Bermuda. *J. Geophys. Res.*, 68: 5545–5558.

Sasai, Y., 1966. The anomalous behaviour of geomagnetic variations of short period in Japan and its relation to the subterranean structure. The 11th report. (Spectral analysis of geomagnetic disturbances). *Bull. Earthquake Res. Inst.*, 44: 167–178.

Sasai, Y., 1967. Spatial dependence of short-period geomagnetic fluctuations on Oshima island (1). *Bull. Earthquake Res. Inst.*, 45: 137–157.

Sato, R., 1967. Attenuation of seismic waves. *J. Phys. Earth*, 15: 32–61.

Sato, R. and Espinosa, A. F., 1967. Dissipation in the earth's mantle and rigidity and viscosity in the earth's core determined from waves multiply reflected from the mantle–core boundary. *Bull. Seismol. Soc. Am.*, 57: 829–856.

Satô, Y., 1955. Analysis of dispersed surface waves by means of Fourier transform (I). *Bull. Earthquake Res. Inst.*, 33: 33–48.

Satô, Y., 1956a. Analysis of dispersed surface waves by means of Fourier transform (II). *Bull. Earthquake Res. Inst.*, 34: 9–18.

Satô, Y., 1956b. Analysis of dispersed surface waves by means of Fourier transform (III). *Bull. Earthquake Res. Inst.*, 34: 131–138.

Satô, Y., 1958. Attenuation, dispersion, and the wave guide of the G wave. *Bull. Seismol. Soc. Am.*, 48: 231–251.

Satô, Y., 1960. Synthesis of dispersed surface waves by means of Fourier transform. *Bull. Seismol. Soc. Am.*, 50: 417–426.

Satô, Y., 1964. Soft core spectrum splitting of the torsional oscillation of an elastic sphere and related problems. *Bull. Earthquake Res. Inst.*, 42: 1–10.

Satô, Y., Usami, T. and Landisman, M., 1963a. Spectrum, phase and group velocities of the theoretical seismograms and the idea of the equivalent surface source of disturbance. *Geophys. J. R. Astron. Soc.*, 8: 1–11.

Satô, Y., Takeuchi, H., Nishimura, E. and Nakagawa, I., 1963b. Free oscillation of the earth observed by gravimeters installed in Kyoto, Japan. *Bull. Earthquake Res. Inst.*, 41: 699–703.

Savage, J. C., 1965. The effect of rupture velocity upon seismic first motions. *Bull. Seismol. Soc. Am.*, 55: 263–275.

Savage, J. C., 1967. Spectra of S-waves radiated from bilateral fracture. *Bull. Seismol. Soc. Am.*, 57: 39–54.

Savage, J. C. and Hasegawa, H. S., 1965. A two-dimensional model study of the directivity function. *Bull. Seismol. Soc. Am.*, 55: 27–45.

Savarenskii, E. F., Fedorov, S. A. and Gogichaishvili, B. V., 1963. The determination of the actual motion of the ground and its spectrum from a seismogram. *Izv. Acad. Sci. USSR, Geophys. Ser.*, 9: 818–822 (Engl. ed.).

Savarensky, E. F., Proskurjakova, T. A. and Voronina, E. V., 1967. On microseism phase velocities and the directions to the excitation source. In: H. Jensen (Editor), *European Seismological Commission, Copenhagen Assembly 1966*, pp.347–356.

Savino, J., Sykes, L. R., Liebermann, R. C. and Molnar, P., 1971. Excitation of seismic surface waves with periods of 15 to 70 seconds for earthquakes and underground explosions. *J. Geophys. Res.*, 76: 8003–8020.

Savino, J., McCamy, K. and Hade, G., 1972a. Structures in earth noise beyond twenty seconds—a window for earthquakes. *Bull. Seismol. Soc. Am.*, 62: 141–176.

Savino, J. M., Murphy, A. J., Rynn, J. M. W., Tatham, R., Sykes, L. R., Choy, G. L. and McCamy, K., 1972b. Results from the high-gain long-period seismograph experiment. *Geophys. J. R. Astron. Soc.*, 31: 179–203.

Savit, C. H., Brustad, J. T. and Sider, J., 1958. The moveout filter. *Geophysics*, 23: 1–25.

Saxton, J. A. (Editor), 1969. Spectra of meteorological variables. *Radio Sci.*, 4: 1099–1387.

Schaub, Yu. B., 1961. The separation of geophysical anomalies from a background of intense interference. *Izv. Acad. Sci. USSR, Geophys. Ser.*, 6: 589–592 (Engl. ed.).

Scheidegger, A. E., 1960. General spectral theory for the onset of instabilities in displacement processes in porous media. *Geofis. Pura Appl.*, 47: 41–54.

Schenk, V., 1971. Attenuation coefficients of the maximum amplitude and the spectral amplitude of stress waves in non-elastic zones of explosive sources. *Pure Appl. Geophys.*, 90: 61–69.

Schevill, W. E., Backus, R. H. and Hersey, J. B., 1962. Sound production by marine animals. In: M. N. Hill (Editor), *The Sea*, Vol. 1. Interscience, pp.540–566.

Schick, R., 1968. Untersuchungen über die Bruchausdehnung und Bruchgeschwindigkeit bei Erdbeben mit kleinen Magnituden ($M < 4$). *Z. Geophys.*, 34: 267–286.

Schick, R., 1970. A method for determining source parameters of small-magnitude earthquakes. *Z. Geophys.*, 36: 205–224.

Schiff, A. and Bogdanoff, J. L., 1967. Analysis of current methods of interpreting strong-motion accelerograms. *Bull. Seismol. Soc. Am.*, 57: 857–874.

Schnabel, P., Seed, H. B. and Lysmer, J., 1972. Modification of seismograph records for effects of local soil conditions. *Bull. Seismol. Soc. Am.*, 62: 1649–1664.

Schneider, G., 1961. Mikroseismik-Ausbreitung in Nord- und Mitteleuropa. *Z. Geophys.*, 27: 118–135.

Schneider, W. A. and Backus, M. M., 1964. Ocean-bottom seismic measurements off the California coast. *J. Geophys. Res.*, 69: 1135–1143.

Schneider, W. A., Farrell, P. J. and Brannian, R. E., 1964. Collection and analysis of Pacific ocean-bottom seismic data. *Geophysics*, 29: 745–771.

Schott, F., 1971. On horizontal coherence and internal wave propagation in the North Sea. *Deep-Sea Res.*, 18: 291–307.

Schroeder, M. R., 1968. Period histogram and product spectrum: New methods for fundamental-frequency measurement. *J. Acoust. Soc. Am.*, 43: 829–834.

Schroeder, M. R. and Atal, B. S., 1962. Generalized short-time power spectra and autocorrelation functions. *J. Acoust. Soc. Am.*, 34: 1679–1683.

Schule, J. J., Jr., Simpson, L. S. and DeLeonibus, P. S., 1971. A study of fetch-limited wave spectra with an airborne laser. *J. Geophys. Res.*, 76: 4160–4171.

Scollar, I., 1970. Fourier transform methods for the evaluation of magnetic maps. *Prospezioni Archeol.*, 5: 9–41.

Scott, J. R., 1969. Some average wave lengths on short-crested seas. *Q. J. R. Meteorol. Soc.*, 95: 621–634.

Seidl, D., Müller, St. and Knopoff, L., 1966. Dispersion von Rayleigh-Wellen in Südwestdeutschland und in den Alpen. *Z. Geophys.*, 32: 472–481.

Seiwell, H. R., 1949. The principles of time series analysis applied to ocean wave data. *Proc. Natl. Acad. Sci.*, 35: 518–528.

Sen, A. K., 1968. A theory of geomagnetic micropulsations II. *J. Geomagn. Geoelectr.*, 20: 245–261.

Sengupta, S. and Ganguli, D. K., 1971. On attenuation and spectral analysis of Rayleigh waves in alluvium and lateritic soil near small blasts. *Pure Appl. Geophys.*, 90: 70–77.

Seyduzova, S. S., 1970. Frequency analysis and problems in investigations of near earthquake recordings. *Izv. Acad. Sci. USSR, Phys. Solid Earth*, 6: 347–352 (Engl. ed.).

Shapiro, H. S. and Silverman, R. A., 1960. Alias-free sampling of random noise. *J. Soc. Ind. Appl. Math.*, 8: 225–248.

Shapiro, R. and Ward, F., 1960. The time-space spectrum of the geostrophic meridional kinetic energy. *J. Meteorol.*, 17: 621–626.

Shapiro, R. and Ward, F., 1963. The kinetic energy spectrum of meridional flow in the mid-troposphere. *J. Atmos. Sci.*, 20: 353–358.

Shapiro, R. and Ward, F., 1966. Three peaks near 27 days in a high-resolution spectrum of the international magnetic character figure, C_i. *J. Geophys. Res.*, 71: 2385–2388.

Sharma, B. and Geldart, L. P., 1968. Analysis of gravity anomalies of two-dimensional faults using Fourier transforms. *Geophys. Prospect.*, 16: 77–93.

Sharma, B., Geldart, L. P. and Gill, D. E., 1970. Interpretation of gravity anomalies of dike-like bodies by Fourier transformation. *Can. J. Earth Sci.*, 7: 512–516.

Shaub, Yu. B., 1963. The use of correlation analysis for the evaluation of geophysical data. *Izv. Acad. Sci. USSR, Geophys. Ser.*, 4: 358–364 (Engl. ed.).

Shaw, D. M. and Donn, W. L., 1964. Sea-level variations at Iceland and Bermuda. *J. Mar. Res.*, 22: 111–122.

Shaw, L., Paul, I. and Henrikson, P., 1969. Statistical models for the vertical deflection from gravity-anomaly models. *J. Geophys. Res.*, 74: 4259–4265.

Sherwood, J. W. C. and Spencer, T. W., 1962. Signal-to-noise ratio and spectra of explosion-generated Rayleigh waves. *Bull. Seismol. Soc. Am.*, 52: 573–594.

Shima, E., 1962. Modifications of seismic waves in superficial soil layers as verified by comparative observations on and beneath the surface. *Bull. Earthquake Res. Inst.*, 40: 187–259.

Shima, E., 1969. Vibration characteristics of subsoil layers in downtown Tokyo during the earthquakes. *Bull. Earthquake Res. Inst.*, 47: 145–163.

Shima, E., McCamy, K. and Meyer, R. P., 1964. A Fourier transform method of apparent velocity measurement. *Bull. Seismol. Soc. Am.*, 54: 1843–1854.

Shimamura, H., 1969. Model study on core–mantle boundary structure. *J. Phys. Earth*, 17: 133–168.

Shimozuru, D., Kamo, K. and Kinoshita, W. T., 1966. Volcanic tremor of Kilauea volcano, Hawaii, during July–December, 1963. *Bull. Earthquake Res. Inst.*, 44: 1093–1133.

Shimozuru, D., Miyazaki, T., Gyoda, N. and Matahelumual, J., 1969. Volcanological survey of Indonesian volcanoes. Part 2. Seismic observation at Merapi volcano. *Bull. Earthquake Res. Inst.*, 47: 969–990.

Shimshoni, M., 1967. The determination of periods in seismic records. *Bull. Seismol. Soc. Am.*, 57: 1347–1354.

Shimshoni, M., 1968a. Improved accuracy in the determination of periods in seismic records. *Suppl. Nuovo Cimento, Ser. I*, 6: 160–165.

Shimshoni, M., 1968b. The separation to two close hidden periodicities. *J. Phys. Earth*, 16 (Spec. Issue): 195–202.

Shimshoni, M., 1971. On Fisher's test of significance in harmonic analysis. *Geophys. J. R. Astron. Soc.*, 23: 373–377.

Shimshoni, M. and Ben-Menahem, A., 1970. Computation of the divergence coefficient for seismic phases. *Geophys. J. R. Astron. Soc.*, 21: 285–294.

Shlien, S. and Toksöz, M. N., 1970. A clustering model for earthquake occurrences. *Bull. Seismol. Soc. Am.*, 60: 1765–1787.

Shonting, D. H., 1968. Autospectra of observed particle motions in wind waves. *J. Mar. Res.*, 26: 43–65.

Shopland, R. C. and Kirklin, R. H., 1969. Application of strain seismographs to the discrimination of seismic waves. *Bull. Seismol. Soc. Am.*, 59: 673–689.

Shopland, R. C. and Kirklin, R. H., 1970. Application of a vertical strain seismograph to the enhancement of P-waves. *Bull. Seismol. Soc. Am.*, 60: 105–124.

Shugart, T. R., 1944. Frequency discrimination in the low-velocity zone. *Geophysics*, 9: 19–28.

Siebert, M., 1961. Atmospheric tides. *Adv. Geophys.*, 7: 105–187.

Siedler, G., 1971. Vertical coherence of short-periodic current variations. *Deep-Sea Res.*, 18: 179–191.

Silverman, B. A., 1968. The effect of spatial averaging on spectrum estimation. *J. Appl. Meteorol.*, 7: 168–172.

Silverman, D., 1967. The digital processing of seismic data. *Geophysics*, 32: 988–1002.

Simonenko, T. N. and Roze, E. N., 1967. On the structure of the magnetic field of the earth. *Izv. Acad. Sci. USSR, Phys. Solid Earth*, 8: 525–530 (Engl. ed.).

Simons, R. S., 1968. A surface wave particle motion discrimination process. *Bull. Seismol. Soc. Am.*, 58: 629–637.

Simpson, J. H., 1969. Observations of the directional characteristics of sea waves. *Geophys. J. R. Astron. Soc.*, 17: 93–120.

Singh, S. J., Ben-Menahem, A. and Shimshoni, M., 1972. Theoretical amplitudes of body waves from a dislocation source in the earth, 1. Core reflections. *Phys. Earth Planet. Inter.*, 5: 231–263.

Siskind, D. E. and Howell, B. F., Jr., 1967. Scale-model study of refraction arrivals in a three-layered structure. *Bull. Seismol. Soc. Am.*, 57: 437–442.

Sitaraman, V., 1970. Spectra and cospectra of turbulence in the atmospheric surface layer. *Q. J. R. Meteorol. Soc.*, 96: 744–749.

Skorik, L. A., 1969. Spectra of induced seismic accelerations in perceptible earthquakes. *Izv. Acad. Sci. USSR, Phys. Solid Earth*, 4: 249–253 (Engl. ed.).

Slaucitajs, L., 1964. Variación secular geomagnética en Sudamérica y alrededores con relación a todo el Hemisferio Sur. *Pure Appl. Geophys.*, 59: 75–83.

Slichter, L. B., 1965. Earth's free modes and a new gravimeter. *Geophysics*, 30: 339–347.

Smart, E., 1971. Erroneous phase velocities from frequency–wavenumber spectral sections. *Geophys. J. R. Astron. Soc.*, 26: 247–253.

Smart, E. and Flinn, E. A., 1971. Fast frequency–wavenumber analysis and Fisher signal detection in real-time infrasonic array data processing. *Geophys. J. R. Astron. Soc.*, 26: 279–284.

Smith, F. B., 1961. An analysis of vertical wind-fluctuations at heights between 500 and 5,000 ft. *Q. J. R. Meteorol. Soc.*, 87: 180–193.

Smith, F. B., 1962. The effect of sampling and averaging on the spectrum of turbulence. *Q. J. R. Meteorol. Soc.*, 88: 177–180.

Smith, H. W., Provazek, L. D. and Bostick, F. X., Jr., 1961. Directional properties and phase relations of the magnetotelluric fields at Austin, Texas. *J. Geophys. Res.*, 66: 879–888.

Smith, M. K., 1958. A review of methods of filtering seismic data. *Geophysics*, 23: 44–57.

Smith, P. F., Richard, J. D. and Stephens, F. H., 1955. A technique for the spectral analysis of sound in the ocean. *Trans. Am. Geophys. Union*, 36: 413–418.

Smith, S. D., 1967. Thrust-anemometer measurements of wind-velocity spectra and of Reynolds stress over a coastal inlet. *J. Mar. Res.*, 25: 239–262.

Smith, S. D., 1970. Thrust-anemometer measurements of wind turbulence, Reynolds stress, and drag coefficient over the sea. *J. Geophys. Res.*, 75: 6758–6770.

Smith, S. D., 1972. Wind stress and turbulence over a flat ice floe. *J. Geophys. Res.*, 77: 3886–3901.

Smith, S. D., Banke, E. G. and Johannessen, O. M., 1970. Wind stress and turbulence over ice in the Gulf of St. Lawrence. *J. Geophys. Res.*, 75: 2803–2812.

Smith, S. W., 1963. Generation of seismic waves by underground explosions and the collapse of cavities. *J. Geophys. Res.*, 68: 1477–1483.

Smith, S. W., 1965. Seismic digital data acquisition systems. *Rev. Geophys.*, 3: 151–156.

Smith, S. W., 1966. Free oscillations excited by the Alaskan earthquake. *J. Geophys. Res.*, 71: 1183–1193.

Smith, S. W., 1972. The anelasticity of the mantle. *Tectonophysics*, 13: 601–622.

Smith, S. W. and Kind, R., 1972. Regional secular strain fields in southern Nevada. *Tectonophysics*, 14: 57–69.

Snodgrass, F. E., Munk, W. H. and Miller, G. R., 1962. Long-period waves over California's continental borderland. Part I. Background spectra. *J. Mar. Res.*, 20: 3–30.

Snodgrass, F. E., Groves, G. W., Hasselmann, K. F., Miller, G. R., Munk, W. H. and Powers, W. H., 1966. Propagation of ocean swell across the Pacific. *Philos. Trans. R. Soc. London, Ser. A*, 259: 431–497.

Snyder, R. L. and Cox, C. S., 1966. A field study of the wind generation of ocean waves. *J. Mar. Res.*, 24: 141–178.

Soga, N. and Anderson, O. L., 1967. Elastic properties of tektites measured by resonant sphere technique. *J. Geophys. Res.*, 72: 1733–1739.

Sokolowski, T. J. and Miller, G. R., 1967. Automated epicenter locations from a quadripartite array. *Bull. Seismol. Soc. Am.*, 57: 269–275.

Solomon, S. C. and Toksöz, M. N., 1970. Lateral variation of attenuation of P- and S-waves beneath the United States. *Bull. Seismol. Soc. Am.*, 60: 819–838.

Solov'ev, S. L. and Pustovitenko, A. N., 1964. On the possible reduction in period of longitudinal waves as the depth of the earthquake focus increases. *Izv. Acad. Sci. USSR, Geophys. Ser.*, 6: 508–512 (Engl. ed.).

Southworth, R. W., 1960. Autocorrelation and spectral analysis. In: A. Ralston and H. S. Wilf (Editors), *Mathematical Methods for Digital Computers*. Wiley, pp.213–220.

Spector, A. and Bhattacharyya, B. K., 1966. Energy density spectrum and autocorrelation function of anomalies due to simple magnetic models. *Geophys. Prospect.*, 14: 242–272.

Spector, A. and Grant, F. S., 1970. Statistical models for interpreting aeromagnetic data. *Geophysics*, 35: 293–302.

Spetner, L. M., 1954. Errors in power spectra due to finite sample. *J. Appl. Phys.*, 25: 653–659.

Spitznogle, F. R. and Quazi, A. H., 1970. Representation and analysis of time-limited signals using a complex exponential algorithm. *J. Acoust. Soc. Am.*, 47: 1150–1155.

Spizzichino, A., 1969. Etude expérimentale des vents dans la haute atmosphère. *Ann. Géophys.*, 25: 697–720.

Springer, D., Denny, M., Healy, J. and Mickey, W., 1968. The Sterling experiment: Decoupling of seismic waves by a shot-generated cavity. *J. Geophys. Res.*, 73: 5995–6011.

Srivastava, H. N., Drakopoulos, J. and Terashima, T., 1971. Spectra of seismic waves of Matsushiro micro-earthquakes. *Pure Appl. Geophys.*, 92: 26–35.

Srivastava, S. P. and White, A., 1971. Inland, coastal, and offshore magnetotelluric measurements in eastern Canada. *Can. J. Earth Sci.*, 8: 204–216.

Stabler, C. L., 1968. Simplified Fourier analysis of fold shapes. *Tectonophysics*, 6: 343–350.

Stacey, F. D. and Westcott, P., 1972. The record of a vector proton magnetometer after the earthquake. In: Comm. Alaska Earthquake, Div. Earth Sci., Natl. Res. Council (Editors), *The Great Alaska Earthquake of 1964. Seismology and Geodesy.* Natl. Acad. Sci., pp.523–525.

Starodubrovskaya, S. P., 1964. The physical prerequisites for using the dynamic characteristics of longitudinal reflected waves for tracing layers of variable thickness. *Izv. Acad. Sci. USSR, Geophys. Ser.*, 12: 1049–1058 (Engl. ed.).

Starovoit, O. E., 1971. Spectra of long-period seismic surface waves. *Obs. R. Belg., Comm. A13, Sér. Géophys.*, 101: 71–79.

Sterling, A. and Smets, E., 1971. Study of earth tides, earthquakes and terrestrial spectroscopy by analysis of the level fluctuations in a borehole at Heibaart (Belgium). *Geophys. J. R. Astron. Soc.*, 23: 225–242.

Stern, D., 1962. The low-frequency power spectrum of cosmic-ray variations during IGY. *J. Geophys. Res.*, 67: 2133–2144.

Stewart, R. W., 1967. Mechanics of the air-sea interface. In: K. F. Bowden, F. N. Frenkiel and I. Tani (Editors), *Boundary Layers and Turbulence. The Physics of Fluids Suppl.* American Institute of Physics, pp.47–55.

Stewart, R. W., 1969. Turbulence and waves in a stratified atmosphere. *Radio Sci.*, 4: 1269–1278.

Stilwell, D., Jr., 1969. Directional energy spectra of the sea from photographs. *J. Geophys. Res.*, 74: 1974–1986.

Střeštík, J., 1969a. Properties of spectra of geomagnetic Pi2 pulsations recorded at the Budkov Observatory. *Studia Geophys. Geodaet.*, 13: 42–59.

Střeštík, J., 1969b. Comparison of data on geomagnetic pulsations recorded by various instruments. *Studia Geophys. Geodaet.*, 13: 293–307.

Střeštík, J., 1970. Determination of the original shape of pulsations from their spectra. *Studia Geophys. Geodaet.*, 14: 344–349.

Střeštík, J., 1971. Irregular Pi2 pulsations resolved into a superposition of sinusoidal oscillations. *Studia Geophys. Geodaet.*, 15: 64–75.

Střeštík, J., Prikner, K. and Dobeš, K., 1973. Daily variations of the characteristics of beating-type Pc3 (Bpc3) pulsations. *Studia Geophys. Geodaet.*, 17: 27–35.

Stringer, E. T., 1972. *Techniques of Climatology.* Freeman, 539 pp.

Stuart, R. D., 1969. *An Introduction to Fourier Analysis.* Methuen, 128 pp.

Stuart, W. F., Sherwood, V. and Macintosh, S. M., 1971. The power spectral density technique applied to micropulsation analysis. *Pure Appl. Geophys.*, 92: 150–164.

Stumpff, K., 1937. *Grundlagen und Methoden der Periodenforschung.* Springer, 332 pp.

Stumpff, K., 1940. Ermittlung und Realität von Periodizitäten. Korrelationsrechnung. *Handbuch Geophysik*, Borntraeger, 10(1): 117 pp.

Sudo, K., 1972. The focal process of the Taiwan-Oki earthquake of March 12, 1966. *J. Phys. Earth*, 20: 147–164.

Sugiura, M., 1960. A note on harmonic analysis of geophysical data with special reference to the analysis of geomagnetic storms. *J. Geophys. Res.*, 65: 2721–2725.

Sumner, R. D., 1967. Attenuation of earthquake generated P-waves along the western flank of the Andes. *Bull. Seismol. Soc. Am.*, 57: 173–190.

Sutton, G. H. and Pomeroy, P. W., 1963. Analog analyses of seismograms recorded on magnetic tape. *J. Geophys. Res.*, 68: 2791–2815.

Suyehiro, S., 1962. Deep earthquakes in the Fiji region. *Pap. Meteorol. Geophys.*, 13: 216–238.

Suyehiro, S., 1968. Change in earthquake spectrum before and after the Matsushiro swarm. *Pap. Meteorol. Geophys.*, 19: 427–435.

Suyehiro, S., Furuta, M., Sato, K. and Hirono, T., 1970. Noise attenuation in shallow holes (III). Improvement of signal to noise ratio. *Pap. Meteorol. Geophys.*, 21: 473–487.

Suzuki, S., 1972. Anomalous attenuation of P-waves in the Matsushiro earthquake swarm area. *J. Phys. Earth*, 20: 1–21.

Suzuki, Y. and Sato, R., 1970. Viscosity determination in the earth's outer core from ScS and SKS phases. *J. Phys. Earth*, 18: 157–170.

Syberg, F. J. R., 1972. A Fourier method for the regional-residual problem of potential fields. *Geophys. Prospect.*, 20: 47–75.

Takahasi, R. and Aida, I., 1961. Studies on the spectrum of tsunami. *Bull. Earthquake Res. Inst.*, 39: 523–535.

Takahasi, R. and Aida, I., 1962. Spectral analyses of long-period ocean waves observed at Izu-Ōshima. *Bull. Earthquake Res. Inst.*, 40: 561–573.

Takahasi, R. and Aida, I., 1963. Spectra of several tsunamis observed on the coast of Japan. *Bull. Earthquake Res. Inst.*, 41: 299–314.

Takano, K., 1970. Attenuation of short-period seismic waves in the upper mantle and its regional difference. *J. Phys. Earth*, 18: 171–179.

Takano, K., 1971a. A note on the attenuation of short-period P- and S-waves in the mantle. *J. Phys. Earth*, 19: 155–163.

Takano, K., 1971b. Analysis of seismic coda waves of ultra microearthquakes in the Matsushiro area—A comparison with Parkfield, California. *J. Phys. Earth*, 19: 209–215.

Takano, K. and Hagiwara, T., 1966. Preliminary observation of microearthquakes with a deep well seismometer. *Bull. Earthquake Res. Inst.*, 44: 1135–1148.

Takano, K. and Hagiwara, T., 1968. Observations of microearthquakes with a deep well seismometer (II). *Bull. Earthquake Res. Inst.*, 46: 1293–1300.

Takeuchi, H., Saito, M., Kobayashi, N. and Nakagawa, I., 1962. Free oscillations of the earth observed on gravimeters. *Zisin, Ser. II*, 15: 122–137 (in Japanese).

Tanaka, T., 1966. Study on the relation between local earthquakes and minute ground deformation. Part 2. *Bull. Disaster Prev. Res. Inst. (Kyoto)*, 16(1): 59–67.

Tanaka, T., 1967a. Study on the relation between local earthquakes and minute ground deformation. Part 3. *Bull. Disaster Prev. Res. Inst. (Kyoto)*, 16(2): 17–36.

Tanaka, T., 1967b. Study on the relation between local earthquakes and minute ground deformation. Part 4. *Bull. Disaster Prev. Res. Inst. (Kyoto)*, 17(3): 7–20.

Tanaka, T., 1968. On the effect of atmospheric pressure upon ground tilt. *Bull. Disaster Prev. Res. Inst. (Kyoto)*, 18(2): 23–36.

Tanaka, T., 1969. Study on meteorological and tidal influences upon ground deformations. *Spec. Contrib. Geophys. Inst. Kyoto Univ.*, 9: 29–90.

Tanaka, T., Dutta, T. K. and Kanai, K., 1966. Study of microtremor amplitude spectrum in relation to ground amplification character for sites in southern California. *Bull. Int. Inst. Seismol. Earthquake Eng.*, 3: 21–37.

Tanaka, Y., 1969. A seismometrical study of Izu-Ōshima (II). On the 4 kinds of volcanic tremors at the volcano Mihara-yama. *Pap. Meteorol. Geophys.*, 20: 385–416.

Taner, M. T. and Koehler, F., 1969. Velocity spectra—digital computer derivation and applications of velocity functions. *Geophysics*, 34: 859–881.

Tarr, A. C., 1969. Rayleigh-wave dispersion in the North Atlantic Ocean, Caribbean Sea, and Gulf of Mexico. *J. Geophys. Res.*, 74: 1591–1607.

Taylor, G. I., 1938. The spectrum of turbulence. *Proc. R. Soc. London, Ser. A*, 164: 476–490.

Teng, T.-L., 1968. Attenuation of body waves and the *Q*-structure of the mantle. *J. Geophys. Res.*, 73: 2195–2208.

Teng, T.-L. and Ben-Menahem, A., 1965. Mechanism of deep earthquakes from spectrums of isolated body-wave signals. 1. The Banda Sea earthquake of March 21, 1964. *J. Geophys. Res.*, 70: 5157–5170.

Terashima, T., 1968. Magnitude of microearthquake and the spectra of microearthquake waves. *Bull. Int. Inst. Seismol. Earthquake Eng.*, 5: 31–108.

Thakur, T. R. and Scheidegger, A. E., 1970. Chain model of river meanders. *J. Hydrol.*, 12: 25–47.

Thatcher, W., 1972. Regional variations of seismic source parameters in the northern Baja California area. *J. Geophys. Res.*, 77: 1549–1565.

Thompson, M. C., Jr., Janes, H. B. and Kirkpatrick, A. W., 1960. An analysis of time variations in tropospheric refractive index and apparent radio path length. *J. Geophys. Res.*, 65: 193–201.

Thompson, N., 1962. Intensities and spectra of vertical wind fluctuations at heights between 100 and 500 ft. in neutral and unstable conditions. *Q. J. R. Meteorol. Soc.*, 88: 328–334.

Thompson, N., 1972. Turbulence measurements over the sea by a tethered-balloon technique. *Q. J. R. Meteorol. Soc.*, 98: 745–762.

Thomson, W. T., 1950. Transmission of elastic waves through a stratified solid medium. *J. Appl. Phys.*, 21: 89–93.

Thomson, W. T., 1959. Spectral aspect of earthquakes. *Bull. Seismol. Soc. Am.*, 49: 91–98.

Toksöz, M. N. and Anderson, D. L., 1966. Phase velocities of long-period surface waves and structure of the upper mantle. 1. Great-circle Love and Rayleigh wave data. *J. Geophys. Res.*, 71: 1649–1658.

Toksöz, M. N. and Ben-Menahem, A., 1963. Velocities of mantle Love and Rayleigh waves over multiple paths. *Bull. Seismol. Soc. Am.*, 53: 741–764.

Toksöz, M. N. and Ben-Menahem, A., 1964. Excitation of seismic surface waves by atmospheric nuclear explosions. *J. Geophys. Res.*, 69: 1639–1648.

Toksöz, M. N. and Kehrer, H. H., 1972. Tectonic strain-release characteristics of CANNIKIN. *Bull. Seismol. Soc. Am.*, 62: 1425–1438.

Toksöz, M. N., Ben-Menahem, A. and Harkrider, D. G., 1964. Determination of source parameters of explosions and earthquakes by amplitude equalization of seismic surface waves. 1. Underground nuclear explosions. *J. Geophys. Res.*, 69: 4355–4366.

Toksöz, M. N., Harkrider, D. G. and Ben-Menahem, A., 1965. Determination of source parameters by amplitude equalization of seismic surface waves. 2. Release of tectonic strain by underground nuclear explosions and mechanisms of earthquakes. *J. Geophys. Res.*, 70: 907–922.

Toksöz, M. N., Arkani-Hamed, J. and Knight, C. A., 1969. Geophysical data and long-wave heterogeneities of the earth's mantle. *J. Geophys. Res.*, 74: 3751–3770.

Tolstoy, I. and Clay, C. S., 1966. *Ocean Acoustics*. McGraw-Hill, 293 pp.

Toman, K., 1965. The spectral shifts of truncated sinusoids. *J. Geophys. Res.*, 70: 1749–1750.

Toman, K., 1966. Fourier transform of the sunspot cycle. *J. Geophys. Res.*, 71: 3285–3286.

Tomoda, Y., 1954. A simplified method for harmonic analysis by means of square wave expansion. *Zisin, Ser. II*, 7: 201–208 (in Japanese).

Tomoda, Y. and Aki, K., 1955. Use of the function sin x/x in gravity problems. *Proc. Japan Acad.*, 31: 443–448.

Treitel, S., Clement, W. G. and Kaul, R. K., 1971. The spectral determination of depths to buried magnetic basement rocks. *Geophys. J. R. Astron. Soc.*, 24: 415–428.

Trembly, L. D. and Berg, J. W., Jr., 1968. Seismic source characteristics from explosion-generated P-waves. *Bull. Seismol. Soc. Am.*, 58: 1833–1848.

Trifunac, M. D. 1971a. Response envelope spectrum and interpretation of strong earthquake ground motion. *Bull. Seismol. Soc. Am.*, 61: 343–356.

Trifunac, M. D., 1971b. Zero baseline correction of strong-motion accelerograms. *Bull. Seismol. Soc. Am.*, 61: 1201–1211.

Trifunac, M. D., 1971c. A method for synthesizing realistic strong ground motion. *Bull. Seismol. Soc. Am.*, 61: 1739–1753.

Trifunac, M. D., 1972. Tectonic stress and the source mechanism of the Imperial Valley, California, earthquake of 1940. *Bull. Seismol. Soc. Am.*, 62: 1283–1302.

Trifunac, M. D. and Brune, J. N., 1970. Complexity of energy release during the Imperial Valley, California, earthquake of 1940. *Bull. Seismol. Soc. Am.*, 60: 137–160.

Trifunac, M. D., Udwadia, F. E. and Brady, A. G., 1973. Analysis of errors in digitized strong-motion accelerograms. *Bull. Seismol. Soc. Am.*, 63: 157–187.

Tryggvason, E., 1965. Dissipation of Rayleigh wave energy. *J. Geophys. Res.*, 70: 1449–1455.

Tsai, N. C., 1970. A note on the steady-state response of an elastic half-space. *Bull. Seismol. Soc. Am.*, 60: 795–808.

Tsai, N. C. and Housner, G. W., 1970. Calculation of surface motions of a layered half-space. *Bull. Seismol. Soc. Am.*, 60: 1625–1651.

Tsai, Y.-B., 1972. Use of LP surface waves for source characterization. *Geophys. J. R. Astron. Soc.*, 31: 111–130.

Tsai, Y.-B. and Aki, K., 1969. Simultaneous determination of the seismic moment and attenuation of seismic surface waves. *Bull. Seismol. Soc. Am.*, 59: 275–287.

Tsai, Y.-B. and Aki, K., 1970a. Source mechanism of the Truckee, California, earthquake of September 12 1966. *Bull. Seismol. Soc. Am.*, 60: 1199–1208.

Tsai, Y.-B. and Aki, K., 1970b. Precise focal depth determination from amplitude spectra of surface waves. *J. Geophys. Res.*, 75: 5729–5743.

Tsai, Y.-B. and Aki, K., 1971. Amplitude spectra of surface waves from small earthquakes and underground nuclear explosions. *J. Geophys. Res.*, 76: 3940–3952.

Tsuboi, C., 1954. A new and simple method for calculating the deflections of the vertical from gravity anomalies with the aid of the Bessel Fourier series. *Proc. Japan Acad.*, 30: 461–466.

Tsuboi, C. and Fuchida, T., 1937. Relation between gravity anomalies and the corresponding mass distribution (I). *Bull. Earthquake Res. Inst.*, 15: 636–649.

Tsuboi, C. and Fuchida, T., 1938. Relation between gravity anomalies and the corresponding mass distribution (II). *Bull. Earthquake Res. Inst.*, 16: 273–284.

Tsuboi, C. and Tomoda, Y., 1958. The relation between the Fourier series method and the sin x/x method for gravity interpretations. *J. Phys. Earth*, 6: 1–5.

Tsujiura, M., 1966. Frequency analysis of seismic waves (1). *Bull. Earthquake Res. Inst.*, 44: 873–891.

Tsujiura, M., 1967. Frequency analysis of seismic waves (2). *Bull. Earthquake Res. Inst.*, 45: 973–995.

Tsujiura, M., 1969. Regional variation of P-wave spectrum (1). *Bull. Earthquake Res. Inst.*, 47: 613–633.

Tsujiura, M., 1972. Spectra of body waves and their dependence on source depth. I. Japanese arc. *J. Phys. Earth*, 20: 251–266.

Tsvang, L. R., 1963. Some characteristics of the spectra of temperature pulsations in the boundary layer of the atmosphere. *Izv. Acad. Sci. USSR, Geophys. Ser.*, 10: 961–965 (Engl. ed.).

Tsvang, L. R., Zubkovskii, S. L., Ivanov, V. N., Klinov, F. Ya. and Kravchenko, T. K., 1963. Measurements of some properties of turbulence in the lowest 300 meters of the atmosphere. *Izv. Acad. Sci. USSR, Geophys. Ser.*, 5: 475–481 (Engl. ed.).

Tukey, J. W., 1961. Discussion, emphasizing the connection between analysis of variance and spectrum analysis. *Technometrics*, 3: 191–219.

Tukey, J. W., 1967. An introduction to the calculations of numerical spectrum analysis. In: B. Harris (Editor), *Spectral Analysis of Time Series*. Wiley, pp.25–46.

Turner, C. H. M., 1954. On the concept of an instantaneous power spectrum, and its relationship to the autocorrelation function. *J. Appl. Phys.*, 25: 1347–1351.

Udias, A., 1971. Source parameters of earthquakes from spectra of Rayleigh waves. *Geophys. J. R. Astron. Soc.*, 22: 353–376.

Udías Vallina, A., S. J., 1969. Estudio del mecanismo focal de los terremotos por medio de ondas superficiales. *Rev. Geofís.*, 28: 427–472.

Udías, A. and López Arroyo, A., 1970. Body and surface wave study of source parameters of the March 15, 1964 Spanish earthquake. *Tectonophysics*, 9: 323–346.

Ulrych, T. J., 1971. Application of homomorphic deconvolution to seismology. *Geophysics*, 36: 650–660.

Ulrych, T. J., 1972. Maximum entropy power spectrum of truncated sinusoids. *J. Geophys. Res.*, 77: 1396–1400.

Usami, T., Kotake, Y. and Satô, Y., 1967. Soft core spectrum splitting and related problems of the spheroidal oscillations of an elastic sphere with a homogeneous mantle and core. *Bull. Earthquake Res. Inst.*, 45: 945–961.

Utsu, T., 1966. Variations in spectra of P-waves recorded at Canadian Arctic seismograph stations. *Can. J. Earth Sci.*, 3: 597–621.

Utsu, T., 1971. Seismological evidence for anomalous structure of island arcs with special reference to the Japanese region. *Rev. Geophys. Space Phys.*, 9: 839–890.

Utsu, T., 1972. Aftershocks and earthquake statistics (IV). *J. Fac. Sci. Hokkaido Univ., Ser. 7 (Geophys.)*, 4: 1–42.

Utsu, T. and Okada, H., 1968. Anomalies in seismic-wave velocity and attenuation associated with a deep earthquake zone, 2. *J. Fac. Sci. Hokkaido Univ., Ser. 7 (Geophys.)*, 3: 65–84.

Valenzuela, G. R., Laing, M. B. and Daley, J. C., 1971. Ocean spectra for the high-frequency waves as determined from airborne radar measurements. *J. Mar. Res.*, 29: 69–84.

Valle, P. E., 1949. Sulla misura della velocità di gruppo delle onde sismiche superficiali. *Ann. Geofis.*, 2: 370–376.

Van den Heuvel, E. P. J., 1966. On the precession as a cause of Pleistocene variations of the Atlantic Ocean water temperatures. *Geophys. J. R. Astron. Soc.*, 11: 323–336.

Van der Hoven, I., 1957. Power spectrum of horizontal wind speed in the frequency range from 0.0007 to 900 cycles per hour. *J. Meteorol.*, 14: 160–164.

Van Isacker, J., 1961. Generalized harmonic analysis. *Adv. Geophys.*, 7: 189–214.

Van Mieghem, J., 1961. Zonal harmonic analysis of the Northern Hemisphere geostrophic wind field. *IUGG Monogr.*, 8: 57 pp.

Vere-Jones, D. and Davies, R. B., 1966. A statistical survey of earthquakes in the main seismic region of New Zealand. Part 2—Time series analysis. *N.Z. J. Geol. Geophys.*, 9: 251–284.

Vesanen, E., 1942. Über die typenanalytische Auswertung der Seismogramme. *Ann. Acad. Sci. Fenn., Ser. A, III. Geol.-Geogr.*, 5: 244 pp.

Vesanen, E., 1946. On seismogram types and focal depth of earthquakes in the north Japan and Manchuria region. *Ann. Acad. Sci. Fenn., Ser. A, III. Geol.-Geogr.*, 11: 25 pp.

Vetter, R. C. and Bretschneider, C. L. (Editors), 1963: *Ocean Wave Spectra. Proc. Conf. Natl. Acad. Sci.* Prentice-Hall, 357 pp.

Vigoureux, P. and Hersey, J. B., 1962. Sound in the sea. In: M. N. Hill (Editor), *The Sea*, Vol. 1. Interscience, pp.476–497.

Vinnichenko, N. K., 1970. The kinetic energy spectrum in the free atmosphere—1 second to 5 years. *Tellus*, 22: 158–166.

Vinnichenko, N. K. and Dutton, J. A., 1969. Empirical studies of atmospheric structure and spectra in the free atmosphere. *Radio Sci.*, 4: 1115–1126.

Vinnik, L. P., 1963. The space–time filtration of seismic signals. *Izv. Acad. Sci. USSR, Geophys. Ser.*, 6: 521–527 (Engl. ed.).

Vinnik, L. P., 1967a. Structure of microseisms in the 1 cps range. I. Method of analysis. *Izv. Acad. Sci. USSR, Phys. Solid Earth*, 8: 496–499 (Engl. ed).

Vinnik, L. P., 1967b. The structure of 4-6 second microseisms. *Izv. Acad. Sci. USSR, Phys. Solid Earth*, 10: 640–647 (Engl. ed.).

Vinnik, L. P., 1967c. Structure of microseisms. In: H. Jensen (Editor), *European Seismological Commission, Copenhagen Assembly 1966*, pp.333–346.

Vinnik, L. P. and Pruchkina, N. M., 1964. A study of the structure of short-period microseisms. *Izv. Acad. Sci. USSR, Geophys. Ser.*, 5: 412–419 (Engl. ed.).

Vinnik, L. P., Deniskov, A. S. and Kon'kov, G. D., 1967. Structure of microseisms in the 1 cps range. II. Results of observations. *Izv. Acad. Sci. USSR, Phys. Solid Earth*, 8: 500–504 (Engl. ed.).

Volkov, V. A., 1971. Détermination des caractéristiques de phase des systèmes enregistreurs de marées – gravimètre – galvanomètre. *Marées Terr., Bull. d'Inform.*, 62: 3149–3163.

Von Hann, J. and Süring, R., 1940–1951. *Lehrbuch der Meteorologie.* Keller and Hirzel, 1092 pp.

Von Seggern, D., 1972. Relative location of seismic events using surface waves. *Geophys. J. R. Astron. Soc.*, 26: 499–513.

Von Seggern, D. and Blandford, R., 1972. Source time functions and spectra for underground nuclear explosions. *Geophys. J. R. Astron. Soc.*, 31: 83–97.

Von Seggern, D. and Lambert, D. G., 1970. Theoretical and observed Rayleigh-wave spectra for explosions and earthquakes. *J. Geophys. Res.*, 75: 7382–7402.

Voorhis, A. D., 1968. Measurements of vertical motion and the partition of energy in the New England slope water. *Deep-Sea Res.*, 15: 599–608.

Wada, T., Furuzawa, T. and Ono, H., 1963. Source-mechanism of the Chilean earthquake from spectra of long-period surface waves. *J. Seismol. Soc. Japan*, 16: 181–187 (in Japanese).

Wada, T., Kamo, K., Furuzawa, T. and Onoue, K., 1972. The observation of microtremors correlated with the existence of cracks at the landslide area. *Bull. Disaster Prev. Res. Inst. (Kyoto)*, 21(3): 217–226.

Walden, H. and Rubach, H.-J., 1967. Gleichzeitige Messungen des Seegangs mit nicht-stabilisierten Beschleunigungsschreibern an Orten mit unterschiedlicher Wassertiefe in der Deutschen Bucht. *Dtsche Hydrogr. Z.*, 20: 157–167.

Walker, R. A., Menard, J. Z. and Bogert, B. P., 1964. Real-time, high-resolution spectroscopy of seismic background noise. *Bull. Seismol. Soc. Am.*, 54: 501–509.

Wallace, J. M., 1971. Spectral studies of tropospheric wave disturbances in the tropical western Pacific. *Rev. Geophys. Space Phys.*, 9: 557–612.

Wallace, J. M. and Chang, C.-P., 1969. Spectrum analysis of large-scale wave disturbances in the tropical lower troposphere. *J. Atmos. Sci.*, 26: 1010–1025.

Walsh, J. L., 1923. A closed set of orthogonal functions. *Am. J. Math.*, 45: 5–24.

Walzer, U., 1972a. Comparison of microseisms of Cuban and Central European stations. *Pure Appl. Geophys.*, 95: 89–99.

Walzer, U., 1972b. The distribution of continents and oceans and its relation to mantle convection. *Gerlands Beitr. Geophys.*, 81: 471–480.

Ward, F. W., Jr., 1960. The variance (power) spectra of C_i, K_p, and A_p. *J. Geophys. Res.*, 65: 2359–2373.

Ward, F. and Shapiro, R., 1961a. Solar, geomagnetic, and meteorological periodicities. *Ann. N. Y. Acad. Sci.*, 95: 200–224.

Ward, F. and Shapiro, R., 1961b. Meteorological periodicities. *J. Meteorol.*, 18: 635–656.

Ward, R. W. and Toksöz, M. N., 1971. Causes of regional variation of magnitudes. *Bull. Seismol. Soc. Am.*, 61: 649–670.

Wardell, J., 1970. A comparison of land seismic sources. *Geoexploration*, 8: 205–229.

Webster, F., 1969a. Turbulence spectra in the ocean. *Deep-Sea Res.*, 16 (Suppl.): 357–368.

Webster, F., 1969b. On the representativeness of direct deep-sea current measurements. In: M. Sears (Editor), *Progress in Oceanography*, Vol. 5. Pergamon Press, pp.3–15.

Weichert, D. H., 1971. Short-period spectral discriminant for earthquake-explosion differentiation. *Z. Geophys.*, 37: 147–152.

Weiler, H. S. and Burling, R. W., 1967. Direct measurements of stress and spectra of turbulence in the boundary layer over the sea. *J. Atmos. Sci.*, 24: 653–664.

Weinstein, M. S., 1968. Spectra of acoustic and seismic signals generated by underwater explosions during Chase experiment. *J. Geophys. Res.*, 73: 5473–5476.

Welch, P. D., 1967. The use of the Fast Fourier Transform for the estimation of power spectra: A method based on time averaging over short, modified periodograms. *IEEE Trans. Audio Electroacoustics*, AU-15: 70–73.

Wendell, L. L., 1969. A study of large-scale atmospheric turbulent kinetic energy in wave-number frequency space. *Tellus*, 21: 760–788.

Wenz, G. M., 1962. Acoustic ambient noise in the ocean: Spectra and sources. *J. Acoust. Soc. Am.*, 34: 1936–1956.

Werth, G. C. and Herbst, R. F., 1963. Comparison of amplitudes of seismic waves from nuclear explosions in four mediums. *J. Geophys. Res.*, 68: 1463–1475.

Weston, D. E., 1960. The low-frequency scaling laws and source levels for underground explosions and other disturbances. *Geophys. J. R. Astron. Soc.*, 3: 191–202.

Whitcomb, J. H., 1969. Array data processing techniques applied to long-period shear waves at Fennoscandian seismograph stations. *Bull. Seismol. Soc. Am.*, 59: 1863–1887.

White, P. H., 1969. Cross-correlation in structural systems: Dispersion and nondispersion waves. *J. Acoust. Soc. Am.*, 45: 1118–1128.

White, R. M. and Cooley, D. S., 1956. Kinetic-energy spectrum of meridional motion in the mid-troposphere. *J. Meteorol.*, 13: 67–69.

Whitham, K., 1963. An anomaly in geomagnetic variations at Mould Bay in the Arctic archipelago of Canada. *Geophys. J. R. Astron. Soc.*, 8: 26–43.

Whitham, K. and Andersen, F., 1965. Magneto-telluric experiments in northern Ellesmere Island. *Geophys. J. R. Astron. Soc.*, 10: 317–345.

Whittle, P., 1954. The statistical analysis of a seiche record. *J. Mar. Res.*, 13: 76–100.

Whorf, T., 1972. Teleseismic and earth noise monitoring with the Block-Moore quartz accelerometer. *Geophys. J. R. Astron. Soc.*, 31: 205–238.

Wickens, A. J. and Kollar, F., 1967. A wide range seismogram digitizer. *Bull. Seismol. Soc. Am.*, 57: 91–98.

Wickens, A. J. and Pec, K., 1968. A crust–mantle profile from Mould Bay, Canada, to Tucson, Arizona. *Bull. Seismol. Soc. Am.*, 58: 1821–1831.

Wiener, N., 1959. *The Fourier Integral and Certain of Its Applications*. Dover Publ., 201 pp.

Wiggins, J. H., Jr., 1962. Note on the effect of record balancing upon strong motion earthquake response spectra. *Bull. Seismol. Soc. Am.*, 52: 963–970.

Wiggins, R. A., 1966. ω-k filter design. *Geophys. Prospect.*, 14: 427–440.

Wiggins, R. A. and Miller, S. P., 1972. New noise-reduction technique applied to long-period oscillations from the Alaskan earthquake. *Bull. Seismol. Soc. Am.*, 62: 471–479.

Wiin-Nielsen, A., 1967. On the annual variation and spectral distribution of atmospheric energy. *Tellus*, 19: 540–559.

Williams, R. B., 1968. Horizontal temperature variations in the upper water of the open ocean. *J. Geophys. Res.*, 73: 7127–7132.

Willis, D. E., 1960. Some observations on the attenuation of seismic waves. *Earthquake Notes*, 31(4): 37–45.

Willis, D. E., 1963a. A note on the effect of ripple firing on the spectra of quarry shots. *Bull. Seismol. Soc. Am.*, 53: 79–85.

Willis, D. E., 1963b. Seismic measurements of large underwater shots. *Bull. Seismol. Soc. Am.*, 53: 789–809.

Willis, D. E., 1963c. Comparison of seismic waves generated by different types of source. *Bull. Seismol. Soc. Am.*, 53: 965–978.

Willis, D. E., 1964. Short period spectral measurements of seismic waves in the northeastern U.S.A. *Earthquake Notes*, 35: 1–13.

Willis, D. E., 1965. Variations in compressional waves at teleseismic distances. *J. Geophys. Res.*, 70: 1877–1883.

Willis, D. E. and DeNoyer, J. M., 1966. Seismic attenuation and spectral measurements from the Lake Superior experiment. *Am. Geophys. Union, Geophys. Monogr.*, 10: 218–226.

Willis, D. E. and Johnson, J. C., 1959. Some seismic results using magnetic tape recording. *Earthquake Notes*, 30(3): 21–25.

Willis, D. E. and Wilson, J. T., 1962. Effects of decoupling on spectra of seismic waves. *Bull. Seismol. Soc. Am.*, 52: 123–131.

Willis, D. E., DeNoyer, J. and Wilson, J. T., 1963. Differentiation of earthquakes and underground nuclear explosions on the basis of amplitude characteristics. *Bull. Seismol. Soc. Am.*, 53: 979–987.

Winch, D. E., 1965. Noncyclic variation and Sq. *Pure Appl. Geophys.*, 61: 45–51.

Winch, D. E., Bolt, B. A. and Slaucitajs, L., 1963. Geomagnetic fluctuations with the frequencies of torsional oscillations of the earth. *J. Geophys. Res.*, 68: 2685–2693.

Wirth, H. and Byl, J., 1965. Beobachtung freier Schwingungen der Erde. *Gerlands Beitr. Geophys.*, 74: 14–19.

Wirth, H. and Skalský, L., 1965. Freie Schwingungen des Erdkörpers. *Gerlands Beitr. Geophys.*, 74: 230–232.

Wirth, H., Buchheim, W. and Schneider, M., 1965. Zur Anregung von Eigenschwingungen des Erdkörpers durch das Erdbeben in Alaska am 28.3.1964. *Gerlands Beitr. Geophys.*, 74: 408–412.

Wonnacott, T. H., 1961. Spectral analysis combining a Bartlett window with an associated inner window. *Technometrics*, 3: 235–243.

Wood, D. E., 1964. New display format and a flexible-time integrator for spectral-analysis instrumentation. *J. Acoust. Soc. Am.*, 36: 639–643.

Wood, L. C., 1968. A review of digital pass filtering. *Rev. Geophys.*, 6: 73–97.

Woods, J. D., 1969. On Richardson's number as a criterion for laminar–turbulent–laminar transition in the ocean and atmosphere. *Radio Sci.*, 4: 1289–1298.

Wooldridge, G. and Reiter, E. R., 1970. Large-scale atmospheric circulation characteristics as evident from ghost balloon data. *J. Atmos. Sci.*, 27: 183–194.

Wright, J. K., Carpenter, E. W. and Savill, R. A., 1962. Some studies of the P-waves from underground nuclear explosions. *J. Geophys. Res.*, 67: 1155–1160.

Wu, F. T. 1968. Parkfield earthquake of June 28, 1966: magnitude and source mechanism. *Bull. Seismol. Soc. Am.*, 58: 689–709.

Wu, F. T. and Ben-Menahem, A., 1965. Surface wave radiation pattern and source mechanism of the September 1, 1962, Iran earthquake. *J. Geophys. Res.*, 70: 3943–3949.

Wu, F. T. and Hannon, W. J., 1966. PP and crustal structure. *Bull. Seismol. Soc. Am.*, 56: 733–747.

Wunsch, C., 1972a. Bermuda sea level in relation to tides, weather, and baroclinic fluctuations. *Rev. Geophys. Space Phys.*, 10: 1–49.

Wunsch, C., 1972b. The spectrum from two years to two minutes of temperature in the main thermocline at Bermuda. *Deep-Sea Res.*, 19: 577–593.

Wunsch, C. and Dahlen, J., 1970. Preliminary results of internal wave measurements in the main thermocline at Bermuda. *J. Geophys. Res.*, 75: 5899–5908.

Wunsch, C., Hansen, D. V. and Zetler, B. D., 1969. Fluctuations of the Florida current inferred from sea level records. *Deep-Sea Res.*, 16 (Suppl.): 447–470.

Wyngaard, J. C. and Coté, O. R., 1972. Cospectral similarity in the atmospheric surface layer. *Q. J. R. Meteorol. Soc.*, 98: 590–603.

Wyrtki, K., 1967. The spectrum of ocean turbulence over distances between 40 and 1000 kilometers. *Dtsche Hydrogr. Z.*, 20: 176–186.

Wyrtki, K. and Graefe, V., 1967. Approach of tides to the Hawaiian Islands. *J. Geophys. Res.*, 72: 2069–2071.

Wyss, M. and Hanks, T. C., 1972. The source parameters of the San Fernando earthquake inferred from teleseismic body waves. *Bull. Seismol. Soc. Am.*, 62: 591–602.

Wyss, M. and Molnar, P., 1972. Source parameters of intermediate and deep focus earthquakes in the Tonga arc. *Phys. Earth Planet. Inter.*, 6: 279–292.

Wyss, M., Hanks, T. C. and Liebermann, R. C., 1971. Comparison of P-wave spectra of underground explosions and earthquakes. *J. Geophys. Res.*, 76: 2716–2729.

Yampolsky, A. D., 1960. On the application of harmonic analysis for processing of the hydrologic observations material. *Izv. Acad. Sci. USSR, Geophys. Ser.*, 7: 712–713 (Engl. ed.).

Yanai, M. and Murakami, M., 1970a. A further study of tropical wave disturbances by the use of spectrum analysis. *J. Meteorol. Soc. Japan*, 48: 185–197.

Yanai, M. and Murakami, M., 1970b. Spectrum analysis of symmetric and antisymmetric equatorial waves. *J. Meteorol. Soc. Japan*, 48: 331–347.

Yanai, M., Maruyama, T., Nitta, T. and Hayashi, Y., 1968. Power spectra of large-scale disturbances over the tropical Pacific. *J. Meteorol. Soc. Japan*, 46: 308–323.

Yokosi, S., 1967. The structure of river turbulence. *Bull. Disaster Prev. Res. Inst. (Kyoto)*. 17(2): 1–29.

Yoshikawa, S., Shima, M. and Irikura, K., 1967. Vibrational characteristics of the ground investigated by several methods. *Bull. Disaster Prev. Res. Inst. (Kyoto)*, 16(2): 1–16.

Yoshiyama, R., 1959. Maximum amplitude and epicentral distance. Proposed a theoretical elucidation of empirical formulas and some development. *Bull. Earthquake Res. Inst.*, 37: 389–404.

Yoshiyama, R., 1960. Propagation of surface waves and internal friction. *Bull. Earthquake Res. Inst.*, 38: 361–368.

Yoshizawa, S., Tanaka, T. and Kanai, K., 1968. Some features of strong underground earthquake motions computed from observed surface records. *Bull. Earthquake Res. Inst.*, 46: 667–686.

Yukutake, T., 1962. The westward drift of the magnetic field of the earth. *Bull. Earthquake Res. Inst.*, 40: 1–65.

Yukutake, T., 1971. Spherical harmonic analysis of the earth's magnetic field for the 17th and the 18th centuries. *J. Geomagn. Geoelectr.*, 23: 11–31.

Yukutake, T. and Tachinaka, H., 1968. The non-dipole part of the earth's magnetic field. *Bull. Earthquake Res. Inst.*, 46: 1027–1074.

Zadro, M., 1961. "Power spectrum analysis" delle deviazione della verticale registrate durante l'eclissi totale di Sole del 15 Febbraio 1961. *Atti Conv. Ann., Assoc. Geofis. Ital.*, 11: 63–78.

Zadro, M. B., 1966. Maree terrestri ed effetti di carico. *Boll. Geofis. Teor. Appl.*, 8: 173–195.

Zadro, M. B., 1971. Non-linear effects in the free oscillations of the earth. *Boll. Geofis. Teor. Appl.*, 13: 187–195.

Zadro, M. B. and Caputo, M., 1968. Spectral, bispectral analysis and *Q* of the free oscillations of the earth. *Suppl. Nuovo Cimento, Ser. I*, 6: 67–81.

Zadro, M. B. and Marussi, A., 1967. Polarization and total energy spectra of the eigenvibrations of the earth recorded at Trieste. *Geophys. J. R. Astron. Soc.*, 12: 425–436.

Zadro, M. B. and Poretti, G., 1972. Spectral techniques for the analysis of tidal time series. *Pure Appl. Geophys.*, 95: 18–26.

Zalkan, R. L., 1970. High frequency internal waves in the Pacific Ocean. *Deep-Sea Res.*, 17: 91–108.

Zeevaert, L., 1964. Strong ground motions recorded during earthquakes of May the 11th and 19th, 1962 in Mexico City. *Bull. Seismol. Soc. Am.*, 54: 209–231.

Zelei, A., 1971. On the design of numerical filters. *Ann. Geofis.*, 24: 457–474.

Zetler, B. D., 1960. The effect of instrumental drift on the harmonic analysis of gravity at Washington, D.C. *Boll. Geofis. Teor. Appl.*, 2: 235–237.

Zetler, B. D., 1964. The use of power spectrum analysis for earth tides. *Marées Terr., Bull. d'Inform.*, 35: 1157–1164.

Zetler, B. D. and Cummings, R. A., 1967. A harmonic method for predicting shallow-water tides. *J. Mar. Res.*, 25: 103–114.

Zetler, B. D. and Lennon, G. W., 1967. Some comparative tests of tidal analytical processes. *Int. Hydrogr. Rev.*, 44: 139–147.

Zetler, B. D., Schuldt, M. D., Whipple, R. W. and Hicks, S. D., 1965. Harmonic analysis of tides from data randomly spaced in time. *J. Geophys. Res.*, 70: 2805–2811.

Zetler, B., Cartwright, D. and Munk, W., 1970. Tidal constants derived from response admittances. *Obs. R. Belg., Comm. (Géophys.)*, A9: 175–178.

Zubkovskii, S. L., 1962. Frequency spectra pulsations of the horizontal component of wind velocity in the surface air layer. *Izv. Acad. Sci. USSR, Geophys. Ser.*, 10: 887–891 (Engl. ed.).

Zubkovskii, S. L., 1963. An experimental investigation of the spectra of pulsations in vertical component of the wind velocity in the free atmosphere. *Izv. Acad. Sci. USSR, Geophys. Ser.*, 8: 782–784 (Engl. ed.).

Zverev, S. M., 1962. Frequency features of explosions in deep seismic sounding in deep seas. *Izv. Acad. Sci. USSR, Geophys. Ser.*, 3: 240–244 (Engl. ed.).

Zverev, S. M. and Galkin, I. N., 1966. Observation methods and the scope for an increase of recording distance in deep seismic sounding at sea. *Izv. Acad. Sci. USSR, Phys. Solid Earth*, 9: 555–560 (Engl. ed.).

Zwang, L. R., 1960a. Measurements of temperature pulse frequency spectra in the surface layer of the atmosphere. *Izv. Acad. Sci. USSR, Geophys. Ser.*, 8: 833–838 (Engl. ed.).

Zwang, L. R., 1960b. Measurements of the spectrum of temperature fluctuations in the free atmosphere. *Izv. Acad. Sci. USSR, Geophys. Ser.*, 11: 1117–1120 (Engl. ed.).

SUBJECT INDEX